Geophysical Monograph Series

Including

IUGG Volumes
Maurice Ewing Volumes
Mineral Physics Volumes

Geophysical Monograph 126

The Oceans and Rapid Climate Change

Past, Present, and Future

Dan Seidov
Bernd J. Haupt
Mark Maslin
Editors

American Geophysical Union
Washington, DC

Library of Congress Cataloging-in-Publication Data
The oceans and rapid climate change: past, present, and future / Dan Seidov, Bernd J. Haupt, Mark Maslin, editors.
    p.cm. -- (Geophysical monograph ; 126)
    Includes bibliographical references.
    ISBN 0-87590-985-X (alk. paper)
    1. Paleoclimatology. 2. Climatic changes. 3. Ocean circulation. I. Seidov, Dan, 1948- II. Haupt, Bernd J. III. Maslin, Mark. IV. Series

QC884.O34 2001
555.6'09'01--dc21   2001045701

ISBN 0-87590-985-X
ISSN 0065-8448

CONTENTS

CONTENTS

PREFACE

Until a few decades ago, scientists generally believed that significant large-scale past global and regional climate changes occurred at a gradual pace within a time scale of many centuries or millennia. A secondary assumption followed: climate change was scarcely perceptible during a human lifetime. Recent paleoclimatic studies, however, have proven otherwise: that global climate can change extremely rapidly. In fact, there is good evidence that in the past at least regional mean annual temperatures changed by several degrees Celsius on a time scale of several centuries to several decades.

Humanity is now faced with the contentious problem of global warming and the potential for catastrophic climatic change. How will the global climate system react to the ever-increasing amounts of anthropogenic carbon dioxide now entering the atmosphere? Although paleoclimatology can provide neither a political nor an economic perspective here, it can enhance our understanding of the global climate system and how it has changed. This understanding of the past is essential for more reliable scenarios of possible future change.

Currently there is evidence that we are on a warming trend, which many scientists suggest is evidence of global warming. There are researchers, however, who argue that the same trend is a more natural adjustment of the global climate system recovering from the Little Ice Age, the last major climate deterioration that ended 100 years ago. Be that as it may, if we are in a period of enhanced global warming, what consequences might follow from the melting of Arctic or Antarctic ice? Or is it possible, in a period of general global warming, that the conveyor belt of deep water will be "switched off," plunging Europe into a new Little Ice Age?

Although such questions are complex, there are basic features of the climate system that can help to narrow our search for answers and dramatically improve our understanding of how climate works. One key approach is to use paleoclimatology to assess the dynamics of the climate system on decadal to millennial time scales using both the geologic record and climate models. In particular, this approach brings into focus the world ocean as the most important element controlling climate system dynamics on large time scales.

By its thermal, freshwater and dynamic impacts, the ocean affects our environment in many ways. The effects of the ocean impact are evident in ocean-atmosphere interactions, sea-ice dynamics, and sea level changes. When measured in decadal to millennial time scales, however, the ocean proves most important for climate change via its enormous heat capacity and its capability to redistribute heat by ocean currents. Although such time intervals are too long for the atmosphere to be dynamically important, they are too short in comparison with the changes of Earth orbital parameters, ice sheet dynamics, or tectonic activity. In addition, we must consider the heart beat of global climate, the thermohaline ocean circulation, which seems to be surprisingly sensitive to relatively small changes of freshwater in the high latitudes. It has also been speculated that anthropogenic global warming may threaten the stability of this oceanic overturning and thus present a considerable challenge to human society.

New evidence from ice cores and deep-sea sediments also suggests that the climates of the two hemispheres may not be synchronized in their response to climate forcing. These findings, which question the role of the thermohaline circulation even further, point to other questions as well: can the thermohaline circulation be altered by surface impacts in the northern and southern hemispheres; what role, if any, does thermohaline play in the previously observed synchronicity; and how can the thermohaline alter those impacts?

Our research, as a result, is multifaceted if focused on charting feedbacks in the climate systems as facilitated by ocean circulation in conjunction with other climatic elements. For example, we know that the engine of long-term climate change, the deep-ocean circulation system, can change dramatically in response to freshwater impacts that are usually associated with the melting of some elements of the cryosphere, either as melting sea ice or ice sheet surges. These and related subjects are the major focus of the present volume.

By focusing on ocean dynamics and the ocean-atmosphere-cryosphere coupling during geologically rapid climate changes, the authors of this volume address the fundamental unresolved issues in paleoclimatology, including factors that may cause or influence strong long-term internal variability of the climate system. The volume provides discussion of proxy data, hypotheses, modeling and synthesizing papers that address key aspects of rapid climate change, ocean circulation variability, and sea-surface conditions during the Late Quaternary.

To structure the material within the volume, we have taken a linear approach. We begin by reviewing what we know about geologically abrupt climate changes of the Pleistocene, and then gradually move from data to ideas, hypotheses, modeling, and finally to perspectives that we may infer from data and modeling analyses.

The book combines discussions of paleoclimatic and paleoceanographic reconstructions with the modeling of past and future, colder and warmer climates, and gives a basis for synthesizing the results and gaining a new momentum in studying millennial-scale ocean and climate variability. In this capacity, the volume is targeted to scientists, researchers, graduate students, and others interested in paleoclimatology, paleoceanography, and the future of the climate system.

This volume delineates work in progress, and does not present ultimate solutions to the problems addressed; rather it proposes initial answers to the key questions summarized above. The first part, Data and Climate Models: Windows to the Past, presents data interpretation, hypotheses based on data analyses, and ideas that may shed new light on many aspects of past climates and their evolution. The second part, Ocean and Climate Models: Bridges From Past to Future, features climate system models that focus on the role of the world ocean in the past and possible future climate change, starting with simpler, ocean-only models that unroll into full, three-dimensional models of the entire climate system. Importantly, most papers emphasize the role of global ocean links, especially between the Northern and Southern Hemispheres.

The book can be considered a continuation of the AGU volume, Mechanisms of Global Climate Change (Clark, P. U., S. R. Webb and L. D. Keigwin, Editors, AGU, Washington, DC, 1999), but in no way repeats that volume. In fact, the gallery of results presented here provides a unique vision of what new observations and modeling can offer for assessing the role of the different components of the climate long-term alteration, and especially the still poorly understood southern oceans connection.

The volume derives from presentations and discussions at two special sessions devoted to rapid climate changes on decadal to millennial time scale. The first took place at the AGU Fall Meeting, San Francisco (December 1999), the second at the Ocean Science Meeting, San Antonio (February 2000). Presenters responded enthusiastically when asked to contribute to the volume. To further enhance the project, several other researchers, who contributed to the discussions, were asked to join the project, and many agreed.

As editors, we are very grateful to all authors, who worked hard to meet the deadline and did everything they could to make this project a success. We appreciate the help, patience, and expertise of AGU staff, who worked diligently to publish the book. We are especially thankful to Allan Graubard, our acquisitions editor, whose support, encouragement, and advice was crucial for finalizing the entire project. We are grateful to the reviewers who devoted so much of their time and effort helping to improve the volume.

Dan Seidov
Earth System Science Center
Penn State University

Bernd J. Haupt
Earth System Science Center
Penn State University

Mark Maslin
Department of Geography
University College London

Ocean Currents of Change: Introduction

Eric J. Barron and Dan Seidov

EMS Environment Institute, Pennsylvania State University, University Park, Pennsylvania

Scientists have traditionally assumed that climate change occurred on a timescale far in excess of a single human lifespan. In fact, significant global and regional climate changes were thought to occur incrementally over many centuries or millennia. Recently, however, our view of climate change has altered. Paleoclimatic studies now show that the global climate can change quite rapidly. There is substantive evidence that in the past, at least from a regional perspective, mean annual temperatures changed by several degrees Celsius over a few centuries or even decades. Evidence from the instrumented record and from climate models also indicates that humans have the potential to alter the Earth's climate significantly within one century – although the actual sensitivity of the climate system to the growing concentration of anthropogenic carbon dioxide and other greenhouse gases in the atmosphere is highly uncertain. In this regard, paleoclimatology offers perspectives that can enhance our understanding of how the global climate system changes. If we are to develop scenarios of possible climate change with greater reliability from the near to far future, the understanding provided by paleoclimatology will play a significant role.

Needless to say, the attribution of the temperature changes that occurred over the last century to a specific set of factors is the subject of considerable debate. The suggestion that the current warming is predominately related to human emissions of greenhouse gases is countered by a recognition that the climate system is also recovering from the Little Ice Age, the last major climate deterioration. Whichever view, in whole or in part, proves correct, both raise important questions about the role of the ocean in climate change. If we are now in a period of enhanced global warming, for instance, what consequences might we face with the melting of ice in the Arctic or Antarctica? Is it possible that global warming will present unexpected, even counterintuitive consequences, such as altering the deep–water conveyor belt and plunging Europe into a new Little Ice Age? Are the causative factors of the Little Ice Age linked to the internal variability of the Earth system? Answering such questions is difficult. Each requires that we considerably improve our understanding of the climate system, and how it works.

Paleoclimatology provides us with a key perspective here. Combining the geologic record with climate models offers a powerful tool by which to assess climate system dynamics on decadal to millennial timescales. More specifically, we are now able to focus our research on the role of the world ocean, a principal element in governing climate system dynamics on these time scales.

Numerical models provide much of the foundation for projecting future climate change. For more than three decades, increasingly complex computer simulations have assessed the potential impact of anthropogenic greenhouse gases on climate. The attempts to look into the geologic past through a computer terminal have a shorter history. Applications of numerical models used in present–day atmosphere and ocean simulations in paleoclimate studies began slightly more than two decades ago, e.g., *Gates*, [1976]; *Barron* [1983]; *Kutzbach and Guettner* [1984]; *Manabe and Broccoli* [1985]; *Seidov* [1986]; *Bryan and Manabe* [1988]; *Barron and Peterson* [1989]; *Maier-Reimer et al.* [1990], to name just a few examples (the special issue of *Paleoceanography* prefaced by *Crowley* [1990] and *Crowley and North* [1991] provide more detail on early paleoceanographic modeling). Early efforts demonstrated that simulations of past climates and ocean circulation patterns have the potential to provide a number of valuable insights into the behavior of the climate system. More advanced and sophisticated paleoclimate and paleoceanographic modeling studies are now yielding significant information about present and possible future climate tendencies.

Earth system history provides a unique contribution to understanding climate change that is principally different from efforts that focus solely on modeling the future. The

The Oceans and Rapid Climate Change: Past, Present, and Future
Geophysical Monograph 126

geologic record contains abundant information that records global changes on a variety of spatial and temporal scales. In particular, the marine record provides a wealth of evidence on past climate changes. Thus, computer models of past climate change on the millennial time scale can be at least partly verified against geologic data – an option that does not exist for predicted future change. Moreover, paleoclimatology provides an important lesson by demonstrating that climate changes, externally and internally driven, can occur extremely quickly, even approaching human time scales. Much of the irregularity and abruptness of these climate changes can be attributed to the ocean. Thus, ocean currents are true "winds of change," the source of significant climate alterations. For this reason, the last two decades have witnessed an explosion of paleo-reconstructions and modeling of the deep ocean and the climatic links associated with the glacial–to–interglacial transition of the Pleistocene. These last two million years of climate history contain a record of substantial climate fluctuation between warmer and colder states with numerous apparent instabilities. The role of the ocean in these fluctuations is still not entirely understood.

The ocean, with its thermal, freshwater and dynamic impacts, controls our environment in a variety of ways. These controls are evident in the nature of ocean–atmosphere interactions, sea–ice dynamics, and in sea level changes. The ocean's role is thought to be dominant on decadal to millennial timescales because of its enormous heat capacity and its capability to redistribute heat by ocean currents. Although these time intervals are too long for the atmosphere to play a dynamically important role, they are too short to associate with changes in the Earth's orbital parameters, ice sheet dynamics, or tectonic activity. In addition, we must consider the growing evidence that the thermohaline circulation represents one of the key pacemakers of global climate, given its apparent sensitivity to relatively small changes of freshwater in the high latitudes. Further support of its importance stems from the speculation that anthropogenic global warming may threaten the stability of this oceanic overturning, potentially providing a considerable challenge to human societies.

In his marvelous new book, *The Two–Mile Time Machine*, Richard Alley points out that "most paleoclimatologists spend their time looking at ocean sediments" [*Alley*, 2000]. Most popular oceanographic books, of course, focus on surface ocean currents, using common examples such as the Gulf Stream to capture the significance of the ocean in climate. However, most ocean modelers direct substantial attention to the so–called meridional overturning streamfunction, a mathematical abstraction that describes the volume of water that circulates in the vertical

plane. Present–day oceanography recognizes that the meridional overturning is an intrinsic and powerful mechanism by which the ocean imposes vital thermal control over the Earth's climate. Slow thermohaline circulation, which is driven by density contrasts between low and high latitudes and between the surface and deep layer of the ocean, appears to be the most important link in the climate system on decadal to centennial time scale, and perhaps on even longer time intervals, as this circulation is responsible for the lion part of meridional oceanic poleward heat transport. The key focus of this volume is thus "oceanic overturning" with studies drawn from ice core research, marine sediments, and oceanic modeling.

New evidence from ice cores and deep-sea sediments shows that the climate of the two hemispheres may not be synchronized in their response to climate forcing (e.g., *Blunier et al.* [1998]). These findings add new questions concerning the role of the ocean thermohaline circulation. What is the potential for differential impacts involving both the northern and southern hemisphere? What governs the observed synchronicity? How do changes in freshwater fluxes alter their impacts? In response, we focus on the feedbacks that link ocean circulation with other elements of the climate system. The engine of long–term climate change, the deep–ocean circulation system, can dramatically change in response to freshwater impacts that are usually associated with melting of some elements of the cryosphere, either as sea ice melting or ice sheets surges.

The discovery of cold deepwater in the equatorial regions was followed by early notions that large–scale deep ocean motion was caused by density contrasts (these findings can be traced back to the eighteenth and early nineteen centuries — see details in *Gill* [1982]). Because motion in the abyss is predominantly geostrophic, with weak and slow turbulent mixing in the deep layers, the water density can be modified essentially only at the sea surface in contact with the atmosphere. Hence, the ocean global thermohaline circulation depends on how intensive the buoyancy flux is across the sea surface in the high latitudes. An understanding of how this oceanic overturning works was first provided by the Stommel–Arons theory [*Stommel and Arons*, 1960], which describes a scheme of a dipole high–latitudinal sinking of dense water in the northern North Atlantic in the Northern Hemisphere and in the Weddell and Ross Seas in the Southern Hemisphere (the most recent and enlightening explanation of this keystone theory is given by *Pedlosky* [1996]). Certainly the most important element in this scheme is the formation of deep water: North Atlantic Deep Water (NADW) produced in the deep convection sites in the Nordic Seas and northern North Atlantic, and Antarctic Deep Water (AABW) formed around Antarctica but primarily in the Weddell Sea. Thus, the first understanding of bi–polarity of the thermohaline circulation origin had been put forth.

Another important issue concerns the role of freshwater. Understanding the importance of salinity in ocean dynamics goes back to the Goldsbrough model [*Goldsbrough*, 1933], which suggests that, theoretically, the ocean circulation on a sphere can be maintained by evaporation and precipitation only, or, equivalently, by density contrasts caused by surface freshwater fluxes altering salinity distributions (also see a discussion of the Goldsbrough model in *Stommel* [1957]). However, real advances in our comprehension of the role of salinity in climate are provided in the pioneering work of *Bryan* [1986] followed by key studies of the stability of the ocean thermohaline circulation (e.g., *Manabe and Stouffer* [1988]; *Marotzke et al.* [1988]; *Weaver et al.* [1991]). A thermohaline circulation driven by buoyancy fluxes across the sea surface was shown to be sensitive to high–latitudinal density variations and to have more than one stable regime. In essence, salinity has, unlike temperature, no simple restoring feedbacks in the ocean–atmosphere system and therefore can be one of the crucial elements responsible for nonlinearity of climate dynamics, including bifurcation of the circulation regime (e.g., *Rahmstorf* [1995a]).

The growing interest in paleoclimate modeling, and in an improved understanding of the role of freshwater in climate change, coincided with the development of new concepts about global ocean circulation. Based on extensive observational, theoretical and modeling efforts, a new vision of the global interhemispheric and interoceanic water transport in the deep-ocean flow system began to emerge about a decade ago (e.g., *Gordon* [1986]; *Broecker and Denton* [1989]; *Cox* [1989]). This new image of deep-ocean currents as a major player has been transformed into the concept of a "global ocean conveyor" connecting the most remote ocean regions and being most sensitive to high-latitudinal freshwater fluxes, and thus to high-latitudinal salinity fluctuations (a "salinity conveyor belt" according to *Broecker* [1991]). Further progress in understanding of freshwater impacts in the high latitudes has been achieved through a number of studies (e.g., in modeling effort of *Weaver et al.* [1991; *Maier-Reimer et al.* [1993]; *Manabe and Stouffer* [1995]; *Rahmstorf* [1995b] to name just a few).

The current paradigm of the global ocean thermohaline conveyor is that it is driven by the formation of the NADW, with surface poleward currents compensating the NADW outflow (e.g., a review in *Gordon et al.* [1992] and a discussion in *Boyle and Weaver* [1994]). As the NADW crosses the equator, there must be a compensating flow at

the surface that maintains continuity of the ocean circulation. The compensating northward warm surface flow crosses the equator and thus provides cross–equatorial heat transport to the high latitudes in the Northern Hemisphere in the Atlantic sector. This cross-equatorial warm flow allows northern Europe to enjoy a far warmer climate than the countries in the Southern Hemisphere at the same distance from the equator. Thus, the high latitudes are potent regulators of the impact of the ocean on global climate (e.g., *Stocker et al.* [1992]; *Stocker* [1994]; *Sakai and Peltier* [1995]; *Broecker et al.* [1999]; *Schmittner and Stocker* [1999]; *Wang and Mysak* [2000]; *Ganopolski and Rahmstorf* [2001]).

One of the most noticeable attributes of present-day climate is its hemispheric asymmetry, with a warm ocean surface in the northern North Atlantic, a moderately cool northern North Pacific and a much colder Southern Ocean (e.g., *Weyl* [1968]; see also the most recent discussion of this issue in *Weaver et al.* [1999]). The salinity contrasts between different ocean regions, and most notably between a saltier Atlantic and fresher Pacific, is a signature of a global abyssal connection that may be responsible for the climate state in which we are living. Therefore, a change in this connection may be at least partly responsible for past and possibly future long–term climate fluctuations that are not externally driven (for example, due to tectonics or change of the Earth's orbit). The bi-polar character of the deep-ocean circulation – a prominent feature of the thermohaline engine – may be helpful in understanding these issues.

The idea of the so-called bi-polar ocean seesaw [*Broecker*, 1998; *Stocker*, 1998; *Broecker*, 2000] (see also a discussion in this volume [*Seidov et al.*, 2001; *Stocker et al.*, 2001]) is the most recent and one of the most exciting additions to the Stommel-Arons scheme and to the whole global conveyor paradigm. The bi-polar seesaw is a fluctuating meridional overturning regime driven by two deep-water sources — the NADW in the north, and the AABW in the south — responding primarily to high-latitudinal freshwater fluctuations and therefore presumably linked to the major glacial cycles of the Pleistocene. However, it has been shown that changes in the intensity of the AABW do not impose a direct impact on the deep–ocean thermal regime. AABW variability controls deep–ocean warming or cooling via NADW, which has been shown to vary, at least in some cases, in counter–phase with the AABW fluctuations (*Seidov et al.* [2001], this volume). The bi–polar seesaw idea appears to be a conceivable mechanism of some of the glacial-to-interglacial fluctuations linked to the global ocean conveyor variability. It may also imply that the observed north-south asymmetry of the millennial-scale climate variability, including the Little Ice Age, was caused by a flip-flopping of the deep-ocean circulation regime [*Broecker*, 2000].

This introduction is not intended to be a comprehensive listing of the many very important and enlightening publications on the role of the oceans in climate change. A Herculean amount of work would be needed to outline the state of science for either observations or modeling efforts addressing this topic. Rather, our intent is to give a sense of the excitement and advances associated with the main issue of the volume — the role of the ocean in fast climate change in the past and future. This volume is a rather balanced presentation of both observational and modeling directions within this theme. Hundreds of references in individual papers within the volume give a broad array of relevant publications supporting the interesting conclusions reached by the authors within the volume. The papers themselves provide an equally broad perspective on the core ideas of the ocean's impact on climate, and specifically on the role of deep-ocean in climate change. Thus, this introductory text reflects only the views outlined above that demonstrate why so much research attention is needed on the role of the ocean in understanding past and possible future climate changes. In essence, we argue that it is time to think along the lines of "ocean currents of change".

Acknowledgements. Discussions with Bernd J. Haupt, Mark Maslin and Allan Graubard, which were helpful in improving the introductory text, are much appreciated.

REFERENCES

Alley, R.B., 2000, The two-mile time machine: ice cores, abrupt climate change, and our future. Princeton University Press, Princeton, NJ, 229 pp.

Barron, E.J., 1983: A warm, equable Cretaceous: the nature of the problem, *Earth-Science Reviews*, 19: 305-338.

Barron, E.J. and W.H. Peterson, 1989: Model simulations of the Cretaceous ocean circulation, *Science*, 244: 684-686.

Blunier, T., J. Chappellaz, J. Schwander, A. Daellenbach, B. Stauffer, T.F. Stocker, D. Raynaud, J. Jouzel, H.B. Clausen, C.U. Hammer and S.J. Johnsen, 1998: Asynchrony of Antarctic and Greenland climate change during the last glacial period, *Nature*, 394: 739-743.

Boyle, E. and A. Weaver, 1994: Conveying past climates, *Nature*, 372: 41-42.

Broecker, W., 1991: The great ocean conveyor, *Oceanography*, 1: 79-89.

Broecker, W.S., 1998: Paleocean circulation during the last deglaciation: A bipolar seesaw?, *Paleoceanography*, 13: 119-121.

Broecker, W.S., 2000: Was a change in thermohaline circulation responsible for the Little Ice Age?, *Proc. Nat. Acad. Sci.*, 97(4): 1339-1342.

Broecker, W.S. and G.H. Denton, 1989: The role of ocean atmosphere reorganizations in glacial cycles, *Geochimica Cosmochimica Acta*, 53: 2465-2501.

Broecker, W.S., S. Sutherland and T.-H. Peng, 1999: A possible 20th century slowdown of Southern Ocean deep water formation, *Science*, 286: 1132-1135.

Bryan, F., 1986: High-latitude salinity effects and interhemispheric thermohaline circulations, *Science*, 323: 301-304.

Bryan, K. and S. Manabe, 1988: Ocean circulation in warm and cold climates, In: M.E. Schlesinger (Editor), *Physically-Based Modelling and Simulation of Climateic Change - Part II*. Kluwer Academic Publishers, New York, pp. 951-966.

Cox, M., 1989: An idealized model of the world ocean, Part I: The global-scale water masses, *Journal of Physical Oceanography*, 19: 1730-1752.

Crowley, T.J., 1990: Foreword, *Paleoceanography*, 5(3): p. 297.

Crowley, T.J. and G.R. North, 1991, Paleoclimatology. New York, Oxford Univ. Press, 339 pp.

Ganopolski, A. and S. Rahmstorf, 2001: Stability and variability of the thermohaline circulation in the past and future: a study with a coupled model of intermediate complexity, *Geophysical Monograph (This Volume)*, D. Seidov, B.J. Haupt and M. Maslin (Editors), American Geophysical Union, Washington, D.C.

Gates, W.L., 1976: the numerical simulation of ice-age climate with a global general circulation model, *Journal of Atmospheric Sciences*, 33: 1844-1873.

Gill, E.G., 1982, Atmosphere-ocean dynamics, International Geophysical Series, 30. Academic Press, San Diego, 666 pp.

Goldsbrough, G.R., 1933: Ocean currents produced by evaporation and precipitation, *Proc. R. Soc. London Ser. A*, 141: 512-517.

Gordon, A., 1986: Interocean exchange of thermocline water, *Journal of Geophysical Research*, 91: 5037-5046.

Gordon, A.H., S.E. Zebiak and K. Bryan, 1992: Climate variability and the Atlantic Ocean, *Eos, Transactions, American Geophysical Union*, 79: 161,164-165.

Kutzbach, J.E. and P.J. Guettner, 1984: The sensitivity of monsoon climates to orbital parameter changes for 9000 years B.P.: Experiments with the NCAR general circulation model, In: A. Berger, J. Imbrie, J. Hays, G.J. Kukla and B. Saltzman (Editors), *Milankovitch and climate*. D. Reidel, Dordrecht, Netherlands, pp. 801-820.

Maier-Reimer, E., U. Mikolajewicz and T. Crowley, 1990: Ocean general circulation model sensitivity experiment with an open central American isthmus, *Paleoceanography*, 5: 349-366.

Maier-Reimer, E., U. Mikolajewicz and K. Hasselmann, 1993: Mean circulation of the Hamburg LSG OGCM and its sensitivity to the thermohaline surface forcing, *Journal of Physical Oceanography*, 23: 731-757.

Manabe, S. and A.J. Broccoli, 1985: The influence of continental ice sheets on the climate of an ice age, *Journal of Geophysical Research*, 90D: 2167-2190.

Manabe, S. and R.J. Stouffer, 1988: Two stable equilibria of a coupled ocean-atmosphere model, *Journal of Climate*, 1: 841-866.

Manabe, S. and R.J. Stouffer, 1995: Simulation of abrupt change induced by freshwater input to the North Atlantic Ocean, *Nature*, 378: 165-167.

Marotzke, J., P. Welander and J. Willebrand, 1988: Instability and multiple steady states in a meridional-plane model of the thermohaline circulation, *Tellus*, 40A: 162-172.

Pedlosky, J., 1996, Ocean Circulation Theory. Springer, New York, 453 pp.

Rahmstorf, S., 1995a: Bifurcations of the Atlantic thermohaline circulation in response to changes in the hydrological cycle, *Nature*, 378: 145-149.

Rahmstorf, S., 1995b: Multiple convection patterns and thermohaline flow in an idealized OGCM, *Journal of Climate*, 8: 3027-3039.

Sakai, K. and W.R. Peltier, 1995: A simple model of the Atlantic thermohaline circulation: Internal and forced variability with paleclimatological implications, *Journal of Geophysical Research*, 100: 13455-13479.

Schmittner, A. and T.F. Stocker, 1999: The stability of the thermohaline circulation in global warming experiments., *Journal of Climate*, 12: 1117-1133.

Seidov, D., B.J. Haupt, E.J. Barron and M. Maslin, 2001: Ocean bi-polar seesaw and climate: Southern versus northern meltwater impacts, *Geophysical Monograph (This Volume)*, D. Seidov, B.J. Haupt and M. Maslin (Editors), American Geophysical Union, Washington, D.C.

Seidov, D.G., 1986: Numerical modeling of the ocean circulation and paleocirculation, In: K. Hsü (Editor), *Mesozoic and Cenozoic oceans. Geodynamics Series.* AGU, Washington, D.C., pp. 11-26.

Stocker, T.F., 1994: The variable ocean, *Nature*, 367: 221-222.

Stocker, T.F., 1998: The seesaw effect, *Science*, 282: 61-62.

Stocker, T.F., R. Knutti and J.-K. Plattner, 2001: The future of the thermohaline circulation - a perspective, , *Geophysical Monograph (This Volume)*, D. Seidov, B.J. Haupt and M. Maslin (Editors), American Geophysical Union, Washington, D.C.

Stocker, T.F., D.G. Wright and W.S. Broecker, 1992: The influence of high-latitude surface forcing on the global thermohaline circulation, *Paleoceanography*, 7: 529-541.

Stommel, H., 1957: Survay of ocean current theory, *Deep Sea Research*, 4: 149-184.

Stommel, H. and A.B. Arons, 1960: On the abyssal circulation of the world ocean, II, An idealized model of the circulation pattern and amplitude in the oceanic basins, *Deep Sea Research*, 6: 217-233.

Wang, Z. and L.A. Mysak, 2000: A simple coupled atmosphere-ocean-sea-ice-land surface model for climate and paleoclimate studies, *Journal of Climate*, 13: 1150-1172.

Weaver, A.J., C.M. Bitz, A.F. Fanning and M.M. Holland, 1999: Thermohaline circulation: high latitude phenomena and the difference between the Pacific and Atlantic, *Annual Review of Earth and Planetary Sciences*, 27: 231-285.

Weaver, A.J., E.S. Sarachik and J. Marotzke, 1991: Freshwater flux forcing of decadal and interdecadal oceanic variability, *Nature*, 353: 836-838.

Weyl, P.K., 1968: The role of the oceans in climatic change: a theory of ice ages, *Meteorological Monographs*, 8: 37-62.

E.J. Barron, EMS Environment Institute, Pennsylvania State University, University Park, PA 16802. (eric@essc.psu.edu)

D.Seidov, EMS Environment Institute, Pennsylvania State University, University Park, PA 16802. (dseidov@essc.psu.edu)

Section I

Data and Climate Models: Windows to the Past

Synthesis of the Nature and Causes of Rapid Climate Transitions During the Quaternary

Mark Maslin

Environmental Change Research Centre, University College London, London , UK

Dan Seidov

EMS Environment Institute, Pennsylvania State University, University Park, Pennsylvania

John Lowe

Centre for Quaternary Research, Royal Holloway College, University of London, UK

The last few million years have been punctuated by many abrupt climate transitions many of them occurring on time-scales of centuries or even decades. In order to better understand our current climate system we need to understand how these past climate transitions occurred. In this study we examine and review the paleoclimate proxies and modeling results for each of the key climate transitions in the Quaternary period. These are identified as: 1) Onset of Northern Hemisphere Glaciation (ONHG), 2) glacial-interglacial cycles, 3) Mid-Pleistocene Revolution (MPR), 4) Heinrich events and glacial Dansgaard-Oeschger (D-O) cycles, 5) last glacial-interglacial transition (LGIT) and the Younger Dryas, 6) interglacial climate transition such as the Intra-Eemian cold event and Holocene D-O cycles. For each climate transition the current theories of causation are critically examined and our own synthesis based on current knowledge is put forward. Most of these transitions appear to be threshold changes where by external forcing combined with internal feedbacks leads to a change in the state of global climate. We argue that bifurcation within the climate system means that it is easier for the global climate to go through these thresholds than it is to return to its previous state. Although this does not necessarily make climate change irreversible, it may provide a mechanism, which facilitates the locking of the climate system into a new equilibrium state. We suggest that the evidence indicates that long-term climate change occurs in sudden jumps rather than incremental changes; which does not bode well for the future.

INTRODUCTION

Rapidity of Global Climate Change

Until a few decades ago it was generally thought that significant large-scale global and regional climate changes occurred gradually over a time scale of many centuries or millennia. Hence the climate shifts were assumed to be scarcely perceptible during a human lifetime. The tendency of climate to change abruptly has been one of the most surprising outcomes of the study of earth history, especially from the polar ice core records for the last 150,000 years [e.g., *Taylor et al.* 1993; *Dansgaard et al.*, 1993; *Alley*, 2000]. Some and possibly most pronounced climate changes (involving, for example, a regional change in mean annual temperature of several degrees Celsius) occurred on a time scale of a few centuries, but frequently within a few decades, or even just a few years.

The decadal-time-scale transitions would presumably have been quite noticeable to humans living at such times [e.g., *deMenocal*, 2001]. For instance *Hodell et al.* [1995] and *Curtis et al.* [1996] consider the possible importance of climate change on the collapse of the Classic period of Mayan civilization. It has also been suggested that alternating wet and dry periods influenced the rise and fall of coastal and highland cultures of Ecuador and Peru [*Thompson and Mosley-Thompson*, 1987; *Thompson*, 1989; *Thompson et al.*, 1995]. The emergence of crop agriculture in the Middle East corresponds very closely with a sudden warming event marking the beginning of the Holocene [*Wright*, 1993], while the global collapse of the urban civilizations coincided with the deterioration of climate around 4,300 BP [*Peiser*, 1998; *Cullen et al.*, 2000]. On longer time scales, the evolution and migration of modern humans has also been linked to climatic changes in Africa [e.g., *de Menocal*, 1995; *Vrba et al.*, 1996; *Wilson et al.*, 2000]. These sudden stepwise climate transitions are also a disturbing scenario to be borne in mind when considering the effects that humans might have on the present climate system through the rapid generation of greenhouse gases. Judging by what we can learn from records of the past, conditions might gradually building to a 'break point' or threshold following which a dramatic change in the global climate system might occur over just a decade or two. The actual trigger for this transition, because of the threshold, may be quite innocuous. In this overview we summarize the current ideas on the causes of rapid climate changes during the last 2.5 million years, their impacts on the modern climate system and implications for our anthropogenic 'forcing' of the global climate system. In many cases we are still at the stage of speculating what may have been the underlying cause of these transitions and thus there is still a huge amount of work to be done in paleoclimatology.

Modes of Climate Change

Global climate changes are responses to external or internal forcing mechanisms. A good example of an external forcing mechanism is the changing orbital parameters which alter the net radiation budget of the Earth, while an example of an internal forcing mechanism is the carbon dioxide content of the atmosphere which modulates the greenhouse effect. We can abstract the way the global climate system responds to an internal or external forcing agent by examining four different scenarios:

Linear and synchronous response (Figure 1a). In this case the forcing produces a direct response in the climate system whose magnitude is in proportion to the forcing.

Muted or limited response (Figure 1b). In this case the forcing may be extremely strong, but the climate system in someway buffered and therefore has very little response.

Delayed or non-linear response (Figure 1c). In this case the climate system may have a slow response to the forcing or is in some way buffered at first. After an initial period then the climate system responds to the forcing but in a non-linear way.

Threshold response (Figure 1d). In this case initially there is no or very little response in the climate system to the forcing, however all the response takes place in a very short period of time in one large step or threshold. In many cases the response may be much larger than one would expect from the size of the forcing and this can be referred to as a response over-shoot.

Though these are purely theoretical models of how the climate system can respond they are important to keep in mind when reviewing global climate transitions. Moreover these scenarios can be applied to the whole range of spatial and temporal scales. This review hopefully illustrates how a study of paleoclimatology helps to distinguish between these possible scenarios and establishing how regional and global climate responded to different forcing mechanisms in the past. Examples of abrupt major climate transitions and cycles are reviewed, ranging from the initiation of Northern Hemisphere Glaciation and the start of the Quaternary, to the higher (on geological time scale) frequency climate changes that characterize the Holocene. Figure 2 shows on a the most important major climatic phenomena that occurred during the Quaternary period log time-scale. These phenomena are the main focus of the review.

An added complication when assessing the causes of climate changes is the possibility that climate thresholds contain bifurcations. This means the forcing required to go one way through the threshold is different from the reverse. This implies that once a climate threshold has occurred it is a lot more difficult to reverse it. This bifurcation of the climate system has been inferred from ocean models which mimic the impact of freshwater on deep-water formation in the North Atlantic [e.g., *Rahmstorf*, 1996]. However such an irregular relationship can be ap-

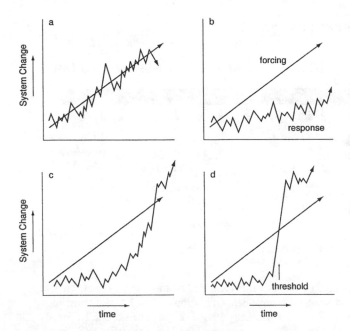

Figure 1. Schematic of the four alternative responses of the global climate system to internal or external forcing: a) Linear and synchronous response. In this case the forcing produces a direct response in the climate system whose magnitude is in proportion to the forcing. b) Muted or limited response. In this case the forcing may be extremely strong, but the climate system in someway buffered and therefore has very little response. c) Delayed or non-linear response. In this case the climate system may have a slow response to the forcing or is in some way buffered at first. After an initial period then the climate system responds to the forcing but in a non-linear way. d) Threshold response. In this case initially there is no or very little response in the climate system to the forcing, however all the response takes place in a very short period of time in one large step or threshold. In many cases the response may be much larger than one would expect from the size of the forcing and this can be referred to as a response overshoot.

plied to any forcing mechanism and the corresponding response of the global climate system. Figure 3 demonstrates this bifurcation of the climate system and shows that there can be different relationships between climate and the forcing mechanism, depending on the direction of the threshold. This is very common in natural systems for example in cases where inertia or the shift between different states of matter need to be overcome. Figure 3 shows that in case A and B the system is reversible, but in case C it is not. In case C the control variable must increase to more than it was in the previous equilibrium state to get over the threshold and return the system to it pre-threshold state. Let us consider this in terms of the salinity of the

North Atlantic versus the production of North Atlantic Deep Water (NADW), as we know that adding more freshwater to the North Atlantic hampers the production of salty cold and hence heavy deep water. In Case A, changing the salinity of the North Atlantic has no effect on the amount of NADW produced. In Case B reducing the salinity reduces the production of NADW, however if the salt is replaced then the production of NADW returns to it previous, pre-threshold level. In case C, reducing the North Atlantic salinity reduces the production of NADW. However simply returning the salt does not return the NADW production to the normal level. Because of the bifurcation, a lot more salt has to be injected to bring back the NADW production to its previous level. It may be that the extra amount of salt required is not possible within the system and hence this makes the system theoretically irreversible. The major problems we face when looking at both past and future climate change whether a bifurcation may occur and whether evolution of the system is reversible (Figure 3).

When discussing the behavior of climatic forcing the response time of the different parts of the Earth climate system must also be considered. Figure 4 shows the response times of the different internal systems that vary from hundreds of millions of years to days. The complication is that all of these processes are constantly changing but at different rates. They will also have different responses to external or internal forcing as suggested in Figure 1. In this paper we review how the history of Quaternary climatic change has been influenced by a wide range of possible forcing mechanisms from the long term tectonic forcing to the much short human time scale changes in the atmospheric and ocean circulation patterns, the impacts of which could be detected within a human lifetime (Figure 4). The paper is arranged along a time-line starting with the onset of the Quaternary period and its characteristic great 'ice ages' [Wilson et al., 2000].

I. ONSET OF NORTHERN HEMISPHERE GLACIATION.

I.A. TIMING.

The earliest recorded onset of significant global glaciation during the last 100 Ma was the widespread continental glaciation of Antarctica at about 34 Ma [e.g., *Hambrey et al.*, 1991; *Breza and Wise*, 1992; *Miller et al.*, 1991; *Zachos et al.*, 1992; 1996; 1999; 2001]. In contrast the earliest recorded glaciation in the Northern Hemisphere is between 10 and 6 Ma [e.g., *Jansen et al.*, 1990; *Wolf and Thiede*; 1991; *Jansen and Sjøholm*, 1991; *ODP Leg 151*, 1994; *Wolf-Welling et al.*, 1995]. Marked expansion of continental ice sheets in the Northern Hemisphere was the

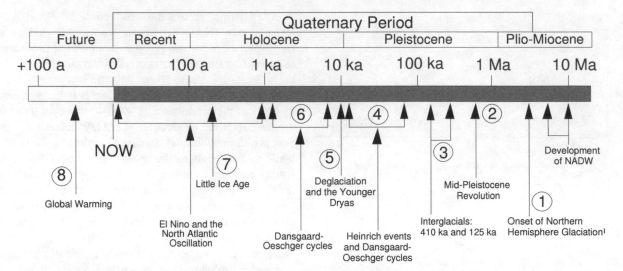

Figure 2. Log time-scale cartoon, illustrating the most important climate events identified by this study in the Quaternary Period [adapted from *Adams et al.*, 1999]. 1. Onset of Northern Hemisphere Glaciation (3.2 to 2.5 Ma) ushering in the strong glacial-interglacial cycles which are characteristic of the Quaternary Period. 2. Mid-Pleistocene Revolution when glacial-interglacial cycles switched from 41 ka, to every 100 ka. The external forcing of the climate did not change, thus, the internal climate feedback's must have altered. 3. The two closest analogs to the present climate are the interglacial periods at 420-390 ka (Oxygen isotope stage 11) and 130-115 ka (Oxygen isotope stage 5e, also known as the Eemian). 4. Heinrich events and Dansgaard-Oeschger cycles (see Text) 5. Deglaciation and the Younger Dryas events 6. Holocene Dansgaard-Oeschger cycles (see Text) 7. Little Ice Age (1700 AD) the most recent climate event which seemed have occur throughout the Northern Hemisphere. 8. Anthropogenic Global Warming

culmination of a longer term, high latitude cooling, which began with this late Miocene glaciation of Greenland and the Arctic and continued through to the major increases in global ice volume around 2.5 Ma [*Prell*, 1984; *Keigwin*, 1986; *Ruddiman et al.*, 1986a and b; *Sarnthein and Tiedemann*, 1989; *Tiedemann et al.*, 1994].

Evidence from orbitally tuned Ocean Drilling Program records (Sites, 609, 610, 642, 644, 552, 882, 887] *Maslin et al.* [1998a] suggested that this long term cooling led to three key steps in the glaciation of the Northern Hemisphere (see Figure 5): a) Eurasian Arctic and Northeast Asia were glaciated at approximately 2.74 Ma, b) glaciation of Alaska at 2.70 Ma and c) the significant glaciation of the North East American continent at 2.54 Ma. Sea level changes, inferred from the benthic foraminiferal oxygen isotope records obtained from ODP Sites 659 [*Tiedemann et al.*, 1994] and 846 [*Shackleton et al.*, 1995, see Fig. 5] supports this and suggests that the two most important stages in the glaciation of the Northern Hemisphere were the maturing of the Eurasian-Northeast Asian ice sheets (oxygen isotope stage 110 or G6) and the proto-Laurentide ice sheet on the eastern North American continent (oxygen isotope stage 100). This step-like nature of the ice rafting records may, however, conceal a more gradual process of

ice build-up, as revealed by the progressive ^{18}O enrichment of benthic isotope records [*Tiedemann et al.*, 1994; *Shackleton et al.*, 1995]. This is because the ice-rafting records indicate only when the continental ice sheets were mature enough to impinge on the adjacent oceans. Dramatic changes, however, are observed in each ocean basin when ice-rafting first occurs.

I.b. Possible Causes of the Onset of Northern Hemisphere Glaciation.

The predominant theories to explain the Onset of Northern Hemisphere Glaciation have focused on major tectonic events and their modification of both atmospheric and ocean circulation [*Hay*, 1992; *Raymo*, 1994a; *Maslin et al.*, 1998a]. For example the uplift and erosion of the Tibetan-Himalayan plateau [*Ruddiman and Raymo*, 1988; *Raymo*, 1991, 1994b], the deepening of the Bering Straits [*Einarsson et al.*, 1967] and/or the Greenland-Scotland ridge [*Wright and Miller*, 1996], and emergence of the Panama Isthmus [*Keigwin*, 1978, 1982; *Keller et al.*, 1989; *Mann and Corrigan*, 1990; *Haug and Tiedemann*, 1998].

Ruddiman and Raymo [1988], *Ruddiman et al.* [1989], and *Ruddiman and Kutzbach* [1991] suggested that the initiation of Northern Hemisphere glaciation was caused

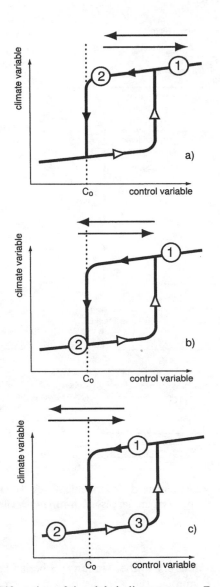

Figure 3. Bifurcation of the global climate system. For example the control variable could be North Atlantic salinity and the climate variable could be production of NADW. a) An insensitive system and the control variable does not vary greatly with large changes in the control variable. b) The control variable drops beneath the critical threshold point C0 (point 2) and thus there is a major change in the climate variable, however by return the control variable to its original state, the system is reversible and the climate variable returns to its original point 1. c) The control variable drops beneath the critical threshold point C0 (point 2), however returning the control variable to its original state does not reverse the change and the climate variable remains at point 3. An additional change to the control variable is required to overcome the bifurcation and return the climate variable back to point 1. Hence, if this additional change is not possible within the system, the threshold becomes an irreversible one. It should be notes that returning the system back to its initial boundry contions by going beyond point 3 may take the system back to a new state of equilibrium that is different from the initial state.

by progressive uplift of the Tibetan-Himalayan and Sierran-Coloradan regions. This may have altered the circulation of atmospheric planetary waves such that summer ablation was decreased, allowing snow and ice to build-up in the Northern Hemisphere. There is much speculation as to whether: orography [*Charney and Eliasson*, 1949; *Bolin*, 1950], differential heating of land and sea surfaces [*Sutcliffe*, 1951; *Smagorinsky*, 1953], or a combination of both [*Trenberth*, 1983], control the structure and direction of the planetary waves. However, most of the Himalayan uplift occurred much earlier between 20 Ma and 17 Ma [*Copeland et al.*, 1987; *Molnar and England*, 1990] and the Tibetan Plateau reached its maximum elevation during the late Miocene [*Harrison et al.*, 1992, *Quade et al.*, 1989]. *Raymo et al.* [1988], *Raymo* [1991, 1994b] and *Raymo and Ruddiman* [1992] then suggesting that the uplift caused a massive increase in tectonically driven chemical weathering in the late Cenozoic. They argue that carbonation of rainwater removes CO_2 from the atmosphere and forms a weak carbonic acid solution. Dissociated H^+ ions in the acidified rainwater by hydrolysis leads to enhanced chemical weathering of rocks. Only weathering of silicate minerals makes a difference to atmospheric CO_2 levels, as weathering of carbonate rocks by carbonic acid returns CO_2 to the atmosphere. Bi-products of hydrolysis reactions affecting silicate minerals are biocarbonates ($HCO_3.$) anions and calcium cations. These, when washed into the oceans, are metabolised by marine plankton and are converted to calcium carbonate. The calcite skeletal remains of the marine biota are ultimately deposited as deep-sea sediments and hence lost from the global biogeochemical carbon cycle for the duration of the life cycle of the oceanic crust on which they were deposited. Consequently, atmospheric CO_2 could have been depleted causing a cooling of the global climate and thus the OHNG. This theory, however, suffers from a number of major draw-backs, 1) there is debate whether the Strontium (Sr) isotope data can be used as evidence of continental weathering [*Kirshnaswami et al.*, 1992; *Berner and Rye*, 1992; *Huh and Edmund*, 1998], 2) there seems to be no obvious negative feedback mechanism to prevent a complete depletion of the relatively small reservoir of atmospheric CO_2 [e.g., *Berner*, 1994; *Compton and Mallinson*, 1996, *Raymo*, 1994b] and 3) there is now evidence that suggest that there was no decrease in atmsopheric carbon dioxide during the Miocene [*Pagani et al.*, 1999; *Pearson and Palmer*, 2000].

A second key tectonic control invoked as a trigger for the ONHG is the closure of the Pacific-Caribbean gateway. *Haug and Tiedemann* [1998] suggest it began to emerge at 4.6 Ma and finally closed at 1.8 Ma. However, there is still considerable debate on the exact timing of the closure [*Burton et al.*, 1997; *Vermeij*, 1997; *Frank et al.*, 1999].

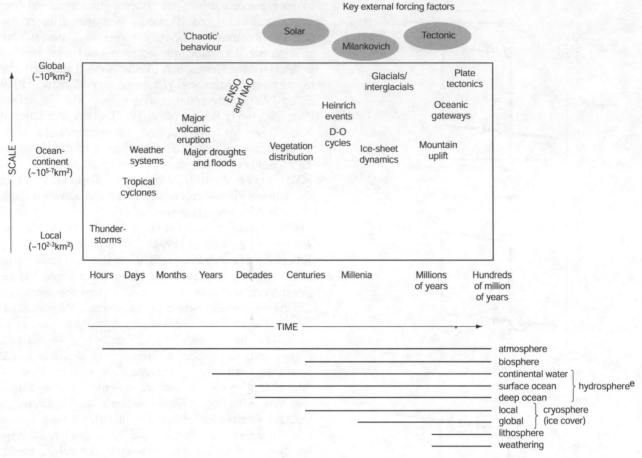

Figure 4. The spatial and temporal dimensions of the Earth's climate system plotted on logarithmic scales. The key forcing functions and response time of different section of the global climate system are also shown. D-O cycles = Dansgaard-Oeschger cycles.

The closure of the Panama gateway however causes a paradox [*Berger and Wefer*, 1996] as there is considerable debate whether it would have helped or hindered the intensification of North Hemisphere glaciation. The reduced inflow of Pacific surface water to the Caribbean would have increased the salinity of the Caribbean. This would have increased the salinity of water carried northward by the Gulf Stream and North Atlantic Current, thus enhancing deepwater formation [*Mikolajewicz et al.*, 1993; *Berger and Wefer*, 1996]. Increased deep-water formation could have worked against the initiation of Northern Hemisphere glaciation as it enhances the oceanic heat transport to the high latitudes and would have opposed ice sheet formation (see Figure 6). This enhanced Gulf Stream would also have pumped more moisture northward, stimulating the formation of ice sheets [*Mikolajewicz et al.*, 1993; *Berger and Wefer*, 1996]. *Driscoll and Haug* [1998] also argue that this increased moisture supply would have enhanced

freshwater delivery to the Arctic via Siberian Rivers. This freshwater input to the Arctic would have facilitated sea ice formation (fresher water has higher freezing point), increased albedo and isolated the high heat capacity of the ocean from the atmosphere. So it would have acted as a negative feedback on the efficiency of the "deep water conveyor belt" heat pump [*Broecker*, 1997a].

Tectonic forcing alone cannot explain the fast changes of both the intensity of glacial-interglacial cycles and mean global ice volume. It has, therefore, been suggested that changes in orbital forcing may have been an important mechanism contributing to the gradual global cooling and the subsequent rapid intensification of Northern Hemisphere glaciation [*Lourens and Hilgen*, 1994; *Maslin et al.*, 1995a; 1998a; *Haug and Tiedemann*, 1998]. This theory extends the ideas of *Berger et al.* [1993] by recognizing distinct phases during the Pleistocene and late Pliocene, characterized by the relative predominant strength of the

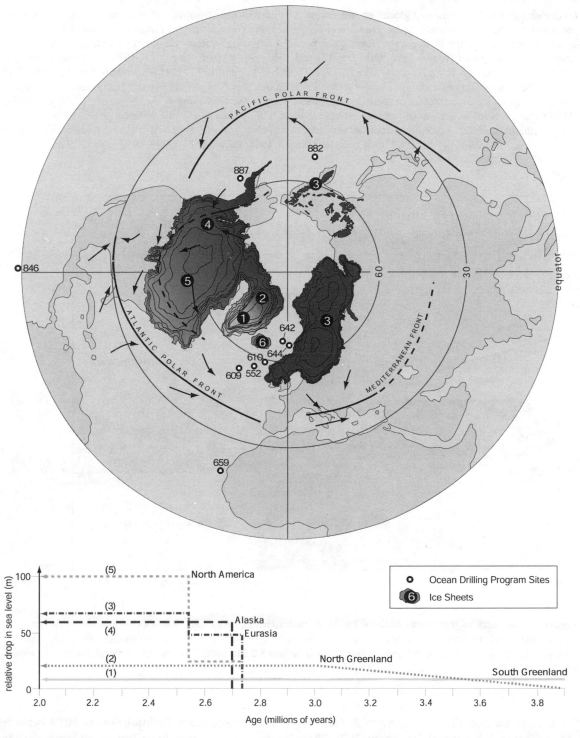

Figure 5. Timing of the intensification of Northern Hemisphere Glaciation [adapted from *Maslin et al.*, 1998a]. Map shows the extent of the Northern Hemisphere ice sheets during the Last Glacial Maximum [*CLIMAP*, 1976, 1981]. The separate ice sheets are identified by different colours. The **modern** climatic fronts are also shown to represent the possible atmospheric circulation of the pre-glacial Pliocene. The graph shows the estimated time at which each ice sheet was large enough to reach the edge of the continents and thus release icebergs, compared with its relative effect on the sea level, estimated from the change in the global $\delta^{18}O$ signal [*Tiedemann et al.*, 1994; *Shackleton et al.*, 1995]. Ice-

land seems to have had repeated glacial episodes from the mid-Pliocene onwards, but no clear dates are yet available for when this became sustained [*Einarsson and Albertsson*, 1988; *Geirsdottir and Eiriksson*, 1994]. The data for the onset of ice rafting of the different ice sheets come from the following sources: Southern Greenland - coarse fraction analysis [e.g., *Wolf and Thiede*, 1991: ODP Leg 151 and 152, 1994; *Wolf-Welling et al.*, 1995], Northern Greenland - coarse fraction analysis [e.g., *Wolf and Thiede*, 1991: ODP Leg 151 and 152, 1994; *Wolf-Welling et al.*, 1995], Eurasian Arctic - lithic fragment counts [*Jansen et al.*, 1990; *Jansen and Sjøholm*, 1991] and Northeast Asia - magnetic susceptibility and coarse fraction analysis [*Haug*, 1995; *Haug et al.*, 1995a and b; *Maslin et al.*, 1995a and 1998a], Alaska and the Northwest coast of America - magnetic susceptibility [*Rea et al.*, 1993; *Maslin et al.*, 1995a and 1998a] and Northeast America - calcium carbonate, magnetic susceptibility, stable isotopes, ostracodes and coarse fraction analysis [e.g., *Shackleton et al.*, 1984; *Ruddiman and Raymo*, 1988; *Raymo et al.*, 1989, 1992; *Cronin et al.*, 1996].

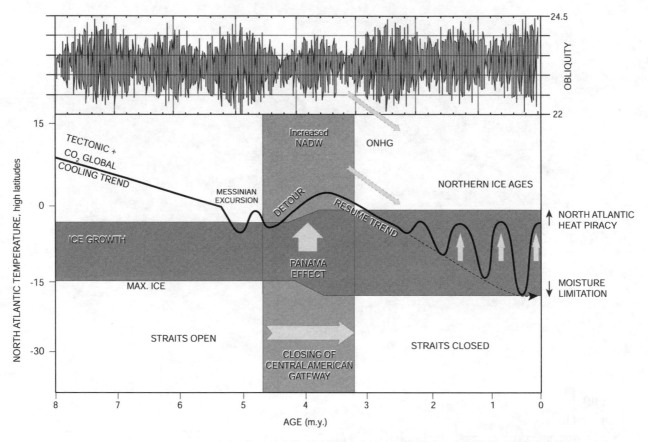

Figure 6. Summary of the causes of the Onset of Northern Hemisphere Glaciation (ONHG). Note the detour of the cooling trend and the expansion of the temperature range of ice growth caused by the closure of the central American gateway. Note also the kick in obliquity which occurs between 3.2 Ma and 2.5 Ma which drives the continued cooling of the climate system and the ONHG [adapted from *Berger and Wefer*, 1996].

different orbital parameters during each interval. *Maslin et al.* [1995a, 1998a] and *Haug and Tiedemann* [1998] have suggested that the observed increase in the amplitude of orbital obliquity cycles, from 3.2 Ma onwards, may have increased the seasonality of the Northern Hemisphere, thus initiating the long-term global cooling trend (see Figure 6). The subsequent sharp rise in the amplitude of precession

and consequently in insolation at 60°N between 2.8 Ma and 2.55 Ma may have forced the rapid glaciation of the Northern Hemisphere. This theory is supported by simulation of the Northern Hemisphere ice-sheet volume variation made by *Li et al.* [1998] with the LLN 2-D model [*Galleé et al.*, 1991, 1992]. In these experiments, ice volume fluctuations were forced by insolation variations

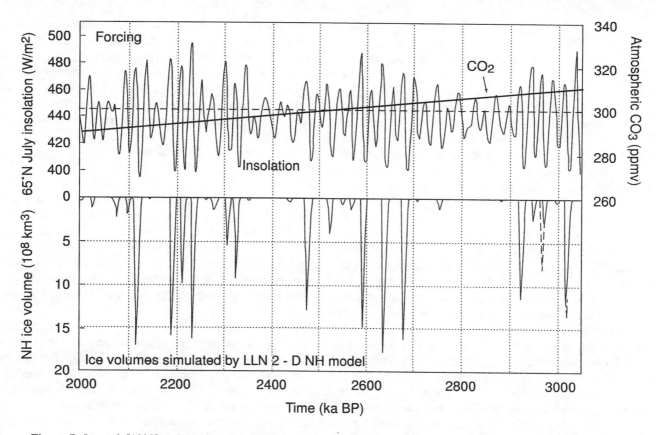

Figure 7. *Li et al.* [1998] and *Maslin et al.* [1998a] clearly showed from modelling results that Northern Hemisphere ice sheets could build-up around 2.5 Ma by only the variations in the orbital parameters. The model, however, due to the lack of climatic feedbacks fails to sustain the ice sheets after the initial forcing has been removed. The July 65°N insolation variation [*Loutre and Berger*, 1993] (solid line, upper panel), the tectonically-induced linear CO_2 concentrations [*Saltzman et al.*, 1993] (dashed line, upper panel), and the simulated Northern Hemisphere ice sheet volume with the LLN 2-D model (solid line, lower panel), from 3.05 Ma to 2 Ma BP [*Li et al.*, 1998] are shown.

[*Loutre and Berger*, 1993] with the assumption of a linearly decreasing atmospheric CO_2 concentration [*Saltzman et al.*, 1993]. Three major periods of glaciation between 2 Ma and 3 Ma were simulated, each one corresponding to the periods of large amplitude in the insolation signal and match with the three key steps in the intensification of Northern Hemisphere Glaciation (Figure 7).

I.c. Synthesis.

We still do not know what caused the Northern Hemisphere to glaciate some two and half million years ago. A plausible theory could be that the Tibetan uplift caused long term cooling during the late Cenozoic. The closure of the Panama Isthmus then may have delayed the onset of Northern Hemisphere Glaciation but ultimately provide the moisture which allowed intensive glaciation to develop at warmer high latitude temperatures (see Figure 6). The global climate system seems to have reached a threshold at about 3 Ma, when orbital configuration may have pushed global climate across this threshold, building all the major Northern Hemisphere ice sheets in a little over 200 kyrs.

II. GLACIAL-INTERGLACIAL CYCLES

II.a. Feedback Mechanisms: the Conventional View.

Glacial-Interglacial cycles are forced by changes in the Earth's orbital parameters. These cycles are, however, not caused by Earth's orbital parameters but rather the Earth's climate system feedback mechanisms. An illustration of this is that the insolation received at the critical 65°N was the same 18,000 during the LGM as it is today [*Berger and Loutre*, 1991; *Laskar*, 1990]. This section will not discuss the orbital parameters and their various influences as that has been done many times in great detail [e.g., *Milank-ovitch*, 1949; *Hays et al.*, 1976; 1984; *Ruddiman and McIntyre*, 1981; *Imbrie et al.*, 1992; 1993; *Berger and Loutre*, 1991; *Laskar*, 1990, see *Wilson et al.*, 2000 for clear review of the different orbital parameters and Appendix I). However, to illustrate the interaction of the global climate system feedback mechanisms these are discussed first in the context of glacial-interglacial and then rapid climate transitions. We present the discussion of the general consensus about the feedback mechanisms, followed by discussion of the alternatives. The insolation changes received by the Northern Hemisphere temperate zone is thought to be critical for glacial-interglacial cycles, because the Southern Hemisphere is presumably unable to drive glacial-interglacial as the expansion of the ice sheets is limited by the Southern Ocean. The critical factor is total summer insolation, as ice sheets surplus must survive the summer melting. So both the glaciation and deglaciation feedback mechanisms start with changes in summer 65°N insolation (Figure 8). The conventional view is that as ice starts to build-up on the northern continents it produces it own sustainable environment primarily by increasing the albedo [e.g., *Hewitt and Mitchell*, 1997]. The next critical stage is when the ice sheets, particularly the Laurentide become big enough to deflect the atmospheric planetary waves [*COHMAP*, 1988]. This changes the storm path across the North Atlantic Ocean and prevents the Gulf Stream penetration as far north as today. This surface ocean change and the increased melt-water in the Nordic Seas and Atlantic Ocean ultimately reduces the production of deep water [e.g., *Broecker*, 1991]. This in turn reduces the amount of warm water pulled northwards. All of which leads of increased cooling in the Northern Hemisphere and expansion of the ice sheets.

There are then secondary feedback mechanisms which also help drive the system towards the maximum possible glacial conditions. These include changes in the carbon cycle which reduces both atmospheric carbon dioxide and methane [e.g., *Chapellaz et al.*, 1993; 1997; *Brook et al.*, 1996]. Sigman and *Boyle* [2000] provide an excellent review of all the main controls on the carbon cycle and the possible causes of reduced glacial atmospheric carbon dioxide. They speculate that glacial-interglacial changes in atmospheric carbon dioxide may be primarily driven by changes in oceanography in the Southern Ocean. This could have altered the nutrient supply to the surface waters, hence surface water productivity. Increased glacial surface water productivity would have down-drawn atmospheric carbon dioxide into the surface water to produce organic matter through photosynthesis. However, the controls on the glacial-interglacial carbon cycle are still very poorly understood. Glacial periods by their very nature

have drier conditions which reduces atmospheric water vapor. For example Lea et al. [2000] provides clear evidence that the water vapor production of the equatorial Pacific zone was greatly curtailed during the last five glacial periods. All three, CO_2, CH_4 and water vapor are crucial greenhouse gases and any reduction in them leads to general global cooling (Figure 8), which in turn furthers glaciation.

These feedbacks are prevented from becoming a run away affect by 'moisture limitation'. As the warm surface water is forced further and further south, supply of the moisture that is required to build ice sheets decreases. Moreover it is very clear that ice sheets are naturally unstable [*Dowdeswell et al.*, 1999] and these feedbacks are constantly altering direction during the whole glacial period, hence it takes 80 ka to achieve the maximum ice extent during the LGM and that period is characterized by rapid oscillation such as the Heinrich events and the D-O cycles.

The natural instability of ice sheets means that deglaciation is much quicker than glaciation. In the case of Termination I it lasts a maximum of 4 ka including the Younger Dryas period. The increase in summer 65°N insolation [e.g., *Imbrie et al.*, 1993] leads to the initial melting of the Northern ice sheets. This raises sea level that then undercuts the ice sheets adjacent to the oceans, which in turns raises sea level. This sea level feedback mechanism is extremely rapid [*Fairbanks*, 1989]. Once the ice sheets are in retreat then the other feedback mechanisms discussed for glaciation are thrown into reverse (see Figure 9) including a massive increase in the greenhouse gases CO_2, CH_4 and water vapor. These feedbacks are prevented from becoming a run away affect by the limit of how much heat the North Atlantic can steal from the South Atlantic to maintain the interglacial deep water overturning rate.

IIb. Feedback Mechanisms: Possible New View?

The latest twist concerning orbital forcing and the feedback mechanisms has been shown by *Shackleton* [2000]. *Shackleton* [2000] suggests from a detailed tuning of the Antarctic Vostok ice core and deep ocean records that at the 100,000 year period, atmospheric carbon dioxide, Vostok air temperatures and deep water temperatures are in phase with orbital eccentricity; whereas ice volume lags these other variables. This suggests that orbitally induced changes in the carbon cycle and hence atmospheric carbon dioxide is the primary feedback causing the 100 ka glacial-interglacial cycles and not the conventional ice sheet dominated view. So in the proposed glaciation feedbacks carbon dioxide could be the primary means of transferring the orbital forcing into the global climate system. It also suggests that Northern Hemisphere ice sheet size is not the

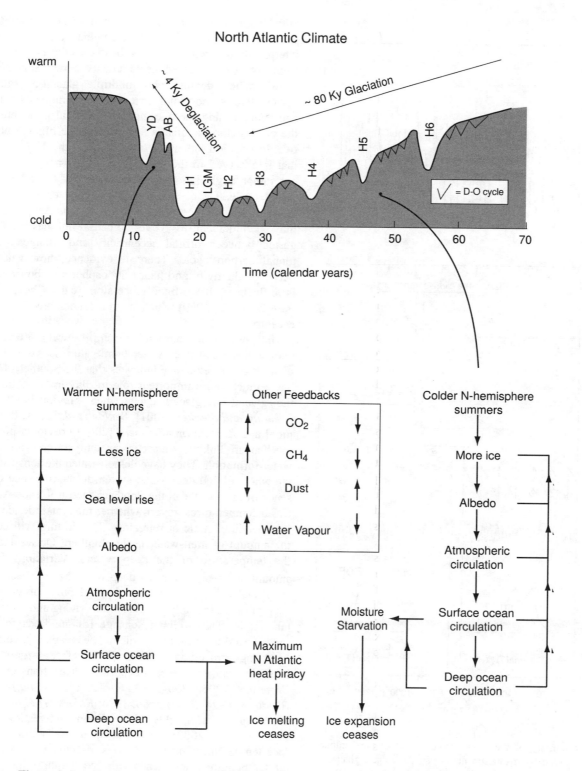

Figure 8. Detailed break down of the Mid-Pleistocene Revolution by *Mudelsee and Stattegger* [1997] demonstrating that there is a delay of 200 ka between the significant increase in global ice volume and the start of the 100 ka glacial-interglacial cycles.

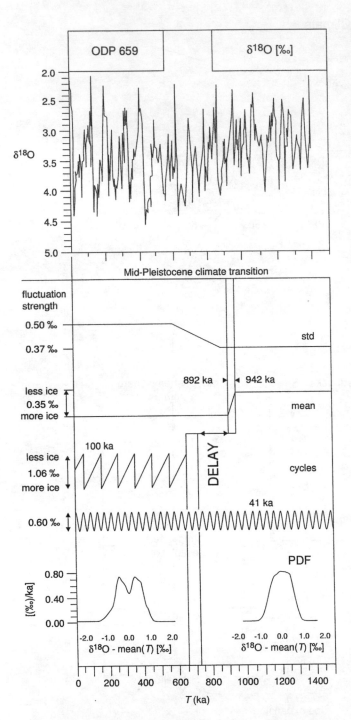

Figure 9. Summary of the conventional view of the feedback mechanism forced by insolation at 65°N which drive glaciation and deglaciation. H = Heinrich events, LGM = Last Glacial Maximum, AB = Allerød Bølling Interstadial, YD = Younger Dryas and D-O = Dansgaard-Oeschger cycles.

initial control on deep water circulation and hence temperatures. However, it must be remember that this new interpretation is reliant on the robustness of the age models from both deep sea sediments and ice cores [Shackleton, 2000]. When dealing with multiple glacial-interglacial cycles there are no radiometric dating techniques available and hence tuning the age models to orbital parameters is the only method available and this can be highly subjective. In addition, others have argued that the 100 ka cycle that is observed in the spectral records is not eccentricity but rather deglaciation triggered every 4th or 5th precessional cycle. Hence the correlation between eccentricity and the paleoclimate proxies found by *Shackleton* [2000] may be coincident. Moreover no causal link has been suggested between orbital eccentricity and changes in the global carbon cycle. There is evidence, however, that ocean productivity and hence atmospheric carbon dioxide is influenced by orbital precession [e.g., *Cane*, 1998; *Beaufort et al.*, 1999] which is in turn modulated by orbital eccentricity.

The possible in-phase relationship between Vostok air temperatures and deep water temperature is easier to explain as recent evidence indicates that the Southern Ocean has a much more important role to play in the control of the glacial-interglacial cycles than previously thought [*Keeling and Stephen*, 2001]. *Seidov et al.* [2001; this volume] and *Keeling and Stephens* [2001] provide a possible mechanism linking Antarctic air temperatures and deep water formation. They have demonstrated the sensitivity of the whole global deep water system to the presence or absence of melt-water in the Southern Ocean. Hence Antarctic air temperatures govern whether the Antarctic ice sheet is expanding, static or retreating. This in turn will control the amount of melt-water in the Southern Ocean and thus the temperature of the deep waters. Variations in the amount and temperature of deep water has a direct effect on global climate. From the physical point of view, the deep ocean is the only candidate for driving and sustaining internal long-term climate change (of hundreds to thousands years) because of its volume, heat capacity, and inertia. Long-term climate change is, therefore, largely regulated by the processes of oceanic heat transfer [e.g., *Broecker* 1995, *Keigwin et al.*, 1994, *Shaffer and Bendtsen*, 1994; *Jones et al.*, 1996; *Rahmstorf et al.*, 1996; *Seidov and Maslin*, 1999]. An additional interesting link is that *Lea et al.* [2000] found that equatorial Pacific sea surface temperature and salinity was in-phase with Antarctic air temperatures and hence leds ice volume changes by about 3,000 years.

So we can speculate that changes in the global carbon cycle, Antarctic air temperatures and equatorial climate could independently respond to orbital forcing. Alternatively one or several of these variables controls the others, for example changes of atmospheric carbon dioxide could control equatorial and Antarctic temperatures. The climate changes over Antarctic then in turn could influence the deep water system which then could influence the pull of the Gulf Stream northward to the formation sites of the North Atlantic Deep Water (NADW). These changes along with the local critical changes in the 65°N insolation instigate either build or collapse of the Northern Hemisphere ice sheets.

II.c. Synthesis

The thought provoking paper by *Shackleton* [2000] forces us to re-assess the conventional view of what causes glacial-interglacial cycles. It provides a major challenge, does Northern Hemisphere ice volume or the global carbon cycle primarily control glacial-interglacial cycles and what is the involvement of orbital forcing? However, before we throw out the conventional view, which has stood the test of time, and embrace a new view there are a large number of problems that need to be addressed. We suggest the primary one is whether a 100 ka cycle represents orbital eccentricity or rather deglaciations triggered by everyother 4th or 5th precessional cycle. This is, thus, an exciting time for paleoclimatology when even the basic view of what causes glacial-interglacial cycles may be wrong.

III. THE MID-PLEISTOCENE REVOLUTION

III.a. A Switch to Non-linear Global Climate System?

The Mid-Pleistocene Revolution (MPR) is the term used to denote a marked prolongation and intensification of the global glacial-interglacial climate cycles which occurred between 900 and 650 Ka [*Berger and Jansen*, 1994]. Prior to the MPR, since at least the onset of Northern Hemisphere glaciation (~2.75 Ma), the global climate conditions appear to have responded linearly to the obliquity (41 ka) orbital periodicity [*Imbrie et al.*, 1992], i.e., the glacial-interglacial cycles occur with a frequency of 41 ka. After about 800 ka glacial-interglacial cycles become more pronounced and occur with a frequency of approximately 100 ka. Hence MPR marks a switch in the timing of glacial-interglacial cycles from 41 ka to 100 ka [*Berger and Jansen*, 1994]. *Imbrie et al.* [1993] call this the "100 ka problem", as only the last 8 glacial-interglacial cycles have occurred with this frequency and increased intensity. The cyclicity and amplitude of the orbital parameters which force long term global climate change, e.g., eccentricity (~100 ka), obliquity (~41 ka) and precession (~21/19 ka),

do not vary during the MPR [*Berger and Loutre*, 1991]. This suggests the MPR is an internal re-organization of the feedback mechanisms that translate orbital forcing into global climate change. Moreover, the 100 ka eccentricity signal is by far the weakest of the orbital parameters and has been thought too be phase-lagged to force directly a 100 ka global climate cyclicity. The MPR, thus could be a change from a linearly-forced climate system to the non-linearly forced climate system which we have at present [*Imbrie et al.*, 1993]. This is, however, assuming that the 100 ka cyclicity that is observed for the last 8 glacial-interglacial cycles reflects an eccentricity signal. It has been argued that the spectral peak of 100 ka is not associated with eccentricity which has two spectral peaks near 100 ka and one at 413 ka; rather the spectrum is caused by deglaciations being triggered by every fourth or fifth precessional cycle [*Raymo*, 1997; *Ridgwell et al.*, 1999]. Hence, the MPR could be an internal adjustment from a obliquity dominated system to a precession dominated system.

During the MPR not only does timing of glacial-interglacial cycles increase but there is also an amplitude increase of the variations of global ice volume. The main causes for this ice volume increase are prolonged glacial periods, evidence of which is provided by high resolution oxygen isotope data from deep sea cores [*Pisias and Moore*, 1981; *Prell*, 1982; *Shackleton et al.*, 1988; *Tiedemann et al.*, 1994; *Berger and Jansen*, 1994; *Raymo et al.*, 1997; *Mudelsee and Stattegger*, 1997]. The MPR, therefore, marks a dramatic sharpening of the contrast between warm and cold periods. *Mudelsee and Stattegger* [1997] used advanced methods of time-series analysis to review the deep sea evidence spanning the MPR and summarized the salient features (Figure 9). The first transition occurs between 942 and 892 ka when there is a significant increase in global ice volume, however, the 41 ka climate forcing continues. This situation persists until about 650-725 ka when the climate system finds a two modal solution and the strong 100 ka climate cycles begin [*Mudelsee and Stattegger*, 1997]. A number of causes have been suggested for the MPR and the subsequent 100 ka 'non-linear' world; these are all based on a threshold model (Figure 1d) and include:

a) Critical size of the Northern Hemisphere ice sheets. *Imbrie et al.* [1993] suggested that the North Hemisphere ice sheets may have reached a critical size during the MPR allowing them to respond non-linearly to eccentricity.

b) Global cooling trend. It has been suggested that long term cooling through the Cenozoic instigated a threshold which allowed the ice sheets to become large enough to ignore the 41 ka orbital forcing and to survive between 80 and 100 ka [*Abe-Ouchi*, 1996; *Raymo et al.*, 1997]

c) Erosion of regolith beneath the Laurentide ice sheet. *Clark and Pollard* [1998] suggested that the constant erosion of regolith beneath the Laurentide ice sheet would eventually provide more bedrock for the ice sheet to rest on and hence it would be more stable allowing it to survive longer than the 41 ka driving force.

d) Orbital inclination. A fourth possible orbital parameter has been identified, orbital inclination, which is the 100 ka variation in the angle of the Earth's plane of orbit compared to the average orbit of the solar system [*Muller and MacDonald*, 1997; *Farley and Paterson*, 1995]. It has been argued that this might have changed the total amount of radiation received and could cause the Earth to pass through increased amounts of dust from outer space (Interplanetary Dust Particles, IDPs) which would encourage glacial-interglacial alternations on 100 ka periodicity. It now appears, however, that this process is not likely to have been of influence on ice ages on Earth [*Kortenkamp and McDermott*, 1998]. Moreover, the possibility that the 100 ka cycle is caused by deglaciation every 4th or 5th precessional cycle removes the need for a 100 ka forcing mechanism.

e) Greenland-Scotland submarine ridge. *Denton* [2000] has envisioned a MPR mechanism based on the uplift of the Greenland-Scotland submarine ridge ca. 950 ka BP (caused by burst of tectonic activity along the Iceland mantle plume) which led to a southward shift of the area of deep-water production from the Arctic to the Nordic seas. This would have had a number of effects on oceanic circulation, especially during times of expanded ice sheets, so that once ice-sheet expansion had commenced, it became much more difficult for the salinity conveyor to re-set into an interglacial mode. In other words, after 950 ka, it became easier for ocean circulation to lock into glacial than interglacial mode, and only occasional highs in the amplitude of the eccentricity-driven precession signal lead to short warm events.

f) Carbon cycle and Atmospheric CO_2. It has been suggested that the reduction of atmospheric CO_2 may have brought the global climate to a threshold allowing it to respond non-linearly to orbital forcing [*Mudelsee and Stattegger*, 1997; *Raymo et al.*, 1997; *Berger et al.*, 1999]. It is also important to consider the latest results of *Shackleton* [2000]. He suggests that the 100 ka eccentricity, atmospheric carbon dioxide, Vostok (Antarctica) air temperatures and deep water temperatures are in phase; whereas ice volume lags these other variables. This is an important finding, if confirmed, as it suggests that atmospheric carbon dioxide and not Northern Hemisphere ice volume may be the primary driving force of the 100 ka glacial-interglacial cycles. The MPR could, thus, have been caused by changes in the internal response of the global carbon cycle to orbital forcing. *Sigman and Boyle* [2000] suggest the Southern Ocean as one of the most important regions modulating atmospheric carbon dioxide. So *Shackleton* [2000] opens the possibility that the Southern Ocean may have become more sensitive to orbital forcing and hence modulated the global carbon cycle. However, what caused this increased sensitivity and whether carbon dioxide changes alone could maintain extended glacial periods is still unknown. The links between Antarctic air temperatures, the Southern Ocean and deep water temperatures is discussed in more details in Sections II IVc.

III.b. Synthesis

At the moment there is no consensus on how and why the MPR occurred. The only real agreement is that it involved a significant internal reorganization of the climate system. The MPR could have been a shift to climate system with a non-linear response to orbital eccentricity [*Imbrie et al.*, 1993] or it may have been a shift from an obliquity to a precession dominated system. *Shackleton* [2000] work provides new ideas for the causation of the MPR including changes in the sensitivity of the global carbon cycle and/or the Southern Ocean to orbital forcing. These ideas have yet to be fully investigated.

IV. RAPID CLIMATE CHANGES IN GLACIAL PERIODS

IV.a. Heinrich Events and Dansgaard-Oeschger Cycles.

Heinrich events are intense and quasi-periodic ice rafting pulses which are thought to originate primarily from the Laurentide ice sheet [e.g., *Ruddiman*, 1977; *Heinrich*, 1988; *Broecker et al.*, 1992; *Bond et al.*, 1992; *Francois and Bacon*, 1993; *Grousset et al.*, 1993; 2000; *Andrews et al.*, 1994; *Gwiazda et al.*, 1996 a and b, *Ramussen et al.*, 1997; *Andrews*, 1998; *Clark et al.*, 1999]. These events occurred against the general background of unstable glacial climate, and represent the brief expression of the most extreme glacial conditions around the North Atlantic. The Heinrich events are evident in the Greenland ice cores as a further 3-6°C drop in temperature from the already cold glacial climate [*Bond et al.*, 1993; *Dansgaard et al.*, 1993]. The Heinrich events have been found to have a global impact [*Clark et al.*, 1999] with evidence for coeval climate changes described from as far field as: South America [*Lowell et al.*, 1995], North Pacific [*Kotilainen and Shackleton*, 1995], Santa Barbara Basin [*Behl and Kennett*, 1996], Arabian Sea [*Schulz et al.*, 1998], China [*Porter and An*, 1995] and the South China Sea [*Wang et al.*, 1999]. Around the North Atlantic much colder conditions are found [*Grimm et al.*, 1993; *Thouveny et al.*, 1994] see

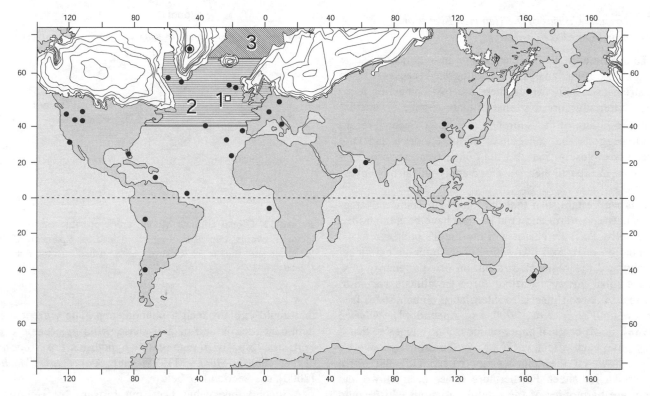

Figure 10. Map showing sites where paleo-records have linked the local climate and environmental changes to the Heinrich events. Number 1 is the location of the original *Heinrich* [1988] study. Number 2 is the North Atlantic which has been covered extensively by a number of studies [e.g., *Bond et al.*, 1992; 1993; Sarnthein et al., in press] and Number 3 is an extensive studied Nordic Seas that has demonstrated that the Greenland and Fenno-Scandinavian ice sheet seems to surge at different times to the Heinrich events [e.g., *Sarnthein et al.*, 1995; *McCabe and Clark*, 1998; *Dowdeswell et al.*, 1999].

Figure 10) and in the North Atlantic Ocean huge armadas of melting icebergs reduce sea surface temperatures by another 1-2°C and reduce the surface salinity by up to 4‰ [*Maslin et al.*, 1995b; *Cortijo et al.*, 1997; *Chapman and Maslin*, 1999; *Vidal et al.*, 1999].

Detailed studies of the sequence of events in ocean sediments and ice cores showed that Heinrich events occurred on an average of 7200 ± 2400 calendar years [Sarnthein et al., in press) in the time interval between about 70 and 10.5 ka. In between the Heinrich events there are much higher frequency, but smaller amplitude events occurring at about every 1500 years, which are referred to as Dansgaard-Oescheger events or cycles [*Dansgaard et al.*, 1993; *Bond et al.*, 1997].

Dansgaard-Oeschger (D-O) cycles were first identified in the Greenland ice core records [*Dansgaard et al.*, 1993]. A succession of short lived 'warm' events or Greenland interstadials (GIS) characterize the ice core records of the last glacial episode and are numbered down from the 'Bolling' as GIS 1 to GIS 24 at 105 ka. The D-O stadials

or cold sections of the D-O cycle have been found in the ice rafting records of the North Atlantic [e.g., *Bond and Lotti*, 1995]. These stadials are characterized by increase meltwater and ice rafted debris originating from Iceland and East Greenland [*Bond and Lotti*, 1995; *Sarnthein et al.*, in press]. These 1500-year cycles seem to persist into the Holocene interglacial [*Bond et al.*, 1997; 1999; *Campbell et al.*, 1998; see below] as well as during earlier glacial and interglacial stages [*Oppo et al.*, 1998; *Hodge*, 2000]. A precise time scale for the duration and sequence these millennial-scale climate events is uncertain, though the events themselves do not appear to have lasted longer than a few centuries [e.g., *Dowdeswell et al.*, 1995; *Zahn et al.*, 1997; *Sarnthein et al.*, in press; van *Kreveld et al.*, 2000; *Grousset et al.*, 2000, in press; *Scourse et al.*, 2000]. Furthermore, the transitions marking the beginnings and ends of each Heinrich and D-O event are particularly rapid, probably lasting no more than a few decades, indicating that climate responded in a step-like series of jumps to whatever driving mechanism initiated them.

IV.b. Possible Causes of the Heinrich Events and Dansgaard-Oeschger Cycles.

To examine the possible causes of these millennial-scale events, we first analysis the suggested causes of the Heinrich events for which there are two competing theories: externally forced global climate change or internal periodic failure of the Laurentide ice sheet. Second, recent work suggests that the causes of the Heinrich and D-O events are closely tied and the connection between them and how the causal theories above apply will also be investigated.

It is well established that the long-term climate fluctuations between warm and cold periods have been controlled externally. Periodic changes in the Earth's orbit and in the inclination of its axis of rotation affect both the net global receipt and the regional distribution of solar energy flux. These solar energy variations force the climate system to alternate between glacial to interglacial climate states [see review in *Wilson et al.*, 2000, and Appendix I). *Heinrich* [1988], in his original paper on ice rafting suggested that it may have been minor variations in the orbital parameters which lowered the solar energy received by the North American ice sheet. Furthermore it has been shown that there are harmonics of the orbital variations which could corresponding to the timing of the Heinrich events [Broecker, 1994; Hagleberg et al., 1994]. These variations could have lowered the heat input or increased the moisture supply to the Laurentide ice sheet causing it to expand, pushing more ice out into the ocean (see Figure11). The global climate model was first put forward by *Broecker* [1994] in response to the work of *Lowell et al.* [1995] who showed that the glaciers of the Chilean Andes in South America expanded at the same time as the Heinrich events. This evidence was interpreted as indicating that ice masses in both the northern and southern hemispheres expanded and contracted synchronously, which in turn would imply the imposition of some over-riding global forcing mechanism. However, the detailed sequence of events has been shown to be much more complex than this. First, *Bond* [1995] has argued that, for the North Atlantic at least, each Heinrich event is immediately preceded by a short episode of climate cooling, so that climate change actually preceded ice-sheet response. This interpretation was based on planktonic formainifera $\delta^{18}O$ records so could also indicate a increase in salinity of the surface waters. Moreover, *Grousset et al.* [2000] see no evidence of the pre-event cooling. Second, there is growing evidence to suggest that ice-mass changes in the two hemispheres during the last glacial cycle were significantly out of phase [e.g., *Blunier et al.*, 1998], which has led to a number of theories about the possible driving mechanisms

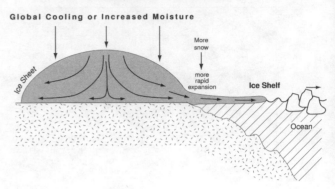

Figure 11. Global Climate Model is one possible cause of the Heinrich events. Global cooling could lead to expansion of the ice sheets and thus more icebergs being pushed into the Atlantic Ocean.

that could explain such a phenomenon. The current predominant idea is one of anti-phase shifts in oceanic currents (the 'bipolar climate seesaw hypothesis') of *Broecker* [1998, 2000], *Stocker* [1998] and *Seidov and Maslin* [2000], see Section IVc below.

A second theory has been put forward by *MacAyeal* [1993a,b] suggesting that the surges were caused by internal instabilities of the Laurentide ice sheet. This ice sheet rested on a bed of soft unconsolidated sediment, when it is frozen it does not deform, and behaves like cement, and so would have been able to support the weight of the growing ice sheet. As the ice sheet expanded the geothermal heat from within the Earth's crust together with heat released from friction of ice moving over ice, was trapped by the insulating effect of the overlying ice. This 'duvet' effect allowed the temperature of the sediment to increase until a critical point when it thawed. When this occurred the sediment became soft, and thus lubricated the base of the ice sheet causing a massive outflow of ice through the Hudson Strait into the North Atlantic (see Figure 12). This, in turn, would lead to sudden loss of ice mass, which would reduce the insulating effect and lead to re-freezing of the basal ice and sediment bed, at which point the ice would revert to a slower build-up and outward movement. According to *MacAyeal* [1993a, b] this system of progressive ice build-up, melting and surge, followed by renewed build up has an approximate periodicity of 7,000 years, which compares with the interval between the last six Heinrich events which have been calculated by *Sarnthein et al.* [in press] to be an average of 7200 ± 2400 calendar years.

The *MacAyeal* [1993] ice surge hypothesis as this seems to explain many of the details of Heinrich events, espe-

MacAyeal Model

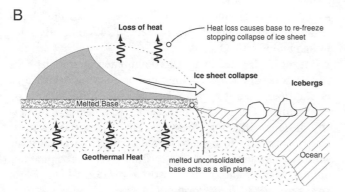

A

Ice Flow
Friction

Bigger icesheet the greater
insolation effect and quicker heat
builds up

Ice Sheet (Insulation traps heat)

Frozen Unconsolidated Base

Ocean

Geothermal Heat

B

Loss of heat

Heat loss causes base to re-freeze
stopping collapse of ice sheet

Ice sheet collapse

Icebergs

Melted Base

Geothermal Heat

melted unconsolidated
base acts as a slip plane

Ocean

Figure 12. MacAyeal or free oscillating ice sheet collapse model is one possible cause of the Heinrich events. It is suggested that geothermal and fractional heat builds up at the base of the ice sheet and causes it to melt triggering the ice sheet collapse.

cially the rapid onset of these events which is evident in sediments records [*Dowdeswell et al.*, 1995]. X-ray photographs of some deep sea cores show that the uppermost tier of fossil burrows are preserved in the sediment deposited immediately preceding the Heinrich events [*McCave*, 1995]. This implies that the ice-rafted debris must have been deposited almost instantly, i.e., within a few years, since it is usual for these fossil burrows to be obliterated by the action of subsequent ocean floor animals. The Heinrich events, therefore, began with a huge and rapid input of sediment which is consistent with an ice sheet surge. The idea of self-destabilizing ice has a further advantage over the 'Global climate model' since the response time of large 3 km thick ice sheets to external forcing likely to be very slow (on the order of thousands of years) which does not fit well with the evidence of the shear rapidity of these events.

Heinrich events, however, can not be considered in isolation as they are part of the general instability that characterizes glacial periods. Hence, the third possible

cause of the Heinrich events which has been suggested is the destabilization of the Laurentide ice sheets triggered by sea level changes induced by the 1,500 year D-O cycles. *Sarnthein et al.* [in press] and *van Kreveld* [2000] present evidence that suggest that ice surges from Iceland and East Greenland could have produced significant increases in sea level, possibly big enough to start under cutting the Laurentide icesheet and thus precipitating a full Heinrich event. Support for this theory comes from the detailed analysis of the sources of ice rafted material in the Heinrich layers. From the detailed break down of H1, H2, H4 and H5 it has been shown that European ice rafted debris was deposited in the North Atlantic upto 1,500 years before the Laurentide material [*Grousset et al.*, 2000, in press; *Scourse et al.*, 2000]. Referring back to the climate change scenarios suggested at the beginning of this paper, it is possible that a critical sea level threshold is exceeded once in every three Dansgaard-Oeschger cycles, precipitating the collapse of the Laurentide ice sheet. This raises a further issue of what causes the Dansgaard-Oeschger cycles?

Again the arguments for causation fall in to the two camps; internal vs external focring. There are two possible internal explainations. This first is a simple MacAyeal type binge-purge process controlling the East Greenland icesheet with a periodicity if 1,500 years [*Sarnthein et al.*, in press; *van Kreveld et al.*, 2000]. Successive failures of the Greenland ice sheet would have a small effect on sea level to the critical point when it influenced the Laurentide ice sheet [*Sarnthein et al.*, in press; *van Kreveld et al.*, 2000]. The second possible internal mecahnism could be an internal oscillation of the deep water system driven by alternating sea ice extent in the Southern Ocean and ice sheet surges in the North Atlantic. This idea and the bipolar seesaw are discussed at length in Section IVd. However, *Scource et al.* [2000] argue for an external climate forcing as suggested by *Broecker* [1994] and Hagleberg et al. [1994]. They have excellent deep-sea core evidence that the British ice sheet surged both before and at the same time as Heinrich event 2. The precursor event occurred 700 to 1000 years before the Heinrich event and hence *Scource et al.* [2000] argue it is to young to be the previous D-O cold event. They believe it shows the different ice mass balance response time of each ice sheet surrounding the North Atlantic due to their varying size. Hence the much smaller British ice sheet responds first and up to a thousand years later the Laurentide ice sheet follows. However, as *Bond et al.* [1997; 1999] suggests that D-O cycles have a frequency of 1,479±532 years, then the precursor event described by *Scource et al.* [2000] which could have been 1000 years before Heinrich event 2, may indeed be the prior D-O cold event.

IV.c. Bipolar Climate Seesaw and Deep Water Circulation

One of the most important finds in the study of millennial-scale climate events is the apparent out of phase climate response of the two Hemispheres [*Blunier et al.*, 1998]. *Blunier et al.* [1998] concluded that Antarctic climate changes were significantly out of step with those in the North Atlantic on the basis of $\delta^{18}O$ data (a proxy for temperature) obtained from the Byrd and Vostok ice-cores. They were able to achieve this by correlating the methane records in the ice cores in both Hemispehere, this has its own inherent problems and drawbacks which are discussed more fully in Section Va. Support for *Blunier et al.* [1998] has been found in well dated deep-sea sediment records and has shown that the sea surface conditions in the tropical and South Atlantic Ocean are out of phase with those in the Northern North Atlantic [*Vidal et al.*, 1999; *Ruhleman et al.*, 1999; *Huls and Zahn*, 2001]. If the climate of the two Hemispheres is out of phase then it suggest the global climate system acts as like a seesaw, with each Hemisphere taking turns to drive the system. This 'bipolar seesaw' can be controlled by the atmosphere [e.g., *Webb et al.*, 1997; *Broecker*, 1998], the ocean [*Stocker*, 1994; 1998], or a combination of these two controls [*Broecker*, 1998]. *Stocker* [1998] suggests a simple mechanical analogue that depicts the ocean behavior as a seesaw driven by either high-latitudinal, or near-equatorial sea-surface perturbations, or both.

The deep ocean is, however, the only candidate for driving and sustaining internal long-term climate change (of hundreds to thousands years) because of its volume, heat capacity, and inertia [e.g., *Labeyrie et al.*, 1992; *Stocker et al.*, 1992; *Weaver*, 1994; *Broecker* 1995, *Keigwin et al.*, 1994, *Shaffer and Bendtsen*, 1994; *Jones et al.*, 1996; *Manabe and Stouffer*, 1988; 1995, 1997; *Rahmstorf*, 1994; 1995; *Rahmstorf et al.*, 1996; *Seidov and Haupt*, 1999; *Seidov and Maslin*, 1996; 1999]. In the North Atlantic, the north-east trending Gulf Stream carries warm and relatively salty surface water from the Gulf of Mexico up to the Nordic seas. Upon reaching this region, the surface water has sufficiently cooled that it becomes dense enough to sink into the deep ocean. The 'pull' exerted by this dense sinking maintains the strength of the warm Gulf Stream, ensuring a current of warm tropical water into the North Atlantic that sends mild air masses across to the European continent [e.g., *Schmitz and McCartney*, 1993; *Schmitz*, 1995; *Rahmstorf et al.*, 1996]. Formation in the North Atlantic Deep Water (NADW) can be weakened by two processes: 1) the atmospheric polar front preventing the North Atlantic Drift from travelling so far north which reduces the amount of cooling that occurs and hence its capacity to sink, as happened during the last glacial period,

or 2) the input of freshwater in the form of melt-water, river discharge, or increased precipitation. If NADW formation is diminished, the weakening of the warm Gulf Stream would cause colder European winters [e.g., *Broecker*, 1995]. However, the Gulf Stream does not give markedly warmer summers in Europe. So a reduction in the Gulf Stream in itself does not in itself explain why summers also became colder during glacial periods.

In contrast in the Southern Ocean the Antarctic Bottom Water (AABW) is formed in coast polnyas where out-blowing Antarctic winds push sea ice away from the continental edge and super cool the exposed surface waters. This leads to more sea ice formation and brine rejection, producing the coldest and saltiest water in the world. AABW flows around Antarctic and penetrates the North Atlantic flowing under the lighter NADW and also the Indian and Pacific Oceans.

For rapid climate chnages to be initiated there must be a trigger for a sudden 'switching off' or a strong decrease in rate of deep water formation in either the North Atlantic or the Southern Ocean; this must be due to a decrease in density or dedensification [*Broecker*, 2000] of surface waters. Such a decrease in density would result from changes in salinity (addition of fresh water from rivers, precipitation, or melt water), and/or increased temperatures [e.g., *Dickson et al.*, 1988; *Rahmstorf et al.*, 1996].

Seidov and Maslin [2000] and *Seidov et al.* [2001 and this volume] have suggested that the asynchrony during *Blunier et al.* [1998] A1 and A2 events (which correspond approximately to H4 and H5 or D-O cycle 8 and 12, see Figure 13) can be explained by variations in relative amount of deepwater formation in the two Hemispheres and thus by the resulting heat piracy. Figure 14 shows from model results of the direction and quantity of heat piracy that occurs during different periods. During modern or interglacial conditions the North Atlantic steals heat from the Southern Hemisphere, this North Atlantic heat piracy (Figure 14) maintains a strong Gulf Stream and prevents ice sheet build-up in the Northern Hemisphere. During a Heinrich event when meltwater is injected into the North Atlantic, heat is transported southward cross the equator (Figure 14). This South Atlantic Ocean counter heat piracy causes the Southern Hemisphere oceans to warm, while the Northern Hemisphere oceans cool (Figure 13.1). When the iceberg armada ceases, the freshwater cap on NADW formation is removed and northward cross-equatorial heat transport kicks in. This North Atlantic Ocean heat piracy warms the Northern Hemisphere while cooling down the Southern Hemisphere (Figure 13.2). If the glacial boundary conditions re-assert themselves, the North and South Atlantic Oceans come back to an almost perfect balance (Figure 13.3).

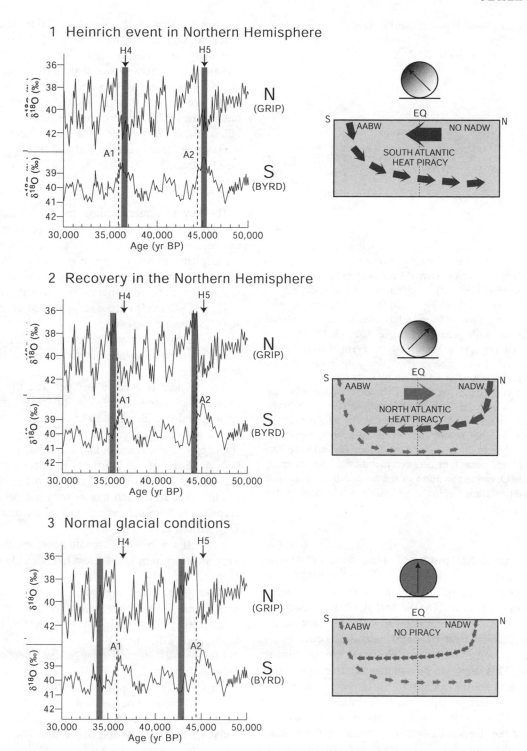

Figure 13. The left panel shows Greenland (GRIP) and Antarctic (Byrd) ice core $\delta^{18}O$ temperature proxy records on the age models generated by *Blunier et al.* [1998]. Three timeslices are shown illustrated on the *Blunier et al.* [1998] records with solid bars; 1. Heinrich event, 2. Northern Hemisphere climate recovery post Heinrich event and 3. Normal glacial conditions. On the right panel is a schematic view of oscillating meridional ocean overturning and associated regimes of heat transports in the Atlantic Ocean. This shows which Hemisphere is controlling the deep water circulation and thus inter-Hemispheric heat transport [adapted from *Seidov and Maslin*, in press].

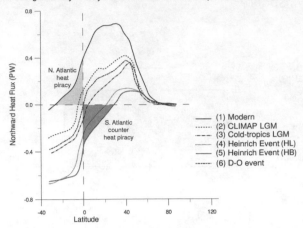

Glacial-interglacial asymmetry of the northward heat transport in the North Atlantic

Figure 14. Atlantic Ocean poleward heat transport (positive indicates a northward movement) as given by the ocean circulation model [*Seidov and Maslin* 1999; 2000] for the following scenarios: (1) present-day (warm interglacial) climate; (2) last glacial maximum (LGM) with generic CLIMAP data; (3) "Cold Tropics" LGM scenario; (4) a Heinrich-type event driven by the meltwater delivered by icebergs from decaying Laurentide Ice sheet; (5) a Heinrich-type event driven by meltwater delivered by icebergs from decaying Barrens Shelf Ice sheet or Scandinavian ice sheet; (6) A general Dansgaard-Oeschger (D-O) meltwater event confined to the Nordic Seas. Note that the total meridional heat transport can only be correctly mathematically computed in the cases of either cyclic boundary conditions (as in Drake Passage for the global ocean), or between meridional boundaries, as in the Atlantic Ocean to the north of the tip of Africa. Therefore the northward heat transport in the Atlantic ocean is shown to the north of 30°S only.

The most important results of these computer simulations is that these scenarios, does NOT require a substantial change in the heat transported by the upper ocean currents. The imbalance between NADW and AABW productions, i.e. between the deep-ocean flows, is the primary control on cross-equatorial heat transport, and thus could be the sole agent responsible for the observed seesaw climate oscillations. This seems to be counter-intuitive as it is usually assumed that only changes of the upper-ocean currents can affect the heat transport to high latitudes. However *Seidov and Maslin* [1999] and *Seidov et al.* [this volume] suggest that the deep-ocean currents could be the ultimate internal mechanism capable of reversing cross-equatorial heat transport within the required time scale of the oscillations. The simple rule is that when the NADW subsides, the AABW picks up and affects the heat transport regime in the opposite way. Though the model experiments described above a based purely by the input of meltwater into

the North Atlantic, *Seidov et al.* [2001; see also this volume] has also demonstrate that the Southern Ocean is also extremely sensitive to meltwater inputs. *Kanfoush et al.* [2000] have clearly shown that there are large iceberg discharges which seem to occur after the Heinrich events. Hence we suggest the true picture of the bipolar seesaw is one driven by both hemispheric sources and that the oscillation of this deep water system may alone be sufficient to explain the 1,500 year glacial and interglacial D-O cycles (see Figure 15).

IV.d. Synthesis

The key to understanding millennial scale fluctuations during glacial periods is not to separate the different events, nor to ignore the role played by the Southern Hemisphere. In attempting to explain the complex patterns of abrupt climate change during the last glacial stage, several important points emerge from recent investigations.

1.   D-O cycles are quasi-periodic or irregular in timing (1,500 ± 500years).

2.   Heinrich events occur at irregular intervals (7,200 ±2,400 years) and appear to triggered by a threshold mechanism.

3.   There is a close and occasionally sensitive interrelationship between the state of ice sheets (i.e., whether relative stable with a frozen base or sensitive to small environmental changes), ocean surface temperature and salinity, deep water formation and sea level.

4.   Each of the Northern Hemisphere ice sheets during the last glacial episode have surged or collapsed at different times, thus each has its own independent rhythm of response to internal processes and external forcing [e.g., *Dowdeswell et al.*, 1999].

5.   If the bipolar climate seesaw did occur then the deep water system is the most likely candidate to control it (Figure 15).

Having reviewed the possible causes of the Heinrich events and D-O cycles, below we provide our own combined theory (Figure 15). We suggest that the meltwater associated with the D-O cold event could reduces the NADW, which in turns changes the direction of the Hemispheric heat piracy and thus the Southern Hemisphere warms up. This increase warmth in the Southern Hemisphere at some point triggers a melting of sea ice and/or a surging of the Antarctic ice sheets. This reverse the process reducing the AABW and hence strengthening the NADW. This in turn warms the North Atlantic and triggers another D-O melting event. Eventually after a certain number of D-O cycles the sea level rise assocaited with them would be significant to undercut the Laurentide ice sheet, causing a Heinrich event. The Heinrich event would then also influence the Southern Hemisphere and cause a further bipolar

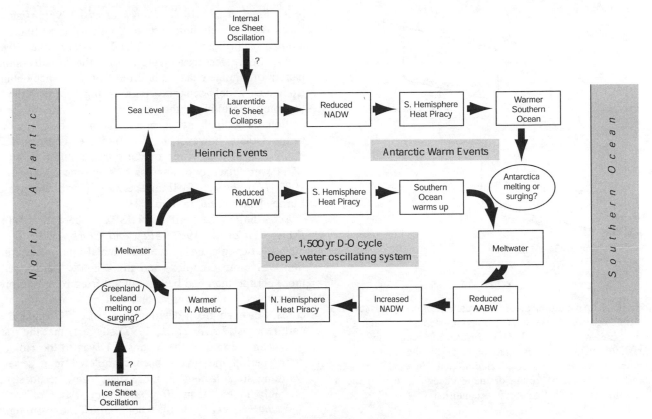

Figure 15. Possible "deep-water oscillatory system" explaining the Glacial and Interglacial Dansgaard-Oeschger cycles [adapted from Maslin and Seidov, submitted]. Meltwater in either the North Atlantic or Southern ocean can start the system going. The meltwater caps the sites of deep water formation in that particular Hemisphere and reduces the deep water formation. This increases heat piracy to the other Hemipshere, which warms that Hemisphere and after hundreds or thousands of years instigates ice sheet melting, thus reversing the whole processes. The system is quasi-periodic because of the oscillatory nature of the system. the stochastic element which is observed in the records [Alley et al., this volume] comes from the different internal dynamics of each ice sheet. The link between the Dangaard-Oeschger ~1,500 year cycles and the Heinrich events is their successive affect on sea level. At a certain point the sea level has risen enough to undercut the Laurentide ice sheet and precipitate a full Heinrich event.

climate swing. Hence once initiated this deep water oscillating system could continue without the need for any external influences.

We can add to this scenario both a stochastic [*Alley et al.*, this volume] and another quasi-periodic element, which is the internal dynamics of the ice sheets (see Figure 15). Which combined with the deep-water oscillating system would produce 1) stable periods when the majority of ice sheet are expanding and 2) occasional periods of very strong instability with episodes of major multiple ice sheet collapse. This seems to be supported by the evidence as the magnitude and timing of each D-O and Heinrich event seem to be different. This combined theory also avoids the

pervasive "chicken-and-egg" problem, which is the attempt to interpret geological records to establish precisely which component of the system is the 'instigator' of any one event, and there is of course no need for there to be only one trigger for any specific event.

One key point to note is that global climate seems to be reversible during the D-O cycles, returning to similar conditions to prior to the event (Figure 16). This is not true for the Heinrich events, though each is followed by a climatic rebound in the North Atlantic, each Heinrich events seems to drive global climate towards evermore severe glacial conditions, so we suggest Heinrich events could be one branch of the climate system's bifurcation (Figure 16).

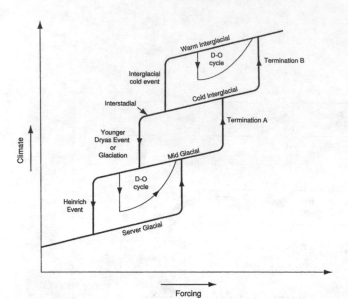

Figure 16. Adaptation of the bifurcation Figure 3 to show the possible relationship between global climate and the forcing factors. Showing that the D-O (Dansgaard-Oeschger) cycles are threshold events but it is relatively easy for the climate system bounce back. In contrast glaciation and Heinrich events are thresholds which require a dramatic change in the forcing to bring the climate back to previous conditions.

V. RAPID CLIMATIC CHANGES DURING THE LAST GLACIAL-INTERGLACIAL TRANSITION

V.a. The Nature of the Last Glacial-Interglacial Transition (LGIT)

The last glacial-interglacial transition (LGIT, ca. 14-9 ^{14}C ka BP) was characterized by a series of extremely abrupt climatic changes [e.g., *Walker*, 1995; *Lowe et al.*, 1994; 1995; 1999; *Alley and Clark*, 1999]. The two most prominent features being rapid warming at 14.7 k GRIP (ss08c chronology, see Lowe et al., in press] ice-core years BP (the start of the "Bølling" episode, or the GI-1 event in the Greenland ice-core isotope stratigraphy of *Björck et al.*, 1998 and *Walker et al.*, 1999] and the well-known period of severe cooling referred to as the Younger Dryas Stadial (GS-1), dated to between 12.7 and 11.5 ka GRIP ice-core yrs BP. These features are easily recognisable in a wide range of proxy records from sites throughout the northern hemisphere [e.g., *Dansgaard et al.*, 1989; *Lowe and Walker*, 1994; *Walker et al.*, in press a; *Alley and Clark*, 1999]. There is also a growing body of evidence which suggests that parts of the Southern Hemisphere may have experienced a broadly similar sequence of climatic events

[e.g., *Bard et al.*, 1997; *Bjorck et al.*, 1996; *Alley and Clark*, 1999], including parts of South America [papers in *Sugden*, 2000] and New Zealand [*Denton and Hendy*, 1994]. Some researchers argue that the broadly similar pattern of climate changes inferred from evidence obtained from such widely-separated regions, together with the fact that they approximate the same interval, points to the operation of globally-synchronous climate forcing mechanisms. Others, however, have provided evidence that may imply that at least some of the climate shifts during the LGIT were diachronous across NW Europe [*Coope et al.*, 1998; *Witte et al.*, 1998] and between North America and Europe [*Elias et al.*, in prep.].

According to the GRIP and GISP2 ice-core records [e.g., *Dansgaard et al.*, 1993; *Alley and Clark*, 1999], the sequence of events during the last glacial-interglacial transition was more complex, and the climatic transitions far more abrupt, than had been assumed from prior interpretations based on continental and marine stratigraphical records. For example, a detailed study of the Greenland ice-core records [*Taylor et al.*, 1997] suggests that the transition from the end of the Younger Dryas to the Holocene warm period may have been completed in a series of warming steps lasting no more than a few decades, most probably in less than 50 years [*Alley and Clark*, 1999; *Severinghaus et al.*, 1998]. About half of the warming was concentrated into a single short step that lasted no more than 15 years. By comparison with the abrupt warming at the start of the 'Bølling' and the Holocene, cooling from the warm 'Bølling' to the Younger Dryas appears to have been a more gradual process, involving a series of cooling steps and minor climatic oscillations, extending over a period in excess of 1500 years.

By comparison with ice-core records, the temporal resolution afforded by most continental and/or marine sequences is much more limited, while the problems inherent in the radiocarbon time-scale severely hamper attempts to effect high-precision correlations between the sequences [e.g., *Lowe and Walker*, 2000]. Hence, despite the fact that the sequence and pattern of environmental changes during the LGIT can be reconstructed in much greater detail than is the case for earlier interstadial events [e.g., *Lowe and Walker*, 1994; *Lundqvist et al.*, 1995; *Renssen and Isarin*, 1998; *Coope et al.*, 1998], the vast majority of LGIT records obtained from marine and terrestrial sediment sequences remain inadequate, in terms of their temporal and spatial resolution, for the sorts of scientific questions now being posed. There are notable exceptions, however, where varved sequences have accumulated, which allow reconstructions to be made at decadal to annual-scale resolution [e.g., *Hughen et al.*, 1998, *Litt et al.*, in press] and *Von Grafenstein et al.*, 1999]. In these instances, the rates of environmental change are comparable with those suggested by the ice-core records, but the degree to which these were synchronous with events in the ice-core records is difficult to establish.

The extent to which abrupt climate shifts during the LGIT were globally synchronous, or not, is pivotal to our understanding of global climate mechanisms, not only for this particular period, but also for the earlier episodes of abrupt climatic change, such as D-O and Heinrich events (see Section IV). A major debate has emerged on the subject, which revolves around whether short-term changes in oceanic circulation or in atmospheric gas content constituted the main driving force behind the short-lived climatic perturbations that characterized the LGIT. Central to the argument are detailed comparisons between the Antarctic and Greenland ice cores, which are the most detailed palaeoclimatic archives we have available for this period. The ice-core records are 'wiggle-matched' using, for example, methane concentration data, which are assumed to reflect contemporaneous atmospheric values, and to be globally synchronous. Comparisons of independent climatic signals [e.g., stable isotope and snow accumulation data) have generated conflicting ideas about the degree to which climatic changes in Greenland were synchronous with those in Antarctica. Thus, for example, *Blunier et al.* [1998] concluded that Antarctic climate changes were significantly out of step with those in the North Atlantic (Figure 17) on the basis of $\delta^{18}O$ data (a proxy for temperature) obtained from the Byrd and Vostok ice-cores. The data suggest that cooling (the Antarctic Cold Reversal) occurred in Antarctica during the period of warming in the North Atlantic region referred to as the 'Bølling-Allerød Interstadial' (or GI-1). They also suggest that the severely cold Younger Dryas (GS-1) Stadial was a period of rapid warming in Antarctica (Figure 17). By contrast, the Taylor Dome $\delta^{18}O$ record appears to indicate a sequence of climate changes in Antarctica remarkably coherent with the Greenland Summit ice-core records [*Steig et al.*, 1998; Figure 17). However the stratigraphy used in the *Steig et al.* [1998] work has subsequently been questioned [*Mulvaney et al.*, 2000] putting in doubt their conclusions.

In turn, these comparisons have fuelled speculation about the global mechanisms that could account for either synchronous or non-synchronous inter-hemispheric climate changes. Prominent among the emerging ideas the *Broecker* [1998] suggestion of a 'bipolar seesaw behaviour in thermohaline circulation', discussed above, as a mechanism which explain anti-phase relationships between the two hemispheres. By contrast, *Denton et al.* [1999] argue for synchronous climate shifts between the two hemispheres, driven by changes in atmospheric greenhouse gas content [e.g., methane, carbon dioxide, atmospheric mois-

ture content), which would be more consistent with the suggestions of *Shackleton* [2000].

It is difficult to test these alternative hypotheses robustly, because of the following limitations in the interpretation of the stratigraphical records. (1) The extent to which the ice-sheet palaeoclimatic records are representative of broader, hemispheric climatic changes is difficult to establish, especially for the southern hemisphere, where few detailed records of the LGIT have yet been developed. (2) Correlation between polar ice cores based on synchronous changes in global methane has one major drawback. This is that each ice core has a unique relationship between the age of the ice and the age of the gas it contains. Hence assumptions of temperature of ice formation, ice accumulation and ice thinning function have to be taken into account to model this difference which can be up to thousands of years. This means that the methane record can be moved substantial up or down compared with the ice $\delta^{18}O$ 'temperature' record. So in fact both Greenland and Antarctic ice core methane records can be seen as on sliding time scales, and the best estimate of the gas age has then been used to fix the $\delta^{18}O$ temperature proxy to infer either synchronous or non-synchronous inter-hemispheric climate changes. (3) 'Wiggle-matching' between stratigraphical records using the assumption that comparable trends in selected proxy data reflect synchronous events is a circular argument. Each record needs to be dated precisely, and independently. (4) There are differences in detail in the chronologies established for the various ice-core records, especially for pre-Holocene intervals [*Lowe et al.*, in press]. (5) Even if it can be established that the polar ice sheets responded in a synchronous manner to some common forcing mechanism during the LGIT, the degree to which this was also the case on the continents and in the oceans needs to be tested, and not simply assumed. While much headway has been made in recent years in developing a rigorous framework for establishing the degree of synchroneity between ice-sheet, marine and continental records in the North Atlantic region [*Peteet*, 1995; *Walker et al.*, in press b; *Lowe et al.*, in press], precise comparisons are not yet possible, due mainly to the problems inherent in the radiocarbon dating of pre-Holocene events.

V.b. Possible Causes of Abrupt Climate Oscillations During the LGIT

What can be said, therefore, about the causes of the very abrupt climatic changes during the LGIT? It has long been thought that deglaciation was dominated by orbitally induced ice sheet reduction resulting in shifts in oceanic circulation, which controlled sea surface temperatures in the North Atlantic through shifting the latitudinal position of the North Atlantic Polar Front [e.g., *Ruddiman et al.*, 1977;

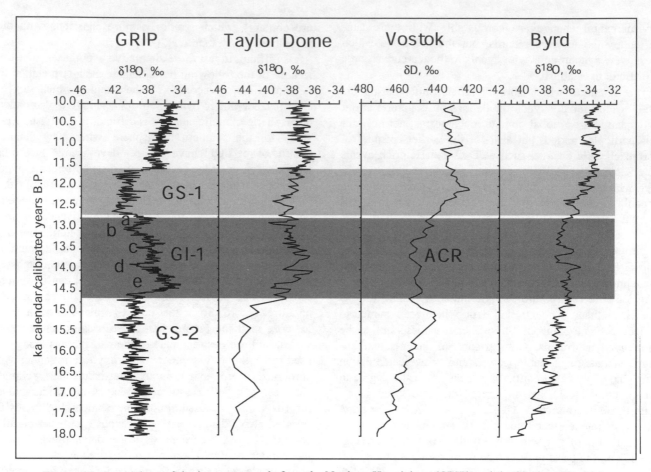

Figure 17. Comparison of the ice core records from the Northern Hemsiphere (GRIP) and the Southern Hemsiphere (Taylor Dome, Vostok and Byrd) for the last glacial-interglacial transition (LGIT). Shaded regions show where it has been suggested that the climate of the North and South Hemisphere was out of phase. GS 1 = Younger Dryas or glacial stadial 1, GI 1 = Bolling-Allerod Interstadial or glacial interstadial 1, GS 2 = glacial stadial 1 and ACR = Antarctic Cold Reversal. It should be noted that the stratigraphy used in the *Steig et al.* [1998] work has subsequently been questioned [*Mulvaney et al.*, 2000] putting in doubt their conclusions that the climate of Antarctic was in phase with Greenland.

Ruddiman and McIntyre, 1981; *Hughen et al.*, 1996]. The colder episodes, especially the very pronounced Younger Dryas cooling [*Dansgaard et al.*, 1989], are then periods imprinted on this trend to warmer interglacial conditions. For example during the Younger Dryas marine records from all parts of the North Atlantic show significant increases in freshwater and ice-rafted debris input. This has led to the suggestion that the Younger Dryas was caused by a sudden influx cold fresh meltwater from the North American and European ice sheets [*Kennett*, 1990; *Bond et al.*, 1992; *Teller et al.*, 1995; *Fronval et al.*, 1995], which was sufficient to cap deep water formation in the North Atlantic. *Berger and Jansen* [1995] suggest it may have only been meltwater diverted through the Gulf of St. Law-

rence which was enough to stop deep water formation. It seems almost axiomatic that the driving mechanism for the Younger Dryas episode, for example, must be some form of internal mechanism, such as oceanic circulation change, since this episode of very pronounced cooling occurred at a time of summer radiation maximum in Milankovitch calculations.

The crucial question that remains to be answered, however, is whether these short episodes of cold-water influx not only displaced the North Atlantic Polar Front southwards to bring devastatingly cold conditions to many parts of the northern hemisphere, but also had synchronous cooling effects in the Southern Hemisphere. Or as suggested by the bipolar seesaw theory there should have been a general warming in the Southern Hemisphere. There is, as yet, no clear consensus on this issue, with arguments for and against synchronous inter-hemispheric developments during the LGIT. Further issues that need to be resolved are whether the very short-lived interstadial and stadial episodes that occurred during the LGIT which were triggered by oceanic changes, or by other factors contributed in some way, or even acted as the catalyst for a cascade of changes that brought about significant climatic effects. These contributing or possible controlling factors include the rapid rise in the greenhouse gases, atmospheric carbon dioxide, methane and water vapor, which are known to have been globally synchronized [*Meeker et al.*, 1997; *Fuhrer and Legrand*, 1997].

VI. RAPID CLIMATE CHANGES WITHIN INTERGLACIALS

VI.a. Marine Oxygen Isotope Stage 5e (the Last Interglacial)

The last interglacial (also called the Eemian in Europe) has often been viewed as a close counterpart of the present (Holocene) interglacial stage: sea surface temperatures were similar, and sea level and atmospheric carbon dioxide levels were possibly higher than pre-industrial times [e.g., *Imbrie and Imbrie*, 1992; *Linsley*, 1996]. Assuming that there is a general similarity between these two warm periods, the Eemian has been used to predict the duration of the present interglacial, and also to study the possibility of sudden climate variability occurring within the next few centuries or millennia. However, it has been shown that during the Holocene and the Eemian both the orbital forcing [e.g., *Berger and Loutre*, 1989] and the climate response such as deep water circulation [e.g., *Duplessy et al.*, 1984] were different. Hence many new studies have concentrated on Marine Oxygen Isotope 11, as a possible analogue for the Holocene since it had very similar orbital characteristics to the Holocene.

VI.b. Controversy Over the Timing of the Last Interglacial.

The Eemian or Marine oxygen Isotope Stage (MIS) 5e interglacial (Fig. 18) began sometime between 130-140 ka [e.g., *Martinson et al.*, 1987; *Sarnthein and Tiedemann*, 1990; *Sowers et al.*, 1993; *Szabo et al.*, 1994; *Stirling et al.*, 1995; *Kukla*, 2000] with a warming phase (of uncertain duration) taking the earth out of an extreme glacial phase, into conditions warmer than today [*Frenzel et al.*, 1992]. Warming into the Eemian may have occurred in two major steps, similar to the last deglaciation [*Sarnthein and Tiedemann*, 1990; *Seidenkrantz et al.*, 1993; 1996].

Though it was named after a warm-climate phase seen in the terrestrial pollen record of the Netherlands [e.g., *Zagwijn*, 1963, 1975], the first generally accepted numerical dates for the start of the Eemian came from astronomic tuning of benthic foraminiferal oxygen isotope data [e.g., *Shackleton*, 1969; *Martinson et al.*, 1987]. The age and duration of the warm Stage 5e, however, is still under discussion [*Frenzel and Bludau*, 1987; *Kukla*, 2000]. Work on deep-sea sediments [*Imbrie et al.*, 1993; *Sarnthein and Tiedemann*, 1990; *Maslin et al.*, 1998b] and corals [*Szabo et al.*, 1994; *Stirling et al.*, 1995; *Slowey et al.*, 1996] suggests that rapid warming could have started as early as 132 ka, while work on the Antarctic Vostok ice core suggests a possible initiation at 134 ka [*Jouzel et al.*, 1993; 1996]. Studies of an Alaskan site (the Eve Interglaciation Forest Bed) suggest that the warming probably postdates 140 ka, because a tephra layer, which is thought to underlie this bed, has been dated to 140 +/-20 ka BP, though the precise relationship between the tephra and the organic bed, and of the age of the organic sediments, remains uncertain [*S. Elias*, pers. comm.]. Uranium-Thorium dated records from a continental karst sediment in the southwestern USA [Devil's Hole; *Winograd et al.*, 1988; 1992; 1997], however, suggest a much earlier start of warming at about 140 ka.

There has been much discussion about the reliability of dating of the marine isotope stages when compared with the Devil's Hole record [e.g., *Imbrie et al.*, 1993], but new radiometric Protactinium-231 dating has apparently confirmed that both records are reliable [*Edwards et al.*, 1997]. We must thus consider the possibility that warm conditions did not last for the same amount of time throughout the world. For instance, comparison of various land records suggests that warming may have occurred at different times in the Alps and in northern France [*de Beaulieu and Reille*, 1989].

Winograd et al. [1997] date the ending of the warm period at about the same time as that suggested by the SPECMAP time scale, which means that the duration of the Eemian warm period according to the Devil's Hole record was significantly longer than that suggested by the marine record (25 ka vs. 17 ka). We need to consider the possibility that warm intervals as seen in pollen records have a longer duration than periods of high sea level and low ice volume in the marine record: *Kukla et al.* [1997; 2000] and *Tzedakis et al.* [1997], for instance, suggest that the Eemian (as reflected in the pollen records) started at about 130 ka, but ended much later than the end of MIS 5e, and that the duration of land interglacials thus is indeed longer than the period of low ice volume.

If the complexity outlined above is accepted, then it seems that for thousands of years warm 'interglacial' type conditions in the mid-latitudes on land could have been occurring at the same time as much colder ocean conditions and expanded Arctic ice sheets. If so the Eemian, it appears, could have been a strange beast quite unlike our present interglacial phase, which began with a rapid and largely simultaneous warming all around the world. This confusion over the nature and duration of the Eemian adds to the difficulty in making simple, general comparisons with our present interglacial, and in interpreting the significance of some of the events seen in the marine and ice core records.

VI.c. Evidence of Climate Instability During the Eemian

Initial evidence from the GRIP ice core [Dansgaard et al., 1993; Taylor et al., 1993] suggested that the Eemian was punctuated by many short-lived extreme cold events. The cold events seemed to last a few thousand years, and the magnitude of cooling was similar to the difference between glacial and interglacial. Furthermore, the shifts between these warm and cold periods seemed to be extremely rapid, possibly occurring over a few decades or less.

A second ice core (GISP2) from the Greenland ice cap [Grootes et al., 1993] provided an almost identical climate record for the last 110 ka, shortly after the end of the Eemian. GISP2 also contains marked changes in the isotope values throughout the deeper section that are not consistent with the GIRP record [Grootes et al., 1993]. Significantly, in GISP2 steeply inclined ice layers occur in this lower portion of the core, indicating that the ice has been disturbed, and that we cannot distinguish simple tilting from folding or slippage that would juxtapose ice of very different ages [Boulton, 1993]. For this reason, the deeper GISP2 record has been interpreted as containing interglacial and glacial ice of indeterminate age, due to the effects of ice tectonics [Grootes et al., 1993; Alley et al., 1997a]. It has been confirmed that the deeper parts of the GRIP ice core record (referred to above), including the crucial Eemian sequence, also tilted [e.g., Grootes et al., 1993; Taylor et al., 1993; Boulton, 1993; Chapellaz et al., 1997; Hammer et al., 1997; Steffensen et al., 1997], with Johnsen et al. [1995, 1997] reporting layers tilted up to 20° within the Marine Isotope Stage 5c (110 ka).

Evidence to support for the occurrence of cold Eemian events was obtained from lake records from continental Europe [de Beaulieu and Reille, 1989, Guiot et al., 1993], in particular the Massif Central in France [Thouveny et al., 1994] and Bispingen inGermany [Field et al. 1994]. However, these records suggest only that the Eemian climate was more variable than that of the Holocene, not the more extreme departures to near-glacial conditions seemingly indicated by the ice cores records. Oceanic records from the North Atlantic [McManus et al., 1994; Adkins et al.,

1997; Oppo et al., 1997; Chapman and Shackleton, 1999] and the Bahamas Outer Ridge [Keigwin et al., 1994] indicate very little or no climatic variability during the Eemian, but a cooling appears to be present in oceanic records from the Sulu Sea [Indonesia; Linsley 1996]. In contrast to this, records from both the Nordic Seas and west of Ireland show a cooling and freshening of the North Atlantic in the middle of the Eemian somewhere between 120 and 123 ka [Cortijo et al., 1994; Fronval and Jansen, 1996]. These Nordic records show highly variable surface water conditions throughout the Eemian period. Records from slightly further south on the north-west European shelf sediments, suggest a similar picture of cold intervals during the Eemian [Seidenkrantz et al., 1995].

Evidence for a single sudden cool event during the Eemian is also present in pollen and lake records in central Europe [Field et al., 1994; Thouveny et al., 1994], from loess sedimentology in central China [Zhisheng and Porter, 1997] and from ocean sediment records from the eastern sub-tropical Atlantic [Maslin et al., 1996; 1998b; Maslin and Tzedakis, 1996] and the re-interpretation of the Nordic Sea records [Cortijo et al., 1994; Fronval and Jansen, 1996] by Maslin and Tzedakis [1996]. Further support for the existence of an intra-Eemian cooling event comes indirectly from coral reef records. High precision U-series coral dates from western Australia indicate that the main global episode of coral reef building during the last interglacial period (dated between 130 and 117 ka) was confined to just a few thousand years between 127 to 122 ka [Stirling et al., 1995], thus ending at the beginning of the Intra-Eemian cold event at about 122 ka (Figure 18). There is also supporting evidence for a cold dry period between 120-121 from a high precision U-series dated speleothem from NW England [Hodge, 2000].

Overall, there is a growing consensus that there was for at least one cold dry event near the middle of the Eemian, at about 120-122 ka. It is characterized by a change in circulation patterns in the North Atlantic, decline in Atlantic surface temperatures by several degrees, and by opening up of the west European forests to give a mixture of steppe and trees. This intra-Eemian cold phase was less dramatic than had been suggested by the variability in the ice core records, but still a major climatic change. Evidence from a high resolution marine core record at Site ODP 658 [Maslin and Tzedakis, 1996] suggests that this event may have been driven by a brief reduction in NADW formation (Figure 18). Afterwards climate recovered, but conditions did not return to the full warmth of the early Eemian 'optimum'. We could view this two stage Interglacial in terms of bifurcation as the intra-Eemian cold drives the climate system into a cooler phase which it can not get out because insolation is already dropping (Figure 19).

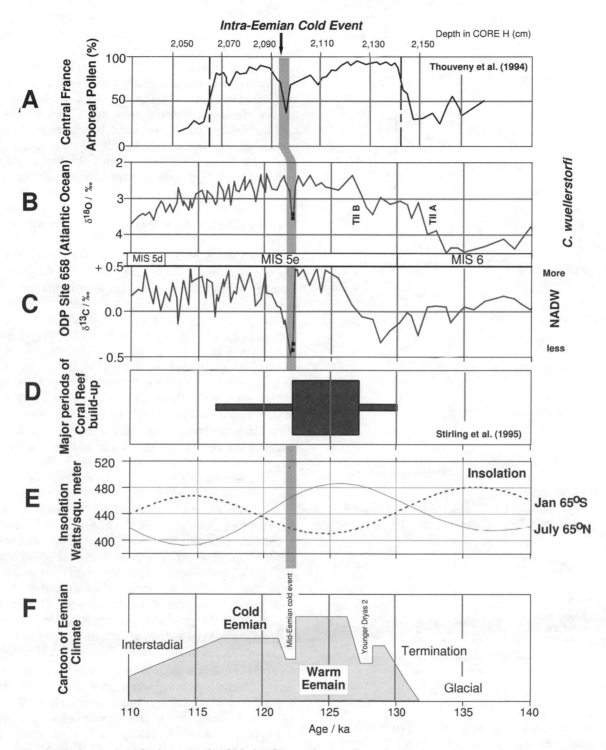

Figure 18. Comparison for key records of Marine Oxygen Isotope Stage 5e. Shaded bar illustartes the possible Mid-Eemian cold event in the records. A summary cartoon illustrating the major changes in climate between 135 ka and 110 ka is given at the bottom [adapted from *Adams et al.*, 1999].

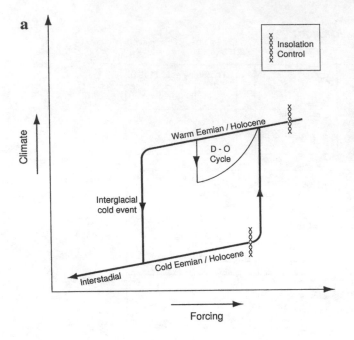

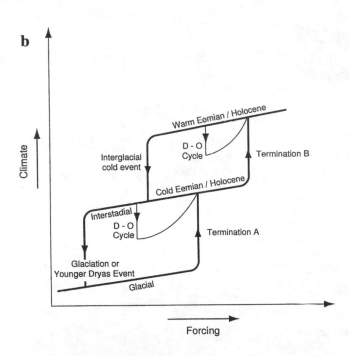

Figure 19. Adaptation of the bifurcation Figure 3 to show the possible relationship between global climate and the forcing factors. A) Shows the possible relationship between a significant Interglacial cold event and the insolation control as an explanation for mild beginnings but colder ends of interglacial periods, see Figure 18. B) Extension of A) to include the Terminations, and Younger Dryas – type events.

VI.d. Holocene

The ice-core records initially suggested the Holocene to be largely complacent as far as climate variability is concerned [e.g., *Dansgaard et al.*, 1993]. This view is being progressively eroded. Long-term trends indicate an early to mid-Holocene climatic optimum with a cooling trend in the late Holocene. Superimposed on this are several distinct oscillations or climate steps which appear to be of widespread significance (8.2 ka, 5.5-5.3 ka and 2.5 ka) see Figure 20. These events now seem to be part of the millenial scale quasi-periodic climate changes, similar to the D-O cycles, and are characteristic of the Holocene [*O'Brien et al.*, 1996; *Bond et al.*, 1997; *Bianchi and McCave*, 1999; *deMenocal et al.*, 2000; *Giraudeau et al.*, 2000].

The periodicity of these Holocene D-O cycles is the subject of much debate. Initial analysis of the Greenland ice core and North Atlantic sediment records have found cycles at approximately the same 1,500 (±500) year rhythm as that found for the last glacial period [*O'Brien et al.*, 1996; *Mayewski et al.*, 1997; *Bond et al.*, 1997; *Campbell et al.*, 1998; *Bianchi and McCave*, 1999; *Chapman and Shackleton*, 2000]. Subsequent analyses have also found both a strong 1,000 year and 550 year cycle [*Chapman and Shackleton*, 2000]. These short cycles have also been recorded in the residual Δ14C data derived from dendrochronology calibrated bidecadal tree-ring measurements spanning the last 11,500 years. The general conclusion is that the Holocene does contain climate variations on the millennial-scale. In general, the coldest points of each of the Holocene millennial-scale cycles surface temperatures of the North Atlantic were about 2-4 °C cooler than at the warmest part [*Bond et al.*, 1997; *deMenocal et al.*, 2000].

The first Holocene event at 8200 ka is the most striking sudden cooling event during the Holocene [*Bjorch et al.*, 1996], giving widespread cool, dry conditions lasting perhaps 200 years before a rapid return to climates warmer and generally moister than the present. This event is clearly detectable in the Greenland ice cores, where the cooling seems to have been about half-way as severe as the Younger Dryas-to-Holocene difference [*Alley et al.*, 1997a; *Mayewski et al.*, 1997]. Records from North Africa across Southern Asia suggest more arid conditions involving a failure of the summer monsoon rains [e.g., *Sirocko et al.*, 1993]. Cold and/or aridity also seems to have hit northern most South America, eastern North America and parts of NW Europe [*Alley et al.*, 1997a and b].

In the middle Holocene at approximately 5,500 ka there seems be a sudden and widespread shift to drier or moister conditions [e.g., *Dorale et al.*, 1992]. The dust and SST records off NW Africa shows that the African Humid Pe-

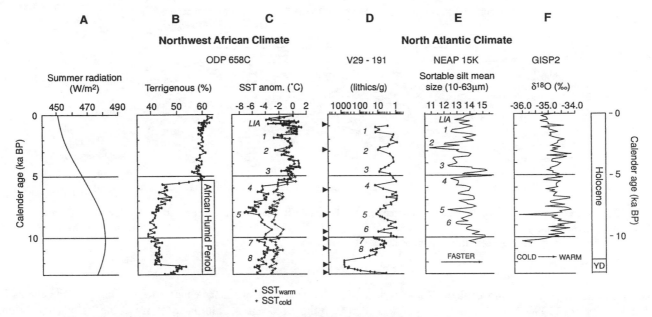

Figure 20. Comparison of summer insolation for 65°N, with Northwest African climate [*deMenocal et al.*, 2000] and North Atlantic climate [V29-191, *Bond et al.*, 1997; NEAP 15K, *Bianchi and McCave*, 1999; GISP2, *O'Brien et al.*, 1995]. Note the similarity of events labeled 1 to 8 and the Little Ice Age (LIA).

riod, when much of subtropical West Africa was vegetated, lasted from 14.8 to 5.5 ka [*deMenocal et al.*, 2000, see Figure 20]. At 5.5 ka there is a 300 year transition to much drier conditions in West Africa (Figure 20). This mid-Holocene shift also corresponds with the decline of the elm (*Ulmus*) in Europe at about 5700 ka and hemlock (*Tsuga*) in North America about 5300 ka. Both vegetation changes were initially attributed to specific pathogen attacks [*Rackham*, 1980; *Peglar*, 1993], but they may be more connected to climate deterioration [*Maslin and Tzedakis*, 1996]. This step to colder and drier conditions could also correspond to the similar change that is observed in the MOIS 5e (Eemian) records. There is also evidence for a strong cold and arid event occurring about 4 ka across the North Atlantic, northern Africa and southern Asia [*Bradley and Jones*, 1992; *O'Brien et al.*, 1996; *Bond et al.*, 1997; *Bianchi and McCave*, 1999; *deMenocal et al.*, 2000; *Cullen et al.*, 2000, see Figure 20]. This cold arid event coincides with the collapse of a large number of the major urban civilizations including: the Old Kingdom in Egypt, the Akkadian Empirer in Mesopotamia, the Early Bronze Age societies of Anatolia, Greece, Israel, the Indus Valley civilization in India, the Hilmand civilization in Afganistan and the Hongshan culture of China [*Peiser*, 1998; *Cullen et al.*, 2000]. Hence these relatively small changes in Holocene climate may have had immense influence on humanity.

VI.e. Little Ice Age (LIA)

The most recent Holocene D-O cold event is the Little Ice Age. This event is really two cold periods, the first of these cold periods follows the 1000 year long Medieval Warm Period and is often referred to as the Medieval Cold Period (MCP) or LIA b [*deMenocal et al.*, 2000]. The MCP played a role in extinguishing Norse colonies on Greenland and caused famine and mass migration in Europe [e.g., *Barlow et al.*, 1997]. It started gradually before 13th century and ended in the middle of the 17th century [*Bradley and Jones*, 1992]. There was then brief respite and a return to milder conditions. Then the second cold period kicked in, this is more classically referred to as the Little Ice Age [*LIA* b, *deMenocal et al.*, 2000] and lasted from the middle of the 18th century to the end of the 19th century. It is thought that the Little Ice Age may have been the most rapid and largest change in the climate of the North Atlantic region during the Holocene according to ice core and deep sea sediment records [*O'Brien et al.*, 1996; *Mayewski et al.*, 1997; *deMenocal et al.*, 2000]. The Little Ice Age events are characterized by a drop of 0.5-1°C in Greenland temperatures [*Dahl-Jensen et al.*, 1998] and sea surface temperature drop of 4°C off the coast of west Africa [*deMenocal et al.*, 2000] and 2°C off the Bermuda Rise [*Keigwin*, 1996]. One question that still remains to be answered is whether the Little Ice Age was a global

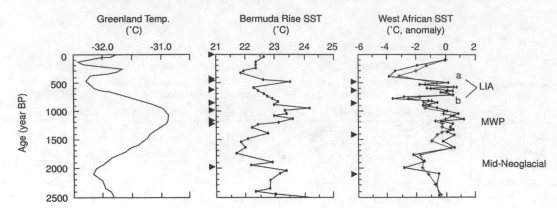

Figure 21. Comparison of Greenland temperatures, the Bermuda Rise sea surface temperatures [*Keigwin*, 1996] and west African and a sea surface temperature [*deMenocal et al.*, 2000] for the last 2.5 ka. LIA = Little Ice Age, MWP = Medieval Warm Period. Solid triangles indicate radiocarbon dates.

or only North Atlantic climate change [*Thompson et al.*, 1986; *Bond et al.*, 1999].

VI.f. Possible Causes of Millennial Climate Fluctuations During Interglacials

Bianchi and McCave [1999] have shown that during the Holocene there have been regular reductions in the intensity of NADW which they link to the Dansgaard-Oeschger cycles identified by *O'Brien et al.* [1995] and *Bond et al.* [1997]. Dansgaard-Oeschger cycles [*Hodge*, 2000] and a major cold event have also been found in the previous interglacial [*Maslin et al.*, 1998b]. There are two possible reasons for the millennial scale changes observed in the intensity of the NADW, instability solely in the North Atlantic region or the bipolar climate seesaw.

First there could be an intrinsic millennial instability in the North Atlantic region. Hence, the amount of freshwater input into the surface waters varies on the millennial time scale to produce the observed reductions. There are a number of possible reasons for this including 1) internal instability of the Greenland ice sheet producing periods of enhanced amounts of melting icebergs, 2) cyclic increases in precipitation over the Nordic Seas due to the North Atlantic storm tracks penetration further north 3) freshwater pulses from the Labrador Sea or 4) changes in surface currents allowing a great import of fresher water from the Pacific, possibly due to reduction in sea ice in Arctic Ocean.

The second possible cause is that the suggested glacial intrinsic millennial-scale bipolar seesaw also operates during interglacial periods (see Figure 15). *Seidov et al.* [2001; this volume] demonstrate that the deep water oscillator described in Figure 15 could be just as valid for interglacial periods. We suggest this is the most plausible ex-

planation at present because the similarity of the events in the Holocene are similar to those in the last glacial period. The defining evidence that is still missing is an out of phase relationship between the climate records of the North Atlantic and the Southern Ocean during the Holocene.

On a more radical note *Wunsch* [2000], provides a very different explanation for the pervasive 1500 year cycle seen in both deep sea and ice core, glacial and interglacial records. *Wunsch* [2000] suggests that the extremely narrow spectral lines (less than two bandwidths) that have been found at about 1500 years in many paleo-records may be due to aliasing. As the 1500 year peak appears precisely at the period predicted as a simple alias of the seasonal cycle inadequately (under the Nyquist criterion) sampled at integer multiples of the common year. When *Wunsch* [2000] removes this peak from the Greenland ice core data [*Mayewski et al.*, 1997] and deep-sea spectral records [*Bianchi and McCave*, 1999] climate variability appears as expected to be a continuum process in the millennial band. This work suggests that finding a cyclicity of 1500 years in a data set may not represent the true occurrence of the millennial scale events in the record. It also supports the case that both Heinrich events and Dansgaard-Oeschger cycles are quasi-periodic with many different and possibly stochastic influences on their occurrence.

THE FUTURE

Future Dramatic Decadal Time-scale Climate Transitions?

From present understanding of Quaternary climate change, we know there are certainly large climatic transitions which have occurred on the time-scale of individual human lifetimes. For example the end of the Younger Dryas and various D-O cold events during the Holocene. Many other substantial shifts in climate took at most a few decades or centuries (Heinrich events, various stages during the last deglaciation). It is clear that despite a huge amount of data being produced there are still many problems with understanding the mechanisms controlling century to millennial scale global climate changes [e.g., *Rind and Overpeck*, 1993, see Figure 22]. The most important finding of this synthesis, however, is the fundamental importance of the deep-water system in controlling rapid global climatic change. Ours is a unique system as the NADW only developed during the Miocene and did not really become important until the closure of the Panama gateway. Since then it has been put forward as a major control on the ONHG, MPR, glacial-interglacial cycles, Heinrich events, glacial and interglacial D-O cycles.

Despite our growing understanding of the causes of rapid climate shift it is still difficult to quantify the future risks of sudden switches in the deep-water system. This is because of the possibility of NADW bifurcation [*Rahmstorf et al.*, 1996] and it is not known where on the threshold figure our current climate system resides (Figure 3). The possibility of deep water variations do appear to be real and relatively small-scale changes in North Atlantic salinity have been observed and studied in the last few decades [*Dickson et al.*, 1988]. Fluctuations in surface water characteristics and precipitation patterns in that region vary on decadal time scales with variations in the strength of high-pressure areas over the Azores and Iceland [North Atlantic Oscillation; *Hurrell*, 1995, 1996], providing an observed apparent link between salinity and climate fluctuations. The fear is that relatively small anthropogenic changes in high-latitude temperature as a result of increased concentrations of greenhouse gases might effect the Nordic Seas in the following ways: increased precipitation, increased freshwater runoff and/or meltwater from the adjacent continents or reduced sea ice formation [*Rahmstorf et al.*, 1996]. All of which would reduce the surface water salinity and thus the potential to form deep water. Some scenarios in which atmospheric carbon dioxide levels are allowed to rise to several times higher than at present result in increased runoff from rivers entering the Arctic Basin, and a rapid weakening of the Gulf Stream, resulting in colder conditions (especially in winter) across much of Europe [*Broecker*, 1997b].

There is however a second twist, the bipolar seesaw that has been discovered in the palaeoclimate records [*Broecker*, 1998; 2000; *Stocker*, 1998; *Seidov and Maslin*, 2000]. If indeed the NADW formation is reduced in the future cooling the North Atlantic region there could be a

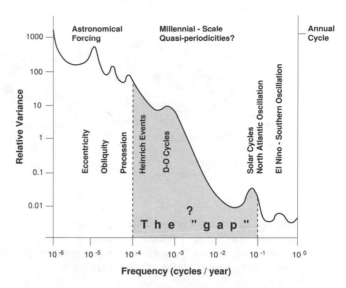

Figure 22. Spectrum of Climate Variance showing the climatic cycles for which we have good understanding and the 'gap' between 100s and 1000s of years for which we still do not have adequate understanding of the causes.

corresponding by delayed warming in the Southern Ocean. Recent modeling results, however, also suggest that the Southern Ocean is as sensitive if not more so to meltwater inputs than the North Atlantic region [*Seidov et al.*, 2001; this volume]. Meltwater input to the Southern ocean has been shown to dramatically reduce AABW production and increase deep water temperatures. This in turn would allow the NADW to strengthen warming the Northern Hemisphere. So currently we have two completely opposite scenarios which will make a significant difference to the inter-Hemisphere heat balance, which influences the whole monsoonal system. The only difference is whether global warming causes the most significant freshwater input to occur in the North Atlantic region or the Southern ocean.

To paraphrase W.S. Broecker; 'Climate is an ill-tempered beast, and we are poking it with sticks'.

Acknowledgments. We would like to thank all those colleagues who over the years have provided their insights into paleoclimatology. We would especially like to thank the reviewers, Ellen Thomas, Rainer Zahn and Francis Grousset whose careful, detailed and challenging reviews improved this paper immeasurably. We would also like to thank C. Pyke, N. Mann, E. McBay of the Department of Geography, Drawing Office for the help with the figures. We are also grateful to Bernd J. Haupt for his help in improving the manuscript.

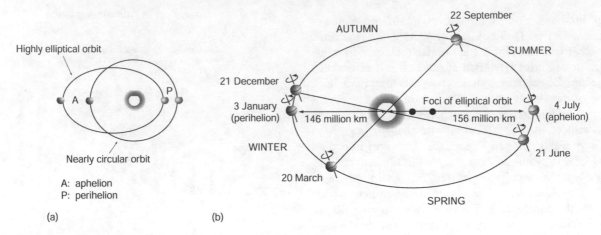

Figure 23. Changes in the shape of the Earth's orbit around the sun. (a) The shape of the orbit changes from a near circular to elliptical. The position along the orbit when the Earth is closest to the Sun is termed the perihelion and the position when it is furthest from the Sun aphelion. (b) The present-day orbit and its relationship to the seasons, solstices and equinoxes [after *Wilson et al.*, 2000].

APPENDIX I: ORBITAL FORCING; THE BASICS [ADAPTED FROM *WILSON ET AL.*, 2000 AND *LOWE AND WALKER*, 2000]

Eccentricity (Figure 23)

The shape of the Earth's orbit changes from near circular to an ellipse over a period of about 100 ka with a long cycle of about 400 ka (in detail there are two distinct spectral peaks near 100 ka and one at 413 ka). Descibed another way the long axis of the ellipse varies in length over time. Today, the Earth is at its closest (146 million km) to the sun on January 3rd: this position is known as perihelion. On July 4th it is at most distance from the sun (156 million km) at the aphelion. Changes in eccentricity cause only very minor variations, approximately 0.03% in the total annual insolation, but can have significant seasonal effects. If the orbit of the Earth were perfectly circular there would be no seasonal variation in solar insolation. Today, the avaerage amount of radiation received by the Earth at perihelion is ~ 351 Wm² reducing to 329 Wm² at aphelion, a difference of more than 6%. At times of maximum eccentricity over the last 5 Ma this difference could have been as large as 30%. *Milankovitch* [1949] suggested that the northern ice sheets are more likely to form when the Sun is more distant in summer, so that each year some of previous winters snow can survive. At a certain size the ice sheets start influencing the local climate and there are positive feedbacks which are discussed more fully in Section IIa. As the intensity of solar radiation reaching the Earth diminishes as the square of the planet's distance, global insolation falls at the present time by near 7% between January and July. A situation that is more favouable for snow surviving in the Northern rather than Southern Hemisphere. The more elliptical the shape of the orbit becomes, the more the season will be exaggerated in one hemisphere and moderated in the other. The other effect of eccentricity is to modulate the precession effects, see below.

Obliquity (Figure 24)

The tilt of the Earth's axis of rotation with respect to the plane of its orbit (the plane of the ecliptic) varies between 21.8° and 24.4° over a period of 41 ka. It is the tilt of the axis of rotation that gives us the seasons. Because in summer the hemisphere is tilted towards the Sun and is warmer because it recieves more than 12 hours of sunlight and the Sun is higher in the sky. At the same time the opposite hemisphere is tilted away from the Sun and is in winter as it colder as it recieves less than 12 hours of sunlight and the Sun is lower in the sky. Hence the greater the obliquity the greater the difference between summer and winter. As *Milankovitch* [1949] suggested the colder the Northern Hemisphere summers the more like ice sheets are to build-up. This is why there seems to be a straight forward explaination for the glacial-interglacial cycle prior to the Mid-Pleistocene Revolution wheich occur every 41 ka.

Precession (Figure 24)

There are two components of precession: that relating to the elliptical orbit of the Earth and that related to its axis of rotation. The Earth's rotational axis moves around a full

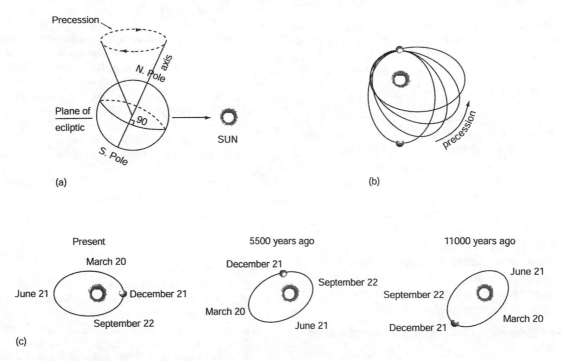

Figure 24. The components of the precession of the equinoxes. (a) The precession of the Earth's axis of rotation. (b) The precession of the Earth's orbit. (c) The precession of the equinoxes [after *Wilson et al.*, 2000].

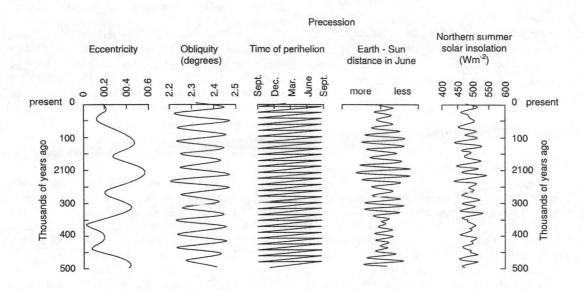

Figure 25. Variations in the Earth's orbital parameters: eccentricity, obliquity and precession and the resultant Northern Hemisphere 65°N insolation for the last half million years [after *Wilson et al.*, 2000].

circle, or precesses every 27 ka (Fig 24 a). This is similar to the gyrations of the rotational axis of a toy spinning top. Precession causes the dates of the equinoxoes to travel around the sun resulting in a change in the Earth-Sun distance for any particular date, for example Northern Hemisphere summer (Fig. 24 c). While the precession of the Earth's orbit is shown in Figure 24 c, which has a periodicity of 105 ka and changes the time of year when the Earth is closest to the Sun (perihelion).

It is the combination of the different orbital parameters that results in the classically quoted precessional periodicities of 23 ka and 19 ka. Combining the precession of the axis of rotation plus the precessional changes in orbit produces a period of 23 ka. Combining the shape of the orbit i.e., eccentricity, and the precession of the axis of rotation results in a period of 19 ka. These two periodicities combine so that perihelion coincides with the summer season in each hemisphere on average every 21.7 ka, resulting in the precession of the equinoxes.

Combining Eccentricity, Obliquity and Precession (Figure 25)

Combining the effects of eccentricity, obliquity and precession provides the means of calculating the insolation for any latitude back through time [e.g., *Milankovitch*, 1941; *Berger*, 1976; 1979; 1988; 1989; *Berger and Loutre*, 1991]. The maximum change in solar radiation in the last 600 ka (see Figure 25) is equivalent to reducing the amount of summer radiation received today at 65°N to that received now over 550 km to the north at 77°N. Which in simplistic terms brings the current glacial limit in mid-Norway down to the latitude of Scotland. However, the key factor to note is that each of the orbital parameters has a different effect with changing latitude. For example obliquity has increasing influence the higher the latitude, while precession has its largest influence in the tropics.

REFERENCES

Abe-Ouchi, A., Quaternary transition: A bifurcation in forced ice sheet oscillations? *Eos Trans. AGU*, 77 (46), Fall Meet. Suppl., F415, 1996.

Adams, J, M.A. Maslin and E. Thomas, Sudden climate transitions during the Quaternary, *Progress in Physical Geography*, 23, 1, 1-36, 1999.

Adkins J. F., Boyle E. A., Keigwin, L., and Cortijo E., Variability of the North Atlantic thermohaline circulation during the last interglacial period. *Nature*, 390, 154-156, 1997.

Alley, R. B., *The two mile time machine*, Princeton University Press, Princton, USA, pp229, 2000.

Alley, R.B. and P.U. Clark, The deglaciation of the Northern Hemisphere, *Ann. Rev. Earth Planet. Sci.*, 27, 149-182, 1999.

Alley, R. B., Gow, A. J., Meese, D. A., Fitzpatrick, J. J., Waddington, E. D., and Bolzan, J. F., Grain-scale processes, folding, and stratigraphic disturbance in the GISP2 ice core. *Journal of Geophysical Research*, 102, 26,819-26,829, 1997a.

Alley, R. B., Mayewski, P. A., Sowers, T., Stuiver, M., Taylor, K. C., and Clark, P. U., Holocene climatic instability: a prominent, widespread event 8200 yr. ago. *Geology*, 25, 483-486, 1997b.

Alley, R.B., S. Anandakrishnan, P. Jung, and A. Clough, Stochastic resonance in the North Atlantic: Further insights (2001; this volume).

Andrews, J. T., Abrupt chnages (Heinrich events) in later Quaternary North Atlantic marine environments, *JQS*, 13, 3-16, 1998.

Andrews, J. T., Erlenkeuser, H., Tedesco, K., Aksu, A. E., and Jull, A. J. T., Late Quaternary (Stage 2 and 3) meltwater and Heinrich events, Northwest Labrador Sea, *Quaternary Research*, 41, 26-34, 1994.

Bard, E. et al., Interhemispheric synchrony of the last deglaciation inferred from alkenone palaeothermomtry, *Nature*, 385, 707-710, 1997.

Barlow, L. K., Sadler, J. P., Ogilvie, A. E. J., Buckland, P. C., Amorosi, T., Ingimundarson, J. H., Skidmore, P., Dugmore, A. J., and McGovern, T, H., Interdisciplinary investigations of the end of the Norse western settlement in Greenland, *The Holocene*, 7, 489-499, 1997.

Behl, R. J., and Kennett, J. P., Brief interstadial events in the Santa Barbara basin, NE Pacific, during the past 60 kyr. *Nature*, 379, 243-246, 1996.

Berger, A., Obliquity and precession for the last 5,000,000 years. *Astronomy and Astrophys*ics, 51, 127-135, 1976.

Berger, A., Insolation signatures of Quaternary climatic changes, *IL Nuoco Cimento*, 2C, 63-87, 1979.

Berger, A., Milankovitch theory and climate. *Review of Geophysics*, 26, 624-657, 1988.

Berger, A., Pleistocene climatic variability at astronomical frequencies. *Quaternary International*, 2, 1-14, 1989.

Berger, A. and Loutre, M.F., Insolation values for the climate of the last 10 million years. *Quat. Sci. Rev.*, 10, 297-317, 1991.

Berger, A., Li, X., Loutre, M.F., Modelling northern hemisphere ice volume over the last 3 Ma. *Quat. Sci. Rev.*, 18, 1-11, 1999.

Berger, W.H. and Jansen, E., Mid-Pleistocene climate shift: the Nansen connection. In: Johannessen et al (Eds). *The Polar Oceans and their role in shaping the global environment*. Geophys Monogr., 84, 295-311, 1994.

Berger, W. H., and Jansen, E., Younger Dryas episode: ice collapse and super-fjord heat pump. In: The Younger Dryas, S. R. Troelstra, J. E. van Hinte, and G. M. Ganssen, eds., Koninklijke Nederlandse Akademie van Wetenschappen, Afdeling Natuurkunde, Eerste Reeks, 44, 61-105, 1995.

Berger, W.H. and G. Wefer, Expeditions into the Past: Paleoceanographic studies in the South Atlantic, In Wefer et al. (editors) *The South Atlantic: present and past ciculation*, Spinger-Verlag, Berlin, 363-410, 1996.

Berger, W.H., Bickert, T., Schmidt, H. and Wefer, G., Quaternary oxygen isotope record of pelagic foraminiferas: Site 806,

Ontong Java Plateau. In Berger, W.H., Kroenke, L.W., Mayer, L.A. et al., *Proc. ODP, Sci. Results.*, 130: College Station, TX (Ocean Drilling Program), 381-395, 1993.

Berger, W. H., Bickert, T., Jansen, E., Wefer, G., and Yasuda, M., The central mystery of the Quaternary ice age, *Oceanus*, 36, 53-56, 1993.

Berner, R.A., Geocarb II: A revised model of atmospheric CO_2 over the phanerozoic time. *American Journal of Science*, 294, 56-91, 1994.

Berner, R. A., and Rye, D. M., Calculation of the Phanerozoic strontium isotope record of the oceans from a carbon cycle model. *American Journal of Science*, 292: 136-148, 1992.

Bianchi, G.G. and I.N. McCave, Holocene periodicity in North Atlantic climate and deep-ocean flow south of Iceland, *Nature*, 397, 515-513, 1999.

Björck, S., Walker, M.J.C., Cwynar, L., Johnsen, S.J., Knudsen, K.L., Lowe, J.J., Wohlfarth, B. and INTIMATE Members, An event stratigraphy for the Last Termination in the North Atlantic based on the Greenland Ice Core record: a proposal by the INTIMATE group. *Journal of Quaternary Science*, 13, 283-292, 1998.

Björck, S., Kromer, B., Johnsen, S., Bennike, O., Hammarlund, D., Lemdahl, G., Possnert, G., Rasmussen, T. L., Wohlfarth, B., Hammer, C. U., and Spurk, M., Synchronized terrestrial-atmospheric deglacial records around the North Atlantic. *Science*, 274, 1155-1160, 1996.

Blunier, T., Chappellaz, J., Schwander, J., Dällenbach, Stauffer, B., Stocker, T.F., Raynaud, D., Jouzel, J., Clausen, H.B., Hammer, C.U. and Johnsen, S.J., Asynchrony of Antarctic and Greenland climate change during the last glacial period. *Nature*, 384, 739-743, 1998.

Bolin, B., On the influence of the earth's orography on the general character of the westerlies. *Tellus*, 2, 184-195, 1950.

Bond, G., Climate and the conveyor *Nature*, 377, 383-4, 1995

Bond, G., Heinrich, H., Broecker, W., Labeyrie, L., McManus, J., Andrews, J., Huon, S., Jantschik, R., Clasen, S., Simet, C., Tedesco, K., Klas, M., Bonani, G., and Ivy, S., Evidence for massive discharges of icebergs into the North Atlantic Ocean during the last glacial period. *Nature*, 360, 245-249, 1992.

Bond, G. C., Broecker, W., Johnsen, S., McManus, J., Labeyrie, L., Jouzel, J., and Bonani, G., Correlations between climate records from North Atlantic sediments and Greenland ice. *Nature*, 365, 143-147, 1993.

Bond, G. C., and Lotti, R., Iceberg discharges into the North Atlantic on millennial time scales during the last deglaciation. *Science*, 267, 1005-1010, 1995

Bond, G., Showers, W., Cheseby, M., Lotti, R., Almasi, P., deMenocal, P., Priore, P., Cullen H., Hajdas I. & Bonani G., A pervasive millenial-scale cycle in North Atlantic Holocene and glacial climates. *Science*, 278, 1257-1265, 1997.

Bond, G., W. Showers, M. Elliot, M. Evans, R. Lotti, I. Hajdas, G. Bonani, and S. Johnson, The North Atlantic's 1-2 kyr climate rhythm: relation to Heinrich events, Dansgaard/Oeschger cycles and the Little Ice Age, In *Mechanisms of global climate change at the millennial time scale* (Clark, Webb and Keigwin editors) Geophysical Monograph, 112, 35-58, 1999.

Boulton, G. S., Two cores are better than one, *Nature*, 366, 507-508, 1993.

Bradley, R. S., and Jones, P., The Little Ice Age. *The Holocene*, 3, 367-376, 1992.

Breza, J. and Wise, S.W., Lower Oligocene ice-rafted debris on the Kergulen Plateau, Evidence for East Antarctic continental glaciation, *Proc. Ocean Drill. Program Sci. Results*, 120, 161-178, 1992.

Broecker, W. S., The great ocean conveyor, *Oceanography*, 1: 79-89, 1991.

Broecker, W. S., Massive iceberg discharges as triggers for global climate change, *Nature*, 372, 421-424, 1994.

Broecker W. S., *The glacial world according to Wally*. Eldigio press, 1995.

Broecker, W. S., Thermohaline circulation, the Achilles heel of our climate system: will man-made CO_2 upset the current balance? *Science*, 278, 1582-1588, 1997a.

Broecker W. S., Will Our Ride into the Greenhouse Future be a Smooth One? *GSA Today*, 7, 1-7, 1997b.

Broecker, W., 1998: Paleocean circulation during the last deglaciation: a bipolar seesaw? *Paleoceanography*, 13, 119-121.

Broecker, W.S., Was a change in thermohaline circulation responsible for the Little Ice Age?, *Proc. Nat. Acad. Sci.*, 97(4): 1339-1342, 2000.

Broecker, W. S., Bond, G., Klas, M., Clark, E., and McManus, J., Origin of the North Atlantic's Heinrich events. *Climate Dynamics*, 6, 265-273, 1992.

Brook, E. J., Sowers, T., and Orchardo, J., Rapid variations in atmospheric methane concentration during the past 110,000 years. *Science*, 273, 1087-1091, 1996

Burton, K. W., Ling, H.-F., and O'Nions, K. R., Closure of the Central American Isthmus and its effect on deep-water formation in the North Atlantic. *Nature*, 386, 382-385, 1997.

Cane, M. A., A Role for the Tropical Pacific. *Science*, 282, 59-61, 1998.

Campbell, I. D., Campbell, C., Apps, M. J., Rutter, N. W., and Bush, A. B. G., Late Holocene ~1500 yr. periodicities and their implications, *Geology*, 26, 471-473, 1998.

Chappellaz J. M., Blunier T., Raynaud D., Barnola J. M., Schwander J. and Stauffer B., Synchronous changes in atmospheric CH_4 and Greenland climate between 40 and 8 kyr BP. *Nature*, 366, 443-445, 1993.

Chappellaz, J., Brook, E., Blunier, T., and Malaizé, B., CH_4 and $\delta^{18}O$ of O_2 records from Antarctic and Greenland ice: a clue for stratigraphic disturbance in the bottom part of the Greenland Ice Core Project and the Greenland Ice Sheet Project 2 ice cores. *Journal of Geophysical Research*, 102, 26,547-26,557, 1997.

Chapman, M., and M.A. Maslin, Low latitude forcing of meridional temperature and salinity gradients in the North Atlantic and the growth of glacial ice sheets, *Geology*, 27, 875-878, 1999.

Chapman, M.R. and N.J. Shackleton, Evidence of 550 year and 1000 years cyclicities in North Atlantic pattern during the Holocene, *Holocene*, 10, 287-291, 2000.

Charney, J.G. and Eliassen, A., A numerical method for predicting the perturbations of the middle-latitude westerlies. *Tellus*, 1, 38-54, 1949.

Clark, P. and Pollard, D., Origin of the middle Pleistocene transition by ice sheet erosion of regolith. *Paleoceanography*, 13, 1-9, 1998.

Clark, P., R. Webb and L. Keigwin (editors), *Mechanisms of global climate change at the millennial time scale*, Geophysical Monograph 112, AGU Washington, pp394, 1999.

Comiso, J. C., and Gordon, A. L., Recurring polynyas over the Cosmonaut Sea and the Maud Rise. *Journal of Geophysical Research*, 92, 2819-2833, 1987.

Compton, J. S. and Mallinson, D.J., Geochemical consequences of increased late Cenozoic weathering rates and the global CO_2 balance since 100 Ma, *Paleoceanography*, 11, 431-446, 1996.

Cortijo E., Duplessy, J.-C., Labeyrie, L., Leclaire, H., Duprat, J., and van Weering, T. C. E., Eemian cooling in the Norwegian Sea and North Atlantic Ocean preceding continental ice-sheet growth. *Nature*, 372, 446-449, 1994.

Cortijo E.,et al., Changes in sea surface hydrology associated with Heinrich event 4 in the North Atalntic Ocean between 40° and 60°N, *Earth and Planet. Sci. Lett.*, 146, 29-45, 1997.

Copeland, P., Harrison, T.M., Kidd, W.S.F., Ronghua, X. and Yuquan, Z., Rapid early Miocene acceleration of uplift of the Gagdese Belt, Xizang (southern Tibet), and its bearing on accomodation mechanisms of the India-Asia collision. *Earth and Planet. Sci. Lett.*, 86, 240-252, 1987.

Cronin, T.M., Raymo, M.E. and Kyle, K.P. Pliocene (3.2-2.4 Ma) ostracode faunal cycles and deep ocean circulation, North Atlantic Ocean. *Geology*, 24, 695-698, 1996.

Cullen, H.M., et al., Climate change and the collapse of the Akkadian Empire: Evidence from the deep sea. *Geology*, 28, 379-382, 2000.

Curtis, J. H., Hodell, D. A., and Brunner, M., Climate variability on the Yucatan Peninsula (Mexico) during the past 3500 years, and implications for Maya cultural evolution. *Quaternary Research*, 46, 47-47, 1996.

Dahl-Jensen, D., et al., Past temperatures directly from the Greenland ice sheet, *Science*, 282, 268-271, 1998.

Dansgaard, W., White, J. W. C., and Johnsen, S. J., The abrupt termination of the Younger Dryas climate event. *Nature*, 339, 532-534, 1989.

Dansgaard, W., Johnson, S. J., Clausen, H. B., Dahl-Jensen, D., Gundenstrup, N. S., Hammer, C. U., Hvidberg, C. S., Steffensen, J. P., Sveinbjörnsdóttir, Evidence for general instability of past climate from a 250-kyr ice-core record. *Nature*, 364, 218-220, 1993.

de Beaulieu, J.-L., and Reille, M., The transition from temperate phases to stadials in the long upper Pleistocene sequence from les Echets (France), *Palaeogeography, Palaeoclimatology, Palaeoecology*, 72, 147-159, 1989.

de Menocal, P., Plio-Pleistocene African Climate. *Science*, 270, 53-59, 1995.

de Menocal, P., J. Ortiz, T. Guilderson and M. Sarnthein, Coherent High- and Low- latitude climate variability during the Holocene warm period. *Science*, 288, 2198-2202, 2000.

de Menocal, P. Cultural Reponses to climate change during the Late Holocene, *Science*, 292, 667-672 (2001).

Denton, G.H. and Hendy, C.H., Younger Dryas age advance of Franz Josef Glacier in the Southern Alps of New Zealand. *Science*, 264, 1434-1437, 1994.

Denton, G.H., Heusser, C.J., Lowell, T.V., Moreno, P.I., Andersen, B.G., Heusser, L.E., Schlüchter, C. and Marchant, D.R., Interhemispheric linkage of paleoclimate during the last glaciation. *Geografiska Annaler*, 81A, 107-153, 1999.

Denton, G. H., Does asymmetric thermohaline-ice-sheet oscillator drive 100,000-yr cycles, *Journal Quaternary Science*, 15, 301-318, 2000.

Dickson, R. R., Meincke, J., Malmberg, S. A., and Lee, A. J., The "Great Salinity Anomaly" in the northern North Atlantic, 1968-1982. *Progress in Oceanography*, 20, 103-151, 1988.

Dorale, J. A., Gonzalez, L. A., Reagan, M. K., Pickett, D. A., Murrell, M. T., and Baker, R. G., A high-resolution record of Holocene climate change in speleothem calcite from Cold Water Cave, Northeast Iowa. *Science*, 258, 1626-1630, 1992.

Dowdeswell, J., M.A. Maslin, J. Andrews and I.N. McCave, Estimation of the timing of the Heinrich events using realistic calculations of the maximum outflow of icebergs from the Laurentide icesheet. *Geology*, 23, No. 4, 301-304, 1995.

Dowdeswell, J.,A. Elveroi, J.T. Andrews and D. Hebbeln, Asynchronous deposition of ice rafted layers in the Nordic Seas and North Atlantic Ocean, *Nature*, 400, 248-351, 1999.

Driscoll, N.W. and G.H. Haug, A short cut in thermohaline circulation: a cause for Northern hemisphere Glaciation. *Science*, 282, 1998.

Duplessy, J. C., Shackleton, N. J., Matthews, R. K., Prell, W. L., Ruddiman, W. F., Caralp, M., and Hendy, C., $\delta^{13}C$ record of benthic foraminifera in the last interglacial ocean: implications for the carbon cycle and the global deep water circulation. *Quaternary Research*, 21, 225-243, 1984.

Edwards, R. L., Cheng, H., Murreil, M. T., and Goldstein, S. J., Protactinium-231 dating of carbonates by thermal ionization mass spectrometry: implications for quaternary climate change. *Science*, 276, 782-786, 1997.

Einarsson, T. and Albertsson, K.J., Glaciation of Iceland. *Phil. Trans. R. Soc. Lond.*, B 318, 227-234, 1988.

Einarsson, T., Hopkins, D.M. and Doell, R.R., The stratigraphy of Tjornes, northern Iceland, and the history of the Bering Land Bridge. In Hopkins, D.M. (Ed.), *The Bering Land Bridge*, California, (Stanford University Press), 312-325, 1967.

Fairbanks, R.G., A 17,000 year glacio-eustatic sea level record: influence of glacial melting rates on the Younger Dryas event and deep-ocean circulation, *Nature*, 342, 637-642, 1989.

Farley, K. and D. B. Patterson, A 100 kyr periodicity in the flux of extraterrestrial ^{3}He to the sea floor, *Nature*, 378, 600-603, 1995.

Fichefet, T., S. Hovine and J.-C. Duplessy, A model study of the Atlantic thermohaline circulation during the Last Glacial Maximum, *Nature*, 372: 252-255, 1994.

Field M, Huntley B. and Muller H., Eemian climate fluctuations observed in a European pollen record. *Nature*, 371, 779-783, 1994.

François, R., and Bacon, M. P., Heinrich events in the North Atlantic: radiochemical evidence, *Deep-Sea Research I*, 41, 315-334, 1993.

Frank, M., Reynolds, B. C., and O'Nions, R. K., Nd and Pb isotopes in Atlantic and Pacific before and after closure of Panama, *Geology*, 27, 1147-1150, 1999.

Frenzel, B., Pécsi, M., and Velichko, A. A., *Atlas of Paleoclimates and Paleoenvironments of the Northern Hemisphere, Late Pleistocene - Holocene*. Geographical Research Institute. Hungarian Academy of Sciences, Budapest, Gustav Fischer Verlag, New York, 153 pp, 1992.

Frenzel, B., and Bludau, W., On the duration of the interglacial transition at the end of the Eemian interglacial (deep sea stage 5e): botanical and sedimentological evidence. in *Abrupt Climatic Change: Evidence and Implications*, W. H. Berger and L. D. Labeyrie, eds., 151-162, 1987.

Fronval, T., and Jansen, E., Rapid changes in ocean circulation and heat flux in the Nordic seas during the last interglacial period. *Nature*, 383, 806-810, 1996.

Fronval, T., Jansen, E., Bloemendal, J. & Johnsen, S., Oceanic evidence for coherent fluctuations in Fennoscandian and Laurentide ice sheets on millennium timescales, *Nature*, 374, 443-446, 1995.

Fuhrer, K., and Legrand, M., Continental biogenic species in Greenland Ice core Project core: tracing back the biomass history of the North American continent. *Journal of Geophysical Research*, 102, 26,735-26,745, 1997.

Galleé, H., van Ypersele, J.P., Fichefet, Th., Tricot Ch. and Berger, A., Simulation of the last glacial cycle by a coupled, sectorially averaged climate-ice sheet model, 1. The climate model. *Journal of Geophysical Research*, 96, 13,139-13,161, 1991.

Galleé, H., van Ypersele, J.P., Fichefet, Th., Tricot Ch. and Berger, A., Simulation of the last glacial cycle by a coupled, sectorially averaged climate-ice sheet model, 2. Response to insolation and CO_2 variations. *Journal of Geophysical Research*, 97, 15,713-15,740, 1992.

Geirsdottir, A. and Eiriksson, J., Growth of an intermittent ice sheet in Iceland during the late Pliocene and early Pleistocene. *Quat. Res.*, 42, 115-130, 1994.

Giraudeau, J., Cremer, M., Mantthe, S., Labeyrie, L., Bond, G., Coccolith evidence for instability in surface circulation south of Iceland during Holocene times. *Earth and Planetary Science Letters*, 179, 257-268, 2000.

Grimm, E. C., Jacobson, G. L., Watts, W. A., Hansen, B. C. S., and Maasch, K., A 50,000 year record of climate oscillations from Florida and its temporal correlation with the Heinrich events. *Science*, 261, 198-200, 1993.

Grootes, P. M., Stuiver, M., White, J. W. C., Johnsen, S., and Jouzel, J., Comparison of oxygen isotope records from the GISP2 and GRIP Greenland ice cores, *Nature*, 366, 552- 554, 1993.

Grousset, F. E., Labaeyrie, L., Sinko, J. A., Cremer, M., Bond, G. , Duprat, J., Cortijo, E., and Huon, S., Patterns of ice-rafted detritus in the glacial North Atlantic (40°-55°N), *Paleoceanography*, 8, 175-192, 1993.

Grousset F.E., Pujol C., Labeyrie L., Auffret G., Boelaert A., Were the North Atlantic Heinrich Events triggered by the behavior of the European ice sheets? *Geology*, 28(2):123-126, 2000.

Grousset F.E., et al. Zooming in on Heinrich layers. Paleoceanography, in press.

Guiot , J. L., de Beaulieu, J. L., Chedaddi, R., David, F., Ponel, P., and Reille, M., The climate in Western Europe during the last glacial/interglacial cycle derived from pollen and insect remains, Palaeogeography, Palaeoclimatology, *Palaeoecology*, 103, 73-93, 1993.

Gwiazda, R. H., Hemming, S. R., and Broecker, W. S., Tracking the sources of icebergs with lead isotopes: the provenance of ice-rafted debris in Heinrich layer 2. *Paleoceanography*, 11, 77-93, 1996a.

Gwiazda, R. H., Hemming, S. R., and Broecker, W. S., Provenance of icebergs during Heinrich event 3 and the contrast to their sources during other Heinrich events. *Paleoceanography*, 11, 371-378, 1996b.

Hagelberg, T., Bond, G., and deMenocal, P., Milankovitch band forcing of sub-Milankovitch climate variability during the Pleistocene. *Paleoceanography*, 9: 545-558, 1994.

Hambrey, M.J., Ehrmann, W.U. and Larsen, B., Cenozoic glacial record of the Prydz Bay Continental Shelf, East Antarctica, *Proc. Ocean Drill. Program Sci. Results*, 119, 77-132, 1991.

Hammer, C., Mayewski, P. A., Peel, D., and Stuiver, M., Preface. *Journal of Geophysical Research*, 102, 26,315-26,316, 1997.

Harrison, T.M., Copeland, P., Kidd, W.S.F. and Yin, A., Raising Tibet, *Science*, 255, 663-670, 1992.

Haug, G.H. and R. Tiedemann, Effect of the formation of the Isthmus of Panama on Atlantic Ocean thermohaline circulation, *Nature*, 393, 673-675, 1998.

Hay, W. The cause of the late Cenozoic Northern Hemisphere Glaciations: a climate change enigma. *Terra Nova*, 4, 305-311, 1992.

Hays, J.D., J. Imbrie, N.J. Shackleton. Variations in the Earth's orbit: Pacemaker of the Ice Ages. *Science*, 194, 1121-1132, 1976.

Hays, J. D., Martinson, D. G., McIntyre, A., Mix, A. C., Morley, J. J., Pisias, N. G., Prell, W. L., Shackleton, N. J., The orbital theory of Pleistocene climate: support from a revised chronology of the marine ^{18}O record. in: A. Berger, J. Imbrie, J. Hays, G. Kukla, B. Saltzman, eds., *Milankovich and Climate*, Reidel, Dordrecht, 269-306, 1984.

Heinrich, H., Origin and consequences of cyclic ice rafting in the northeast Atlantic Ocean during the past 130,000 years. *Quaternary Research*, 29, 142-152, 1988.

Hewitt,T., W. Mitchell, Radiative forcing and response of a GCM to ice age boundary conditions, cloud feedback and climate sensitivity, *Climate Dynamics*, 13, 821-834, 1997.

Hodell, D. A., Curtis, J. H., and Brenner, M., Possible role of climate in the collapse of Classic Maya civilization. *Nature*, 375, 391-394, 1995.

Hodge, E., Climate variability in the last interglacial: New evidence using high-resolution speleothem stable isotope records from Lancaster hole caves, North Yorkshire, England, MSc thesis, University of London, p50, 2000.

Hughen, K. A., Overpeck, J. T., Paterson, L. C., and Trumbore, S., Rapid climate changes in the tropical Atlantic Ocean during the last deglaciation. *Nature*, 230, 51-54, 1996.

Hughen, K.A., Overpeck, J.T., Lehman, S.J., Kashgarian, M., Southon, J., Peterson, L.C., Alley, R. and Sigman, D.M., Deglacial changes in ocean circulation from an extended radiocarbon calibration. *Nature*, 391, 65-68, 1998.

Huh, Y., and Edmond, J. M., On the interpretation of the oceanic variations in 87Sr/86Sr as recorded in marine limestones. *Proc. Indian Acad. Sci. (Earth Planet. Sci.)*, 107 (4): 293-305, 1998.

Huls, M. and R. Zahn, Millennial-scale sea surface temperature variability in the western tropical North Atalntic from planktonic formainiferal census counts, *Paleoceanography*, 15, 659-678, 2000.

Hurrell, J. W., Decadal trends in the North Atlantic Oscillation: Regional temperatures and precipitation. *Science* 269, 676-679, 1995.

Hurrell, J. W., Influence of variations in extratropical wintertime teleconnections on Northern Hemisphere temperature, *Geophysical Research Letters*, 23, 665-668, 1996:

Imbrie, J., Berger, A., Boyle, E. A., Clemens, S. C., Duffy, A., Howard, W. A., Kukla, G., Kutzbach, J., Martinson, D. G., McIntyre, A., Mix, A. C., Molfino, B., Morley, J. J., Peterson, L. C., Pisias, N. G., Prell, W. G., Raymo, M. E., Shackleton, N. J., and Toggweiler, J. R., On the Structure and Origin of Major Glaciation Cycles. 2. The 100,000 year cycle. *Paleoceanography*, 8, 699-735, 1993.

Imbrie, J., Boyle, E., Clemens, S., Duffy, A., Howard, W., Kukla, G., Kutzbach, J., Martinson, D., McIntyre, A., Mix, A., Molfino, B., Morley, J., Peterson, L., Pisias, N., Prell, W., Raymo, M., Shackleton, N., and Toggweiler, J., One the Structure and Origin of major Glaciation Cycles. 1. Linear responses to Milankovitch forcing. *Paleoceanography*, 7, 701-738, 1992.

Imbrie, J., Mix, A. C., and Martinson, D. G., Milankovitch theory viewed from Devil's Hole. *Nature*, 363, 531-533, 1993

Jansen, E., Sjoholm, J., Bleil, U. and Erichsen, J.A., Neogene and Pleistocene glaciations in the Northern hemisphere and late Miocene - Pliocene global ice volume fluctuations: evidence from the Norwegian Sea. In Bleil, U., and Thiede, J. (Eds.), *Geological history of the polar oceans: Arctic versus Antarctic*, Netherlands, Kluwer Academic Publishers, 677-705, 1990.

Jansen, E. and Sjoholm, J., Reconstruction of glaciation over the past 6 Myr from ice-borne deposits in the Norwegian Sea, *Nature*, 349, 600-603, 1991.

Johnsen, S. J., Clausen, H. B., Dansgaard, W., Gundenstrup, N. S., Hammer, C. U., and Tauber, H., The Eem stable isotope record along the GRIP ice core and its interpretation. *Quaternary Research*, 43, 117-124, 1995.

Johnsen, S. J., Clausen, H. B., Dansgaard, W., Gundenstrup, N. S., Hammer, C. U., Andersen, U., Andersen, K. K., Hvidberg, C. S., Dahl-Jensen, D., Steffensen, J. P., Shoji, H.,

Sveinbjörnsdóttir, A. E., White, J., Jouzel, J., and Fisher, D., The $\delta^{18}O$ record along the Greenland Ice Core Project deep ice core and the problem of possible Eemian climate instability. *Journal of Geophysical Research*, 102, 26,397-26,409, 1997.

Jones, P. D., Bradley, R. S. and Jouzel, J., eds., *Climatic variations and forcing mechanisms over the last 2,000 years*. Springer Verlag, Berlin, 1996.

Jouzel, J., Barkov, N. I., Barnola, J. M., Bender, M., Chapellaz, J., Genthon, C., Kotlyakov, V. M., Lipenkov, V., Lorius, C., Petit, J. R., Raynaud, D., Raisbeck, G., Ritz, C., Sowers, T., Stievenard, M., Yiou, F., and Yiou, P., Extending the Vostok ice-core record of palaeoclimate to the penultimate glacial period, *Nature*, 364, 407-411, 1993.

Jouzel, J., Waelbroeck, C., Malaize, B., Bender, M., Petit, J. R., Stievenard, M., Barkov, N. I., Barnola, J. M., King, T., Kotlyakov, V. M., Lipenkov, V., Lorius, C., Raynaud, D., Ritz, C., and Sowers, T., Climatic interpretation of the recently extended Vostok ice records. *Climate Dynamics*, 12, 513-521, 1996.

Kanfoush, S.L. et al., Millennial-scale instability of the Antarctic ice sheet during the last glaciation. *Science*, 288, 1815-1819, 2000.

Keeling, R.F. and B.B. Stephens, Antarctic sea ice and the control of Pleistocene climate instability, *Paleoceanography*, 16, 112-131, 2001.

Keigwin, L.D., Pliocene closing of the Isthmus of Panama, based on biostratigraphic evidence from nearby Pacific Ocean and Caribbean cores, *Geology*, 6, 630-634, 1978.

Keigwin, L.D., Pliocene paleoceanography of the Caribbean and east Pacific: Role of panama uplift in late Neogene times, *Science*, 217, 350-353, 1982.

Keigwin, L.D., Pliocene stable isotope record of DSDP Site 606: sequential events of ^{18}O enrichment beginning at 3.1 Ma. In Kidd, R.B., Ruddiman, W.F. and Thomas, E., et al., *Init. Repts. DSDP*, 94, Washington, (U.S. Govt. Printing Office), 911-920, 1986.

Keigwin, L., The Little Ice Age and Medieval Warm Period in the Sargasso Sea, *Science*, 274, 1504-1508, 1996.

Keigwin, L., Curry, W. B., Lehman, S. J., and Johnsen, S., The role of the deep ocean in North Atlantic climate change between 70 and 130 kyr ago, *Nature*, 371, 323-329, 1994.

Keller, G., Zenker, C.E. and Stone, S.M.. Late Neogene history of the Pacific-Caribbean gateway, *J. of South Am. Earth Sci.*, 2, 73-108, 1989.

Kennett , J. P. (ed.): , Special Section: The Younger Dryas Event, *Paleoceanography*, 5, 891-1041, 1990.

Kortenkamp, S. J., and Dermott, S. F., A 100,000 year Periodicity in the accretion rate of interplanetary dust. *Science*, 280, 874-876, 1998.

Kotilainen, A.T., and Shackleton, N.J., Rapid climate variability in the North Pacific Ocean during the past 95,000 years. *Nature*, 377: 323-326, 1995.

Krishnaswami, S., Trivedi, J. R., Sarin, M. M., Ramesh, R., and Sharma, K. K., Strontium isotoipes and Rubidium in the Ganga-Brahmaputra river system: weathering in the Himalaya, fluxes to the Bay of Bengal and contributions to the evolution of oceanic 86Sr/87Sr. *Earth Planet Sci Lett.*, 109: 243-253, 1992.

Kukla, G., McManus, J., Rousseau, D.-D., and Chuine, I., How long and how stable was the last interglacial? *Quaternary Science Reviews,* 16, 605-612, 1997.

Labeyrie, L. D., Duplessy, J.-C., Duprat, J., Juillet-Leclerc, A., Moyes, J., Michel, E., Kallel, N., and Shackleton, N. J., Changes in the vertical structure of the North Atlantic Ocean between glacial and modern times. *Quaternary Science Reviews*, 11, 401-413, 1992.

Laskar, J., The chaotic motion of the solar system: A numerical estimate of the chaotic zones, *Icarus*, 88, 266-291, 1990.

Lea, D.W., Pak, D., and Spero, H., Climate impact of late Quaternary Equatorial sea surface temperature variations, *Science*, 289, 1719-1724, 2000.

Linsley, B. K., Oxygen-isotope record of sea level and climate variations in the Sulu Sea over the past 150,000 years. *Nature*, 380, 234-237, 1996.

Li, X-S., A. Berger, M-F. Loutre, M.A.Maslin, G.H. Haug and R. Tiedemann, Simulating late Pliocene Northern Hemisphere climate with the LLN 2-D model, *Geophysical Research Letters*, 25, 915-918, 1998.

Litt, T., Brauer, A., Goslar, T., Merkt, J., Balaga, K., Müller, H., Ralska-Jasiewiczowa, M., Stebich, M. & Nedendank, J.F.W. Correlation and synchronisation of Lateglacial continental sequences in northern central Europe based on annually-laminated lacustrine sediments. *Quaternary Science Reviews*, 19, (in press).

Lourens, L.J. and Hilgen, F.J., Chapter 9: Long-period orbital variations and their relation to Third-order Eustatic cycles and the onset of major glaciations - 3.0 million years ago. In Astronomical forcing of Mediterranean climate during the last 5.3 million years. (Ph.D. thesis) Univ. of Utrecht, Utrecht, Holland, 199-206, 1994.

Loutre, M.F. and Berger, A., Sensibilite des parametres astro-climatiques au cours des 8 derniers millions d'annees. *Scientific Report 1993/4. Institut d'Astronomie et de Geophysique G. Lemaitre,* Universite Catholique de Louvain, Louvain-la-Neuve, 1993.

Lowe, J.J. and Walker, M. J. C., *Reconstructing Quaternary Environments*. Longman, NY., 1984.

Lowe, J.J. and Walker, M.J.C., Radiocarbon dating the last glacial interglacial transition (ca. 14-9 ^{14}C ka BP) in terrestrial and marine records: the need for new quality assurance protocols. *Radiocarbon*, 42, 53-68, 2000.

Lowe, J.J., Birks, H.H., Brooks, S.J., Coope, G.R., Harkness, D.D., Mayle, F.E., Sheldrick, C., Turney, C. and Walker, M.J.C., The chronology of palaeoenvironmental changes during the Last Glacial-Holocene transition: towards an event stratigraphy for the British Isles. *Journal of the Geological Society of London*, 156, 397-410, 1999.

Lowe, J.J, Ammann, B., Birks, H.H., Björck, S., Coope, G.R., Cwynar, L., De Beaulieu, J-L., Mott, R.J., Peteet, D.M. & Walker, M.J.C., Climatic changes in areas adjacent to the North Atlantic during the last glacial-interglacial transition (14-9 ka BP): a contribution to IGCP-253. *Journal of Quaternary Science,* 9, 185-198, 1994

Lowe, J.J., Coope, G.R., Harkness, D.D., Sheldrick, C. & Walker, M.J.C., Direct comparison of UK temperatures and Greenland snow accumulation rates, 15-12,000 calendar years ago. *Journal of Quaternary Science,* 10, 175-180, 1995.

Lowe, J.J., Hoek, W. & INTIMATE Group, Inter-regional comparisons of climatic reconstructions for the Last Termination: recommended protocol for assessments of radiocarbon chronologies. *Quaternary Science Reviews* (in press)

Lowell, T. V., Heusser, C. J., Andersen, B. G., Moreno, P. I., Hauser, A., Heusser, L. E., Schluchter, C., Marchant, D. R., and Denton, G. H., Interhemispheric correlation of late Pleistocene glacial events, *Science*, 269, 1541-1549, 1995.

Lundqvist, J., Saarnisto, M. & Rutter, N. (eds.), IGCP-253 - Termination of the Pleistocene. *Quaternary International*, 28, 201 pp, 1995.

MacAyeal, D. R., A Low-Order Model of the Heinrich Event Cycle. *Paleoceanography*, 8, 767-773, 1993a.

MacAyeal, D. R., Binge/Purge oscillations of the Laurentide Ice Sheet as a cause of the North Atlantic's Heinrich Events, *Paleoceanography*, 8, 775-784, 1993b.

Manabe, S. and R.J. Stouffer, Two stable equilibria of a coupled ocean-atmosphere model, *Journal of Climate*, 1: 841-866, 1988.

Manabe, S. and R.J. Stouffer, Simulation of abrupt change induced by freshwater input to the North Atlantic Ocean, *Nature*, 378: 165-167, 1995.

Manabe, S. and R. Stouffer, Coupled ocean-atmosphere model response to freshwater input: Comparison to Younger Dryas event. *Paleoceanography*, 12, 321-336, 1997.

Mann, P. and Corrigan, J., Model for late Neogene deformation in Panama. *Geology*, 18, 558-562, 1990.

Martinson, D. G., Pisias, N. G., Hays, J. D., Imbrie, J., Moore, T. C. Jr., and Shackleton, N. J., Age dating and the orbital theory of ice ages: development of a high-resolution 0 to 300,000 years Chronostratigraphy, *Quaternary Research*, 27, 1-29, 1987.

Maslin, M., and Tzedakis, C., Sultry last interglacial gets sudden chill. *EOS*, 77, 353- 354, 1996.

Maslin, M.A., G. Haug, M. Sarnthein, R. Tiedemann, H. Erlenkeuser and R. Stax, Northwest Pacific Site 882: The initiation of major Northern Hemisphere Glaciation, *ODP Leg 145 Scientific Results Volume*, 315-329, 1995a.

Maslin, M. A., Shackleton, N. J., and Pflaumann, U., Surface water temperature, salinity and density changes in the northeast Atlantic during the last 45,000 years: Heinrich events, deep-water formation, and climatic rebounds. *Paleoceanography*, 10, 527-543, 1995b.

Maslin, M. A., Sarnthein, M., and Knaack, I.-J., Subtropical eastern Atlantic climate during the Eemian. *Naturwissenschaften*, 83, 122-126, 1996.

Maslin M.A., X-S. Li, M-F. Loutre and A. Berger. The contribution of orbital forcing to the progressive intensification of Northern Hemisphere Glaciation, *Quaternary Science Review*, 17, No. 4-5, 411-426, 1998a.

Maslin, M. A., Sarnthein, M., Knaack, J.-J., Grootes, P., and Tzedakis, C., Intra-interglacial cold events: an Eemian-Holocene comparison. In: Cramp, A., MacLeod, C.J., Lee, S. and Jones, E.J.W. (eds) Geological Evolution of Ocean Basins: Results from the Ocean Drilling Program. *Geological Society of London Special Publication*, 131, 91-99, 1998b.

Mayewski, P. A., Meeker, L. D., Twickler, M. S., Whitlow, S., Yang, Q., Lyons, W. B., and Prentice, M., Major features and forcing of high-latitude northern hemisphere atmospheric circulation using a 110,000-year-long glaciochemical series. *Journal of Geophysical Research*, 102, 26,345-26,365, 1997.

McCabe, A. M., and Clark, P. U., Ice-sheet variability around the North Atlantic Ocean during the last deglaciation, *Nature*, 392, 373-377, 1998.

McCave I.N., Sedimenatry processes and the creation of the straigraphic record in the late Quaternary North Atlantic Ocean. *Royal Society Philosophical transactions, Series B*, 348, 229-241, 1995.

McManus, J. F., Bond, G. C., Broecker, W. S., Johnsen, S., Labeyrie, L., and Higgins, S., High resolution climate records from the North Atlantic during the last interglacial, *Nature*, 371, 326-329, 1994.

Meeker, L. D., Mayewski, P. A., Twickler, M. S., Whitlow, S. I., and Meese, D., A 110,000 year history of change in continental biogenic emissions and related atmospheric circulation inferred from the Greenland Ice Sheet Project Ice Core, *Journal of Geophysical Research*, 102, 26,489-26,505, 1997.

Mikolajewicz, U., Maier-Reimer, E., Crowley, T.J. and Kim, K.Y., Effect of Drake and Panamanian gateways on the circulation of an ocean model. *Paleoceanography*, 8, 409-427, 1993.

Milankovitch, M.M., Kanon der Erdbestrahlung und seine Anwendung auf das Eiszeitenproblem. *Royal Serbian Sciences, Spec. pub.* 132, *Section of Mathematical and Natural Sciences*, 33, Belgrade, pp.633, 1949 (Canon of Insolation and the Ice Age Problem, English translation by Israel Program for Scientific Translation and published for the U.S. Department of Commerce and the National Science Foundation, Washington D.C., 1969).

Miller, K.G., Wright, J.D. and Fairbanks, R.G., Unlocking the ice house: Oliogcene-Miocene oxygen isotope, eustacy, and margin erosion, *J. Geophys. Res.*, 96, 6829-6848, 1991.

Molnar, P. and England, P., Late Cenozoic uplift of mountain ranges and global climate change: chicken or egg? *Nature*, 346, 29-34, 1990.

Mudelsee, M., and Schulz, M., The Mid-Pleistocene climate transition: onset of 100 ka cycle lags ice volume build up by 280 ka. *Earth and Planetary Science Letters*, 151, 117-123, 1997.

Mudelsee, M. and Stattegger, K., Exploring the structure of the mid-Pleistocene revolution with advance methods of time-series analysis. *Geol. Rundsch.*, 86, 499-511, 1997.

Muller, R. A., and MacDonald, G. J., Glacial cycles and atsronomical forcing. *Science*, 277, 215-218, 1997.

Mulvaney R, Rothlisberger R, Wolff EW, Sommer S, Schwander J, Hutteli MA, Jouzel J., The transition from the last glacial period in inland and near-coastal Antarctica, *Geophysical Research Letters*, 27 (17): 2673-2676, 2000.

O'Brien, S. R., Mayewski, A., Meeker, L. D., Meese, D. A., Twickler, M. S., and Whitlow, S. I., Complexity of Holocene climate as reconstructed from a Greenland ice core. *Science*, 270, 1962-1964, 1996.

ODP Leg 151 Scientific Party, Exploring Artic History Through Scientific Drilling. *Eos*, 75 (25), 281-286, 1994.

Oppo, D. W., Horowitz, M., and Lehman, S. J., Marine evidence for reduced deep water production during Termination II followed by a relatively stable substage 5e (Eemian), *Paleoceanography*, 12, 51-63, 1997.

Oppo, D. W., J. F. McManus, J. L. Cullen, Abrupt Climate Events 500,000 to 340,000 Years Ago: Evidence from Subpolar North Atlantic Sediments. *Science*, 279 1335-1338, 1998.

Peglar, S. M., The mid-Holocene *Ulmus* decline at Diss Mere, Norfolk, UK: a year-by-year pollen stratigraphy from annual laminations. *The Holocene*, 3, 1-13, 1993

Peiser, B.J., Comparative analysis of late Holocene environmental and social upheaval: evidence for a disaster around 4000 BP. In Natural Catastrophes during Bronze Age Civilisations (Eds: Peiser, B.J., Palmer, T, and Bailey, M), *BAR International series*, 728, 117-139, 1998.

Peteet, D. M., Global Younger Dryas, *Quaternary International*, 28, 93-104, 1995.

Pisias N.G., and Mooore, T.C., The evolution of Pleistocene climate: a time series approach. *Earth Planet*, 52, 450-458, 1981.

Porter, S. and Z. An, Correlations between climate events in the North Atlantic and China during the last glaciation, *Nature*, 375, 305-308, 1995

Prell, W., Oxygen isotope stratigraphy for the Quternary of Hole 502B. *Initial Reports DSDP*, Vol. 68, US Govt. Printing Ofice, Washington DC, 455-464, 1992.

Prell, W. L., Covariance patterns of foraminifera $\delta^{18}O$: An evaluation of Pliocene ice-volume changes near 3.2 million years ago, *Science*, 226, 692-694, 1984.

Quade, J., Cerling, T. E. and Bowman, J. R., Development of Asian monsoon revealed by marked ecological shift during the latest Miocene in northern Pakistan. *Nature*, 342, 163-165, 1989.

Rackham O., Ancient Woodland: its history, vegetation and uses in England. Arnold, London, 1980.

Rahmstorf, S., Rapid climate transitions in a coupled ocean-atmosphere model. *Nature*, 372, 82-85, 1994.

Rahmstorf, S., Bifurcations of the Atlantic thermohaline circulation in response to changes in the hydrological cycle; *Nature*, 378: 145-149, 1995.

Rahmstorf, S., Marotzke, J., and Willebrand, J., Stability of the Thermohaline Circulation, In: *The Warmwatersphere of the North Atlantic Ocean*, W. Kraus, ed., Gebrüder Bornträger, Berlin, 129-157, 1996.

Ram, M., and Koenig, G., Continuous dust concentration profile from the pre-Holocene ice from the Greenland Ice Sheet Project 2 ice core: dust stadials, interstadials, and the Eemian. *Journal of Geophysical Research*, 102, 26,641-26,648, 1997.

Rasmussen T. L., van Weering T. C. E. and Labeyrie L., Climatic instability, ice sheets and ocean dynamics at high northern latitudes during the last glacial period (58-10 KA BP). *Quaternary Science Reviews*, 16, 73-80, 1997.

Raymo, M.E., Geochemical evidence supporting T.C. Chamberlin's theory of glaciation, *Geology*, 19, 344-347, 1991.

Raymo, M.E., The Himalayas, organic carbon burial and climate change in the Miocene, *Paleoceanography*, 9, 399-404, 1994a.

Raymo, M.E., The initiation of Northern Hemisphere glaciation, *Annu. Rev. Earth Planet. Sci.*, 22, 353-383, 1994b.

Raymo, M.E., The timing of major climate terminations, Paleoceanography, 12, 577-585, 1997.

Raymo, M.E. and Ruddiman, W.F., Tectonic forcing of late Cenozoic climate. *Nature*, 359, 117-122, 1992.

Raymo, M.E., Ruddiman, W.F. and Froelich, P.N., Influence of late Cenozoic mountain building on ocean geochemical cycles, *Geology*, 16, 649-653, 1988.

Raymo, M.E., Ruddiman, W.F., Backman, J., Clement, B.M. and Martinson, D.G., Late Pliocene variations in Northern Hemisphere ice sheet and North Atlantic deep water circulation. *Paleoceanography*, 4, 413-446, 1989.

Raymo, M., Oppo, D., and Curry, W., The mid-Pleistocene climate transition: A deep sea carbon isotope perspective. *Paleoceanography*, 12, 546-559, 1997.

Rea, D.K., Basov, I.A., Janecek, T.R., Palmer-Julson, A. et al., *Proc. ODP, Init. Repts.*, 145, College Station, TX (Ocean Drilling Project), 1993.

Renssen, H. & Isarin, R.F.B., Surface temperature in NW Europe during the Younger Dryas: a GCM simulation compared with temperature reconstructions. *Climate Dynamics*, 14, 33-44, 1998.

Ridgwell, A. A. Watson, and M. Raymo, Is the spectral signiture of the 100 kyr glacial cycle consistent with a Milankovitch origin, *Paleoceanography*, 14, 437-440, 1999

Ruddiman, W. F., Late Quaternary deposition of ice-rafted sand in the subpolar North Atlantic (lat. 40° to 65°N). *Geological Society of America Bulletin*, 88, 1813-1827, 1977.

Ruddiman, W.F. & McIntyre, A., The North Atlantic ocean during the last deglaciation. *Palaeogeography, Palaeoclimatology, Palaeoecology*, 35, 145 – 214, 1981

Ruddiman, W.F., Sancetta, C.D. & McIntyre, A., Glacial/interglacial response rate of subpolar North Atlantic waters to climatic change: the record left in deep-sea sediments. *Philos. Trans. R. Soc. Lond. B*, 280, 119 – 142, 1977.

Ruddiman, W.F. and McIntyre, A., Oceanic mechanisms for amplification of the 23,000-year ice volume cycle, *Science*, 212, 617-627, 1981.

Ruddiman, W.F. and Raymo, M.E., Northern Hemisphere climate regimes during the past 3 Ma: possible tectonic connections. *Phil. Trans. R. Soc. Lond.*, B 318, 411-430, 1988.

Ruddiman, W.F., et al., Late Miocene to Pleistocene evolution of climate in Africa and the low-latitude Atlantic - overview of Leg 108 results. In Ruddiman, W.F., Sarnthein, M., Baldauf, J., et al., *Proc. ODP, Sci. Results.*, 108, College Station, TX (Ocean Drilling Program), 463-487, 1989.

Ruddiman, W.F. and Kutzbach, J.E., Plateau uplift and climatic change. *Sci. Am.*, 264, 66-75, 1991.,

Ruddiman, W.F., McIntyre, A. and Raymo, M., Paleoenvironmental results from North Atlantic sites 607 and 609. In Kidd, R.B., Ruddiman, W.F., and Thomas, E., et al., *Init. Repts. DSDP*, 94, Washington, (U.S. Govt. Printing Office), 855-878, 1986a.

Ruddiman, W.F., Raymo, M. and McIntyre, A., Matuyama 41,000-year cycles: North Atlantic Ocean and Northern Hemisphere Ice Sheets. *Earth Planet. Sci. Lett.*, 80, 117-129, 1986b.

Ruddiman, W.F., Sarnthein, M., Baldauf, J., et al.,*Proc. ODP, Sci. Results.*, 108, College Station, TX (Ocean Drilling Program), 1989.

Ruhlemann, C., S. Mulitza, P.J. Muller, G. Wefer and R. Zahn, Tropical Atlantic warming during conveyor shut down, *Nature*, 402, 511-514, 1999.

Rind, D., and Overpeck, J., Hypothesized causes of decade-to-century-scale climate variability: climate model results. *Quaternary Science Reviews*, 12, 357-374, 1993.

Saltzman B., Maasch, K.A. and Verbitsky, M.Y., Possible effects of anthropogenically-increased CO_2 on the dynamics of climate: implications for ice age cycles. *Geophysical Research Letters*, 20, 1051-1054, 1993.

Sarnthein, M. and Tiedemann, R., Toward a high resolution stable isotope stratigraphy of the last 3.4 million years: Sites 658 and 659 off northwest Africa. In Ruddiman, W.F., Sarnthein, M., Baldauf, J., et al., *Proc. ODP, Sci. Results*, 108, College Station, TX (Ocean Drilling Program), 167-187, 1989.

Sarnthein, M. and Tiedemann, R., Younger Dryas-style cooling events at glacial terminations I-VI at ODP Site 685: associated benthic $\delta^{13}C$ anomalies constrain meltwater hypothesis. *Paleoceanography*, 5, 1041-1055, 1990.

Sarnthein et al., Fundermental modes and abrupt changes in North Atlantic circulation and climate over the last 60 kyr, In *The northern North Atlantic: A changing environment*, Schafer et al. (edditors), Spinger-Valag, New York, in press.

Schmitz, W.J., Jr., On the interbasin-scale thermohaline circulation, *Reviews of Geophysics*, 33: 151-173, 1995.

Schmitz, W.J., Jr. and M.S. McCartney, On the North Atlantic circulation, *Reviews of Geophysics*, 31: 29-49, 1993.

Schulz H., von Rad, U. and Erlenkeuser H., Correlation between Arabian Sea and Greenland Climate oscillations of the past 110,000 years. *Nature*, 393, 54-57, 1998.

Scourse, J.D. , I.R. Hall, I.N. McCave, J.R. Young, and C. Sugdon, The origin of Heinrich layers: evidence from H2 for European precurser events, *EPSL*, 5596, 1-9, 2000.

Seidenkrantz, M.-S., Benthic foraminiferal and stable isotope evidence for a "Younger Dryas-style" cold spell at the Saalian-Eemian transition, Denmark. *Palaeogeography, Palaeoclimatology, Palaeoecology*, 102, 103-120, 1993.

Seidenkrantz, M.-S., Kristensen, P., and Knudsen, K. L., Marine evidence for climatic instability during the last interglacial in shelf records from northwest Europe. *Journal of Quaternary Science*, 10 , 77-82, 1995.

Seidenkrantz, M.-S., Bornmalm, L., Johnsen, S. J., Knudsen, K. L., Kuijpers, A., Lauritzen, S.-E., Leroy, S. A. ., Mergeai, I., Schweger, C., and van Vlet-Lanoe, B., Two-step deglaciation at the oxygen isotope stage 6/5e transition: the Zeifen-Kattegat climate oscillation. *Quaternary Science Reviews*, 15, 63-75, 1996.

Seidov, D. and B.J. Haupt, Last glacial and meltwater interbasin water exchanges and sedimentation in the world ocean. *Paleoceanography*, 14, 760-769, 1999.

Seidov, D. and Maslin, M. A., Seasonally ice-free glacial Nordic seas without deep water ventilation. *Terra Nova*, 8, 245-254, 1996.

Seidov, D. and M. Maslin, North Atlantic Deep Water circulation collapse during the Heinrich events, *Geology*, 27: 23-26, 1999.

Seidov, D. and M. Maslin, Atlantic Ocean heat piracy and the bipolar climate see-saw during Heinrich and Dansgaard-Oeschger events, *JQS*, 2001, in press.

Seidov, D., E.J. Barron and B.J Haupt: Meltwater and the global ocean conveyor: Northern versus southern connections, *Global and Planetary Change*, 2001, in press.

Seidov, D., E. Barron, B. J. Haupt, and M.A. Maslin, 2001, Ocean bi–polar Seesaw and climate: Southern versus northern meltwater impacts, *Geophysical Monograph (This Volume)*, D. Seidov, B.J. Haupt and M. Maslin (Editors), American Geophysical Union, Washington, D.C.

Severinghaus. J. P., Sowers, T., Brook, E. J., Alley, R. B., and Bender, M. L., Timing of abrupt climate change at the end of the Younger Dryas interval from thermally fractionated gases in polar ice. *Nature*, 391, 141-146, 1998.

Shackleton, N. J., The last interglacial in the marine and terrestrial records, *Proceedings of the Royal Society, London, B*, 174, 135-154, 1969.

Shackleton, N.J., The 100,000 year Ice Age cycle identified and found to lag temperature, carbon dioxide and orbital eccentricity, *Science*, 289, 1989-1902, 2000.

Shackleton, N.J., Imbrie, J., and Pisias, N.G., the evolution of oceanic oxygen-isotope variability in the North Atlantic over the past 3 million years. *Phil. Trans. R. Soc. Lond. B*, 318, 679-686, 1988.

Shackleton, N.J., et al., Oxygen isotope calibration of the onset of ice-rafting and history of glaciation in the North Atlantic Region, *Nature*, 307, 620-623, 1984.

Shackleton, N.J., Hall, M.A. and Pate, D., Pliocene stable isotope stratigraphy of ODP Site 846. In Mayer, L., Pisias, N. and Janecek, T., et al., *Proc. ODP, Sci. Results*, 138, College Station, TX (Ocean Drilling Program), 337-355, 1995.

Shaffer, G., and Bendtsen, J., Role of the Bering Strait in controlling North Atlantic ocean circulation and climate. *Nature*, 367, 354-357, 1994.

Sigman, D., and E. Boyle, Glacial/interglacial variations in atmospheric carbon dioxide, *Nature*, 407, 859 – 869, 2000.

Sirocko, F., Sarnthein, M., Erlenkeuser, H., Lange, H., Arnold, M., and Duplessy, J.-C., Century-scale events in monsoonal climate over the past 24,000 years. *Nature*, 364, 322-324, 1993.

Slowey, N. C., Henderson, G. M., and Curry, W. B., Direct U-Th dating of marine sediments from the two most recent interglacial periods. *Nature*, 383, 242-244, 1996.

Smagorinsky, J., The dynamical influence of large-scale heat source and sinks on the quasi-stationary mean rotations of the atmosphere. *Q. Jlr. Met. Soc.*, 79, 342-366, 1953.

Sowers, T., Bender, M., Labeyrie, L., Martinson, D., Jouzel, J., Raynaud, D., Pichon, J. J., and Korotevich, Y. S., A 135,000 year Vostok-SPECMAP common temporal framework. *Paleoceanography*, 8, 737-764, 1993.

Steffensen, J. P., Clausen, H. B., Hammer, C. U., LeGrand, M., and De Angelis, M., The chemical composition of cold events within the Eemian section of the Greenland Ice Core Project ice core from Summit, Greenland. *Journal of Geophysical Research*, 102, 26,747-26,753, 1997.

Steig, E.J., Brook, E.J., White, J.W.C., Sucher, C.M., Bender, M.L., Lehman, S.J., Morse, D.L., Waddington, E.D. and Clow, G.D., Synchronous climate changes in Antarctica and the North Atlantic. *Science*, 282, 92-95, 1998.

Stirling, C. H., Esat, T. M., McCulloch, M. T., and Lambeck, K., High-precision U-series dating of corals from Western Australia and implications for the timing and duration of the last interglacial. *Earth and Planetary Science Letters*, 135, 115-130, 1995.

Stocker, T.F., The variable ocean, *Nature*, 367: 221-222, 1994.

Stocker, T.F., The seesaw effect, *Science*, 282: 61-62, 1998.

Stocker, T.F., D.G. Wright and W.S. Broecker, The influence of high-latitude surface forcing on the global thermohaline circulation, *Paleoceanography*, 7, 529-541, 1992.

Sugden, D. (ed.), Quaternary Climate Change and South America, Special Issue of *Journal of Quaternary Science*, 15, 2000.

Sutcliffe, R.C., Mean upper-air contour patterns of the Northern Hemisphere - the thermal-synoptic viewpoint. *Q. Jlr. Met. Soc.*, 77, 435-440, 1951.

Szabo, B. J., Ludwig, K. R., Muhs, D. R., and Simmons, K. R., Thorium-230 ages of corals and duration of last interglacial sea-level highstand on Oahu, Hawaii, *Science*, 266, 93-96, 1994.

Taylor, K. C., Alley, R. B., Doyle, G. A., Grootes, P. M., Mayewski, P. A., Lamorey, G. W., White, J. W. C., and Barlow, L. K., The 'flickering switch' of late Pleistocene climate change. *Nature*, 361, 432-436, 1993.

Taylor, K. C., P. A. Mayewski, R. B. Alley, E. J. Brook, A. J. Gow, P. M. Grootes, D. A. Meese, E. S. Saltzman, J. P. Severinghaus, M. S. Twickler, J. W. C. White, S. Whitlow, G. A. Zielinski, The Holocene-Younger Dryas Transition Recorded at Summit, Greenland. *Science*, 278 825-827, 1997.

Teller, J.T., History and drainage of large ice-dammed lakes along the Laurentide ice sheet. *Quaternary International*, 28, 83-92, 1995.

Thompson, L. G., Ice-core records with emphasis on the global record of the last 2000 years. In: *Global Changes of the Past*, R. Bradley, ed., UCAR/Office of Interdisciplinary Earth Studies, Boulder, CO (USA), 2, 201-223, 1989.

Thompson, L. G., and E. Mosley-Thompson, Evidence of abrupt climatic damage during the last 1,500 years recorded in ice cores from the tropical Quelccaya ice cap, Peru. in *Abrupt Climatic Change: Evidence and Implications*, W. H. Berger and L. D. Labeyrie, eds., 99-110, 1987.

Thompson, L. G., Mosley-Thompson, E., Davis, M. E., Lin, P.-N., Henderson, K. A., Cole-Dai, J., Bolzan, J. F., and Liu, K.-B., Late glacial stage and Holocene tropical ice core records from Huascara, Peru. *Science,* 269, 46-50, 1995.

Thompson, L. G., E. Mosley-Thompson, W. Dansgaard, and P. M. Grootes, The "Little Ice Age" as recorded in the stratigraphy of the tropical Quelccaya ice cap. *Science,* 234, 361-364, 1986.

Thouveny, N., de Beaulieu, J. L., Bonifay, E., Creer, K. M., Guiot, J., Icole, M., Johnsen, S., Jouzel, J., Reille, M., Williams, T., and Williamson, D., Climate variations in Europe over the past 140 kyr deduced from rock magnetism, *Nature,* 371, 503-506, 1994.

Tiedemann, R., Sarnthein, M. and Shackleton, N.J., Astronomic timescale for the Pliocene Atlantic $\delta^{18}O$ and dust flux records of ODP Site 659. *Paleoceanography,* 9, 619-638, 1994.

Trenberth, K.E., Interactions between orographically and thermally forced planetary waves, *J. Atm. Sci.,* 40, 1126-1153, 1983.

Tzedakis, P. C., Andrieu, V., de Beaulieu, J.-L., Crowhurst, S., Folieri, M., Hooghiemstra, H., Magri, D., Reille, M., Sadori, L., Shackleton, N. J., and Wijmstra, T. A., Comparison of terrestrial and marine records of changing climate of the last 500,000 years. *Earth and Planetary Science Letters,* 150, 171-176, 1997.

van Kreveld, S. et al., Potential links between surging ice sheets, circulation changes and the Dansgaard-Oeschger cycles in the Irminger Sea 60-18 kyr. *Paleoceanography,* 15, 425-442, 2000.

Vermeij, G. J., Strait answers from a twisted Isthmus. Paleobiology, 23 (2): 263-269, 1997.

Vidal, L., R. Schneider, O. Marchal, T. Bickert, T. Stocker, G. Wefer, Link between the North and South Atlantic duirng the Heinrich events of the last glacial period, *Climate Dynamics,* 159, 909-919, 1999.

Von Grafenstein, U., Erlenkauser, H., Brauer, A., Jouzel, J. & Johnsen, S. J., A mid-European decadal isotope-climate record from 15,500 to 5,000 years BP. *Science,* 284, 1654-1657, 1999.

Vrba E., Denton G. H., Partridge T. C. and Burckle L. H. (editors) *Palaeoclimate and Evolution, with emphasis on human origins.* Yale University Press, New Haven, pp367, 1996.

Walker, M.J.C., Climatic changes in Europe during the Last Glacial/Interglacial Transition. *Quaternary International,* 28, 63-76, 1995.

Walker, M.J.C., Björck, S., Lowe, J.J., Cwynar, L.C., Johnsen, S., Knudsen, K.-L., Wohlfarth, B. and INTIMATE group, Isotopic 'events' in the GRIP ice core: a stratotype for the late Pleistocene. *Quaternary Science Reviews,* 18, 1143-1150, 1999.

Walker, M.J.C., Björck, S. & Lowe, J.J. (eds) *Records of the Last Termination Around the North Atlantic.* Special Xvth INQUA Congress Issue (Durban, 1999), *Quaternary Science Reviews,* in press a.

Walker, M.J.C., Bryant, C., Coope, G.R., Harkness, D.D., Lowe, J.J. & Scott, E.M., Towards a radiocarbon chronology for the Late-glacial in Britain. *Radiocarbon,* in press b

Wang, L. et al., East Asian monsoon climate during the late Pleistocene: High resolution sediment records from the South China Sea, Marine *Geology,* 156, 245-284, 1999.

Weaver, A.J. and T.M.C. Hughes, Rapid interglacial climate fluctuations driven by North Atlantic ocean circulation. *Nature,* 367, 447-450, 1994.

Webb, R.S., D. Rind, S. Lehmann, R. Healy, and D. Sigman, Influence of ocean heat transport on the climate of the last glacial maximum, *Nature.* 385, 695-699, 1997.

Wilson, R.C.L., S.A. Drury, J.A. Chapman, *The Great Ice Age; Climate change and life.* Routledge, London, pp. 288, 2000.

Winograd, I. J., Szabo, B. J., Coplen, T. B., and Riggs, A. C., A 250,000-year climatic record from Great Basin vein calcite: implications for Milankovich theory. *Science,* 242, 1275-1280, 1988.

Winograd, I. J., Coplen, T. B., Landwehr, J. M., Riggs, A. C., Ludwig, K. R., Szabo, B. J., Kolesar, P. T., and Revesz, K. M., Continuous 500,000-year climate record from vein calcite in Devil's Hole, Nevada, *Science,* 258, 255-260, 1992.

Winograd I. J., Landwehr J. M., Ludwig K. R., Coplan T. B. and Riggs A. C., Duration and structure of the past four interglaciations. *Quaternary Research,* 48, 141-154, 1997.

Witte, H.J.L., Coope, G.R., Lemdahl, G. & Lowe, J.J., Regression coefficients of thermal gradients in northwestern Europe during the last glacial-Holocene transition using beetle MCR data, *Journal of Quaternary Science,* 13, 435-446, 1998.

Wolf, T.C.W. and Thiede, J., History of terrigenous sedimentation during the past 10 my in the North Atlantic (ODP-Leg's 104, 105, and DSDP-Leg 81), *Marine Geology,* 101, 83-102, 1991.

Wolf-Welling, T.C.W., Thiede, J., Myhre, A.M. and Leg 151 Shipboard scientific party, Bulk sediment parameter and coarse fraction analysis: Paleoceanographic implications of Fram Strait Sites 908 and 909, ODP Leg 151 (NAAG), *Eos Transactions,* 76 (17), suppl., p166, 1995.

Wright H. E. Jr., Environmental determinism in Near Eastern Prehistory. *Current Anthropology,* 34, 458-469, 1993.

Wright, J. D. and Miller, K.G., Control of North Atlantic Deep Water circulation by the Greenland-Scotland Ridge, *Paleoceanography,* 11, 157-170, 1996.

Wunsch, C., On sharp spectral lines in the climate record and the millennial prak, *Paleoceanography,* 15, 417-424, 2000.

Yiou, P., Fuhrer, K., Meeker, L. ., Jouzel, J., Johnsen, S., and Mayewski, P. A., Paleoclimatic variability inferred from the spectral analysis of Greenland and Antarctic ice core data. *Journal of Geophysical Research,* 102, 26,441-26,453, 1997.

Zachos, J.C., Breza, J. and Wise, S.W., Early Oligocene ice-sheet expansion on Antarctica, Sedimentological and isotopic evidence from Kerguelen Plateau, *Geology,* 20, 569-573, 1992.

Zachos, J.C., Quinn, T.M. and Salamy, K.A., High-resolution (10^4 years) deep-sea foraminiferal stable isotope records of the Eocene-Oligocene climate transition, *Paleoceanography,* 11, 251-266, 1996.

Zachos, J. C., Opdyke, B. N., Quinn, T. M., Jones, C. E., and Halliday, A. N., Early Cenozoic Glaciation, Antarctic weathering and seawater 87Sr/86Sr; is there a link? *Chemical Geology,* 161: 165-180, 1999.

Zachos, J.C., M. Pagani, L. Sloan, E. Thomas, and K. Billups, Trends, rhythms and aberrations in global climate 65 Ma to present, *Science*, 292, 673-678 (2001).

Zagwijn, W. H., Pleistocene stratigraphy in the Netherlands, based on changes in vegetation and climate, Nederlands Geologisch en Mijnbouwkundig Genootschap, Verhandelingen, *Geologische Serie*, 21-2, 173-196, 1963.

Zagwijn, W. H., Variations in climate as shown by pollen analysis, especially in the lower Pleistocene of Europe. In: *Ice Ages: Ancient and Modern*, A. E. Wright and F. Moseley, eds., Seel House Press, Liverpool, UK, 137-152, 1975.

Zahn, R. et al., Thermohaline instability in the North Atlantic during meltwater events stable isotope and detritus records from core SO75-26KL, Portuguese margin. *Paleoceanography*, 12, 696-710, 1997.

Zhisheng A. & Porter S. C., Millenial-scale climatic oscillations during the last interglaciation in central China. *Geology*, 25, 603-606, 1997.

J. Lowe, Centre for Quaternary Research, Department of Geography, Royal Holloway, University of London, Egham, Surrey TW20 0EX, UK. (j.lowe@rhbnc.ac.uk)

M. Maslin, Environmental Change Research Centre, Department of Geography, University College London, 26 Bedford Way, London. WC1H 0AP, UK. (mmaslin@ucl.ac.uk)

D. Seidov, EMS Environment Institute, Pennsylvania State University, University Park, PA 16802. (dseidov@essc.psu.edu)

The Big Climate Amplifier
Ocean Circulation-Sea Ice-Storminess-Dustiness-Albedo

Wallace S. Broecker

Lamont-Doherty Earth Observatory of Columbia University, Palisades, New York

Regardless of their origin, the abrupt and large shifts in climate associated with the Dansgaard-Oeschger and Younger Dryas cold events must involve a large amplifier. In this paper, I propose that this amplifier involves the lofting of dust and sea salt into the atmosphere and in turn the Earth's albedo. Further, I postulate that the delivery of these substances to the atmosphere is strongly influenced by the magnitude of the wintertime latitudinal thermal gradient, the stronger the gradient, the higher the frequency of intense storms. It is further proposed that the thermal gradient is geared to the position of the ice front and that, at least in the sea, the position of this front is modulated by the bipolar seesaw. During glacial time, this alternation in the strength of the conveyor circulation leads to sizable changes in the position of the sea-ice front and in so doing provides the needed amplification. During interglacials, when sea-ice coverage is small, the power of this amplifier is greatly reduced.

The millennial duration events which punctuated much of the last period of glaciation remain an enigma. Not only were the magnitude of the shifts in Northern Hemisphere climate extremely large but they were also abrupt. Temperature shifts in Bermuda surface waters were 4 to 5°C [*Sachs and Lehman*, 1999] and in Greenland probably 6 to 9°C. The dust and sea salt content of Greenland ice shifted abruptly back and forth by a factor of three (see Figure 1). Ventilation of oxygen-starved portions of today's main oceanic thermocline off California [*Behl and Kennett*, 1996] and Pakistan [*Schulz et al.*, 1998] underwent major changes. In order to generate these large and sudden shifts, a powerful amplifier must have been operative. I propose that this amplifier involved changes in the extent of sea ice coverage driven by seesawing of the ocean's thermohaline circulation. Shifts from dominant deep water formation in the northern Atlantic to dominant deep water formation in the Southern Ocean altered the steepness of the northern hemisphere's latitudinal thermal gradient and in turn its at-

mospheric storminess. The stronger the thermal gradient, the greater the amount of dust and sea salt lofted into the atmosphere. The excess dust and aerosols not only reflected away sunlight but also increased the availability of cloud condensation nuclei and thereby created more numerous and hence more reflective water droplets and ice crystals. The net result was an increase in albedo and hence a cooling.

During times of glaciation, both Polar Regions were far colder than today [*Broecker and Denton*, 1990]. Although acceptable proxies are as yet in their formative stages [*de Vernal et al.*, 2000], these coolings must have led to sizable expansions of sea ice cover. Through borehole thermometry, it has been documented that on the average during the last glacial period the air temperature over Greenland was 15°C colder than today [*Cuffey et al.*, 1995; *Cuffey and Clow*, 1997]. It seems unlikely that a cooling of this magnitude could have been achieved unless sea ice cover around Greenland was far more extensive than today's. Further, as the Bolling Allerod–Younger Dryas oscillation appears to call for shifts in the mode of deep water formation perhaps involving a seesaw action between the northern Atlantic and the Southern Ocean [*Broecker*, 1998; *Hughen et al*, 1998; *Stocker*, 1998], it is likely that the ex-

The Oceans and Rapid Climate Change: Past, Present, and Future
Geophysical Monograph 126

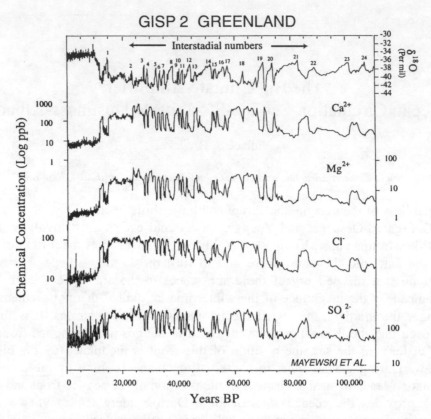

Figure 1. Record in the GISP 2 Greenland ice core [Mayewski et al., 1997] of two cations largely supplied by dust and two anions largely supplied by sea salt aerosols. For comparison is shown the ^{18}O record. The numbers designate the warm phases of the Dansgaard-Oeschger cycles (i.e., 2-20) and the Bolling Allerod warm (i.e., 1).

tent of excess sea ice in the northern Atlantic was larger during intervals when the Atlantic's thermohaline circulation was 'off' than when it was 'on'. Models suggest that the ocean's thermohaline circulation is subject to bifurcation [*Rahmstorf*, 1994]. If so, these reorganizations would have occurred abruptly and since sea ice formation and melting take place on the time scale of years to decades, the extent of sea ice coverage would rapidly adjust to the new circulation pattern.

The boundary of sea ice is meteorologically important because it separates that portion of the atmosphere subject to intense winter colds from that portion kept 'warm' by the release of heat from the open sea. Thus a consequence of shifts in sea ice extent would be changes in the temperature gradient between the warm tropics and the frigid ice edge. The further equatorward the ice margin, the steeper the thermal gradient in the Temperate Zone.

It is well documented that atmospheric dustiness in the Earth's temperate zones was far higher during times of glaciation. Not only was the dust content of glacial-age ice in polar regions thirty or so times higher than that in interglacial ice, but glacial age deposits of loess (i.e., wind-blown

silt) blanketed parts of central Europe [*Kukla*, 1975], China [*Kukla*, 1987], United States [*Muhs and Bettis*, 2000], and Argentina [*Imbellone and Terruggi*, 1993]. As Holocene analogues of loess are rare, conditions must have been very different during glacial time. One factor influencing the lofting and transport of dust is storminess and, of course, storminess is correlated with temperature gradient. During times when Canada and Scandinavia were covered with continental ice sheets, the winter temperature gradient in the temperate Northern Hemisphere was steeper than now. However, on the time scale of the millennial oscillations, the extent of ice cover is unlikely to have changed very much and, in particular, during the several-decade-duration transitions between the episodes of intense cold and moderate cold, it could not have changed at all. Thus, some other factor must be responsible for the large changes in dust delivery associated with the Dansgaard-Oeschger events and the Bolling Allerod–Younger Dryas oscillation. I propose that these modulations of dust transport were the result of changes in the extent of sea ice coverage.

In a recent paper, *Rosenfeld* [2000] provides observational support that the brightness of clouds is influenced by the availability of condensation nuclei. Using satellite photographs, he shows that where pollution plumes intersect cloud banks, the clouds are more reflective when viewed from above. Hence excess dust and sea salt aerosols could have impacted glacial climate by direct scattering of sunlight and by the enhancement of cloud albedo [*Twomey*, 1974; *Tegen et al.*, 1996; *Yung et al.*, 1996; *Toon*, 2000]. Perhaps together these impacts are large enough to explain a major portion of the glacial cooling [*Yung et al.*, 1996].

Not only was the dust loading of the atmosphere larger during glacial time, but so also was the production of NaCl aerosols (See Figure 1). This observation supports the idea that storminess rather than precipitation or vegetative cover lies at the root of dustiness changes. Indeed, as ocean air is currently deficient in particulates, sea salt aerosols may have been even more important cooling agents than dust grains.

There can be little doubt that a turning up and down of the strength of the Atlantic's conveyor circulation would change the water temperature and hence the extent of sea ice coverage in the northern Atlantic. So if, as many believe, the Dansgaard–Oeschger and Allerod-Younger Dryas oscillations were triggered by the action of the bipolar seesaw [*Broecker*, 1998; *Hughen et al.*, 1998; *Stocker*, 1998], then it is reasonable that these ocean reorganizations resulted in large changes in the extent of ice cover in the northern Atlantic. Thermohaline seesawing may also be responsible for Bond's 1500-year cycle in the composition of ice-rafted debris in the northern Atlantic [*Bond et al.*, 1997]. The pattern of currents during times when the conveyor was 'off' would surely have differed from the pattern when the conveyor was 'on'. Bond's finding could also explain why a seesawing of thermohaline circulation produced very large climate changes during glacial periods, yet only very modest ones during interglacials [*Bond et al.*, 1997]. For flips of the bipolar seesaw would make only small changes in sea ice extent during times of interglaciation (see Figure 2).

If indeed Bond's 1500-year cycle in the composition of ice-rafted debris is a reflection of a seesawing of deep water formation, then the question arises as to what paces these circulation flips. My guess is that the pacing involves the interplay between fresh water transport through the atmosphere and salt transport through the sea. The salt buildup in regions of the surface ocean where evaporation exceeds precipitation and river runoff must be balanced by an export to those regions where fresh water accumulates. In particular, the buildup of salt resulting from the export of fresh water from the Atlantic Ocean and its drainage basin must be compensated by the export of salt from the Atlantic. Currently, this export is via the lower limb of the con-

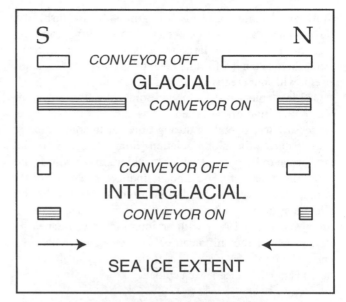

Figure 2. Cartoon showing the contrast in sea ice cover between glacials and interglacials and between times when the conveyor is 'on' and 'off'.

veyor. However, this compensation could well be oscillatory rather than steady. In other words, salt might build up in the Atlantic until its salinity reaches a critical level where deep-water formation in the northern Atlantic is initiated. Then the salt content decreases to the point where the conveyor shuts down. The current rate of fresh water export from the Atlantic has been estimated to be 0.25 ± 0.10 Sverdrups [*Zaucker and Broecker*, 1992]. If not compensated by salt export, this would result in a salinity increase in the Atlantic of about one gram per liter per millennium. Hence the half-cycle time of 750 years is not unreasonable. However, Bond's observations that the average length of these cycles is the same during interglacials as during glacials is admittedly puzzling for one would expect it to change with climate.

I admit that my scenario is highly speculative. In particular, the conjecture that relatively small changes in thermal gradient generate large changes in storminess has, as yet, no firm basis in modeling studies. While most of these simulations do not provide information regarding the frequency of the intense wind events, a recent simulation by *Tegen and Rind* [2000] does address this issue. As the pick up of dust varies with the cube of the wind velocity, it would be necessary for models to correctly simulate extreme events. Deficiencies also exist in our ability to quantitatively relate cooling to atmospheric turbidity. Nevertheless, as scenarios designed to amplify weak forcings into the observed large responses are as rare as hen's teeth, there is a need for this level of speculation. If nothing else, it will spur creative thinking by others.

One last point must be made. Bond's 1500-year cycle in the composition of ice-rafted debris suggests that the bipolar seesaw was operative during times of interglaciation as well as during times of glaciation [Bond et al., 1997]. This may prove to be good news for mankind. In the past, I have made the case that a greenhouse-induced shutdown of the Atlantic's conveyor circulation might cause the Earth's climate system to jump to a quite different state of operation [Broecker, 1997] with serious global consequences. But, if these large shifts can only be triggered in the presence of major sea ice cover, then the magnitude of this threat is much reduced. Mea culpa!

REFERENCES

Behl, R. J., and J. P. Kennett, Brief interstadial events in the Santa Barbara Basin, NE Pacific during the past 60 kyr, *Nature*, 379, 243-246, 1996.

Bond, G., W. Showers, M. Cheseby, R. Lotti, I. Hajdas, and G. Bonani, A pervasive millennial-scale cycle in North Atlantic Holocene and glacial climates, *Science*, 278, 1257-1266, 1997.

Broecker, W. S. Thermohaline circulation, the Achilles heel of our climate system: Will manmade CO_2 upset the current balance? *Science*, 278, 1582-1588, 1997.

Broecker, W. S., and G. H. Denton, The role of ocean—atmosphere reorganizations in glacial cycles, *Quat. Sci. Rev.*, 9, 305-341, 1990.

Broecker, W. S., Paleocean circulation during the last deglaciation: A bipolar seesaw? *Paleoceanography*, 13, 119-121, 1998.

Cuffey, K. M. et al., Large Arctic temperature change at the Wisconsin-Holocene glacial transition, *Science*, 270, 455-458, 1995.

Cuffey, K. M., and G. D. Clow, Temperature, accumulation, and ice sheet elevation in central Greenland through the last deglacial transition, *J. Geophys. Res.*, 102, 26,383-26,396, 1997.

de Vernal, A., C. Hillaire-Marcel, J.-L. Turon, and J. Matthiessen, Reconstruction of sea-surface temperature salinity and sea-ice cover in the northern North Atlantic during the last glacial maximum based on dinocyst assemblages, *Can. J. Earth Sci.*, 37, 725-750, 2000.

Hughen, K. A. et al., Deglacial changes in ocean circulation from an extended radiocarbon calibration, *Nature*, 391, 65-68, 1998.

Imbellone, P. A., and M. I. Terruggi, Paleosols in Loess deposits of the Argentine Pampas, *Quat. International*, 17, 49-55, 1993.

Kukla, G. J., Loess Stratigraphy of Central Europe, in *After the Australopithecines*, edited by K. W. Butzer and G. L. Isaac, pp. 99-188, Mouton Publishers, The Hague, 1975.

Kukla, G., Loess stratigraphy in central China, *Quat. Sci. Rev.*, 6, 191-219, 1987.

Mayewski P. A., et al., Major features and forcing of high-latitude northern hemisphere atmospheric circulation using a 110,000-year-long glaciochemical series, *J. Geophys. Res.*, 102, 26,345-26,366, 1997.

Muhs, R. D., and E. A. Bettis III, Geochemical variations in Peoria Loess of western Iowa indicate paleowinds of midcontinental North America during last glaciation, *Holliday*, 2000.

Rahmstorf, S., Rapid climate transitions in a coupled ocean-atmosphere model, *Nature*, 372, 82-85, 1994.

Rosenfeld, D., Suppression of rain and snow by urban and industrial air pollution, *Science*, 287, 1793-1796, 2000.

Sachs, J. P., and S. J. Lehman, Subtropical North Atlantic temperatures 60,000 to 30,000 years ago, *Science*, 286, 756-759, 1999.

Schulz, H., U. von Rad, and H. Erlenkeuser, Correlation between Arabian Sea and Greenland climate oscillations of the past 110,000 years, *Nature*, 393, 54-57, 1998.

Stocker, T. F., The seesaw effect, *Science* 282, 61-62, 1998.

Tegen, I., A. A. Lacis, and I. Fung, The influence on climate forcing of mineral aerosols from disturbed soils, *Nature*, 380, 419-422, 1996.

Tegen, I., and D. Rind, Influence of the latitudinal temperature gradient on soil dust concentration and deposition in Greenland, *J. Geophys. Res.*, 105, 7199-7212, 2000.

Toon, O. B., How pollution suppresses rain, *Science*, 287, 1763-1764, 2000.

Twomey, S., Polution and the planetary albedo, S. *Atmos. Environ.*, 8, 1251-1256, 1974.

Yung, Y. L., T. Lee, C.-H. Wang, and Y.-T. Shieh, Dust: A diagnostic of the hydrologic cycle during the last glacial maximum, *Science*, 271, 962-963, 1996.

Zaucker, F., and W. S. Broecker, The influence of atmospheric moisture transport on the fresh water balance of the Atlantic drainage basin: GCM simulations and observations, *J. Geophys. Res.*, 97, 2765-2774, 1992.

Broecker, Wallace S., Lamont-Doherty Earth Observatory of Columbia University, Route 9W, Palisades, NY 10964

Stochastic Resonance in the North Atlantic: Further Insights

R.B. Alley

Environment Institute and Department of Geosciences, Pennsylvania State University, University Park, Pennsylvania

S. Anandakrishnan

Department of Geology, University of Alabama, Tuscaloosa, Alabama

P. Jung

Department of Physics and Astronomy, Ohio University, Athens, Ohio

A. Clough

Environment Institute and Department of Geosciences, Pennsylvania State University, University Park, Pennsylvania, and Department of Earth Sciences, University of Leeds, Leeds, UK

The large, abrupt, widespread, millennial changes recorded in many paleoclimatic archives pose a major challenge to our understanding of the climate system. Both periodic and stochastic models have been proposed to explain these events. We have argued that Greenland ice-core data are more consistent with a stochastic-resonance hypothesis. In this model, a combination of a weak periodicity plus "noise" perhaps caused by ice-sheet-related changes in freshwater flux to the north Atlantic produced switches between warm and cold climate modes. Here, we show that the stochastic-resonance hypothesis is consistent with a wider range of previously published data than analyzed before including a north Atlantic marine record and the Byrd Station, Antarctica ice-isotopic record; however, a record of hematite-stained quartz grains in north Atlantic sediment appears more periodic than stochastically resonant.

INTRODUCTION

The enigmatic Dansgaard-Oeschger oscillations and associated events pose significant challenges to our understanding of the climate system. These millennial changes were prominent during the cooling into the most recent ice age and warming from it [e.g., *Johnsen et al.*, 1992; *Alley et al.*, 1993; *Bond et al.*, 1993; *Mayewski et al.*, 1997; *Tay-lor et al.*, 1997; *Severinghaus et al.*, 1998], and appear to have been active through many previous ice ages [*McManus et al.*, 1999]. The changes have been subhemispheric to perhaps global in extent, and involved shifts over decades to perhaps years of one-third to one-half the entire glacial-interglacial amplitude.

No fully convincing model for these changes exists yet, but switches in the mode of operation of the north Atlantic "conveyor-belt" circulation almost certainly were involved [e.g., *Broecker et al.*, 1990; *Sarnthein et al.*, 1994]. The north Atlantic seems to spend most of its time either with vigorous circulation carrying much heat to high latitudes, or with reduced circulation carrying less heat to high lati-

The Oceans and Rapid Climate Change: Past, Present, and Future
Geophysical Monograph 126

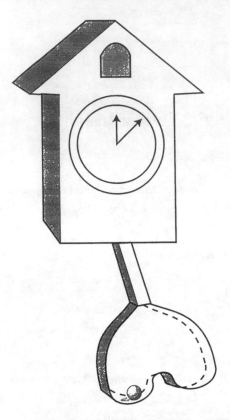

Figure 1a. Cartoon of a system that can exhibit stochastic resonance. A weak periodic oscillation, plus the appropriate random behavior of the jumping bean, combine to produce jumps back and forth between the two cups of the pendulum.

tudes; relatively little time is spent at intermediate levels [e.g., *Bryan*, 1986; *Stocker et al.*, 1992; *Rahmstorf*, 1995]. Switches between these are associated with the millennial Dansgaard-Oeschger oscillations [reviewed by *Alley and Clark*, 1999; *Alley et al.*, 1999]. A third mode of circulation with even less oceanic heat transport is linked to the few-millennial Heinrich events [*Sarnthein et al.*, 1994; *Alley and Clark*, 1999], but is less prominent in Greenland ice-core records and some other records around the north Atlantic than is the Dansgaard-Oeschger oscillation [*Alley et al.*, 1999], on which we concentrate here.

Conflicting evidence exists for causation of the abrupt north Atlantic changes. Some of the prominent coolings, especially the Younger Dryas and the brief cold event about 8200 years ago, occurred immediately after floods of ice-dammed waters from North America [e.g., *Broecker et al.*, 1988; *Barber et al.*, 1999], and other changes may have been related to switches in volume and site of freshwater discharge to the north Atlantic as well [*Licciardi et al.*, 1999]. Because of the major role of freshwater forcing in modeled behavior of north Atlantic circulation, cooling

following freshwater injection is not surprising [e.g., *Rahmstorf*, 1995].

Although the timing of outburst floods probably was linked to climate through its effects on ice-sheet mass balance, flood forcing is unlikely to have produced strongly periodic behavior. The timing of floods should have depended in part on history through isostatic tilting of drainages, and on geologic details of where the low places occurred in drainage divides. Outburst flooding thus would have exhibited significant stochastic elements.

Yet, numerous records show a strong periodicity of about 1500 years associated with the Dansgaard-Oeschger oscillations [*Denton and Karlen*, 1973; *Keigwin and Jones*, 1989; *Bond et al.*, 1997; 1999; *Grootes and Stuiver*, 1997; *Mayewski et al.*, 1997; *Yiou et al.*, 1997]. (As discussed by *Meeker et al.* [in review] and *Alley et al.* [2001], the periodicity is evident in non-Fourier as well as Fourier analyses of the data, is stable against changes in sampling interval, interpolation interval, and other details of the analysis, and so is not an alias of any shorter periodicity such as the annual cycle as suggested by *Wunsch*, [2000].)

It remains possible that there are different types of events, the recent ones triggered by outburst floods, and older ones exhibiting periodicity. However, the more recent changes seem to exhibit approximately the same periodicity as older ones [*Bond et al.*, 1997; *Mayewski et al.*, 1997], and the spatial patterns of climate anomalies of older and younger events appear similar insofar as we can tell [reviewed by *Alley and Clark*, 1999]. Alternatively, the older Dansgaard-Oeschger oscillations, the Younger Dryas and the 8200-year event may be of the same type, but with both stochastic and periodic behavior [*Alley et al.*, 2001].

Systems forced to mode transitions by a combination of a weak periodicity and noise are often studied under the heading of "stochastic resonance" [*Benzi et al.*, 1982]. These ideas were originally developed to explain ice-age cycles, but the simplest version may be more applicable to the Dansgaard-Oeschger oscillations [*Alley et al.*, 2001].

For a thorough review of the ideas of stochastic resonance, see *Gammaitoni et al.* [1998]. For a heuristic explanation, consider a specially designed clock pendulum as shown in Figure 1a, with a stochastic jumping bean in one of the two wells. If the pendulum swings far to each side and the jumping bean is internally quiet, the bean will make transitions back and forth between the wells on each swing, and the bean position will be related simply to the periodicity of the system. On the other hand, if the pendulum swings are small but the jumping bean is quite active, the bean will switch between the wells based on its own stochastic behavior, with little relation to the periodicity of the pendulum. These are periodic and stochastic cases, respectively.

However, if the pendulum amplitude and the jumping-bean activity have appropriate intermediate values, a transition will be likely when the pendulum and the bean work together. Figure 1b shows three numerical simulations of such a periodicity-plus-noise system in which the noise level is less than, approximately equal to, and greater than the optimum for stochastic resonance (following *Gammaitoni et al., 1998*) The optimum stochastic-resonance situation for the usual double-well potential, as shown, occurs when the pendulum swing is not sufficient to cause a transition by itself, and the bean activity exhibits white noise with a mean transition time between cups in the absence of pendulum motion equal to the half-period of the pendulum [*Gammaitoni et al.*, 1998].

In such a situation, the most likely outcome is for the bean to jump left when the pendulum swings right, and vice versa. However, if the bean fails to make a transition when the pendulum is favorable owing to low "noise" level, a transition will be highly unlikely when the phase of the pendulum then becomes unfavorable. Thus, transitions in one direction (say, left to right) will exhibit a favored spacing equal to the period, T, of the pendulum, and to integral multiples nT of the period, but not to half-integral multiples $(2n+1)T/2$.

A simple test for stochastic resonance is to measure the time intervals between successive transitions in a chosen direction, and to form a recurrence histogram of how often different waiting times were observed. The recurrence histogram of the ensemble average of a stochastically resonant system will exhibit peaks at waiting times $T, 2T, 3T$, ..., nT, ..., with exponentially decreasing peak height with increasing n, separated by valleys at waiting times $<T$, $3T/2, 5T/2, ..., (2n+1)T/2, ...$ [*Gammaitoni et al.*, 1998]. In contrast, a periodic system would exhibit a single peak at waiting time T, and a white-noise system would lack significant peaks but exhibit an exponential decrease in number of occurrences with increasing waiting time.

In our previous paper on this topic [*Alley et al.*, 2001], we tested the hypothesis that both signal and noise contributed to millennial oscillations in north-Atlantic climate records using the recurrence histograms of different waiting times between warmings recorded in the ice-isotopic ratios of the GRIP and GISP2 cores from central Greenland. We chose these records because they combined a strong linkage between proxy and climate, good dating, high time resolution of sampling, and great length [e.g., *Cuffey et al.*, 1995; *Alley et al.*, 1997; *Grootes and Stuiver*, 1997; *Jouzel et al.*, 1997; *Johnsen et al.*, 1997; *Meese et al.*, 1997]. In the interest of brevity, our previous paper left a number of questions unanswered, some of which we try to answer here. We provide further insights to the analysis of the records presented in our previous paper, and we

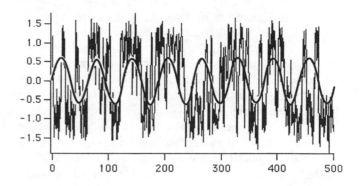
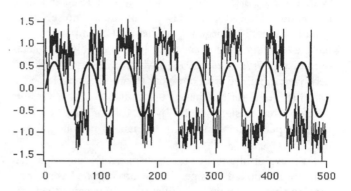
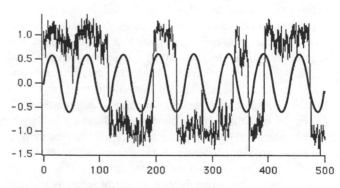

Figure 1b. Numerical simulation of an idealized version of the system in Figure 1a, following Gammaitoni et al. [1998], showing behavior when the noise level is less than (bottom), approximately equal to (middle), and greater than (top) the optimum level for stochastic resonance. The sine-wave forcing is also shown.

show results of applying the same analytical techniques to additional climate records, to provide a clearer view of the strenghts and weaknesses of the stochastic-resonance hypothesis.

METHODS

The methods applied here are the same as in our previous paper [*Alley et al.*, 2001]. We resampled data sets in equal time increments by fitting a spline to the raw data (using the SPLINE function in Numerical Recipes [*Press et al.*, 1989]), sampling the spline in subannual increments (using the related SPLINT function), and then combining those increments to the desired resolution. This resampling is sufficiently coarse that it eliminates any need to correct for diffusive smoothing of the ice-core records [*Johnsen et al.*, 1997]. The resampling also serves as a low-pass filter to remove seasonal cycles, anomalous storms, and other situations in which a variable changed from "low" to "high" and back without recording a persistent mode-switch of the north-Atlantic climate system. (Sub-annual to few-annual ice-isotopic data contain much "noise" unrelated to climate of large regions as well as important "signal"; averaging over decades or longer largely eliminates the noise [*White et al.*, 1973].) Most of the results shown here were obtained with one-year resampling and 150-year lumping. As described in our previous paper, these results are not affected significantly by twofold or larger changes in these parameters, including the use of nonintegral values.

To focus on the millennial Dansgaard-Oeschger oscillations, we removed changes in the orbital band while preserving timing and size of millennial events using the high-pass zero-phase Butterworth filter implemented in the software package MATLAB. As noted in our previous paper, the few-millennial Bond cycles, comprising successively colder millennial Dansgaard-Oeschger oscillations culminating in a Heinrich event [reviewed by *Alley and Clark*, 1999], are sufficiently subtle that they do not dominate our transition analysis of Greenland ice-isotopic ratios; we obtain similar results whether the high-pass cutoff is longer or shorter than the typical Bond-cycle length. Results shown here were obtained with a 7000-year cutoff.

Owing to the pronounced asymmetry of most Dansgaard-Oeschger oscillations, with faster warming than cooling, we chose to analyze only warmings to avoid having event timings depend on the chosen threshold. We did analyze coolings for some records, and found similar results. For the ice-core calcium data, hematite-stained quartz grains, and abundance of *N. pachyderma* (s) analyzed here, we equate warming to a drop in abundance.

We defined a warming transition to have occurred when a paleoclimatic variable rose above its mean value by more than a specified threshold level, d, after falling more than d below the mean. We typically used $d \sim 20\%$ of the standard deviation of the resampled and high-pass-filtered data; much lower values tend toward a noisy limit by giving tiny fluctuations the same importance as major changes, whereas much higher values yield so few transitions that the records are not long enough to provide statistically significant results. We calculated waiting times between successive warmings, and formed histograms of how often different waiting times occurred. For many of the plots, we show sensitivity of the results to d, with the range of d values chosen to show significant changes in the histograms.

In our previous paper, we provided a partial test for statistical significance of the stochastic-resonance hypothesis for the 110,000-year-long ice core records. We first generated 11 million years of a stochastically resonant time series with the same periodicity as inferred for the ice-core data, sufficiently long that additional points had little effect on the recurrence histogram. We then evaluated the effect of time-series length by using the chi-square test to assess the similarity between the recurrence histograms of the parent time series and of 110,000-year-long subsets of that time series. We similarly compared the observed recurrence histogram of the ice-core data to that of the long simulated time series. Greater similarity of the recurrence histogram of the long simulated time series to that of the ice-core data than to some of those of the simulated subsets then is evidence that the data are consistent with the stochastic-resonance hypothesis.

This is a rather stringent test, because measurement errors, dating errors, and any additional active processes would cause the recurrence histogram of a paleoclimatic record of a stochastically resonant system to deviate from the ideal. If dating and measurement errors were well-characterized, one might improve the test by "degrading" the subsets of the synthetic time series by adding appropriate errors; however, especially for dating errors and possible contributions from additional processes, we do not know how to characterize the errors well.

More simply, our inspection of the recurrence histograms of the long synthetic time series and of its subsets (Figure 2) shows that one can recognize potentially stochastically resonant time series rather easily for record length and mean spacing between warmings similar to those of the ice-core records. A recurrence-histogram peak at about 1500 years, and a smaller one at about 3000 years, with valleys flanking and between them and no other large peaks, indicates a system consistent with stochastic resonance. This is strengthened by a third peak at about 4500 years, but consistency is possible with only the first two. Thus, if one sees a large peak on the recurrence histogram, there is a preferred spacing and a potential periodicity in the system; the presence of a second peak at about twice the waiting time of the first indicates the possibility of stochastic resonance.

We note that we began our studies with what we consider to be the best records in terms of dating, climate-recorder transfer function, length, and sampling interval. We thus expect that if the other records analyzed here are recording a stochastically resonant system, they should produce recurrence histograms that differ from that of the long synthetic time series by more than does the recurrence histogram of the ice-isotopic data from Greenland. Because we cannot quantify this well, here we rely primarily on the presence or absence of peaks on the observed histograms near 1500, 3000, and perhaps 4500 years.

RESULTS

We have conducted analyses of ice-core data from central Greenland, including the GRIP and GISP2 ice-isotopic data and the GRIP calcium data. We have also analyzed three north-Atlantic marine records, and the Byrd Station, Antarctica ice-isotopic record. These records were chosen for length, sampling interval, dating accuracy, and prominence in the literature, but clearly are a representative rather than comprehensive selection of available data sets. Here, we summarize and extend our earlier work on ice-isotopic ratios, and present the results of the other analyses.

Greenland Ice-Isotopic Ratios

The main result of our earlier investigations of Greenland ice-isotopic ratios is shown in Figure 2, modified slightly from *Alley et al.* [2001]. As described above, the GRIP ice-isotopic record is compared to a synthetic time-series of a stochastically resonant system one-hundred times as long as GRIP with a periodicity chosen to match that of GRIP, a subset of that synthetic time-series ten times as long as GRIP, and two subsets of that synthetic time series as long as GRIP. Chi-square testing shows that the histogram of the GRIP data bears a stronger resemblance to the long synthetic time-series than do about one-quarter of the GRIP-length subsets of the synthetic time series.

The observed peaks at about 1500, 3000, and 4500 years separated by troughs <1500 years and near 2250 and 3750 years are difficult to explain by other models we tested but are expected for stochastic resonance. (The best-fit period, T, typically falls somewhere between 1450 and 1500 years; we will use 1500 years in most discussions for convenience.) We argued from this that the GRIP data are consistent with the stochastic-resonance model.

Figures 3a and 3b compare the observed GRIP data to the output of a white-noise process with a 1500-year mean waiting time between warmings, and to the outcome of replacing the GRIP observations by randomly generated data and then conducting the same analyses on them. Clearly, there is little relation between the observed histogram and these white-noise possibilities. In the limit of a single periodicity with zero noise, all transitions would occur at time T, giving a single vertical line on the recur-

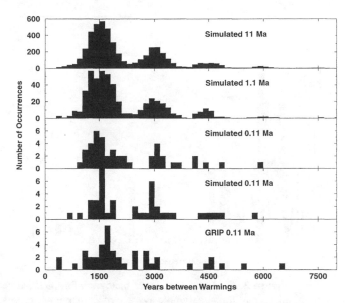

Figure 2. Number of occurrences of different waiting times between warmings for: (2a, top) a simulated stochastically resonant time-series 100x longer than the GRIP data; (2b-d) subsets of that simulated time series that are 10x longer than GRIP (2b, second from top) and of the same length as GRIP (2c and 2d, middle and second from bottom); and (2e, bottom) the observed GRIP ice-core data (150-year samples, high-pass cutoff of 7000 years, and $d=0.25$ or ~20% of the standard deviation of the resampled and high-pass-filtered data). The long simulated series has the expected stochastically resonant pattern. The shorter subsets display deviations from this pattern related to the smaller sample size. The GRIP data are more similar to the long simulated series than is one of the two shorter subsets shown and than are about one-quarter of all GRIP-length subsets examined, based on chi-square testing.

rence histogram at T. We can say with high statistical confidence (>99%) that neither the pure periodic model nor the pure white-noise models are consistent with the data, although as noted above, the stochastic-resonance model is consistent with the data. We showed in our previous paper that consistency of the results with the stochastic-resonance hypothesis is not affected by small changes in any of the parameters in the analysis.

Greenland Calcium Concentrations

Many other Greenland ice-core data sets are available [e.g., *NSIDC*, 1997]. *Ditlevsen* [1999] conducted a similar crossover analysis on the logarithms of calcium concentrations in the GRIP ice core [*Fuhrer et al.*, 1993]. Calcium is an important paleoclimatic indicator [*Fuhrer et al.*, 1993; *Mayewski et al.*, 1997], although the linkage to the climate is somewhat less direct and with more noise than for ice-

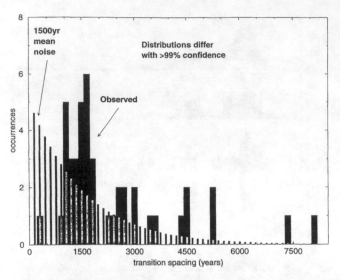

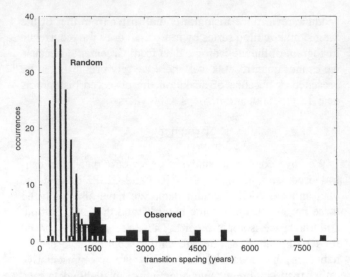

Figure 3. Observed recurrence histogram for GRIP, Greenland ice-isotopic data as in Figure 2, compared to: a) the histogram for an ensemble average of a white-noise process with a 1500-year mean spacing between warmings; and b) the GRIP ice-isotopic data set replaced with random numbers and then treated in the same way as the actual data. The observed data are highly distinct statistically from the white-noise models.

isotopic ratios. *Ditlevsen* [1999] analyzed raw data, with approximately annual resolution through much of the record, and assessed times between transitions across chosen thresholds that were not adjusted for orbital changes in the baseline of the data. *Ditlevsen* [1999] concluded that there is no significant evidence of either millennial periodicity or stochastic resonance in the GRIP calcium data.

However, calcium data are known to record numerous signals and noise, including subannual changes, an annual cycle, and orbital changes (e.g., *Fuhrer et al.*, 1993; *Mayewski et al.*, 1997]. The use of the highest-resolution data by *Ditlevsen* [1999] without low-pass filtering is the most direct treatment possible, but gives equal importance to widely differing signals: a single large dust storm is accorded the same significance as a millennium-long switch in typical dust concentration such as the Younger Dryas event. Also, because the tens-of-millennial orbital signal affected the baseline of the data, the strength of millennial or shorter oscillations needed to cause a transition across a time-invariant threshold is different at different times. We thus are not surprised that the raw data do not yield evidence of periodicity or stochastic resonance in the millennial band.

We have analyzed the GRIP calcium data archived on the Greenland Ice Cores CD-ROM [*NSIDC*, 1997]. We follow *Ditlevsen* [1999] in using the logarithms of calcium concentration. Our analysis of the raw archived data reaches a similar conclusion to that of *Ditlevsen* [1999]. However, if we apply the same resampling and high-pass filtering as for the ice-isotopic ratios, we obtain very dif-

ferent results than *Ditlevsen* [1992], as shown in Figure 4.

The signal is not quite as similar to the stochastically resonant pattern as for the ice-isotopic ratios. However, the scarcity of transitions at short times is evident; the filtering shown would allow transitions as frequently as 300 years, but transitions with waiting times shorter than about 1500

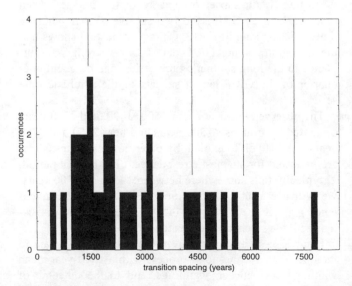

Figure 4. Recurrence histogram for GRIP calcium data, showing a clear peak at about 1500-year spacing and a possible peak at about 3000-year spacing. 7000-year high-pass cutoff, 150-year resampling, and transition threshold set at about 20% of the standard deviation of the resampled and high-pass-filtered data. Lines extending downward are spaced 1470 years apart.

years are greatly underrepresented compared to the expectation of a white-noise process. The weak peak in occurrences at about 3000 years, with relatively fewer occurrences near 2250 and 3750 years, is consistent with the stochastic-resonance hypothesis. We believe that this figure provides support for a preferred spacing of about 1500 years between drops in calcium concentration of the GRIP ice core, and at least some support for stochastic resonance in the calcium data.

North Atlantic Marine Records

We chose to examine three marine records from the north Atlantic Ocean that have figured prominently in discussions of millennial variability. *Bond et al.* [1993] demonstrated a strong tie between ice-core and oceanic records, *McManus et al.* [1999] looked at times older than covered by the Greenland ice-core records, and *Bond et al.* [1999] provided strong support for the hypothesis that the millennial events continued from the ice age into the Holocene with greatly reduced amplitude but similar spacing.

Bond et al. [1993] developed paleoclimatic records including the abundance of the cold-water planktonic foraminifera *N. pachyderma* (s) in core VM23-81, from 54°N, 17°W in the eastern north Atlantic, transferred to the GISP2 time scale [*Meese et al.*, 1997] based on curve matching and radiocarbon control in younger parts. As discussed by *Bond et al.* [1999], this is not purely a record

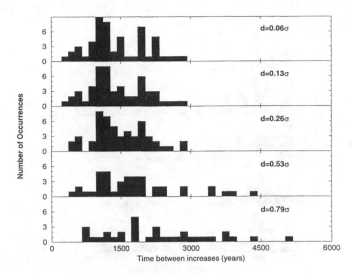

Figure 6. Recurrence histograms for occurrence of hematite-stained quartz grains from *Bond et al.* [1999] betweem 0 ka and 80 ka. 7000-year high-pass cutoff, 150-year resampling, transition threshold *d* varied as a function of the standard deviation of the resampled and high-pass-filtered data as indicated.

of sea-surface temperature, but is related to sea-surface temperature.

Figure 5 shows the recurrence histogram for warmings (drops in abundance *of N. pachyderma* (s)) as a function of the transition threshold, *d*, taken as a fraction of the standard deviation of the resampled (150-year samples) and high-pass-filtered (7000-year cutoff) data set. The stochastic-resonance signal is not as clear as for the ice-core data, but the lack of quick transitions and of transitions near 2250 and 3750 years, together with many transitions near 1500 years or slightly shorter and near 3000 years, is consistent with a stochastic-resonance model. Given the great similarity between ice-core and VM23-81 records [*Bond et al.*, 1993], this result is not surprising, but it is reassuring.

We next consider the record of hematite-stained quartz grains from the same core, VM23-81 (54°N, 17°W) in the eastern north Atlantic [*Bond et al.*, 1999], covering the last 80 ka. Again, dating includes radiocarbon control for younger parts and correlation to the GISP2 timescale older than 26 ka. Sample intervals are typically 100-200 years. Results of crossover analysis on the resampled, high-pass-filtered data are shown in Figure 6. With a high threshold, *d*, one does not obtain significant peaks in the recurrence histogram; with a low threshold, only a single peak appears at about 1500 years, with the possibility of a slight peak near 3000 years.

The frequency of occurrence of hematite-stained quartz grains is some sort of ocean-circulation tracer [*Bond et al.*, 1999]. The transfer function between climate and this

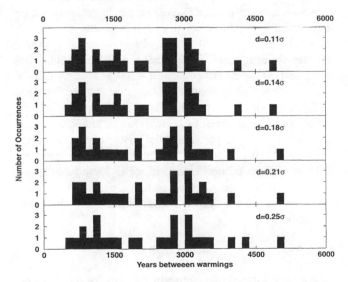

Figure 5. Recurrence histograms for occurrence of decreases in occurrence of cold-loving planktonic foraminifera *N. pachyderma* (s) in the northeastern Atlantic, from *Bond et al.* [1993] between 0 ka and 80 ka. 7000-year high-pass cutoff, 150-year resampling, transition threshold *d* varied as a function of the standard deviation of the resampled and high-pass-filtered data as indicated.

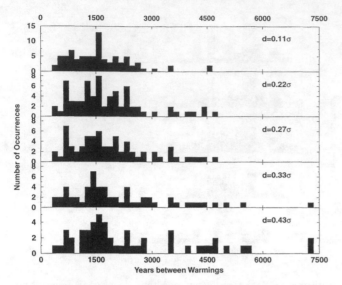

Figure 7. Recurrence histograms for oxygen-isotopic ratios of *Neogloboquadrina pachyderma* (d) from *McManus et al.* [1999] betweem 350 ka and 480 ka. 7000-year high-pass cutoff, 150-year resampling, transition threshold *d* varied as a function of the standard deviation of the resampled and high-pass-filtered data as indicated.

tracer is less direct than for some of the other indicators we consider. However, the clear signal in the data is recording some characteristic of the climate system, and our analysis reinforces that of *Bond et al.* [1999] that there is a preferred spacing of approximately 1500 years between changes associated with warmings (drops in hematite-stained quartz grains). The recurrence histogram for this record is more consistent with a single periodicity of about 1500 years than with stochastic resonance.

McManus et al. [1999] developed several records from the Ocean Drilling Program (ODP) Site 980 core from the Feni abyssal drift at about 55°N, 15°W in the eastern north Atlantic. We have analyzed their oxygen-isotopic data of the planktonic foraminifera *Neogloboquadrina pachyderma* (d). As discussed by *McManus et al.* [1999], this isotopic record provides a useful proxy for sea-surface temperature. We consider only the section with highest sampling resolution between 350 ka and 450 ka.

The 130 ka span of this high-resolution record is slightly longer than the intact portion of the Greenland ice-core records. However, dating of the marine record is by interpolation of an orbitally tuned scale, and so could introduce uncertainty to the estimated duration and spacing of the millennial events, especially if sedimentation rates changed through the millennial events [*McManus et al.*, 1998]. In addition, despite the high time resolution compared to many other records (average 250 years/sample), on average there are only six samples per 1500-year cycle, with some

parts of the record more sparsely sampled. Hence, the combined uncertainties from dating and sampling cast some doubt on any results from our analysis of this core. The need for even longer, more-finely-sampled records is clear.

Application of our analysis produces the results shown in Figure 7 as a function of the threshold, *d*, at which a warming is taken to have occurred. A strong 1500-year preferred spacing is quite clear in the data. The stochastic-resonance pattern is not obvious in this plot, with little evidence for a 3000-year peak, although one cannot reject the stochastic-resonance hypothesis with especially high confidence based on these data owing to dating and sampling issues.

Byrd Station, Antarctica Ice-Isotopic Ratios

The final time-series we consider here is the Byrd Station, Antarctica ice-isotopic record [*Johnsen et al.*, 1972]. This has been placed on the GISP2 time scale using the methane concentrations of air bubbles by *Blunier and Brook* [2001], and extends between 8 ka and 103 ka. Because previous analyses were conducted on north Atlantic warmings, and the Byrd data are proposed to exhibit a seesaw relation with those from Greenland [*Blunier et al.*, 1998; *Bender et al.*, 1999; *Blunier and Brook*, 2001], in Figure 8 we show results for times between coolings (Figure 8a) as well as between warmings (Figure 8b).

The Byrd ice-isotopic data are consistent with the stochastic-resonance pattern. Transitions in the same direction with waiting times shorter than a millennium are rare. A peak occurs near 1500 years, although perhaps shifted slightly younger than 1500 years. A peak near 3000 years is present, and a broad peak near 4500 years may also be present. The slightly shorter waiting time for the first peak in Byrd as compared to GRIP and GISP2 ice-isotopic records could have several explanations including statistical fluctuations, but is perhaps most easily explained as the effect of having a smaller millennial signal in Antarctic than in corresponding Greenland ice-isotopic records [*Alley et al.*, 2001, Figs. 3c, 3d], so that the noise is more important in the Byrd record and produces more short-time transitions.

DISCUSSION

We have formed recurrence histograms for the times between transitions in a chosen direction in paleoclimatic time series, and assessed the similarity of these histograms to those expected for stochastic resonance, white noise, and periodic behavior. We have looked at a range of records, including ice-isotopic ratios from both GISP2 and GRIP in

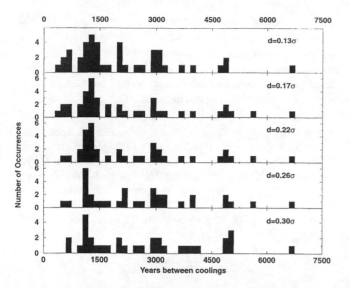
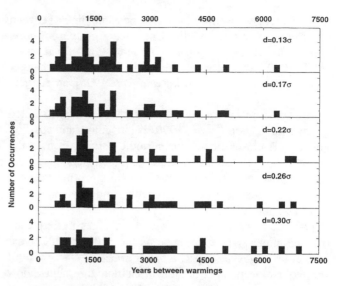

Figure 8. Recurrence histograms for coolings (Figure 8a) and warmings (Figure 8b) in ice-isotopic ratios from the Byrd Station, Antarctica core [*Johnsen et al.,* 1972] as dated by *Blunier and Brook* [2001] between 8 and 103 ka. 7000-year high-pass cutoff, 150-year resampling, transition threshold *d* varied as a function of the standard deviation of the resampled and high-pass-filtered data as indicated.

Greenland and Byrd Station in Antarctica, calcium data from GRIP, and oxygen-isotopic ratios and abundance of hematite-stained quartz grains and cold-water planktonic foraminifera *N. pachyderma* (s) from marine cores in the northeast Atlantic. We find that the ice-isotopic ratios from both Greenland and Antarctica yield patterns fully consistent with stochastic resonance and inconsistent with other simple models, as do the calcium data from Greenland and the *N. pachyderma* (s) abundances. We analyzed the oceanic isotopic record of *McManus et al.* [1999] despite concerns about dating and sample resolution. The result produces strong support for a 1500-year periodicity between 350 and 480 ka, but little support for stochastic resonance; we cannot tell to what extent this argues against stochastic resonance, and to what extent it is an artifact of dating and sampling issues. The hematite-stained quartz grains produce a signal more consistent with a single periodicity.

We note that consistency with the stochastic-resonance pattern does not prove stochastic resonance; almost certainly, other hypotheses can be generated that produce the observed recurrence histograms. However, we were led to the stochastic-resonance hypothesis by consideration of mechanisms, and we developed the recurrence histograms as a test of the hypothesis. In addition, as discussed by *Alley et al.* [2001], other simple models we tested including white noise and simple combinations of periodicities can be ruled out based on our results.

Better tests of the stochastic-resonance hypothesis using our techniques require longer, better-dated records with higher sampling resolution. The available time series typically include only about 40 transitions, too few to reliably define more than two or three peaks in the recurrence histogram. And even if longer records were available, relative dating errors of only a few centuries over many millennia would disrupt any evidence of stochastic resonance at longer waiting times.

Although we have used a digital filter to remove orbital-frequency variability, one can do almost the same by drawing the ice-age cycle "by eye". Our technique thus is not fundamentally spectral, yet we obtain the same approximately 1500-year preferred spacing of events in the paleoclimatic records as found in spectral analyses [e.g., *Mayewski et al.*, 1997; *Yiou et al.*, 1997; *Bond et al.*, 1999]. The persistent appearance of this spacing in numerous climate records of different ages based on very different proxies from disparate regions and analyzed in fundamentally different ways is strong evidence that there is such a periodicity in the climate system. However, the dating of records is not sufficiently accurate to assess whether this 1500-year preferred spacing maintains a constant phase over glacial-interglacial cycles [e.g., *Johnsen et al.*, 1992; *Meese et al.*, 1997]

The 1500-year time-scale of the oscillations is consistent with processes involving the deep ocean [*Broecker et al.*, 1990]. However, tidal processes [*Keeling and Whorf*, 2000], ENSO processes perhaps related to orbital changes [*Cane and Clement*, 1999], and solar forcing [*Mayewski et al.*, 1990] have also been suggested to explain these Dansgaard-Oeschger oscillations. The possible links between the 1500-year cycle and the Little Ice Age [e.g., *Denton and Karlen*, 1973; *Keigwin and Jones*, 1989; *Bond et al.*, 1997; 1999], and between the Little Ice Age and changes in solar activity [e.g., *Crowley and Kim*, 1996], are suggestive.

The apparently periodic behavior of the hematite-stained quartz grains [*Bond et al.*, 1999] is intriguing, and possibly indicates a closer tie to the periodic forcing than to the overall climate response to the combined periodicity and noise. Unfortunately, the precise meaning of this paleoclimatic indicator is not clear; better understanding of it would be helpful.

We cannot easily assess possible changes in frequency of oscillations over time, such as the 550- and 1000-year cyclicities suggested by *Chapman and Shackleton* [2000] for the Holocene. Setting the transition threshold d low enough to capture the small amplitude of Holocene fluctuations gives "noise" in ice-age behavior. The Holocene itself is not long enough to provide robust results if analyzed independentlly, and use of a time-variable threshold is rather ad hoc, but we may investigate it in the future.

CONCLUSIONS

Based on analysis of recurrence histograms for changes in paleoclimatic records, we find that the climate system includes important variability with a spacing of approximately 1500 years in Antarctica as well as Greenland and the north Atlantic. However, less-frequent occurrence of spacings of twice and three times this interval in ice-core records from Greenland and Antarctica and one ocean-core record from the northeast Atlantic shows that if a warming is missed, the next warming maintains the phase of the oscillation through at least two to three times the dominant period.

This observed pattern is consistent with stochastic resonance, in which a weak periodicity plus noise combine to cause mode switches. The noise may be related to changes in freshwater fluxes into the north Atlantic, although other causes are possible; several hypotheses exist for the cause of the periodic variability, but none is widely accepted yet.

Further complicating the picture, the hematite-stained quartz-grain record of *Bond et al.* [1999] appears to be sensitive primarily to the weak periodicity and does not exhibit the stochastic-resonance pattern. The *McManus et al.* [1999] record of northeast-Atlantic planktonic foraminiferal isotopic ratios exhibits the 1500-year peak and additional longer-waiting-time warmings but not obviously at integral multiples of the primary peak; dating and sampling issues may or may not be significant.

Further work focused on assessing both periodic and stochastic elements in the climate system is likely to be useful in improving understanding of the millennial climate changes. Longer, more highly resolved, and better-dated records would be of special value; we remain data-limited in understanding the patterns of climate change.

Acknowledgements We thank the U.S. National Science Foundation for support, the GISP2 Science Management Office and chief scientist Paul Mayewski, the US 109th Air National Guard, the Polar Ice Coring Office, the National Ice Core Lab, the NOAA NGDC Paleoclimatology Program for data archival, Eric Wolff, Minze Stuiver, Larry Wilen, Sigfus Johnsen, Jerry McManus, Gerard Bond, Kirk Maasch, Ed Brook, and other colleagues.

REFERENCES

Alley, R.B., S. Anandakrishnan and P. Jung, Stochastic resonance in the north Atlantic, *Paleoceanography*, 16, 190-198, 2001.

Alley, R.B., and P.U. Clark, The deglaciation of the northern hemisphere: A global perspective, *Ann. Rev. Earth Planet. Sci.*, 27, 149-182, 1999.

Alley, R.B., P.U. Clark, L.D. Keigwin, and R.S. Webb, Making sense of millennial-scale climate change, in *Mechanisms of Millennial-Scale Global Climate Change*, edited by P.U. Clark, R.S. Webb and L.D. Keigwin, pp. 385-394, American Geophysical Union, Washington, DC, 1999 (corrected paste-up at http://www.agu.org/pubs/covers/ASGM112095X.pdf).

Alley, R.B., C.A. Shuman, D.A. Meese, A.J. Gow, K.C. Taylor, K.M. Cuffey, J.J. Fitzpatrick, P.M. Grootes, G.A. Zielinski, M. Ram, G. Spinelli, and B. Elder, Visual-stratigraphic dating of the GISP2 ice core: basis, reproducibility, and applications, *J. Geophys. Res., 102C*, 26367-26381, 1997.

Alley, R.B., D.A. Meese, C.A. Shuman, A.J. Gow, K.C. Taylor, P.M. Grootes, J.W.C. White, M. Ram, E.D. Waddington, P.A. Mayewski, and G.A. Zielinski, Abrupt increase in Greenland snow accumulation at the end of the Younger Dryas event, *Nature, 362*, 527-529, 1993.

Barber, D.C., A. Dyke, C. Hillaire-Marcel, A.E. Jennings, J.T. Andrews, M.W. Kerwin, G. Bilodeau, R. McNeely, J. Southon, M.D. Morehead, and J.M. Gagnon, Forcing of the cold event of 8,200 years ago by catastrophic drainage of Laurentide lakes, *Nature, 400*, 344-348, 1999.

Bender, M.L., B. Malaize, J. Orchardo, T. Sowers, and J. Jouzel, High precision correlations of Greenland and Antarctic ice core records over the last 100 kyr, in *Mechanisms of Millennial-Scale Global Climate Change*, edited by P.U. Clark, R.S. Webb and L.D. Keigwin, pp. 149-164, American Geophysical Union, Washington, DC, 1999.

Benzi, R., G. Parisi, A. Sutera, and A. Vulpiani, Stochastic resonance in climatic change, *Tellus, 34*, 10-16, 1982.

Blunier, T. and E. Brook, Timing of millennial-scale climate change in Antarctica and Greenland during the last glacial period, *Science, 291*, 109-112, 2001.

Blunier, T., J. Chappellaz, J. Schwander, A. Dallenbach, B. Stauffer, T.F. Stocker, D. Raynaud, J. Jouzel, H.B. Clausen, C.U. Hammer, and S.J. Johnsen, Asynchrony of Antarctic and Greenland climate change during the last glacial period, *Nature, 394*, 739-743, 1998.

Bond, G., W. Broecker, S. Johnsen, J. McManus, L. Labeyrie, J. Jouzel, and G. Bonani, Correlations between climate records from North Atlantic sediments and Greenland ice, *Nature, 365*, 143-147, 1993.

Bond, G., W.J. Showers, M. Cheseby, R. Lotti, P. Almasi, P. deMenocal, P. Priore, H. Cullen, I. Hajdas, and G. Bonani, A pervasive millennial-scale cycle in North Atlantic Holocene and glacial climates, *Science, 278*, 1257-1266, 1997.

Bond, G.C., W. Showers, M. Elliot, M. Evans, R. Lotti, I. Hajdas, G. Bonani and S. Johnsen, The North Atlantic's 1-2 kyr climate rhythm: relation to Heinrich events, Dansgaard/Oeschger cycles and the Little Ice Age, in *Mechanisms of Millennial-Scale Global Climate Change*, edited by P.U. Clark, R.S. Webb and L.D. Keigwin, pp. 35-58, American Geophysical Union, Washington, DC, 1999.

Broecker, W.S., M. Andree, W. Wolfli, H. Oeschger, G. Bonani, J. Kennett, and D. Peteet, The chronology of the last deglaciation: Implications to the cause of the Younger Dryas event, *Paleoceanography, 3*, 1-19, 1988.

Broecker, W.S., G. Bond, M. Klas, G. Bonani, and W. Wolfli, A salt oscillator in the glacial Atlantic? 1. The concept, *Paleoceanography, 5*, 469-477, 1990.

Bryan, F., High-latitude salinity effects and interhemispheric thermohaline circulations, *Nature, 323*, 301-304, 1986.

Cane, M., and A. Clement, A role for the tropical Pacific coupled ocean-atmosphere system on Milankovitch and millennial timescales. Part II: Global Impacts, in *Mechanisms of Millennial-Scale Global Climate Change*, edited by P.U. Clark, R.S. Webb and L.D. Keigwin, pp. 373-383, American Geophysical Union, Washington, DC, 1999.

Chapman, M.R., and N.J. Shackleton, Evidence of 550 year and 1000 year cyclicities in North Atlantic pattern during the Holocene, *Holocene, 10*, 287-291, 2000.

Crowley, T.J., and K.-Y. Kim, Comparison of proxy records of climate change and solar forcing, *Geophys. Res. Lett., 23*, 359-362, 1996.

Cuffey, K.M., G.D. Clow, R.B. Alley, M. Stuiver, E.D. Waddington, and R.W. Saltus, Large Arctic temperature change at the glacial-Holocene transition, *Science, 270*, 455-458, 1995.

Denton, G.H., and W. Karlen, Holocene climatic variations—their pattern and possible cause, *Quat. Res., 3*, 155-205, 1973.

Ditlevsen, P.D., Observation of α-stable noise induced millennial climate changes from an ice-core record, *Geophys. Res. Lett., 26*, 1441-1444, 1999.

Fuhrer, K., A. Neftel, M. Anklin, and V. Maggi, Continuous measurements of hydrogen-peroxide, formaldehyde, calcium and ammonium concentrations along the new GRIP ice core from Summit, central Greenland, *Atm. Env. A, 27*, 1873-1880, 1993.

Gammaitoni, L., P. Hanggi, P. Jung, and F. Marchesoni, Stochastic resonance, *Rev. Modern Phys., 70*, 223-287, 1998.

Grootes, P.M., and M. Stuiver, Oxygen 18/16 variability in Greenland snow and ice with 10^{-3}- to 10^5-year time resolution, *J. Geophys. Res., 102*, 26455-26470, 1997.

Johnsen, S.J., H.B. Clausen, W. Dansgaard, N.S. Gundestrup, C.U. Hammer, U. Andersen, K.K. Andersen, C.S. Hvidberg, D. Dahl-Jensen, J.P. Steffensen, H. Shoji, A.E. Sveinbjornsdottir, J. White, J. Jouzel, and D. Fisher, The $\delta^{18}O$ record along the Greenland Ice Core Project deep ice core and the problem of possible Eemian climatic instability, *J. Geophys. Res., 102*, 26397-26410, 1997.

Johnsen, S.J., H.B. Clausen, W. Dansgaard, K. Fuhrer, N. Gundestrup, C.U. Hammer, P. Iversen, J. Jouzel, B. Stauffer, and J.P. Steffensen, Irregular glacial interstadials recorded in a new Greenland ice core, *Nature, 359*, 311-313, 1992.

Johnsen, S.J., W. Dansgaard, H.B. Clausen, and C.C. Langway, Jr., Oxygen isotope profiles through the Antarctic and Greenland ice sheets, *Nature, 235*, 429-434, 1972.

Jouzel, J., R.B. Alley, K.M. Cuffey, W. Dansgaard, P. Grootes, G. Hoffman, S.J. Johnsen, R.D. Koster, D. Peel, C.A. Shuman, M. Stievenard, M. Stuiver, and J. White, Validity of the temperature reconstruction from water isotopes in ice cores, *J. Geophys. Res., 102*, 29471-29487, 1997.

Keeling, C.D., and T.P. Whorf, The 1800-year oceanic tidal cycle: A possible cause of rapid climate change, *Proc. Nat. Acad. Sci., 97*, 3814-3819, 2000.

Keigwin, L.D., and G.A. Jones, Glacial-Holocene stratigraphy, chronology, and paleoceanographic observations on some North Atlantic sediment drifts, *Deep-Sea Res., 36*, 845-867, 1989.

Licciardi, J.M., J.T. Teller, and P.U. Clark, Freshwater routing by the Laurentide Ice Sheet during the last deglaciation, in *Mechanisms of Millennial-Scale Global Climate Change*, edited by P.U. Clark, R.S. Webb and L.D. Keigwin, pp. 177-200, American Geophysical Union, Washington, DC, 1999.

Mayewski, P.A., L.D. Meeker, M.S. Twickler, S. Whitlow, Q. Yang, W.B. Lyons, and M. Prentice, Major features and forcing of high-latitude northern hemisphere atmospheric circulation using a 110,000-year-long glaciochemical series, *J. Geophys. Res., 102*, 26345-26366, 1997.

McManus, J.F., D.W. Oppo, and J. Cullen, A 0.5-million-year record of millennial-scale climate variability in the North Atlantic, *Science, 283*, 971-975, 1999.

McManus, J.F., R.F. Anderson, W.S. Broecker, M.Q. Fleisher, and S.M. Higgins, Radiometrically determined sedimentary fluxes in the sub-polar North Atlantic during the last 140,000 years, *Earth Planet. Sci. Lett., 155*, 29-43, 1998.

Meeker, L.D., P.A. Mayewski, P.M. Grootes, R.B. Alley, and G.C. Bond, Regarding "On sharp spectral lines in the climate record and the millennial peak." by C. Wunsch, *Paleoceanography, 15*, 417-424, submitted.

Meese, D.A., A.J. Gow, R.B. Alley, G.A. Zielinski, P.M. Grootes, M. Ram, K.C. Taylor, P.A. Mayewski, and J.F. Bolzan, The Greenland Ice Sheet Project 2 depth-age scale: methods and results, *J. Geophys. Res., 102*, 26411-26423, 1997.

NSIDC (National Snow and Ice Data Center, University of Colorado at Boulder, and the WDC-A for Paleoclimatology, National Geophysical Data Center, Boulder, Colorado), *The Greenland Summit Ice Cores CD-ROM*, http://www.ngdc.noaa.gov/paleo/icecore/greenland/summit/, 1997.

Press, W.H., B.P. Flannery, S.A. Teukolsky, and W.T. Vetterling, *Numerical Recipes: The Art of Scientific Computing*, Cambridge University Press, Cambridge, UK, 1989.

Rahmstorf, S., Bifurcations of the Atlantic thermohaline circulation in response to changes in the hydrological cycle, *Nature, 378*, 145-149, 1995.

Sarnthein, M., K. Winn, S.J.A. Jung, J.C. Duplessy, L. Labeyrie, H. Erlenkeuser, and G. Ganssen, Changes in east Atlantic deepwater circulation over the last 30,000 years: Eight time slice reconstructions, *Paleoceanography, 9*, 209-267, 1994.

Severinghaus, J.P., T. Sowers, E.J. Brook, R.B. Alley, and M.L. Bender, Timing of abrupt climate change at the end of the Younger Dryas interval from thermally fractionated gases in polar ice, *Nature, 391*, 141-146, 1998.

Stocker, T.F., D.G. Wright, and W.S. Broecker, The influence of high-latitude surface forcing on the global thermohaline circulation, *Paleoceanography, 7*, 529-541, 1992.

Taylor, K.C., P.A. Mayewski, R.B. Alley, E.J. Brook, A.J. Gow, P.M. Grootes, D.A. Meese, E.S. Saltzman, J.P. Severinghaus, M.S. Twickler, J.W.C. White, S. Whitlow, and G.A. Zielinski, The Holocene/Younger Dryas transition recorded at Summit, Greenland, *Science, 278*, 825-827, 1997.

White, J.W.C., L.K. Barlow, D. Fisher, P. Grootes, J. Jouzel, S.J. Johnsen, M. Stuiver, and H.J. Clausen, The climatic signal in the stable isotopes of snow from Summit, Greenland: results of comparisons with modern climate observations, *J. Geophys. Res., 102*, 26425-26439, 1997.

Wunsch, C., On sharp spectral lines in the climate record and the millennial peak, *Paleoceanography, 15*, 417-424, 2000.

Yiou, P., K. Fuhrer, L.D. Meeker, J. Jouzel, S. Johnsen, and P.A. Mayewski, Paleoclimatic variability inferred from the spectral analysis of Greenland and Antarctic ice-core data, *J. Geophys. Res., 102*, 26441-26454, 1997.

R.B. Alley, Environment Institute and Department of Geosciences, Pennsylvania State University, University Park, PA 16802, USA, ralley@essc.psu.edu

S. Anandakrishnan, Department of Geology, University of Alabama, Tuscaloosa, AL 35487, USA

P. Jung, Department of Physics and Astronomy, Ohio University, Athens, OH 45701, USA

A. Clough, School of Earth Sciences, Univesrity of Leeds, Leeds LS2 9JD, UK

Late Holocene (~ 5 cal ka) Trends and Century-scale Variability of N. Iceland Marine Records: Measures of Surface Hydrography, Productivity, and Land/Ocean Interactions

John T. Andrews[1], Gréta B. Kristjánsdóttir[1,2], Áslaug Geirsdóttir[2], Jórunn Hardardóttir[1,2], Gudrún Helgadóttir[3], Árny E. Sveinbjörnsdóttir[4], Anne E. Jennings[1], and L. Micaela Smith[1]

We present data from five piston cores (B997-319, -321, -327, -328, and -330) collected along two of the major troughs on the N. Iceland shelf. These troughs have significantly different hydrographies associated with the influence of the North Iceland Irminger Current (Atlantic Water) and the East Iceland Current (Polar Water). Sediment accumulation rates are between 10 and 20 yrs/cm. Our paper focuses on the last 5000 yrs of record and examines mass accumulation rates (MAR g cm^{-2} 100 yr) for carbonate, total organic carbon, and mass magnetic susceptibility. Carbonate fluxes declined in all cores, furthermore, there is a striking difference between the accumulation in Eyjafjardaráll (averaging ca 0.18 g$_{carbonate}$ cm^{-2}. 100 yr) and Reykjafjardaráll / Húnaflóaáll (averaging 0.58 g$_{carbonate}$ cm^{-2}. 100 yr). The flux of total organic carbon showed a low but persistent increase in the two Eyjafjardaráll cores and lower accumulation rates than the Reykjafjardaráll/Húnaflóaáll sites. Carbonate maximum values are dated from ca 3.8 and 2 cal ka. Correlations between the records are moderately high with r values of between 0.6 and 0.8 with minimum "tweaking". Mass magnetic susceptibility variations proved similar between the fjord and inner shelf sites in the Húnaflói area and differed from the mid-shelf sites. Possible recurring periodicities occurred at around 530 and 360 yrs. These data from N. Iceland reflect an overall reduction in productivity reflecting the summer decrease in insolation at these latitudes, plus shorter-lived events which indicate multicentury variability in hydrographic conditions.

[1]INSTAAR and Department of Geological Sciences, University of Colorado, Boulder, CO

[2]Department of Geosciences, University of Iceland, Reykjavík, Iceland

[3]Marine Research Institute, Reykjavík, Iceland

[4]Science Institute, University of Iceland, Reykjavík, Iceland

The Oceans and Rapid Climate Change: Past, Present, and Future
Geophysical Monograph 126
Copyright 2001 by the American Geophysical Union

INTRODUCTION

Historically, the waters off N. Iceland have recorded significant decadal-scale changes in hydrography and sea-ice extent [e.g. *Bergthorsson*, 1969; *Lamb,* 1979; *Ogilvie,* 1991; *Ogilvie et al.*, 2000], which in turn has affected the adjacent land areas [*Andrews et al., 2001*, in press b; *Caseldine*, 1987; *Eiriksson et al.*, 2000; *Haflidason*, 1983; *Stotter et al.*, 1999]. In this paper we extend the analysis to include a multicentury analysis of the last 5,000 years of record. Our contribution thus extends the recent interest in high-resolution paleoceanographic studies from areas of

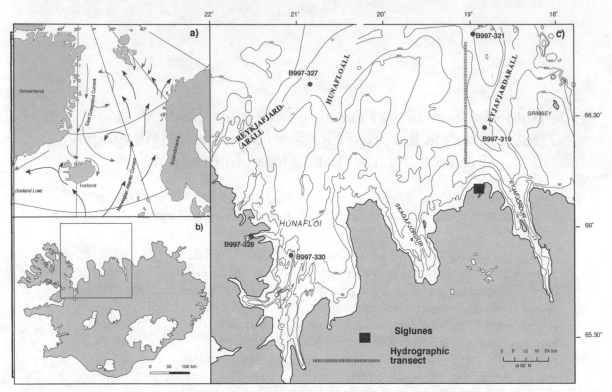

Figure 1. A) Location map and surface currents around Iceland. B) View of Iceland showing the location (square) of our research area (see Figure 1B). C) Bathymetry of N. Iceland and the location of the five B997 cores [*Helgadottir*, 1997]. The location of the Siglunes hydrographic transect of the Marine Research Institute is also shown (see www.hafro.is).

the North Atlantic [cf. *Keigwin*, 1996; *Jennings and Weiner*, 1996].

Our proxies have been measured on a series of cores collected in 1997 (Figure 1) as part of a joint Icelandic/USA marine geology and paleoceanography cruise [*Helgadottir*, 1997] using the RV *Bjarni Sæmundsson* equipped with a 3.5 KHz sub-bottom sounder, a piston core rigged with up to 6 m of core barrel, a 3 m lightweight (10 cm diameter) gravity corer, and small gravity corer designed to sample the water/sediment interface. Grab samples were also taken, and surface samples carefully obtained.

On board the ship we measured the whole-core magnetic susceptibility with a Bartington meter with a sample spacing of 3 cm. The cores were shipped to the University of Colorado where they were stored at 5°C. The cores were split, visually described, photographed, X-radiographed, and sampled for different properties at intervals of between 2 and 5 cm. Samples for AMS ^{14}C dating were primarily selected on the basis of the occurrence of bivalves, visible on X-radiographs [*Andrews et al.*, 2001; *Heier-Nielsen et al.*, 1996]. Most of the dates have been reported earlier in detail [*Smith and Licht*, 2000]. In this paper we used the

CALIB 4.01 program [*Stuiver et al.*, 1986, 1988; *Stuiver and Reimer*, 1986] taking the R as established in the program and allowing a ± 50 yr standard deviation on this estimate. Because our paper focuses on the last 5000 yrs the calibration procedure is more simple than dealing with later intervals which involve both changes in the ocean reservoir correction and major changes in the production of ^{14}C [*Hughen et al.*, 1998; *Voelker and al.*, 1998]. However, we recognize that R can vary temporally and spatially because of the differences in age of Polar versus Atlantic water [*Haflidason et al.*, 2000].

We present data from five cores (Table 1) and concentrate on three proxies for changes in this shelf environment. The Holocene sediments off Iceland [*Andrews et al.*, in press b] are primarily massive, bioturbated muds. In the last 5 cal ka no major lithofacies changes have occurred in the five cores. The proxies are: 1) carbonate weight % or the mass accumulation rate (MAR_{carb}) which we believe is primarily a function of surface [cf.*Giraudeau et al.*, 2000] and benthic productivity, but which could be a measure of dilution by the input of terrestrial sediments ; 2) Total organic carbon (TOC%), which is a more complex function of sur-

Table 1. Information on the five cores discussed in this paper (see Figure 1B for locations).

Core ID	Latitude and Longitude	Water depth (m)	Core length (cm)
Eyjafjardaráll			
B997-319	66°26.53'N & 18°50.82'W	422	227
B997-321	66°53.48'N & 18°58.85'W	484	300
Reykjafjardaráll/Húnaflóaáll			
B997-327	66°38.48' & 20°51.78'W	373	329
B997-328	65°57.41'& 21°32.9'W	96	422
B997-330	65°52.00' & 21° 05.28'W	170	540

face productivity, but also influenced by the possibility of transfers from the land, and degradation of OC during burial [*Jennings et al.*, 2001; *Syvitski et al.*, 1990]; and 3) rock magnetic properties, mass or volume magnetic susceptibility, which is influenced by carbonate and TOC%'s, but also reflects changes in land/sea transport of sediment [*Andrews et al.*, in press a & b.; *Stoner and Andrews*, 1999].

PRESENT OCEANOGRAPHY AND BATHYMETRY OF THE NORTH ICELAND SHELF

The N. Iceland shelf west of 18° W is crossed by two major troughs. To the east, Eyjafjardaráll has maximum depths of > 600 m and is separated from the outer shelf and slope by a sill rising to ca. 420 m (Figure 1C). This trough has distintinctive structural control [*Thors*,1982 ; *Thors and Boulton*, 1991] and late Quaternary faulting is evident in our seismic data [*Kristjansdottir*, 1999]. Húnaflóaáll is shallower and is not silled (Figure 1C). A main branch of the trough enters from the west and is called Reykjafjardaráll. The location and bathymetry influence the hydrography of each trough. The surface flow consists of the North Iceland Irminger Current (NIIC) which brings warm, salty Atlantic Water around the NW Peninsula of Iceland (Figure 1A) [*Hopkins*, 1991; *Malmberg*, 1969; *Stefansson*, 1962]. Offshore, but in years replacing the NIIC, is the colder and fresher water of the East Iceland Current [*Lamb*, 1979; *Olafsson*, 1999; *Thordardottir*, 1977]. CTD casts in 1997 as part of cruise B997 show some differences in the hydrography which can be related to both location (east and west) and bathymetry of the troughs. Figure 2 shows the contrast in water temperatures along the two troughs as of

late July 1997. In Húnaflóaáll a tongue of warm water (~ 6° C) indicates the presence of Atlantic Water at a shallow depth virtually along the length of the trough (Figure 1C & 2), whereas in Eyjafjardaráll this water mass is restricted to the inner shelf. The cold water in Eyjafjardaráll (< 0° C) may represent winter-cooled water which is difficult to exchange because of the seaward sill. Surface salinities in early summer are quite low (not shown) because of rain and run-off from the adjacent snow covered land area. Over the last 50 years, the mean salinity for the Siglunes section (Figure 1C) in June is 34.727‰ [*Olafsson*, 1999].

Reviews of the oceanography of the North Atlantic, and N. Iceland waters in particular, are discussed in depth by several authors [*Belkin et al.*, 1998; *Malmberg*, 1985]. The Marine Research Institute (Iceland) undertakes quarterly hydrographic surveys along standard transects around the Iceland margin (see www.hafro.is). The area included within our purview includes the Siglunes section (Figure 1B & 1C). Olafsson [1999, p. 45] summarized the Siglunes section data for the period 1952-1998AD; these temperature and salinity data represent numerically integrated mean values for June. The series indicate maximum shifts in mean June T°C of 4.8° and 0.7‰ in salinity. There is a strong positive correlation between T°C and S ($r^2 = 0.5$). An important observation is that during winter months sea-ice can develop *in situ* on the N. Iceland shelf when surface salinities fall to 34.7‰ (*Hopkins*, 1991).

The productivity of the N. Iceland surface waters are strongly affected by the hydrography [*Thordardottir*, 1984, 1986]. During years in which the NIIC dominates there is little vertical density stratification and mixing brings nutrients into the photic zone. However, Thordardottir [1977] showed that when Polar Water lies offshore the highly stratified water column results in nutrient depletion and productivity decreased by a factor of 2, and can be as great as a factor of 10.

Because of the East Iceland Current and changes in surface salinity, the N. Iceland shelf can be heavily impacted by sea-ice due to its polar latitude. Indeed, the NW of Iceland, including Húnaflóaáll, is the most severely impacted area of Iceland in terms of sea-ice extent and coverage [*Ogilvie*, 1991; *Sigurdsson*, 1969; *Stefansson*, 1969]. Stotter et al [1999], amongst others, have noted the strong influence of changes in sea-ice extent on the climate of the adjacent coastal areas and on local glacier mass balance (negative correlation between sea-ice extent and weather station data of $r^2 = -0.4$ [*Stotter et al.*, 1999 p. 464]. Since 1850AD the sea-ice index has varied from between 5-10 in the period 1925-1960AD to maximum values of ca. 80 in the 1860's, 1890's, and during the Great Salinity Anomaly of the 1960's [*Stotter et al.*, 1999].

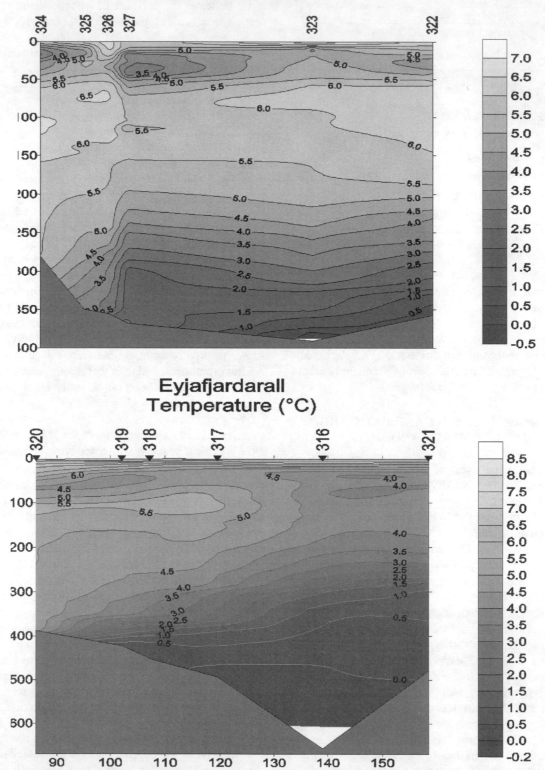

Figure 2. Cross sections showing the water column characteristics along the two troughs based on the CTD casts during the late July, 1997, B997 cruise of *Bjarni Sæmundsson* [*Helgadottir,* 1997]

METHODS

Our major proxies for changes in the environment of N. Iceland waters are carbonate weight % (5 cm intervals), TOC%, and mass magnetic susceptibility (x 10^{-7} m^3/kg). Carbonate weight % was determined by either a Coulometer (USC Model 5012--- with a precision on replicate samples of 0.04 ± 0.22% (n = 34)), or we used an automated WHOI Carbonate device (at a 2 cm interval on some cores). The agreement on replicate runs between these two systems gave an r^2 = 0.97 with a slope of 1.02 ± 0.025. TOC weight % is based on the difference between the USC model 5012 Coulometer determinations of Total Carbon (TC) and Total Inorganic Carbon (TIC)---this is changed to carbonate weight % by multiplying by 8.3333. Calibrations between this procedure and the Walkley-Black titration method for direct determination of TOC results [*Walkley*, 1947] gave strong positive correlations [*Andrews*, 1987].

Discrete mass magnetic susceptibility (mass MS) was measured on all samples at the 5 cm sampling interval using a 10 cc holder and calibrating against a manufactures standard [*Thompson and Oldfield*, 1986; *Walden et al.*, 1999]. Cores 321, 317, and 330 were also processed in U-channels (2 x 2 x 150 cm) through the UC-Davis NSF Cryogenic Magnetometer Facility [*Verosub*, 1999], but the results are not reported here.

Our sediment sampling procedure [cf. *Andrews et al.*, in press b; *Kristjansdottir*, 1999] consists of taking a 10.5 cc sediment sample every 5 cm from which we initially determine wet and dry sediment density, moisture weight %, and then carbonate, TOC, and mass MS. Information on down core variations in dry sediment density are necessary in order to determine mass accumulation rates (MAR).

RESULTS

Elsewhere we have commented on the carbonate records from 328 [*Andrews et al.*, in press a] and 327 and 330 [*Andrews et al.*, in press a], all in Húnaflóaáll, and argued that a) the records are well correlated in terms of their major peaks and troughs, and b) that the variations primarily represent changes in surface productivity, and most likely the production of coccoliths [*Andrews et al.*, in press a; *Giraudeau et al.*, 2000]. Here we add to our previous results by including sites to the east, from the deep Eyjafjardaráll trough [*Kristjansdottir*, 1999] (Figure 1C). Moreover, we present the data as MAR$_{carb}$ in units of g/cm^2.100 yrs plotted against calibrated yrs.

Sediment Accumulation and Mass Accumulation Rates

Radiocarbon dates from our cores [*Andrews et al.*, 2000; *Smith and Licht*, 2000] indicate that over the last few thou-

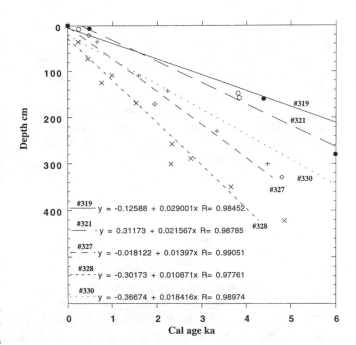

Figure 3. Depth versus calibrated age plots for the five cores. These data were derived from the CALIB program [*Stuiver et al.*, 1998] assuming a R of 0 ± 50 yrs.

sand years the sediment accumulation rate (SAR) is approximately linear with depth (Figure 3) [Andrews et al., in press a]. We use least-squares regression to determine the "best fit" line and use this to develop a time-series for each core. The sense of the errors involved in this process [*Andrews et al.*, 1999] was estimated using a Monte Carlo simulation of the combined counting and calibration errors on the dates. No statistically significant improvement was obtained by using a second-order polynomial fit.

Temporal resolution for these cores at a 5 cm sampling interval varies between 50 and 100 years; given the length of our time-series this indicates that we should be able to resolve frequencies of between 0.00066 and 0.01 cycles/yr, or periodicties of between 100 and 1500 yrs. We use the program "AnalySeries" [*Paillard et al.*, 1996] to quantify the level of agreement between our records [*Crowley*, 1999]. We used the program "Stata^{TM}" for regular statistical analyses of regression and confidence levels [*Stata*, 1999; *Davis*, 1986; *Hamilton*, 1990].

Plotting age/depth relationships indicate that SARs, expressed as the number of years required to accumulate 1 cm of sediment, ranged between ~10 (#328) to 20 yrs/cm (#321). Thus the graph of calibrated age (ka) versus depth (Figure 3) indicates that sedimentation rates differ by a factor of 2 between the more inshore sites and those farther offshore (Figure 1B). However, dry sediment density (dvd) invariably increases with depth, although not monotonically, hence the total MAR (g/cm^2.unit time) tends to decrease toward the surface for a constant SAR. Our measurements indicate that on average density increases with depth by 0.00067 g/cm, thus the total dvd range over our 3-5 m cores is from ca. 0.4 to 0.8 g/cc. We calculated the MAR for carbonate and TOC by:

$$(p_S)(p_i)/(SAR)$$

where (p_S) is the sediment dry weight per unit volume, p_i is the weight fraction, and SAR is the rate of sediment accumulation expressed as yrs/cm. With linear SAR (Figure 3) these results indicate that at any one site, MAR decreases by approximately a factor of 2 over the last 5 cal ky. This has important implications in a simple analysis of weight % data. MAR units were finally calculated as g/cm^2. 100 yr.

We are dealing with net accumulation. Both the carbonate and organic carbon can be subject to removeal by various processes. However, we saw no specific evidence for carbonate dissolution during picking for foraminifera (although not all cores have been processed for this). Although the sediments show evidence for bioturbation on X-radiographs, we do not think that this changes either the trends or individual events.

Mean Values and Variability

Figure 4 shows boxplots [*Velleman and Hoaglin*, 1981] of the distribution of MAR$_{carb}$, MAR$_{TOC}$, and mass MS for the five sites. This plot indicates that substantial differences exist between the two troughs in terms of MAR$_{carb}$ and MAR$_{TOC}$, both in terms of the median value and in the variability (as shown by the width of box plots). The high TOC estimates for the fjord site, #328, may represent an accumulation of TOC from land erosion after the Settlement of Iceland [cf. *Jennings et al.*, 2001; *Andrews et al.*, in press a]. Mass MS differences between the troughs is not as great and #330 stands out because of its low values.

Trends

The records for MAR$_{carb}$ (Figure 5A) show century-scale variability superimposed on a persistent decrease in over the last 5 cal ka. The MAR$_{carb}$ from the five cores all show a progressive decrease in net accumulation over this period. However, there is a distinct difference between the two troughs, with MAR$_{carb}$ varying by a factor of between 2-4,

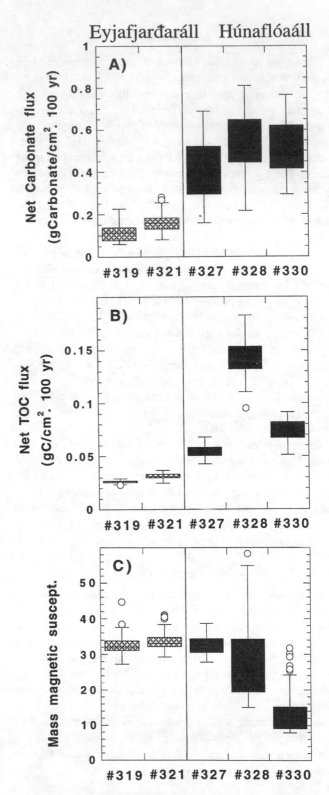

Figure 4. Box plots of MAR for carbonate and total organic carbon, and mass magnetic susceptibility for the five sites over the last ca. 5 cal ka. The box includes 50% of a sample's distribution and the line within the box is the median value. Small circles represent outliers [*Velleman and Hoaglin*, 1981].

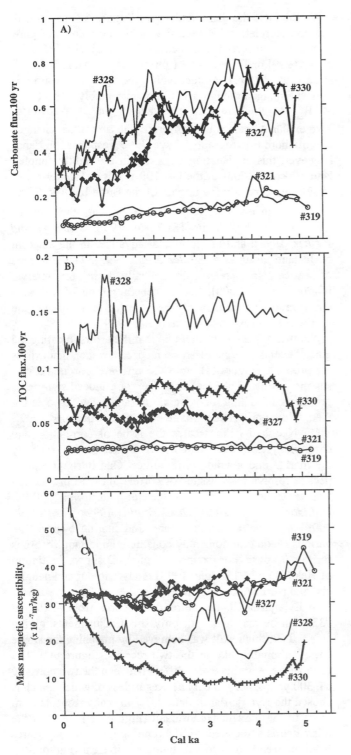

Figure 5. A) Plot of the changes in the mass accumulation rate of carbonate in the fives sites from N. Iceland (Figure 1B). B) Plot of changes in the mass accumulation of TOC. C) Changes in mass magnetic susceptibility at the five sites.

with the net accumulation in Eyjafjardaráll (#319 and #321) being substantially lower (Figure 4). Our estimates of the numbers of foraminifera/g sediment, combined with estimates of average foraminifera weights, indicated that the foraminifera contribute < 10% to MAR$_{carb}$ [*Andrews et al.*, in press a]. Thus we associate the bulk of the carbonate variability with changes in coccolith production during the extensive spring blooms [cf. *Clark and Maynard*, 1986; *Thordardottir*, 1977].

The average rate of decrease per 100-yr, over the last 5 cal ka, varied from 0.002 MAR$_{carb}$ for #321 and #319, to 0.0066 MAR$_{carb}$ for #330 and #328. The largest decrease was seen at #327 which amounted on average to 0.012 (g$_{carbonate}$ cm^{-2}.100 yr). The overall carbonate trend for the last 5 cal ka is similar to that reported form the Gardar Drift, south of Iceland [*Giraudeau et al.*, 2000], where the coccoliths are the prime contributor to the carbonate budget.

MAR$_{TOC}$ on-the-other-hand (Figure 5B) shows a much wider range between the five sites, plus there are differences in the sign of the trends. The Eyjafjardaráll sediments indicate a slight tendency for MAR$_{TOC}$ to increase toward the present. These two sites have relatively low overall net mass accumulations. The two western shelf sites (#327 and #330, Figure 5B) show intermediate MAR$_{TOC}$ values and a tendency for MAR$_{TOC}$ to decrease toward the present. The MAR$_{TOC}$ at the fjord site, #328, is much higher than the four shelf sites. There is a noticeable trend for a decrease in net carbon accumulation over the last 1 cal ka; this appears to exert a major control on an overall tendency for MAR$_{TOC}$ to decrease to the present day. It is important to note that if TOC% is plotted against time then all sites show TOC% increasing toward the present. This is a very different picture than that presented by the MAR$_{TOC}$ data (Figure 5B) and indicates the importance of estimating the mass accumulation rates.

Magnetic mass susceptibility (mass MS, 10^{-7}m^3/kg) is obtained by dividing the dry weight of a unit volume into the volume magnetic susceptibility [*Thompson and Old-field*, 1986]. Because the values are mass corrected they serve a parallel purpose to other MAR units, in this case they are a measure of the magnetite concentration in the sediment. However, the concentration is diluted by diamagnetic materials such as carbon and carbonate, thus changes in mass MS can be largely driven by changes in the net accumulation of organic carbon and carbonate [*Andrews et al.*, in press b; *Stoner and Andrews*, 1999]. The sediments in the two troughs have high mass MS values of between 8 and 60 x 10^{-7}m^3/kg (Figure 4). Sites #328 and #330 have similar trends with a pronounced increase in mass MS over the last 1 cal ka (Figure 5C). These are the two sites closest to the Húnaflói coast and thus more prone to receiving terrestrial sediments. However, this influence may be driven by the decrease in MAR_{carb}. In contrast, #327, #319, and #321 have very similar values (~30 to 40 $x10^{-7}m^3/kg$) and a slight trend toward lower mass MS (magnetite concentrations) over this interval.

Our data from #330 and other cores which extend from 5-10 ka (not shown), indicate that the Holocene carbonate maximum is achieved in the mid-Holocene and is represented in our data by the double MAR_{carb} peaks at ca. 2 and 3.8 cal ka (Figure 5A). Eiriksson et al. [2000] also noted a deterioration in conditions on the N. Iceland shelf during the last few thousand years. Over the last 5 cal ka the July insolation at 65°N has decreased by 23 W m^{-2} [Berger and Loutre, 1991]. This decrease is evident in the ^{18}O records from the Greenland and Renland ice caps [e.g. Johnsen et al., 1992]. An environmental threshold may have been passed ca 5 cal ka as Neoglacial activity is first dated at about that time in N. Iceland [Gudmundsson, 1997; Stotter et al., 1999]. On the East Greenland shelf, rafting of IRD materials suddenly intensifies and becomes a dominant transport mechanism in the last 5-6 cal ka [Andrews et al., 1997].

A reviewer pointed to one element of the discussion which we had ignored, and that was a strong negative relationship between water depth and the masss accumulation rates of carbonate and carbon respectively (r^2 of 0.82 and 0.9). We believe that this represents an "accident" of geography because the shallower Reykjafjardaráll/Húnaflóaáll trough is dominated by Atlantic Water (Figure 2) when compared to the deep and cold Eyjafjardaráll trough.

Correlations and Periodicities

To proceed further we evaluate the correlation between detrended time-series using "AnalySeries" [Paillard et al., 1996] to derive a quantitative measure of agreement [Crowley, 1999]. In an earlier study [Andrews et al., in press a] we showed substantial correlations in the carbonate % data between the three Húnaflóaáll records with r values of between 0.7 and 0.9.

We used this program to correlate significant features of the MAR_{carb} records between any two cores and to adjust the time-scales of the cores. The carbonate records have many points of similarity; especially noteworthy are the two carbonate peaks with estimated ages of ~ 2 and 3.5 cal ka (Figure 6A). In all, 13 or so discrete "events" can be recognized and correlated between these three cores. Fitting the #330 data to the #328 time scale increased the correlation from r=0.54 to r=0.78 (Figure 6A) and results in changes in the #330 depth/age scale which can be easily accommodated within the radiocarbon errors.

We then use the detrended series (Figure 6B) to investigate the correlation between the carbonate records (Figure 6A) and dated Neoglacial records from N. Iceland [Stotter et al., 1999] or IRD records from the North Atlantic [Bond et al., 1997, 1999]. At face value the agreements are not overwhelming although we might reasonably expect low MAR_{carb} to coincide with intervals of glacial re-advance. Two explanations can be proffered. The first is that changes the offshore hydrography have little response on land. However, this explanation does not fit the tightly coupled land/sea connections of the last 100-yrs or so [Stotter et al., 1999], and certainly the impact of the Little Ice Age (LIA) is evident in our data (Figure 6A), and in other records from the North Atlantic [Keigwin, 1996; Jennings and Weiner, 1996], and in the glacial advances in the adjacent N. Iceland mountains [Stotter et al., 1999]. The incidence of sea-ice also increased dramatically during this interval [Ogilvie et al., 2000], hence iceberg rafting of sediment (IRD) might also reasonably be expected to increase in the northern North Atlantic [Bond et al., 1999].

Another explanation might be that the errors in dating the glacial events and the offshore records are such that offsets are probable (Figure 6B, note the agreement around 3 versus the mismatch at 2 cal ka). There are indeed significant problems in dating neoglacial moraines and associating them with appropriate climatic conditions [Andrews, 1999].

On Neoglacial time-scales the issue of correlation and chronology is critical, and needs further investigation by the marine and glacial communities. One intriguing prospect to resolve these issues of correlation is to use the extensive tephra record from Iceland [e.g. Boygle, 1999; Haflidason et al., 2000; Hardardottir, 1999; Wastl et al., 1999] to directly correlate land and marine events. The various Hekla eruptions may constitute important isochrons for this purpose. For example, Figure 6B shows the timing of Hekla 4 and 3 [Boygle, 1999] relative to our and neoglacial and North Atlantic IRD events.

In Eyjafjardaráll, the detrended records from #319 and #321 can be matched with only small adjustments, giving an r = 0.69; this result was achieved by matching the single large carbonate spike in the two records (Figure 6C). The estimated age of this event differs between the two cores by ca. 300 yrs (4.3 or 4.0 cal ka). Regardless of which match is chosen the Eyjafjardaráll data differ in many respects from those to the west in Húnaflóaáll (Figure 6D). Part of the mismatch is associated with chronology, but the departures from the trend are also substantially different (Figure 5A). Part of the offset in chronology might be explained by a difference in the ocean reservoir correction between these two troughs. Haflidason et al. [2000] indicated a 750 yr correction in Eyjafjardaráll at the time of the Saksunarvatn

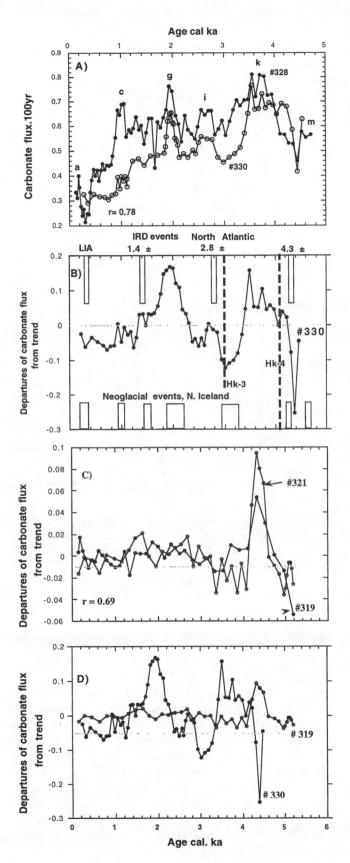

tephra (10.18 ± cal ka), but in Húnaflóaáll this correction is only 400 yrs [unpublished data]. Nevertheless, the Eyjafjardaráll sites do not have the second main peak at ca 2 cal ka and the magnitude of the departures is much reduced. One interpretation of the difference is that Eyjafjardaráll has been much more influenced by the cold water of the East Iceland Current than Húnaflóaáll, where the North Iceland Irminger Current is more dominant, but is also subject to much greater hydrographic variability [Olafsson, 1999] (Figure 2).

Several authors have commented on the presence in North Atlantic sediments of recurrent periodicities in the range of 500 to 1500 yrs [Bianchi and McCave, 1999; Bond et al., 1999; Chapman and Shackelton, 2000]. We use the AnalySeries program [Paillard et al., 1996] to derive a first estimate of significant periodicites in each of the detrended records. Because the SARs are unchanging over the last 5 cal ka our samples are approximately evenly spaced, which can be a concern in spectral analysis [Schulz and Stattegger, 1997]. The data were processed using the Maximum Entropy "compromise solution" in terms of resolution and confidence [Pestiaux et al., 1988]. Because our series are short (n = 100) we developed an empirical basis for the acceptance or rejection of a "significant frequency". The following strategies were adopted: 1) the five input series were all detrended, normalized (mean of 0, standard deviation of ±1.0) and prewhitened); 2) these were evaluated against randomly generated time series of length 100, mean of 0, standard deviation of ±1.0, and prewhitened. An average power spectrum for the 10 random series was computed and we calculated the 90% and 95% upper confidence levels using the appropriate value for t [Hamilton, 1990]. We also introduce another restriction in that for a particular frequency to be judged of interest then it must be noted in more than one of the five time series. We concen-

Figure 6. Correlation estimates of the agreement between the carbonate records from the cores. A) is the agreement between #330 adjusted to the #328 time-scale and giving a measure of correlation of 0.78. B) The oscillations in the MAR_carb data from #330 are plotted against the IRD cycles in the North Atlantic which have a quasi periodicity of ca 1.4 ± ky [Bond et al., 1997], and against the estimated dates on neoglacial advances on N. Iceland [Stotter et al., 1999]. The black dashled lines labelled Hk-3 and Hk-4 represent the timing of Hekla eruptions [Boygle, 1999], C) The fit between #319 and #321 using the #328 time scale. The level of association is r = 0.69 with the peak ca. 4.3 cal ka used to force the correlation. Note that the scale extends to 6 cal ka. D) The difference between the unadjusted detrended MAR_carb records from Húnaflóaáll (#330) and Eyjafjardaráll. Peak k (Figure 6A) in Húnaflóaáll is offset from the major carbonate event in Eyjafjardaráll by ~400 ± yrs.

trate our analyses on the MAR$_{carb}$ records because the mass MS records have a significant imprint of the MAR$_{carb}$ variations [*Stoner and Andrews*, 1999] and the TOC measurements are derived from the difference between TC and TIC measurements, hence are more subject to error. The results for the five cores, and the confidence level, are shown on Figure 7. The most striking aspect of the figure is the large amount of power in all five data sets, which resides in frequencies of < 0.001 cycles/yr; the differences between the MAR$_{carb}$ and the randomly generated series is striking. Remember that these are detrended (linear) time-series, however, it is difficult to attach great confidence to the low frequency spectra because they are only samped two to three times in records of these lengths (5-6 cal ka). Other recurring peridocities include ca. 530 yrs (4 cores), 360 years (4 cores), and 260 years (2 cores), but not all the individual peaks cross the upper 90% limit (Figure 7A & B). We view Figure 7 as an interesting hypothesis that requires verification from higher resolution data and a more rigorous test of assumptions. We are now sampling at 2 cm resolution, and will test an alternative approach to the assumption of stationary time-series which is to use wavelet analysis [*Torrance and Compo*, 1998], and ascertain if there is any evolution in frequencies over the last 5000 yrs.

CONCLUSIONS

The records from five cores from N. Iceland show the two adjacent troughs have strikingly different rates of MAR$_{carb}$ and MAR$_{TOC}$; these differences reflect the difference in oceanography and hydrography between the two cores with the cooling and sinking of warmer Atlantic Water providing higher levels of nutrients in Húnaflóaáll than in the colder waters of Eyjafjardaráll (Figure 2). Over the last 5000 years net mass accumulations of both parameters has decreased; this may be attributable to the decrease in summer insolation over this interval at the critical latitude of 65° N [*Berger and Loutre*, 1991], which is indeed close to the latitude of our sites (Figure 1).

Superimposed on the decrease in MAR are a series of well-defined excursions, which are especially noticeable in the Húnaflóaáll MAR$_{carb}$ data, specifically the two major peaks at ca 2000 ± and 3800 ± cal yrs BP (Figure 6B). The decrease in net accumulation of MAR$_{carb}$ is most striking after 2000 cal yrs BP. The lowest MAR$_{carb}$ for the last 5000 cal yrs BP occurred in the last few hundred years. In the best dated core, #328 [*Andrews et al.,* in press a], this trough firmly dates from the Little Ice Age. An important conclusion from our comparison of these two adjacent troughs (Figure 1B) is that, at least in terms of two proxies, how different the records look (Figures 5&6). This indi-

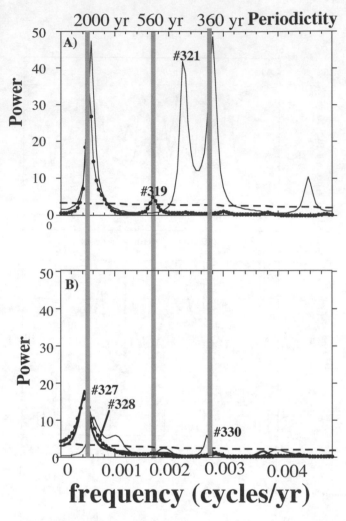

Figure 7. Compromise Maximum Entropy spectra from the five sites compared against the upper 90% probability limit from a series of ten randomly generated time-series of length 100, with means of 0 ± 1. A) shows data from the two sites in Eyjafjardaráll, and B) shows the sites in Húnaflóaáll. The gray vertical lines identify the main periodicties.

cates that in detail (i.e. 50-yr resolution) the two troughs have experienced rather different hydrographies over the last 5 cal ka. The decrease in the net mass accumulation of carbonate and TOC in Eyjafjardaráll is in agreement with the suggestion in Eiriksson et al. [2000, p. 38] which shows a decrease in the relative strength of the North Iceland Irminger Current during this interval. Based on the carbonate data we would hypothesize that conditions may have peaked ca. 4-5 cal ka (Figure 6C). Their [*Eiriksson et.al.,* 2000] cores were also taken in Eyjafjardaráll; they indicate that ice-rafting of sediments has occurred nearly continuously over the last 5-6 cal ka.

Exploratory analysis of the MAR$_{carb}$ time-series of the five cores indicates significant power at millennial and century periodicities, even though the series were detrended, and suggests that periods of around 2000, 530 and 360 yrs might be captured in these data. The lower frequency peaks may, or may not, represent the same forcing as the 1600-1400 year beat reported from other North Atlantic sites using other proxies for paleoceanography [*Bianchi and McCave*, 1999; *Bond et al.*, 1997, 1999], and other multi-century periodicities have been reported similar to ours (Figure 7) [*Chapman and Shackelton*, 2000].

We emphasize that our explicit hypothesize is that the net production of carbonate and organic carbon is driven by changes in temperature and salinity. However, we are cognizant that net productivity might be forced by other processes such as changes in upwelling, or the location of fronts. We are testing the causes of changes in our proxies (Figure 5) by an analysis of the isotopic composition of benthic and planktonic foraminifera, and by developing census data on the composition of the faunas [cf. *Andrews et al.*, in press a; *Kristjansdottir*, 1999]. Changes in the concentration of ice, sea ice or icebergs, is being evaluated by studies of changes in the composition of the sand-size fraction.

Acknowledgements. We appreciate the support of the Marine Research Institute, Iceland, in providing ship time for cruise B997. Sedimentological data analyses were supported by grants from the National Science Foundation (ATM-9531397, OPP-9726510, OCE–98-09001). Radiocarbon dates were obtained from the above grants (University of Arizona) and some were obtained through the Earth System History (ESH) radiocarbon dating initiative involving INSTAAR and WHOI. Other dates were provided through the Science Institute, University of Iceland, through a cooperative agreement with the Aarhus AMS Facility. Kristjánsdóttir's and Geirsdóttir's research was supported by funds from Iceland Science Foundation. We greatly appreciate the positive comments of a reviewer and D. Seidov. PARCS Contribution # 163

REFERENCES

Andrews, J. T., *Downcore variations in carbon content of fjord piston cores and association with sedimentation rates*, Sedimentology of Arctic Fjords Experiment, compilers J.P.M. Syvitski and D. B. Praeg, 1987.

Andrews, J. T., Dating Glacial Events and Correlation to Global Climate Change, in *Quaternary Geochronology. Methods and Applications*, edited by J. S. Moller, J. Sowers and W. R. Lettis, pp. 447-455, American Geophysical Union, Washington, DC, 1999.

Andrews, J. T., L. M. Smith, R. Preston, T. Cooper, and A. E. Jennings, Spatial and temporal patterns of iceberg rafting (IRD) along the East Greenland margin, ca. 68 N, over the last 14 cal.ka, *Journal of Quaternary Science, 12*, 1-13, 1997.

Andrews, J. T., D. C. Barber, and A. E. Jennings, Errors in generating time-series and dating events at late Quaternary (radiocarbon) time-scales: Examples fromBaffin Bay, Labrador Sea, and East Greenland, in *Mechanisms of Global Climate Change at Millennial Time Scales*, edited by P.U. Clark, a. R.S. Webb and L. D. Keigwin, pp. 23-33, American Geophysical Union, Washington, D.C., 1999.

Andrews, J. T., J. Hardardottir, G. Helgadottir, A. E. Jennings, A. Geirsdottir, A. E. Sveinbjornsdottir, S. Schoolfield, G. B. Kristjansdottir, L. M. Smith, K. Thors, and J. P. M. Syvitski, The N and W Iceland Shelf: Insights into Last Glacial Maximum Ice Extent and Deglaciation based on Acoustic Stratigraphy and Basal Radiocarbon AMS dates, *Quaternary Science Review, 19*, 619-631, 2000.

Andrews, J. T., C. Caseldine, N. J. Weiner, and J. Hatton, Late Quaternary (~4 ka) Marine and Terrestrial Environmental Change in Reykjarfjördur, N. Iceland: Climate and/or Settlement?, *Journal Quaternary Sciences, 16*, 135-144, 2001.

Andrews, J. T., G. Helgadottir, A. Geirsdottir, and A. E. Jennings, Century-scale variability in northern North Atlantic productivity and hydrography: Evidence from the N. Iceland margin , *Quaternary Research*, in press a.

Andrews, J. T., R. Kihl, A. E. Jennings, G. B. Kristjansdottir, G. Helgadottir, and L. M. Smith, Contrast in Holocene Sediment Properties on the Shelves Bordering Denmark Strait (64-68°N), North Atlantic, *Sedimentology* , in press b.

Belkin, I. M., S. Levitus, J. Antonov, and S.-A. Malmberg, "Great Salinity Anomalies" in the North Atlantic, *Progress in Oceanography, 41*, 1-68, 1998.

Berger, A., and M. F. Loutre, Insolation values for the climate of the last 10 million years, *Quat. Sci. Rev., 10*, 297-318, 1991.

Bergthorsson, P., An Estimate of Drift Ice and Temperature in Iceland In 1000 years, *Jokull, 19* (Symposium on Drift Ice and Climate), 94-101, 1969.

Bianchi, G. G., and I. N. McCave, Holocene periodicity in North Atlantic climate and deep-ocean flow south of Iceland, *Nature, 397*, 515-517, 1999.

Bond, G., W. Showers, and e. al., A Pervasive Millennial-Scale Cycle in North Atlantic Holocen and Glacial Climates, *Science, 278*, 1257-1266, 1997.

Bond, G. C., W. Showers, M. Elliot, M. Evans, R. Lotti, I. Hajdas, G. Bonani, and S. Johnson, The North Atlantic's 1-2 kyr Climate Rhythm: Relation to Heinrich Eevnts, Dansgaard/Oescher Cycles and the Little Ice Age, in *Mechanisms of Global Climate Change at Millennial Time Scales*, edited by P. U. Clark, R. S. Webb. and L. D. Keigwin, pp. 35-58, American Geophysical Union, Washington, D.C., 1999.

Boygle, J., Variability of tephra in lake and catchment sediments, Svinavatn, Iceland, *Global and Planetary Change, 21*, 129-149, 1999.

Caseldine, C., Neoglacial glacier variations in northern Iceland: Examples from the Eyjafjorour area, *Arctic and Alpine Research, 19*, 296-304, 1987.

Chapman, M. R., and N. J. Shackelton, Evidence of 550-year and 1000-year cyclicities in North Atlantic circulation patterns during the Holocene, *The Holocene, 10*, 297-301, 2000.

Clark, D. K., and N. G. Maynard, Coastal zone color scanner imagery of phytoplankton pigment distribution in Icelandic waters, in *The International Society of Optical Engineering*, 637, pp. 350-357, SPIE, Orlando, FL, 1986.

Crowley, T. J., Correlating high frequency climate variations, *Paleoceanography, 14* (3), 271-272, 1999.

Davis, J. C., *Statistics and data analysis in Geology*, 646 pp., John Wiley & Sons, New York, 1986.

Eiriksson, J., K. L. Knudsen, H. Haflidason, and P. Henriksen, Late-glacial and Holocene paleoceanography of the North Iceland Shelf, *Journal of Quaternary Science, 15*, 23-42, 2000.

Gudmundsson, H. J., A Review of the Holocene Environmental History of Iceland, *Quat. Sci. Rev., 16*, 81-92, 1997.

Giraudeau, J., M. Cremer, S. Manthe, L. Laberrie, and G. Bond, Coccolith evidence for instabilities in surface circulation south of Iceland during Holocene times, *Earth Planet Sci. Lett., 179*, 257-268, 2000.

Haflidason, H., The Marine Geology of Eyjafjordur, North Iceland: Sedimentology, Petrographical and stratigraphical studies, MPhil thesis, University of Edinbugh, 1983.

Haflidason, H., J. Eiriksson, and S. Van Kreveld, The tephrachronology of Iceland and the North Atlantic region during the Middle and Late Quaternary: a review, *Journal of Quaternary Science, 15*, 3-22, 2000.

Hamilton, L. C., *Modern Data Analysis*, 684 pp. pp., Brooks/Cole Publishing Co., Pacific Grive, CA,, 1990.

Hardardottir, J., Late Weichselian and Holocene Environmental History of South and West Iceland as Interpreted from Studies of Lake and Terrestrial Sediments, Doctor of Philosophy thesis, Univ. of Colorado, 1999.

Heier-Nielsen, S., K. Conradsen, J. Heinemeier, K. L. Knudsen, H. L. Nielsen, N. Rud, and A. E. Sveinbjornsdottir, Radiocarbon dating of shells and foraminifera from the Skagen Core, Denmark: Evidence of reworking, *Radiocarbon, 37* (3), 1996.

Helgadottir, G., *Paleoclimate (0 to >14 ka) of W. and NW Iceland: An Iceland/USA Contribution to P.A.L.E., Cruise Report B9-97*, Marine Research Institute of Iceland , 1997.

Hopkins, T. S., The GIN Sea- A synthesis of its physical oceanography and literature review 1972-1985, *Earth Science Reviews, 30*, 175-318, 1991.

Hughen, K. A., J. T. Overpeck, S. J. Lehman, M. Kashgarian, J. Southon, L. C. Peterson, R. Alley, and D. M. Sigman, Deglacial changes in ocean circulation from an extended radiocarbon calibration, *Nature, 391*, 65-68, 1998.

Jennings, A. E., and N. J. Weiner, Environmental change on eastern Greenland during the last 1300 years: Evidence from Foraminifera and Lithofacies in Nansen Fjord, 68°N, *The Holocene, 6*, 179-191, 1996.

Jennings, A. E., J. Hardardottir, R. Stein, A. E. J. Ogilvie, and I. Jonsdottir, Oceanographic Change and Terrestrial Human Impacts in a post 1400 AD record from the Southwest Iceland Shelf., *Climatic Change, 48*, 83-100, 2001.

Johnsen, S., H. B. Clausen, W. Dansgaard, N. S. Gundestrup, M. Hansson, P. Johnsson, P. Steffensen, and A. Sveinbjornsdottir, E., A "deep" ice core from East Greenland, *Meddelelser om Gronland, Geoscience, 29*, 22 pp., 1992.

Keigwin, L. D., The Little Ice Age and Medieval Warm Period in the Sargasso Sea, *Science, 274* (29 Nov.), 1504-1508, 1996.

Kristjansdottir, G. B., Late Quaternary climatic and environmental changes on the North Iceland shelf, MSc thesis, University of Iceland, 1999.

Lamb, H. H., Climatic variations and changes in the wind and ocean circulation: The Little Ice Age in the Northeast Atlantic, *Quaternary Research, 11*, 1-20, 1979.

Malmberg, S.-A., Hydrographic Changes in the Waters Between Iceland and Jan Mayen in the Last Decade, *Jokull, 19* (Symposium on Drift Ice and Climate), 30-43, 1969.

Malmberg, S.-A., The water masses between Iceland and Greenland, *Journal Marine Research Institute, 9*, 127-140, 1985.

Ogilvie, A. E. J., Climatic change in Iceland A.D. c.865 to 1598, *Acta Archaeologica, 61*, 233-251, 1991.

Ogilvie, A. E., L. K. Barlow, and A. E. Jennings, North Atlantic Climate c. A.D. 1000: Millennial Reflections on the Viking Discoveries of Iceland, Greenland and North America, *Weather, 55* (2), 34-45, 2000.

Olafsson, J., Connections between oceanic conditions off N-Iceland, Lake Myvatn temperature, regional wind direction variability and the North Atlantic Oscillation, *Rit Fiskideildar, 16*, 41-57, 1999.

Paillard, D., L. Labeyrie, and P. Yiou, Macintosh Program Performs Time-Series Analysis, *EOS, 77* (39), 379, 1996.

Schulz, M., and K. Stattegger, Spectrum: spectral analysis of unevenly spaced paleoclimatic time series, *Computers and Geosciences, 23* (9), 929-945, 1997.

Pestiaux, P., I. Van der Mersch, and A. Berger, Paleoclimatic variability at frequencies ranging from 1 cycle per 10,000 years to 1 cycle per 1000 years: evidence for nonlinear behavious of the climate system, *Climatic Change, 12*, 9-37, 1988.

Sigurdsson, F. H., Report on Sea Ice off the Icelandic Coasts October 1967 to September 1968, *Jokull, 19* (Symposium on Drift Ice and Climate), 77-93, 1969.

Smith, L. M., and K. J. Licht, *Radiocarbon Date List IX: Antarctica, Arctic Ocean, and the Northern North Atlantic*, INSTAAR, UNiversity of Colorado , 2000.

Stata, 1999, Stata Reference Manuanl, Stata Press, Release 6, College Station, TX (6 volumes).

Stefansson, U., North Icelandic Waters, *Rit Fiskideildar, III. Bind, Vol 3*, 1962.

Stefansson, U., Temperature Variations in the North Icelandic Coastal Area During Recent Decades, *Jokull* (April 19), 18-28, 1969.

Stoner, J. S., and J. T. Andrews, The North Atlantic as a Quaternary magnetic archive, in *Quaternary Climates, Environments and Magnetism*, edited by B. Maher, and R. Thompson, pp. Cambridge University Press, Cambridge, UK, 1999.

Stotter, J., M. Wastl, C. Caseldine, and T. Haberle, Holocene paleoclimatic reconstructions in Northern Iceland: Approaches and Results, *Quat. Sci. Rev., 18*, 457-474, 1999.

Stuiver, M., G. W. Pearson, and T. Braziunas, Radiocarbon age calibration of marine samples back to 9000 cal yr BP, , *28*, 980-1021, 1986.

Stuiver, M., and P. J. Reimer, A computer program for radiocarbon age calibration, , *28*, 1022-1030, 1986.

Stuiver, M., P. J. Reimer, E. Bard, J. W. Beck, K. A. Hughen, B. Kromer, F. G. McCormack, J. v.d. Plicht, and M. Spurk, INTCAL98 Radiocarbon age calibration 24,000-0 cal BP, *Radiocarbon, 40*, 1041-1083, 1998.

Syvitski, J. P. M., K. W. G. LeBlanc, and R. E. Cranston, The flux and preservation of organic carbon in Baffin Island fjords, in *Glacimarine Environments: Processes and Sediments*, edited by J. A. Dowdeswell, and J. D. Scourse, pp. 177-200, The Geological Society of London, Special Publication No. 53, London, 1990.

Thompson, R., and F. Oldfield, *Environmental Magnetism*, 227 pp pp., Allen & Unwin, Winchester, Mass., 1986.

Thordardottir, T., Primary production in North Icelandic Waters in relation to Recent Climatic Change, *Polar Oceans: Proceedings of the Oceanographic Congress*, 655-665, 1977.

Thordardottir, T., Primary Production North of Iceland in relation to Water Masses in May-June 1970-1980, *Council for the Exploration of the Sea, C.M. 1984/L20*, 1-17, 1984.

Thordardottir, T., Timing and Duration of Spring Blooming South and Southwest of Iceland, *NATO ASI Series G-7*, 345-360, 1986.

Thors, K., Shallow seismic Stratigraphy and Structure of the southernmost Part of the Tjornes Fracture Zone, *Jokull, 32* (April), 107-111, 1982.

Thors, K., and G. S. Boulton, Deltas, spits and littoral terraces associated with rising sea level: Late Quaternary examples from northern Iceland, *Mar. Geol., 98*, 99-112, 1991.

Torrence, C., and G. P. Compo, A practical guide to wavelet analysis, *Bulletin American Meteorological Society, 79*, 61-78, 1998.

Velleman, P. F., and D. C. Hoaglin, *Applications, Basics, and Computing of Exploratory Data Analaysis*, 354 pp pp., Duxbury, Boston, 1981.

Verosub, K., Paleomagnetic Dating, in *Quaternary Geochronology. Methods and Applications*, edited by J. S. Noller, J. M. Sowers and W. R. Lettis, pp. 339-356, America Geophysical Union, Washington, DC, 1999.

Voelker, A. H. L., and e. al., Correlation of Marine 14C ages from the Nordic Seas with the GISP2 isotope record: Implications for 14C calibration beyond 25 ka BP, *Radiocarbon, 40* (1), 517-534, 1998.

Walden, J., F. Oldfield, and J. Smith (Eds.), *Environmental Magnetism. A Practical Guide*, 243 pp., Quaternary Research Association, London, 1999.

Walkley, A., A critical examination of a rapid method for determining organic carbon in soils.-effects of variation in digestion conditions of inorganic soil constituents, *Soil Science, 63*, 251-264, 1947.

Wastl, M., J. Stotter, and C. Caseldine, Tephrochronology-A tool for correlating records of Holocene environmental and climatic change in the North Atlantic region, *Geological Society of America Abstract vol, 31*, A315, 1999.

J.T. Andrews, J. Hardardóttir, A.E. Jennings, G.B. Kristjánsdóttir, and L.M. Smith, INSTAAR and Department of Geological Sciences, University of Colorado, Box 450, Boulder, CO 80309, USA.

Á. Geirsdóttir, J. Hardardóttir, G.B. Kristjánsdóttir, Department of Geosciences, University of Iceland, vid Sudurgötu, 101 Reykjavík, Iceland.

G. Helgadóttir, Marine Research Institute, Skúlagata 4, 101 Reykjavík, Iceland.

Á.E. Sveinbjörnsdóttir, Science Institute, University of Iceland, Dunhaga 3, 107 Reykjavík, Iceland.

Changes of Potential Density Gradients in the Northwestern North Atlantic During the Last Climatic Cycle Based on a Multiproxy Approach

Claude Hillaire-Marcel, Anne de Vernal, Laurence Candon, and Guy Bilodeau

GEOTOP, Université du Québec à Montréal, Montréal, Canada

Joseph Stoner

University of Davis in California, Davis, California

A multi-proxy approach was developed to document changes of potential density ($\sigma\theta$) in surface, mesopelagic, and bottom waters of the Labrador Sea during the last climatic cycle. This approach relies on dinocyst transfer functions allowing to estimate sea-surface temperatures and salinities, which were used to calibrate a relationship between $\delta^{18}O$ in calcite and $\sigma\theta$. The $\delta^{18}O$ of epipelagic (*Globigerina bulloides*), deeper-dwelling (*Neogloboquadrina pachyderma* left coiling), and benthic (*Uvigerina peregrina* and *Cibicides wuellerstorfi*) foraminifera, then allowed to extrapolate density gradients between the corresponding water layers. This approach has been tested in surface sediments in reference to modern hydrographic conditions. Its application in cores from the Orphan Knoll area permitted to reconstruct 3 distinct hydrographic configurations for the last 100 ka. i) The modern-like regime established during the deglaciation and reached full stability after 7 ka. It is marked by weak density gradients between the surface and intermediate water masses, allowing winter convection down to a lower pycnocline between intermediate and deep-water masses, i.e., formation of intermediate Labrador Sea Water (LSW). ii) A "full glacial" regime is characterized by variable but generally low density ($\sigma\theta \sim 25\text{-}27$) in the buoyant surface layer overlying a much denser ($\sigma\theta \sim 27.5\text{-}28.5$) water mass occupying the water column down to the sea floor. iii) The third regime occurred sporadically, notably at the end of Heinrich events. It is marked by a particularly low density ($\sigma\theta < 25$) and much deeper surface mixed water layer. Potential density in the intermediate-deep waters reached maximum values at ca. 27 ka ($\sigma\theta \sim 28\text{-}28.5$), then decreased rapidly by almost 2 units between 25.5 and 23 ka. Our data also suggest that temperatures in intermediate-deep waters recorded minimum values between $\sim$ 25 and 15 ka. In brief, during most of the last climatic cycle, a single and dense water mass seems to have occupied the water column below a generally low-density surface water layer, thus preventing deep convection. The last 7 ka allowing continuous formation of LSW constitute a noticeable exception.

The Oceans and Rapid Climate Change: Past, Present, and Future
Geophysical Monograph 126

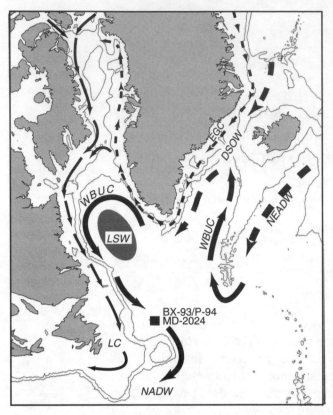

Figure 1. Map of the northwest North Atlantic and location of the coring site. P-94 refers to piston core HU-94-045-094 (50°12.26'N-45°41.14'W; water depth = 3448 m) and BX-93 to box core HU-91-045-093 (50°12.28'N-45°41.15'W; water depth = 3448) [cf. *Hillaire-Marcel et al.*, 1991]; MD-2024 refers to the core collected during the 1995 Images-I cruise on the Marion Dufresne (50°12.40'N-45°41.22'W; water depth = 3539 m). The thin arrows correspond to surface water currents fed by polar waters: EGC (dashed) = East Greenland Current; LC (plain) = Labrador Current. The thick arrows illustrate major deep current trajectories. The dashed thick arrows correspond to deep water mass overflow paths, the North East Atlantic Deep Water (NEADW) and the Denmark Strait Overflow Water (DSOW), respectively; the plain lines show the trajectory of the Western Boundary Undercurrent (WBUC). The hatched area corresponds to the area of intermediate Labrador Sea Water (LSW) formation [cf. *Clarke and Gascard*, 1983; *Lazier*, 1988]. The isobaths correspond to 500 and 2000 meters.

1. INTRODUCTION

Recent ocean model experiments suggest the possibility of a near future reorganization of the thermohaline circulation with the shutdown of one of the two main "pumps" driving the formation of North Atlantic intermediate and deep waters, the one in the Labrador Sea [*Wood et al.*, 1999], where the intermediate Labrador Sea Water (LSW)

mass now forms and rapidly spreads into the North Atlantic [e.g., *Sy et al.*, 1997]. From this viewpoint, information on the status of the LSW during late Quaternary climate-ocean changes is needed to identify the various modes of operation of the North Atlantic thermohaline circulation. Several recent studies provide at least qualitative information on past thermohaline conditions in the northern North Atlantic and on the rate of deep-water formation over the last climate cycle [e.g., *Keigwin et al.*, 1994; *Maslin et al.*, 1995; *Rasmussen et al.*, 1996; *Vidal et al.*, 1997; *Dokken and Jansen*, 1999; *Chapman and Shackleton*, 1999; *Chapman and Maslin*, 1999; *Hillaire-Marcel and Bilodeau*, 2000]. However, quantitative data on past variations of physico-chemical properties in the water column, notably in the Labrador Sea, are still missing or questionable. This poor knowledge of absolute changes in the water mass properties is partly due to misinterpretation of some of the proxies used to reconstruct salinity and temperature in the upper water column. For example, the isotopic composition of planktonic foraminifera such as *Neogloboquadrina pachyderma* left coiled was thought to respond to salinity and temperature in the surface layer. It is now demonstrated that it reflects conditions deeper in the water column, along the pycnocline within the underlying water masses [e.g., *Wu and Hillaire-Marcel*, 1994a; *Kohfeld et al.*, 1996; *Candon et al.*, 1999; *Hillaire-Marcel and Bilodeau*, 2000]. Another reason explaining the lack of accurate information on the upper water column lies in the fact that single proxy approaches have been often used for paleoceanographic reconstruction, whereas multiple tracers are needed to provide the complementary data required to constrain vertical and seasonal gradients in temperature and salinity, in addition to interannual variability. In the present study, micropaleontological and isotopic indicators are combined to reconstruct seasonal to millennial changes in temperature, salinity and potential density from the surface to deep waters of the southern Labrador Sea. Seasonal temperature and salinity gradients in the photic zone are estimated from transfer functions based on dinocyst assemblages [cf. *de Vernal et al.*, 1997, 2000; *de Vernal and Hillaire-Marcel*, 2000]. Temperature and salinity gradients along the pycnocline between surface and intermediate water masses are based on detailed examination and isotopic analysis of foraminifer assemblages with special attention paid to oxygen isotope gradients in response to growth rates and shell density, notably of the epipelagic taxon *Globigerina bulloides* and of the mesopelagic taxon *Neogloboquadrina pachyderma* left coiling. These data are complemented by information on deep waters, from the isotopic composition of benthic foraminifera. We document the reliability of such a methodological approach in reference to box core top assemblages compared to modern

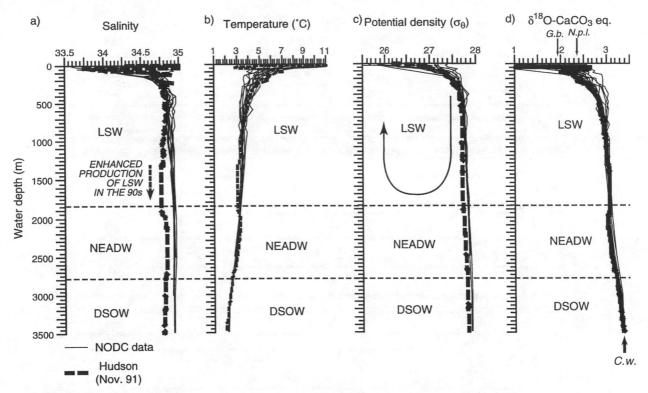

Figure 2. Structure of the modern water column at the Orphan Knoll site, in western North Atlantic.
From left to right: (a) monthly mean salinities from *NODC* [1994] and November 1991 record; (b) monthly mean temperatures from *NODC* [1994] and November 1991 record; (c) corresponding potential density values ($\sigma\theta$); (d) corresponding $\delta^{18}O$ values for a calcite precipitated in isotopic equilibrium with ambient water (from basic equations to be found in section 3.1 of the text), and isotopic composition in foraminiferal assemblages from top of box-core BX-93 (see arrows; *Globigerina bulloides* = G.b.; *Neogloboquadrina pachyderma* left-coiled = N.p.l.; *Cibicides wuellerstorfi.* = C.w.).

hydrographical conditions. The modern data and the reconstructed time series spanning the last 100,000 years are from the Orphan Knoll area, in the southern Labrador Sea (Figure 1). The site is located at the outlet of the Western Boundary Undercurrent (WBUC) into the North Atlantic Ocean, i.e., at the end of the trajectory of bottom-deep water masses throughout the Nordic seas, the Irminger Basin and the Labrador Sea, prior to form the NADW. The Orphan Knoll site therefore provides a key location to monitor temporal changes in the water mass properties of the NADW as well as changes in the rate of formation of intermediate waters (LSW) (see Fig. 2). A chronological time frame for the paleohydrological record extracted from the Orphan Knoll site is based on the development of a high resolution geomagnetic paleointensity-assisted chronostratigraphy that is linked to the Greenland Summit GISP2 ice core official chronology [cf. *Stoner et al.*, 1998, 2000]. This chronostratigraphy permits direct comparison of the paleohydrographic record from the Labrador Sea with other globally derived paleoclimatic time series and

the Greenland ice cores. The resolution of this time-series and the precise stratigraphy allows millennial scale records to be established in relation with high frequency climate oscillations, such as the Dansgaard-Oeschger Oscillations and Heinrich Events [cf. *Bond et al.*, 1993, 1999].

2. MATERIAL AND METHODS

2.1. Core Stratigraphy

The sediments at the Orphan Knoll site have been sampled during the CSS-Hudson cruise 91-045, and during the IMAGES I campaign of the Marion Dufresne in 1995. The box core HU-91-045-093 (cf. BX-93 in Figure 1) represents the upper few tens of centimeters of sediment. The piston cores HU-91-045-094 (cf. P-94 in Fig.1) and MD-2024 are respectively 1100 and 2616 centimeter long; they respectively represent the last 100,000 and 116,000 years of sedimentation [cf. *Hillaire-Marcel and Bilodeau*, 2000; *Stoner et al.*, 2000]. The apparent sedimentation rate of

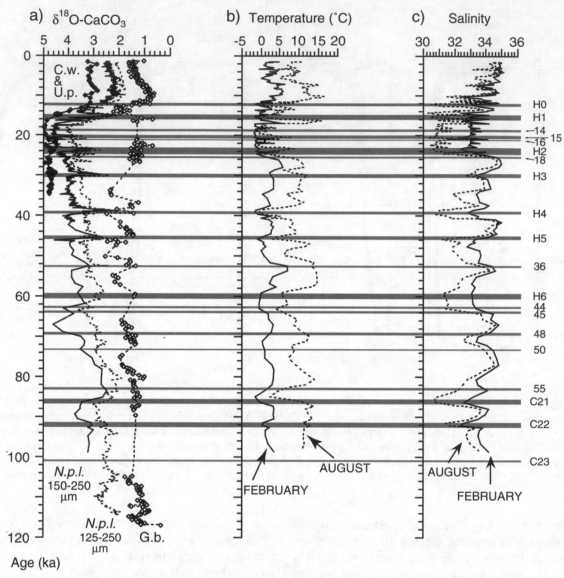

Figure 3. Stratigraphy at the Orphan Knoll site reported against the GISP2 chronology. (a) Oxygen isotope measurements in the planktic foraminifera *Globigerina bulloides* (G.b.) from core MD-2024, and *Neogloboquadrina pachyderma* left coiled (N.p.l.) from core P-94 (size fraction 150-250 μm), or core MD-2024 (size fraction 125-250 μm), and in the benthic foraminifera *Cibicites wuellerstorfi* (C.w.) or *Uvigerina peregrina* (U.p.), both from core P-94 (C.w. is present in the deglacial to Holocene sediment; U.p., in the glacial sediment); C.w. values were corrected by 0.64‰ to account for specific fractionation [*Shackleton*, 1974]; note the ~ +0.39‰ mean shift in δ18O values of N.p.l. in the large size range vs. the smaller one (150-250 μm vs.125-250 μm) [see *Hillaire-Marcel and Bilodeau*, 2000]. (b, c) The August and February temperatures and salinities in the photic zone as reconstructed from transfer functions using dinocysts [cf. *de Vernal et al.*, 2000]. (d) Carbon isotope measurements in G.b. and N.p.l. assemblages (core MD-2024), and in C.w. (core P-94). Detrital event and Heinrich layer age limits and numbering are from *Stoner et al.* [2000].

core MD-2024 is almost twice the one recorded in core P-94 partly due to sediment stretching using the Marion Dufresne long-coring device. However, this coring artifact seems to have had minimum impact on physical and sedimentological properties of the core MD-2024 [cf. *Turon, Hillaire-Marcel et al.*, 1999]. The isotopic stratigraphy in

these two cores has been established on the basis of measurements in the planktonic foraminifer *Neogloboquadrina pachyderma* left coiling (P-94 and MD-2024), *Globigerina bulloides* (MD-2024) and the benthic foraminifera *Cibicides wuellerstorfi* or *Uvigerina peregrina* (P-94). The organic carbon and calcium carbonate content, and the coarse sand percentage in the >120μm fraction have been systematically measured. Parts of these data were published by *Hillaire-Marcel and Bilodeau* [2000] and are available on the GEOTOP web site (www.unites.uqam.ca/geotop/). Cores MD-2024 and P-94 were intensively studied, notably for their physical and magnetic properties [*Stoner et al.*, 1995, 1996, 1998, 2000]. The isotopic, sedimentological, geochemical and paleomagnetic analyses of the two cores contributed to establish unequivocal correlation between the two sequences at the centimeter level [cf. *Hillaire-Marcel and Bilodeau*, 2000; *Stoner et al.*, 1998].

The core MD-2024 geomagnetic paleointensity record has been correlated to the Greenland Summit ice core records, based on the inverse relationship between cosmogenic radionuclide flux and geomagnetic field strength [*Stoner et al.*, 2000]. The paleointensity record from core MD-2024 has also been correlated with a new North Atlantic paleointensity stack (NAPIS-75) [*Laj et al.*, 2000], which is independently tied to the Greenland Summit ice core record, based on a match between magnetic susceptibility cycles and the isotope data in ice [cf. *Kissel et al.*, 1999]. These correlations were used to develop a chronological framework, which permits direct comparison with the ice core paleoclimate record [cf. *Dansgaard et al.*, 1993; *Grootes et al.*, 1997]. The chronology adopted is the GISP2 official chronology [*Meese et al.*, 1994; *Bender et al.*, 1994]. It is of note that a chronological frame consistent with that of *Stoner et al.* [1998, 2000] was independently established on the basis of AMS-^{14}C measurements in monospecific planktonic foraminifer populations for the upper 20 ka of the Orphan Knoll cores. The chronostratigraphy of cores MD-2024 and P-94 is therefore as robust as it can be: the calendar time frame for the last 50 ka is fairly well secured, but there are still uncertainties concerning the absolute chronology for the lower part of the records [cf. discussion in *Stoner et al.*, 2000]. Figure 3 thus illustrates the stratigraphy of cores MD-2024 and P-94 against the GISP2 time scale. The Heinrich layers and other detrital layers are labeled following the *Bond et al.* [1993, 1997, 1999] nomenclature as proposed by *Stoner et al.* [2000] (cf. figure 3).

2.2. Sea-surface Estimates of Temperature and Salinity

The conditions of temperature and salinity in the photic zone are estimated on the basis of transfer functions using dinocyst assemblages and the best analogue method (or modern analogue technique). The reference data set used here includes 371 sites from middle to high latitudes of the North Atlantic Ocean and its subpolar basins [cf. *de Vernal et al.*, 1997; *Rochon et al.*, 1999]. The reference data,

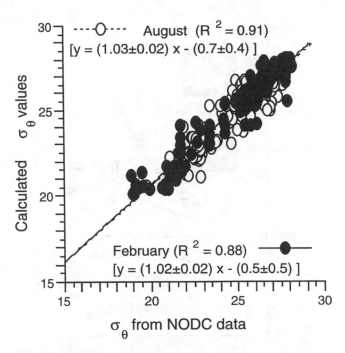

Figure 4. Validation of the transfer function approach to estimate the potential density: σθ values calculated from salinity and temperature estimates for the months of August and February based on dinocyst transfer functions (y-axis), vs. σθ values from NODC data sets at the corresponding sites. The data base includes 371 sites from the northern North Atlantic and adjacent basins [*de Vernal et al.*, 1997]. Uncertainties in the reconstructions are comparable to the interannual variability depicted by NODC data sets.

which come from both oceanic and neritic environments, are representative of a wide range of sea-surface conditions with respect to sea ice cover (from ice-free to quasi-perennial sea ice cover), salinity (20-36) and temperature (from freezing up to 25°C) with various ranges of seasonal amplitude. The dinocyst assemblages permit reconstruction of the parameters needed to calculate potential density in the surface water layer during the warmest and coldest months of the year, i.e., August and February. Validation exercises allow the following degree of accuracy to be determined: ± 1.6°C and ± 1.2°C for the August and February temperatures respectively, and ± 0.7 for salinity in the 25-36 (units) range. The reference hydrographic data set were compiled from the National Ocean Data Center [*NODC*, 1994]. The calibration was done for the surface (0 m) layer. The potential density values (σθ) for the warmest and coldest months were calculated using current equations [see *Gill*, 1982]. Comparison of σθ values derived from temperature and salinity estimates to those of the NODC data set (Fig. 4) yielded reasonably good results. The correlation coefficients (r^2) between estimated and actual

a)
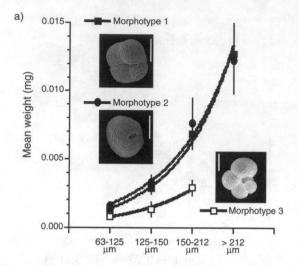

b)
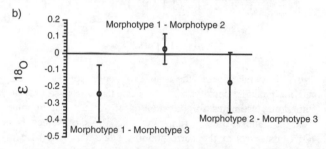

Figure 5. Size and morphotype dependence of isotopic compositions in *Neogloboquadrina pachyderma* left-coiling (N.p.l.). (a) On top, growth curves of N.p.l. morphotypes in box core top sediments from the NW Atlantic [*Candon*, 2000]. These morphotypes are practically identical to those defined by *Kohfeld et al.* [1996] from plankton tow studies in the Northeast Water Polynya area. (b) At bottom, differences in isotopic compositions (for comparable size ranges) between morphotypes; specimens of the morphotype 3 are systematically depleted in ^{18}O vs. the densest, thick-walled and/or encrusted specimens of the other morphotypes. This corresponds to a growth habitat of morphotype 3, on the seasonal pycnocline, in shallower more diluted and warmer waters.

potential density in surface water are 0.88 and 0.91 for February and August, respectively (cf. Fig. 4). Standard deviations between estimated and observed σθ values are ±0.54 and ±0.63, for February and August, respectively. These uncertainties are within the range of the interannual variations observed for the physical properties of surface waters [see also *de Vernal and Hillaire-Marcel*, 2000].

2.3. Reconstruction of Thermohaline Gradients in the Subsurface to Mesopelagic Layers

The thermohaline conditions along the pycnocline from the surface water layer to the underlying water masses of

the upper water column are reconstructed based on isotopic measurements in planktonic foraminifera that are converted into potential density values, using a σθ vs. δ^{18}O value in calcite calibration relationship. We follow here approaches developed notably by *Ravelo and Fairbanks* [1992], *Kohfeld et al.* [1996], *Andreasen and Ravelo* [1997], and more recently by *Faul et al.* [2000]. Our calibration equations are set from August and February temperatures and salinities as reconstructed from the dinocyst transfer functions. Details on boundary conditions, notably the isotopic composition of the fresh and oceanic water end members are provided below.

The subsurface conditions are given by the δ^{18}O of *Globigerina bulloides*, which is known to be an epipelagic taxa, living in the upper 100 meters of the water column [*Bé and Tolderlund*, 1971; *Bé et al.*, 1977]. *G. bulloides* (G.b.) is usually considered to precipitate its calcite in ^{18}O-equilibrium with ambient water. However, experiments by *Spero and Lea* [1996] seem to indicate that its shell could present a slight size-dependent isotopic offset from the equilibrium value, at least in the 16-22°C temperature range, i.e., a range well above that of our study area. Nevertheless, the isotopic analyses of G.b. shells in box-core top sediments of the northwestern North Atlantic have shown δ^{18}O values that correspond to calcite precipitation in summer (~ August temperature and salinity) at a depth of 50-70 meters in the water column [*Candon*, 2000]. This depth range corresponds to the bottom of the mixed layer, towards the top of the seasonal pycnocline. Analyses of G.b. shell density and isotopic composition were also performed in different size fractions in order to identify eventual encrustements and variations in depth habitat during the life cycle. Results reveal no significant change in G.b. shell density with size, invalidating the hypothesis of enhanced encrustement during their life span [*Candon*, 2000]. The isotopic data demonstrate no clear variation with shell size neither, suggesting no significant deepening of their habitat during shell growth.

Deeper along the pycnocline between surface and mesopelagic water masses, thermohaline conditions were reconstructed from the isotopic composition of *Neogloboquadrina pachyderma* left coiling (N.p.l.) assemblages. N.p.l. is known to be a relatively deep dwelling planktonic species, which experiences variations in its depth habitat during its life cycle [e.g., *Bé and Tolderlund*, 1971; *Bé et al.*, 1977; *Sverdlove and Bé*, 1985; *Carstens and Wefer*, 1992; *Carstens et al.*, 1997]. N.p.l. often dominates planktonic foraminifer assemblages of the northern North Atlantic, but shows a wide range of morphological variations [e.g., *Cifelli*, 1973]. Among N.p.l. specimens from the Labrador Sea, there are gradation variations which can be classified into 3 different varieties or morphotypes (Fig. 5) [see also *Plouffe*, 1994; *Kohfeld et al.*, 1996; *Candon*, 2000]. (i) The morphotype 1 corresponds to square shaped specimens characterized by a thick and encrusted shell, with shallow and rough sutures between the chambers; this morphotype is the most abundant and ubiquitous. (ii) The morphotype 2 designates specimens so encrusted that sutures between chambers are barely visible and the aperture is almost close; these specimens are characterized by a rounded to ovoid shape. (iii) The morphotype 3 is represented by shells with a relatively large aperture, and with thin and delicate walls; their outer chambers are well separated by deep sutures (see Fig. 5). The shell density of these N.p.l. morphotypes has been calculated using the mean diameter and mean weight, as measured either using a precision scale or from precisely calibrated CO_2 pressures in the inlet system of the mass spectrometer used for stable isotope analyses (see *Hillaire-Marcel and Bilodeau* [2000] for details on analytical methods for isotopic measurements).

Identification, counts and picking of N.p.l. shells were performed in different size fractions: 63-125, 125-150, 150-212 and > 212 μm for a few key intervals [cf. *Candon*, 2000], and in the 125-250 (MD-2024) or 150-250 μm (P-94) size ranges, throughout the whole sequence. In box core tops from the NW North Atlantic, isotopic measurements for each size indicate that small, likely juvenile, N.p.l. specimens form their calcite towards the top of the seasonal pycnocline, near the habitat of G.b. The largest most incrusted specimens suggest calcite overgrowth deposition at much greater depths, along the pycnocline, as deep as a few hundred meters for open ocean sites characterized by relatively high summer temperatures [*Candon*, 2000]. As a consequence, changes in the size of N.p.l. assemblages used for paleoceanography studies will unavoidably lead to shifts in $\delta^{18}O$ values. For example, at the Orphan Knoll site, *Hillaire-Marcel and Bilodeau* [2000] observed a 0.39‰ mean difference in $\delta^{18}O$ values of N.p.l. specimens in the size ranges 125-250 vs. 150-250 μm, throughout the last climatic cycle. However, the specimens belonging to the most common morphotype of N.p.l., those with the thickest walls and densest tests, do not show significant differences in their ^{18}O contents (Fig. 5b). The lightest shells of the morphotype 3 with thin walled structure are systematically depleted in ^{18}O by approximately 0.2‰ vs. that of the other morphotypes in a given size range, suggesting calcite precipitation slightly above these, along the pycnocline. It seems likely that there is a strong relationship between the shell density, determined by the thickness of its walls, the habitat of N.p.l., and the density gradient along the pycnocline [cf. also *Srinivasan and Kennett*, 1974]. On methodological grounds, one may conclude from this brief overview of the ecological requirements of G.b. and N.p.l. that ^{18}O gradients between them

provide a proxy for the strength of the seasonal pycnocline. This pattern could correspond to a slightly different depth habitat along the summer pycnocline, but more likely, to a slightly different growth season [cf. *Sautter and Thunell*, 1989]. Thus, calcite precipitation occurs during different stages of development of the seasonal pycnocline. In the subarctic North Atlantic, G.b. growth occurs during the warmest months, when the temperature at the base of the mixed layer approaches or exceeds 8°, whereas N.p.l. usually shows a much earlier and longer growing season, i.e., in cooler, more saline and denser waters [e.g., *Bé and Tolderlund*, 1971; see also *Wu and Hillaire-Marcel*, 1994a].

The seasonal and vertical changes in hydrographic conditions at the study site permit to assess on the importance of the seasonality of foraminiferal growth on isotopic compositions of the shells (Fig. 2d). The interannual variability is also illustrated here by the enhancement of LSW production in the Nineties [see also *Rahmstorf*, 1999] shown by the steeper salinity and σθ profiles (Fig. 2a and b) recorded during the CSS-Hudson 1991 cruise [cf. *Lucotte and Hillaire-Marcel*, 1994], compared with the conditions of the preceding decades from NODC compilations.

3. CALIBRATION OF A POTENTIAL DENSITY vs. $\delta^{18}O$-CaCO3 RELATIONSHIP THROUGH TIME

3.1. Basic Equations Used

The equations used to calculate potential density values (σθ) are from *Gill* [1982]. The paleotemperature relation used to transcribe the isotopic composition of calcite vs. PDB (δc) into that of water vs. SMOW (δw) during calcite precipitation, or vice versa, for a given temperature in °C (t) is that of *O'Neil et al.* [1969] and *Shackleton* [1974] (see equations 1 and 2). The offset value of - 0.27‰ in equation 2 stands for conversion into the PDB scale [*Coplen*, 1988].

$$t = 16.9 - 4.38 \, (\delta c - A) + 0.10 \, (\delta c - A)^2 \qquad (1)$$

$$A = (\delta w - 0.27) \qquad (2)$$

The next step consists in setting a relationship between δw and salinity (Sw). In general, these parameters are linked by linear relationships. A noticeable exception is found at sites where sea ice formation results in the production of winter brines, isotopically slightly lighter than the sea water at their origin, and in summer meltwater characterized by a very low salinity but very high $\delta^{18}O$ values [*Bédard et al.*, 1981]. Such effects cannot be totally ignored in Arctic environments [e.g., *Kohfeld et al.*, 1996] but are likely of a lesser importance in subarctic basins.

Thus, we used here a simple linear relationship between the isotopic composition (δw) and the salinity (Sw) of the ambient water during calcite precipitation, based on two end members, i.e., the isotopic composition and salinity of standard oceanic water (δsw and Ssw) and the isotopic composition of local freshwaters (δfw), as indicated below.

$$\delta w = \delta sw - [(\delta fw - \delta sw)*(Sw - Ssw)/Ssw] \qquad (3)$$

For modern conditions, the salinity of the standard ocean end member has been set to 35 and its isotopic composition to 0‰ (i.e., that of SMOW). Local fresh water inputs have been set to a δfw value of −17.8‰ [from *Wu and Hillaire-Marcel*, 1994a] with a salinity of 0.

3.2. Setting of Boundary Conditions for the Last Climatic Cycle

The linkage of stable isotope compositions with potential density ($\sigma\theta$) as well as the transcription of temperature and salinity estimates from transfer functions into isotopic composition for a calcite precipitated in equilibrium with the ambient water (δc), both require the setting of several boundary conditions: δsw, δfw and Ssw. In order to account for temporal changes in these boundary conditions, due to ice sheet growth and decay, independent proxies were used. The benthic isotope record from ODP-Site 677 [from *Shackleton*, 1990], which is often used as a proxy for changes in ocean water volume during glaciations, was directly imported to estimate temporal changes in the isotopic composition of the ocean (δsw). This data set was then adjusted to a modern reference value of 0‰ by subtracting 3.93‰ to the foraminifer δc values (i.e., the isotopic composition of the benthic foraminifera in the core top) to obtain an estimate for paleo-δsw values during the last climatic cycle. The ODP-677 data also provided a first estimate for salinity variations of the standard ocean water end-member. We set an absolute range from 35 to 36 for the ocean salinity between the present interglacial and the LGM, which corresponds to a salinity shift of 1 for a maximum δ^{18}O shift of 1.2‰ at site 677. Thus, the simple relation indicated below was used to constrain the salinity of the seawater end-member.

$$Ssw = 35 + (\delta sw/1.2) \qquad (4)$$

The use of this relation implies some approximations, notably that temporal changes in the isotopic composition of the ice sheets has been neglected. In a similar way, the ODP-Site 677 benthic isotope curve was imported according to the SPECMAP time scale [cf. *Martinson et al.*, 1987], despite the fact that we used the revised GISP2

chronology for our deep sea cores [*Meese et al.*, 1994; *Bender et al.*, 1994]. Small discrepancies exist between the two time scales particularly in the isotopic stage 4-5 part of the records [e.g., *Stoner et al.*, 2000], but the conversion from one to the other would have raised additional problems. Furthermore, the temporal resolution of the ODP-677 curve averages 2 ka, whereas our data set is based on a temporal resolution of 500 to 150 years, depending upon the proxy. It was thus necessary to interpolate the δsw and Sw values between benchmarks provided by the ODP-677 data set. Due to the mixing and renewal times of the ocean, a better resolution for the definition of the marine end member of the mixing equation does not seem necessary.

With respect to temporal changes in the isotopic composition of the freshwater end member (δfw), we also used a relatively simple approach (see equation 5). It is based on the assumption that riverine and meltwater inputs from the continents and ice sheets surrounding the Labrador Sea had temporal variations in their ^{18}O content proportional to those depicted by the GISP2 record [*Grootes and Stuiver*, 1997], i.e., that the latitudinal gradient in δ^{18}O of precipitation also applied during the ice age. A value of ~0.51 that represents the ratio between modern δfw and δGISP2 values is thus applied to evaluate the isotopic composition of the freshwater end member at a given time (t) from the isotopic composition of the ice and that of the ocean.

$$\delta fw = [(\delta fw)modern /(\delta GISP2)modern]*[(\delta GISP2)t-(\delta sw)t]$$

$$\sim 0.51* [(\delta GISP2)t-(\delta sw)t] \qquad (5)$$

3.3. Estimating $\sigma\theta$ From δ^{18}O in Foraminiferal Calcite: the Dinocyst Transfer Function Pathway

Finally, the conversion of δ^{18}O values of foraminiferal calcite into $\sigma\theta$ values requires further explanation. Ideally, the setting by an independent method of one of the parameters governing the paleotemperature equation would be the best way to solve the problem. For example, Mg/Ca ratios in foraminifer shells are more and more used to set independently the growth temperature [e.g., *Nürnberg et al.*, 2000; *Lea et al.*, 1999; *Elderfield and Ganssen*, 2000]. However, the calibration curves for Mg/Ca vs. temperature are poorly constrained in the thermal domain of the present study. In a similar fashion, ecological temperature requirements are not narrow enough to provide the independent parameter needed. For instance, G.b. seems to occupy a large temperature domain [*Bé and Tolderlund*, 1971; *Sautter and Thunell*, 1989], although it may possibly find its minimum temperature limit (~8°C) in the Labrador Sea. Similarly, N.p.l. seems to tolerate a relatively large temperature range, from ~ 0°C in Arctic settings [e.g.,

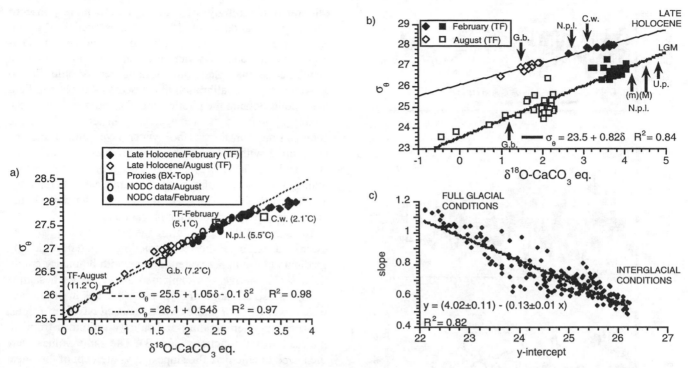

Figure 6. Calibration of potential density vs. δ^{18}O-CaCO3 eq. (a) Comparison of potential density vs. δ^{18}O-CaCO3 values at the study site derived respectively from (i) modern NODC temperature and salinity data sets (August and February), (ii) dinocyst transfer function reconstructions of temperature and salinity of August and February for the late Holocene period (P-94), and (iii) data from box-core top (BX-93; 0-1 cm). The latters include (i) reconstructions of temperature and salinity for August and February, from dinocyst transfer functions, and (ii) one control point provided by δ^{18}O-values in benthic foraminifera and the modern temperature and salinity of the corresponding bottom water. The linear vs. polynomial calibration equations are best fits using both NODC and late Holocene data sets. When projected on the polynomial curve, δ^{18}O-values in planktic foraminifera (G.b. and N.p.l.) allow to constrain the mean growth temperature for the two taxa (between brackets). More details on the validation of the $\sigma\theta$ vs. δ^{18}O-CaCO3 relationship can be found in section 4 of the text. (b) In the intermediate diagram, are shown the linear calibrations based on February vs. August estimates from transfer functions for the late Holocene (0-3 ka) and the LGM intervals, respectively, at the Orphan Knoll site. δ^{18}O-values in planktic (G.b. & N.p.l.) and benthic (U.p. & C.w.) foraminifera are indicated by arrows. δ^{18}O in N.p.l. are reported with respect to both median (m) and mean (M) values. The LGM is defined as the interval encompassed by the H1 and H2 events (see Fig. 3). Arrows correspond to paleo-potential density values that can be reconstructed for the corresponding intervals based on isotopic compositions of foraminifera. (c) The lower diagram reports the slope vs. y-intercepts of the linear relationships calculated from August and February conditions in surface waters (from transfer functions) for the interval 0-100 ka (core P-94). The corresponding equations are used to extrapolate or interpolate $\sigma\theta$ values from isotopic measurements in foraminifera.

Kohfeld et al., 1996] to temperatures as high as 8°C [*Wu and Hillaire-Marcel,* 1994a] and possibly more.

Fortunately, most studies show that $\sigma\theta$ and δc values are linked by simple polynomial relationships, generally sublinear [e.g., *Billups and Schrag,* 2000; *Candon et al.,* 1999; *Candon,* 2000]. Here, we simply used the seasonal thermohaline range (i.e., August-February) in the surface water layer, yielded by dinocyst transfer functions, to establish $\sigma\theta$ vs. δ^{18}O linear calibration equations (Fig. 6). Since G.b. and N.p.l. develop between these two periods, along the pycnocline with the underlying LSW (Fig. 2), these linear

calibration equations should allow interpolating $\sigma\theta$ values from δ^{18}O in G.b. and N.p.l., with a reasonable accuracy. In practical terms, we firstly used temperature and salinity values reconstructed for August and February, from dinocyst transfer functions, to calculate the corresponding interseasonal (i.e., August vs. February) $\sigma\theta$ gradients. In a second step, we used the same salinity and temperature values to calculate δ^{18}O values, accordingly for August and February. Theses two seasonal end-members were then used to calculate the slope and y-intercept of the corresponding $\sigma\theta$ vs. δ^{18}O linear relationship. The parameters of the lin-

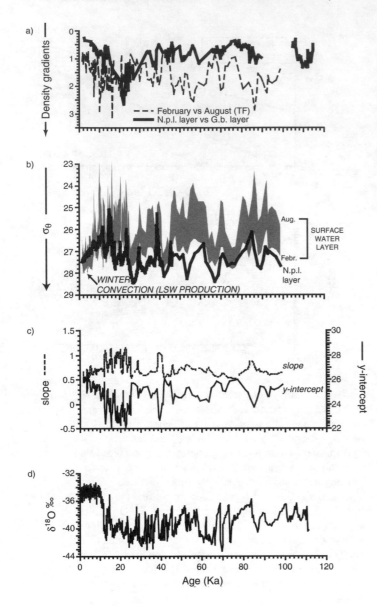

Figure 7. Potential density (σθ) changes in the upper water column throughout the last 100,000 years on the GISP2 time scale. (a) Seasonal σθ gradients in surface waters and differences in σθ values between the G.b. and N.p.l. "layers" (defined on the basis of their bathymetric and seasonal distribution). (b) Potential density in the surface water layer (gray shaded envelope for August to February) calculated from dinocyst transfer function data, versus σθ values in the underlying pycnocline represented by N.p.l. data (N.p.l. layer density is calculated from the equation below). The only episode with a reversal of potential density gradients in winter corresponds to the deglaciation to Holocene interval. Note the stabilization after 7 ka BP. (c) Slope and y-intercep of the calibration equations "σθ = a*δCaCO3 eq. + b", calculated from salinity and temperature reconstructions for February and August using transfer functions (see Fig. 3). (d) GISP2 isotopic record.

ear equation strongly vary through time in response to changes in boundary conditions, as illustrated in Figs. 6b and 6c (see also below, point 4 and 5). One caveat, however, concerns benthic foraminifer data (U.p. or C.w.). One may assume that planktonic foraminifera should fit approximately on a calibration line based on August and February conditions in the photic zone: the end members of the mixing system are the same, notably in terms of isotopic composition, for the surface water layer and along the pycnocline with the underlying water mass, occupied by G.b. and N.p.l. However, this assumption could reveal wrong, for a deep-water mass with drastically distinct age and hydrographic history. Benthic foraminifer data should thus be interpreted with some precautions.

The status of the LSW during the last glaciation has been examined based on the reconstruction of potential density gradients in the upper water column, with boundary conditions set as above. As explained above, at each sample level, a seasonal range of $\delta^{18}O$ and σθ values in the surface water layer (Figs. 7 and 8) was calculated based on August and February temperature and salinity estimates from dinocyst transfer functions (Fig. 3). The same values were then used to calculate the slope and y-intercept of the linear $\delta^{18}O$-σθ relationship for the corresponding time interval (Fig. 7b). Finally, σθ values in the water masses occupied by G.b. and N.p.l. were calculated using this linear equation and their respective $\delta^{18}O$ values. The data set includes dinocyst assemblage counts for 193 samples spanning the last 100 ka (Fig. 3), thus allowing a resolution of ~0.5 ka for the reconstruction of conditions in the surface water layer. A slightly better resolution has been achieved for isotopic measurements with G.b. (280 samples, i.e., ~ 0.35 ka/sample), and a very high resolution was obtained for N.p.l. isotopic measurements (669 samples, i.e., ~ 0.15 ka/sample. For many N.p.l. and G.b. samples, we had thus to interpolated calibration equation parameters (Fig. 7b) from the immediately under- and overlying samples in which dinocyst counts were made.

4. THE LATE HOLOCENE-MODERN SITUATION: VALIDATION OF THE APPROACH

Figure 6a shows a comparison of the σθ vs. $\delta^{18}O$ values calculated for the modern water column with those estimated for the late Holocene interval. The modern values are calculated for the months of August and February using *NODC* [1994] data, and the late Holocene estimates for the same months are based on transfer function runs in core P-094 for the 0-3 ka interval. This time slice has been chosen for the relative stability of its paleoceanographic conditions (see Fig. 3). The two data sets accept the same best fits within standard deviations. Generally, a second-degree polynomial σθ vs. δ¹⁸O relationship yields better correlation coefficients than a linear regression [e.g., *Faul et al.*, 2000]. However, the difference between the polynomial and linear regressions is not much important in most hydrographic situations. As illustrated here (Fig. 6), the only critical departure between the curve and the line is observed for extreme values, well beyond the domain of critical interest.

Proxy data, from BX-93 core top sample (0-1 cm), are also shown in Fig. 6a. One should note, here, that they represent a time interval spanning a few hundred years due to mixing by benthic organisms [see *Wu and Hillaire-Marcel*, 1994b]. Such a time interval is intermediate between those of the NODC (ca. 50 a) and late Holocene (3 ka) data sets. These proxies provide complementary control points. Two are from transfer function reconstruction for February and August. Two others are from δ¹⁸O measurements in planktonic foraminifera. When projected on the above regression line, these δ¹⁸O-values correspond to growth temperatures of 5.5 °C and 7.2°C, respectively for N.p.l. and G.b., which are compatible with their respective ecological requirements. Finally, a control point for benthic conditions is provided by the isotopic composition of C.w. and the modern temperature at the sea floor (i.e., 2.1°C; Fig. 2). The maximum departure of these proxies from the above regression line is less than 0.02 σθ unit.

The relatively good fit of the C.w. control point with the calibration line established from surface water data, suggests that, in terms of isotopic composition and salinity, the DSOW that bathes here the sea floor (Fig. 2) [see *Lucotte and Hillaire-Marcel*, 1994] represents mixing end-members not much different from those characterizing the overlying water masses. Therefore the reservation concerning the interpretation of benthic foraminifera data, made in section 3.3, above, does not apply under the modern hydrographic pattern.

The fact that the data from the modern water column and those from the late Holocene and surface sediment sample fit closely on the same calibration line lead to believe that the proposed approach is appropriate for the reconstruction of paleo-potential density gradients in the water column. In addition, it shows that for any given sample downcore, a simple regression line between σθ estimates for the coldest and warmest months, from transfer functions, can be used to interpolate σθ values for the planktonic foraminifera habitats. With some precaution, this regression line can

Figure 8. Blow up of the changes in potential density in the water column during the last 40 ka on a GISP2 time scale. Detrital event and Heinrich layer age limits and numbering are from *Stoner et al.* [2000]. (a) Potential density of the N.p.l. "layer";

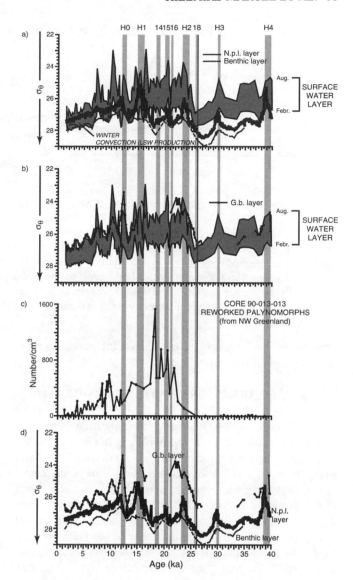

N.p.l. broadly corresponds to winter conditions in surface waters, i.e., it develops deeper than G.b. along the seasonal pycnocline; the only period with February conditions allowing convection to occur corresponds to the deglacial interval, and particularly, the last 7 ka. (b) Potential density of the G.b. "layer" compared with seasonal gradients in surface waters indicated by transfer functions; due to ecological requirements, G.b. likely develops here on top of the pycnocline, during summer months only. (c) Pre-Quaternary reworked palynomophs in a correlative core from the southern Greenland Rise showing the inception, after 25 ka BP, of deep currents carrying them from their source area in NE Greenland. (d) Potential density values in the G.b. and N.p.l. vs. benthic "layers"; note i) the coupling of the N.p.l. and benthic profiles during the glacial interval (and their decoupling during the Holocene); ii) the large oscillations in the N.p.l. "layer" potential density during H-events; iii) the Holocene oscillatory response of the G.b. "layer" and, to a lesser extend, of the benthic layer.

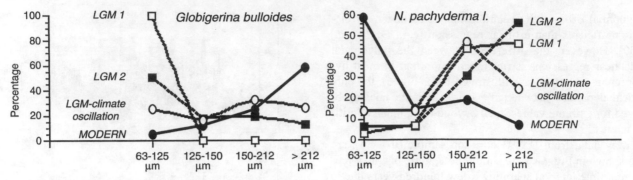

Figure 9. Percentages and mean size of *Neogloboquadrina pachyderma* left coiled (N.p.l.; right diagram) and of *Globigerina bulloides* (G.b.; left diagram) in LGM samples vs. surface sediment samples. The "LGM-Climate oscillation" refers to the major of several (at least 3) short-duration, large-amplitude climate oscillations recorded in LGM sediments from the Labrador Sea [see *Hillaire-Marcel and Bilodeau*, 2000]. LGM1 and LGM2 refer to the time intervals encompassed respectively by H2 and the climate oscillation, and by the latter and H1. The mirror size distributions in G.b. and N.p.l. assemblages of LGM ages, vs. the modern assemblages, illustrate their ecological affinities for respectively warmer and colder waters. The larger, thus denser N.p.l. shells of the LGM (notably of LGM 2) indicate the development of a much stronger pycnocline during the LGM than in the modern Labrador Sea.

also be used to extrapolate $\sigma\theta$ values for bottom waters, using benthic foraminifera data.

5. THE EXTREME SITUATION OF THE LGM

In Figure 6b, the late Holocene calibration equation is compared to that of the LGM, which is the interval of the last climatic cycle characterized by the largest $(\delta fw - \delta sw)$ offset (see section 3 above). The LGM data points correspond to August and February conditions in surface waters for the interval encompassed by the H1-H2 events. They indicate a $\sigma\theta$ vs. $\delta^{18}O$ calibration line with a much steeper slope and lower y-intercept than that of the Holocene, actually the steepest slope and lowest y-intercept that can be calculated throughout the whole climatic cycle (Fig. 6c). Compared to those of the Holocene, the LGM data points are somewhat scattered. This is due to the fact that at least three short-duration but large-amplitude climate oscillations are recorded in the Labrador deep sediments during this interval (Fig. 3a) [see also *Hillaire-Marcel and Bilodeau*, 2000; *Stoner et al*, 2000]. They correspond to dilution and warming pulses in surface waters, recorded by dinocyst assemblages (Fig. 6b), and to deepening of the surface water layer shown by light $\delta^{18}O$ values in N.p.l. (Fig. 3). As a consequence, N.p.l. assemblages from the LGM show a large difference between their median $\delta^{18}O$-value (4.51‰; n = 91) and their mean value (4.20±0.73). Comparatively, U.p. recorded much narrower isotopic fluctuations in its benthic habitat (4.82±0.15; n = 14), and G.b. seems to have developed during exceptionally warm years at the beginning of the LGM (Fig. 3), as shown by its very light mean $\delta^{18}O$-value (1.20±0.23; n = 10).

These isotopic compositions deserve further comments. Firstly, the light $\delta^{18}O$-values in G.b. confirm the very low salinity conditions in surface waters reconstructed from dinocysts (Fig. 3). A buoyant surface water layer was then strongly stratified over a high salinity and cold water mass that occupied the underlying water column down to the sea floor, as shown by the high $\delta^{18}O$-$\sigma\theta$ values depicted by the mesopelagic (N.p.l.) and benthic (U.p.) assemblages (Fig. 6). The median $\delta^{18}O$-value of N.p.l, heavier than the mean value, indicates that N.p.l. developed during intervals with contrasted regimes. Under severe glacial conditions, heavy $\delta^{18}O$-values in N.p.l., almost identical to those recorded by U.p., indicate that N.p.l. developed at the deepest part of a strong pycnocline, on top of the cold and dense water mass then occupying most of the water column. On the contrary, during the short duration climate oscillations of the LGM interval, meltwater pulses resulted in deepening of the pycnocline. N.p.l. then developed slightly higher on the pycnocline, in less saline and possibly slightly warmer waters.

Size distributions in G.b. and N.p.l. assemblages confirm the above interpretation (Fig. 9). Large and heavy N.p.l. shells characterize the most severe LGM episodes (LGM1 & LGM2 in Fig. 9), whereas lighter and smaller shells are found during the largest amplitude climate oscillation recorded of the interval. On the contrary, the more thermophilous species G.b. follows a reverse pattern with lighter shells during LGM1 and LGM2 episodes (Fig. 9). This pattern corresponds to calcite precipitation, respectively towards the deepest part of the pycnocline (for dense N.p.l. shells), and on top of it (for light G.b. shells). It seems therefore that these species behave as "proxidensitometers", for instance with large thick-walled and/or encrusted N.p.l. shells (morphotypes 1 & 2) developing in denser waters than lighter shells of morphotype 3 (Fig. 5).

6. THE STATUS OF THE LSW DURING THE LAST CLIMATIC CYCLE

Figure 7 illustrates density gradient changes in the upper water column throughout the last 100 ka, with reference to the GISP2 paleoclimate record. Very large amplitude oscillations are depicted by seasonal density gradients in the surface water layer (Fig. 7a), likely in response to the modes and rates of meltwater discharge from surrounding ice sheets. A few intervals with reduced seasonal gradients are seen, but except for the Holocene period, they were likely of short duration. By opposition, density gradients between the water layers and growth seasons of G.b. and N.p.l. seem much less variable (Fig. 7a), with a noticeable exception during the LGM interval, when their $\sigma\theta$ values differed by ~ 2.5 units. This suggests that G.b. developed in relatively diluted waters, on top of the pycnocline, whereas N.p.l. occupied a much denser layer, deeper along the pycnocline, as already seen above.

As seen in Figure 7b, several periods depict relatively high-density conditions in surface waters. They correspond to two distinct settings depending upon the climate interval concerned. During the last ice age stricto sensu, the high density intervals in surface waters (~ 78-72, ~ 68-64, 45-40 and 30-25 ka; Fig. 7b) are interpreted as matching ice advance phases, with reduced meltwater supplies, rather than maximum ice extend standstills, when equilibrium conditions between ice accumulation and ice ablation should have led to relatively high meltwater supply rates. During these intervals with dense surface waters, the water layer occupied by N.p.l. remained still denser, as shown by the corresponding $\sigma\theta$ values (Fig. 7b). Winter convection seems thus unlikely (no LSW formation). On the contrary, during the deglacial interval, the trend towards densification of surface waters (Fig. 7b) resulted in $\sigma\theta$ values in February exceeding occasionally those indicated by N.p.l. for the seasonal pycnocline with the underlying water mass. Therefore, thermohaline conditions during this interval were occasionally compatible with winter convection and production of LSW, notably after 7 ka BP (Fig. 7b).

A closer examination of the 35-0 ka interval (Fig. 8a) allows to precise the timing of major hydrographic reorganizations. A major switch in surface water conditions seems to have occurred at the onset of the LGM, between 25.5 and 23 ka. It corresponds to a ~ 3 unit decrease of $\sigma\theta$ values in surface waters with an almost parallel shift in $\sigma\theta$ values in the deeper water layer occupied by N.p.l. Since ~ 23 ka, the general trend has been an increase in potential density of surface waters until the Present. However, N.p.l. still shows major fluctuations with two minimums in the corresponding $\sigma\theta$ values peaking at 15 ka (late H1 event) and 12 ka (late H0), i.e., before and just after the Younger Dryas (YD). They indicate maximum dilution of intermediate waters, likely due to the maximum meltwater discharge pulses of the deglaciation (cf. termination 1A and 1B [Duplessy et al., 1981; Fairbanks, 1989]). The comparison of February conditions in surface waters with those indicated for the deeper "N.p.l. layer" (Fig. 8a) suggests the possibility of intermittent convection and LSW formation as early as 16 ka BP. However, it is only after 7ka BP (Fig. 8a), that surface waters clearly show much denser winter conditions than those recorded deeper, by N.p.l., along the pycnocline. Maximum production of LSW seems therefore to be a specific feature of the mid-late Holocene interval.

As clearly shown by Figure 8b, the good fit between $\sigma\theta$ values from G.b. and those yielded for August, by the dinocyst transfer function, demonstrate that G.b. developed almost exclusively during the warmest months. In opposition, N.p.l. data generally fit closely with February conditions (Fig. 8a). Another striking feature of Fig. 8 (b and d) appears in the strong oscillatory signal depicted by $\sigma\theta$ values for the "G.b. layer", and to a lesser extend, by $\sigma\theta$ values in August, from transfer function data. Since the YD, summer conditions show six major oscillations of relatively large amplitude, possibly in relation changes in iceberg and/or freshwater seasonal discharge rates. They likely correspond to the "pervasive millennial cycle" already observed by Bond et al. [1997] in a few North Atlantic records. Worth of mention is the fact that these oscillations are practically unseen in the "N.p.l. layer" or in winter conditions (Fig. 8b and d). However, as explained below, the benthic foraminifera bathed by the DSOW water mass have recorded at least some of these oscillations (Fig. 8d), leading to believe that the process at their origin may also have had some influence on surface water conditions in the Greenland Sea, where this water mass forms.

7. CONDITIONS IN THE DEEP BASIN: THE BENTHIC FORAMINIFER RECORD

7.1. Changes in the Thermohaline Structure of the Water Column

The benthic $\sigma\theta$ record (Fig. 8b and d) is derived from isotopic measurements in U.p. for glacial sediments, and in C.w. for deglacial to Holocene sediments [see Bilodeau et al., 1994]. It does not extend beyond ~ 35 ka essentially due to the low abundance of shells: benthic productivity was low in the Labrador Sea during the glacial interval [Hillaire-Marcel et al., 1994b], and sedimentation rates were relatively high at Orphan Knoll [e.g., *Stoner et al.*, 1998], this resulting in very low benthic foraminiferal concentrations.

As explained before, the extension of the σθ vs. δ¹⁸O calibration equations to the benthic record could lead to small biases in the estimation of σθ values for intervals when the (δfw – δsw) offset for the bottom water layer could have differed drastically from that of surface waters. The possibility of occasional penetration of Antarctic Bottom Waters in the deep Labrador Sea, during the last ice age, cannot be discarded [see *Bilodeau et al.*, 1994]. However, if proven, such anomalies would not modify significantly our interpretation: the parallel and close fluctuations of σθ values suggested by N.p.l. and benthic foraminifer data, in the glacial part of the sequence notably (Fig. 8d), rather suggest that they both record potential density changes respectively on top and bottom of a single water mass then occupying most of the water column. During a few intervals, the LGM and YD notably, N.p.l. and the benthic foraminifera depict almost similar σθ values (Fig. 8d), suggesting negligible density gradients in the whole water column, below the dilute surface water layer that characterized these climatic episodes.

Both N.p.l. and benthic foraminifera indicate a significant change in the thermohaline properties of the Labrador Sea starting at ~ 25.5 and ending at ~ 23 ka BP. It is marked by a near 2-unit decrease of σθ values throughout the water column (Fig. 8a), suggesting a major reorganization of thermohaline circulation. Independent evidence is provided by palynological data in a core raised from the southern Greenland Rise, at the entry of the WBUC that drives the NADW masses into the Labrador Sea (Fig. 1) [core 90-013-013; see *Hillaire-Marcel et al.*, 1994a]. In this core, reworked palynomorphs (Fig. 8c), originating from pre-Quaternary sedimentary units outcropping along the coastline of NE Greenland, show a contemporaneous concentration increase. This indicates the inception of strong currents that would have carried this material around Greenland, as far south as the coring site and as early as ~ 25 ka.

Another feature of the glacial record also arises from the comparison of the N.p.l. and benthic records (Fig. 8d). At the end of each H-event, and to some extend during the higher frequency climate oscillations of the glacial interval [see *Hillaire-Marcel and Bilodeau*, 2000], σθ values for the "N.p.l. layer", depict a short-duration negative shift of a 1 to 2 σθ unit, decoupling the N.p.l. record from the benthic one (Fig. 8d). Density conditions in surface water show concomitant fluctuations, also matched by peaks in ice rafting deposition [see *Clarke et al.*, 1999], suggesting episodes of iceberg spreading into the NW Atlantic. Fluctuations of potential density of the "N.p.l. layer" indicate

that these episodes were characterized by a deeper pycnocline, due to enhanced meltwater supplies leading to a thickening of the dilute surface water layer (up to a few hundred meters? [See *Hillaire-Marcel and Bilodeau*, 2000]). During such episodes, the habitat of N.p.l. would have been slightly shifted upwards, along the pycnocline. Finally, one should note that the N.p.l. and benthic σθ–curves are decoupled during the Holocene (Fig. 8d), in response to the development of a modern-like pycnocline between the intermediate and deep-water masses.

7.2. Constraints on the age and Amplitude of Maximum Cooling in the Deep Water Layer

The timing and succession of events, as illustrated by σθ records, cannot be readily reconciled with our current understanding of thermohaline circulation changes during the LGM and deglacial interval. The spreading of a low density and shallow surface water layer, between ~ 23 and 17 ka (Fig. 8a), i.e., during the LGM, could explain the contrasted seasonal regime with relatively warm summers but cold winters suggested for surface waters by most northern North Atlantic dinocyst records of the interval [*de Vernal et al.*, 2000]. However, the shift towards lower potential densities in deeper water masses that occurred as early as ~25.3-23 ka (Fig. 8a or d), seems more difficult to interpret. The possibility of a forcing of σθ values in the "benthic and N.p.l. layers" due to the methodological approach used, led us to examine further the benthic record.

In Figure 10, σθ values from the benthic records are compared for the 0-35 ka interval with values calculated from temperatures set arbitrarily at –1.8, 0 and 2°C, i.e., within a range compatible with the ecological requirements of C.w. and U.p., and using the boundary conditions of chapter 3 to constrain δsw and Sw values. During the Holocene interval, the σθ values indicated by C.w. fit with a 2°C growth temperature, i.e., a temperature near that observed today at sea floor at the study site (Fig. 2b). On the contrary, during the interval 25-25 the σθ values from U.p. rather indicate a much lower growth temperature (near – 1.8°C), whereas, during the preceding interval, somewhat intermediate temperatures could be inferred. We then used a reverse modeling approach, tuning temperatures through time, to make the "ecological curve" to fit with the benthic curve. The resulting temperature curve is illustrated Fig. 10a in comparison with isotopic records in benthic foraminifera and N.p.l. assemblages. It suggests Holocene bottom temperatures near the modern ones, possibly peaking at ca. 9 ka BP, and minimum bottom temperatures, between 25 and 15 ka BP. The absolute temperatures calculated, close to freezing point, are perhaps a little too low. Small offsets in the boundary conditions used could perhaps be evoked. Nevertheless, the age of this episode seems to match precisely that of the LGM, which would thus have been characterized by minimum bottom temperatures in the Labrador Sea. Interestingly, the temperature curve reconstructed for the 15-35 ka interval shows absolute minimums matching major shifts recorded by N.p.l. at the end of the H-events that we interpreted as an indication for deepening of the surface water layer (Fig. 10a). We have no immediate explanation for this fit, but it seems to indicate that there is some robustness into the reconstructed bottom temperature curve.

8. SUMMARY AND CONCLUSIONS

Rapid changes in the North Atlantic thermohaline circulation depend upon density gradient in the upper water column, notably in the Greenland and Labrador seas where winter convection allows the formation of deep or intermediate waters. Here, we have documented instabilities in the structure of the water column in the Labrador Sea during the last 100 ka using cores from Orphan Knoll, the outlet site of intermediate and deep waters into the North Atlantic. Proxies for the reconstruction of the upper water column density include transfer function based on dinocyst assemblages that provide constraints for the surface waters as well as a calibration equation used to transcribe isotopic measurements in epipelagic (*Globigerina bulloides*), mesopelagic (*Neogloboquadrina pachyderma*) and benthic foraminifera (*Uvigerina peregrina* and *Cibicides wuellerstorfi*), into potential densities. A special attention has been paid to the size dependence of the oxygen isotope composition in *Neogloboquadrina pachyderma* that provides an insight into salinity and temperature gradients along the pycnocline, between the base of the mixed layer and the intermediate water mass. Paleo-potential density estimates demonstrate large amplitude changes in the structure of the water column that are well beyond the degree of uncertainty intrinsic to the methodological approaches developed here.

A few major features can be retained:

i) Three distinct hydrographical situations are found to have existed throughout the last climatic cycle, but only one of them, which developed during postglacial times, allowed winter convection and formation of Labrador Sea Water to occur. Intermittent winter convection may have occurred as early as 16 ka BP. However, it is only since 7ka BP that it became a permanent feature. Production of LSW seems therefore to characterize essentially the mid-late Holocene interval.

ii) High-density conditions in surface waters are recorded during the following time intervals: ~ 78-72, ~ 68-64, 45-40 and 30-25 ka. They seem out of phase with maximum

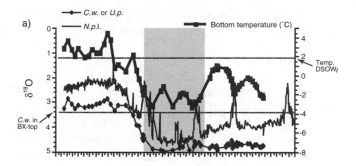

a)

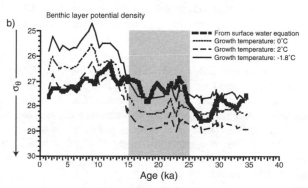

b)

Figure 10. Changes in benthic layer conditions for the 35-0 ka interval. (a) Top: temperature estimates from reverse modeling for the benthic layer (see section 7.2 in the text): note minimum bottom temperatures for the interval 15-25 ka and minimum peaks matching shifts in $\delta^{18}O$ values of N.p.l. assemblages during late H-events. (b) Bottom: potential density curve for C.w. or U.p. calculated from the surface water calibration equation *vs.* potential density estimates for growth temperatures in the range +2/-1.8°C (see text).

ice extend standstills, but rather correspond to ice advance phases, with reduced meltwater supplies.

iii) The most recent major change in the water column properties is dated between ~ 25.5 and ~ 23 ka BP. It corresponds to a ~2 to 3 unit decrease of potential density throughout the water column.

iv) Our interpretation also suggests cold conditions in bottom waters during H-events, and overall minimum bottom temperatures, almost 4°C below the modern ones, during the 25-15 ka BP interval.

v) Between ~ 23 and 17 ka, i.e., during the LGM, the spreading of a low density but shallow surface water layer could have been responsible for contrasted seasonal regimes, with relatively warm summers and cold winters.

vi) Two minimums in the paleo-potential density of the mesopelagic layer are observed at 15 ka (late H1 event) and 12 ka (late H0), respectively, i.e., before and just after the YD interval. They correspond to maximum dilution of the regional intermediate waters, likely to terminations 1A and 1B [*Duplessy et al.*, 1981; *Fairbanks*, 1989].

Over the past 100 ka, the hydrography of the northwestern North Atlantic experienced drastic changes, but we have not yet found past situations similar to the modern regime, suggesting that paleoceanographic data may not necessarily yield analogues to predict future changes. Moreover, our study provides evidence for an extremely sensitive hydrographic system of "fibrillation" that compares to that of records as sensitive as those from Greenland Ice cores. Finally, this study also indicates distinct response times, from millennial time scales for deepening of the pycnocline, to a few years for complete stop of winter convection, which suggest the effect of different forcing mechanisms.

Acknowledgements. This study is a contribution to the Climate System, History and Dynamics project, supported by the National Science and Engineering Research Council of Canada, and to the international Images program. Complementary support by the Fonds pour la Formation de chercheurs et l'Aide à la Recherche of the Quebec Province is acknowledged. Comments from Mark Maslin (University College, London) and an anonymous reviewer helped to improve the manuscript.

REFERENCES

Andreasen, D.J., and A.C. Ravelo, Tropical Pacific Ocean thermocline depth reconstruction for the last glacial maximum, *Paleoceanography*, *12*, 395-413, 1997.

Bé, A.W.H. and D.S. Tolderlund, Distribution and ecology of living planktonic foraminifera in surface waters of the Atlantic and Indian oceans. In: *The Micropaleontology of Oceans*, ed.: B.M. Funnel and W.R. Riedel, Cambridge University Press, Cambridge, 105-149, 1971.

Bé, A.W.H., C. Hemleben, O.R. Anderson, M. Spindler, J. Hacunda, and S. Choy, Laboratory and field observations of living planktonic foraminifera, *Micropaleontology*, *23*, 155-179, 1977.

Bédard, P., C. Hillaire-Marcel, and P. Pagé, ^{18}O-modelling of freshwater inputs in Baffin Bays and Canadian Arctic coastal waters, *Nature*, *293*, 287-289, 1981.

Bender, M. B., T. Sowers, M.-L. Dickson, J. Orchardo, P. Grootes, P. A. Mayewski, and D.A. Messe, Climate correlations between Greenland and Antarctica during the past 100,000 years, *Nature*, *372*, 663-666, 1994.

Billups, K. and D.P. Schrag, Surface ocean density gradients during the Last Glacial Maximum, *Paleoceanography*, *15*, 110-123, 2000.

Bilodeau, G., A. de Vernal, and C. Hillaire-Marcel, Benthic foraminifer assemblages in deep Labrador Sea sediments: relation with deep water mass changes since the deglaciation, *Canadian Journal of Earth Sciences*, *31*, 128-138, 1994.

Bond, G. C., W. Showers, M. Elliot, M. Evans, R. Lotti, I. Hajdas, G. Bonani, and S. Johnson. The North Atlantic's 1-2 kyr climate rhythm: Relation to Heinrich Events, Dansgaard/Oeschger Cycles and the Little Ice Age. In: *Mechanisms of global climate change at millennial time scales*, edited by P.U. Clark, R.S. Webb, L.D. Keigwin, AGU, Washington DC, 35-58, 1999.

Bond, G., W. Showers, M. Cheseby, R. Lotti, P. Almasi, P. de Menocal, P. Priore, H. Cullen, I. Hajdas, G. Bonani, A pervasive millennial-scale cycle in the North Atlantic Holocene and glacial climate, *Science*, *278*, 1257-1266, 1997.

Bond, G., Broecker, W., Johnsen, S., McManus, J., Laberyie, L., Jouzel, J. & Bonani, G., Correlations between climate records from North Atlantic sediments and Greenland ice, *Nature*, *365*, 143-147, 1993.

Candon, L., *Structure des populations et composition isotopique des foraminifères planctoniques dans le nord-ouest de l'Atlantique Nord (actuel versus dernier maximum glaciaire)*, M.Sc. Thesis, Université du Québec à Montréal, Montréal, 2000.

Candon, L., C. Hillaire-Marcel, C., and A. de Vernal, Size-dependent isotopic composition of epi- and meso-pelagic foraminifera versus sea-surface conditions based on dinocyst transfer functions: a probe into the halo-thermocline structure of the North Atlantic Ocean during the LGM. American Geophysical Union, Fall Meeting, San Francisco, Session U-5 (abstracts), 1999.

Carstens, J., and G. Wefer, Recent distribution of planktonic foraminifera in the Nansen Basin, Arctic Ocean, *Deep-Sea Research*, *39*, S507-S524, 1992.

Carstens, J., D. Hebbeln, and G. Wefer, Distribution of planktic foraminifera at the ice margin in the Arctic (Fram Strait), *Marine Micropaleontology*, *29*, 257-269, 1997.

Chapman, M.R., and M.A. Maslin, Low latitude forcing of meridional temperature and salinity gradients in the subpolar North Atlantic and the growth of glacial ice sheets, *Geology*, *27*, 875-879, 1999.

Chapman, M.R., and N.J. Shackleton, Global ice volume fluctuations, North Atlantic ice-rafting events, and deep ocean circulation between 130 and 70 ka, *Geology*, *27*, 795-799, 1999.

Cifelli, R., Observations *on Globigerina pachyderma* (Ehrenberg) and *Globigerina incompta* Cifelli from the North Atlantic, *Journal of Foraminiferal Research*, *3*, 157-166, 1973.

Clarke, G.K., S.J. Marshall, C. Hillaire-Marcel, G. Bilodeau, and C. Veiga-Pires, A Glaciological Perspective on Heinrich Events, In: *Mechanisms of global climate change at millennial time scales*, edited by P.U. Clark, R.S. Webb, L.D. Keigwin, AGU, Washington DC, 243-263, 1999.

Clarke, R.A., and J.-C. Gascard, The formation of the Labrador Sea Water. Part I: large scale processes, *Journal of Physical Oceanography*, *13*, 1764-1778, 1983.

Coplen, T.B. Normalization of oxygen and hydrogen isotope data, *Chemical Geology*, *72*, 293-297, 1988.

Dansgaard, W., S.J. Johnsen, H.B. Clausen, D. Dahl-Jensen, N.S. Gundestrup, C.U. Hammer, C.S. Hvidberg, J.P. Steffensen, A.E. Sveinbjirnsdottir, J. Jouzel, and G. Bond, Evidence for general instability of past climate from a 250-kyr ice-core record, *Nature*, *364*, 218-220, 1993.

de Vernal, A., and C. Hillaire-Marcel, Sea-ice, sea-surface salinity and the halo/thermocline structure in the northern North Atlantic: modern versus full glacial conditions, *Quaternary Science Reviews, 19*, 65-85, 2000.

de Vernal, A., C. Hillaire-Marcel, J.-L. Turon, and J. Matthiessen, Reconstruction of sea-surface conditions in the northern North Atlantic during the last glacial maximum based on dinocyst assemblages, *Canadian Journal of Earth Sciences, 37*, 725-750, 2000.

de Vernal, A., A. Rochon, J.-L. Turon, and J. Matthiessen, Organic-walled dinoflagellate cysts: palynological tracers of sea-surface conditions in middle to high latitude marine environments, *GEOBIOS, 30*, 905-920, 1997.

Dokken, T.M., and E. Jansen, Rapid changes in the mechanism of ocean convection during the last glacial period, *Nature, 401*, 458-461, 1999.

Duplessy, J.-C., G. Delibrias, J.-L. Turon, C. Pujol, and J. Duprat, Deglacial warming of the northeastern Atlantic Ocean. Correlation with the paleoclimatic evolution of the European continent, *Palaeogeography, Palaeoclimatology, Palaeoecology, 35*, 121-144, 1981.

Elderfield, H. and G. Ganssen, Past temperature and $\delta^{18}O$ of surface ocean waters inferred from foraminiferal Mg/Ca ratio, *Nature, 405*, 442-445, 2000.

Fairbanks, R.G., A 17 000-year glacio-isostatic sea level record: influence of glacial melting dates on the Younger Dryas event and deep ocean circulation, *Nature, 342*, 637-642, 1989.

Faul, K., A.C. Ravelo, and M.L. Delanay, Reconstructions of upwelling, productivity, and photic zone depth in the eastern equatorial Pacific ocean using planktonic foraminiferal stable isotopes and abundances, *Journal of Foraminiferal Research, 30*, 110-125, 2000.

Gill, A.E., *Atmosphere-Ocean Dynamics*. International Geophysics Series, volume 30, Academic Press Inc., San Diego, 662 p., 1982.

Grootes, P. M., and M. Stuvier, Oxygen 18/16 variability in Greenland snow and ice with 10^{-3} to 10^5–year time resolution, *Journal of Geophysical Research, 102*, 455-470, 1997.

Hillaire-Marcel, C., and G. Bilodeau, Instabilities in the Labrador Sea water mass structure during the last climatic cycle, *Canadian Journal of Earth Sciences, 37*, 795-809, 2000.

Hillaire-Marcel, C., A. de Vernal, G. Bilodeau, and G. Wu, Isotope Stratigraphy, sedimentation rates and paleoceanographic changes in the Labrador Sea, *Canadian Journal of Earth Sciences, 31*, 63-89, 1994a.

Hillaire-Marcel, C., A. de Vernal, M. Lucotte, A. Mucci, G. Bilodeau, A. Rochon, S. Vallières, and G. Wu, Productivité et flux de carbone dans la mer du Labrador au cours des derniers 40 000 ans, *Canadian Journal of Earth Sciences, 31*, 139-158, 1994b.

Hillaire-Marcel, C., A. de Vernal. S. Vallières, S., and on-board participants, *The Labrador Sea, The Irminger and Iceland basins. CSS-Hudson Cruise 91-045 Report and on-board studies*, Open file Report, Bedford Institute of Oceanography, Dartmouth (NS), Canada, 1991.

Keigwin, L.D., W.B. Curry, S.J. Lehman, and S. Johnsen, The role of the deep ocean in North Atlantic climate change between 70 and 130 kyr ago, *Nature, 371*, 323-326, 1994.

Kissel, C., C. Laj, L. Labeyrie, T. Dokken, A. Voelker, and D. Blamart, Rapid climatic variations during marine isotope stage 3: magnetic analysis of sediments from Nordic Seas and North Atlantic, *Earth and Planetary Science Letters, 171*, 489-502, 1999.

Kohfeld, K.E., R.G. Fairbanks, S.L. Smith, and I.D. Walsh, I.D., *Neogloboquadrina pachyderma* (sinistral coiling) as paleoceanographic tracers in polar oceans: Evidence from Northeast Water Polynya plankton tows, sediment traps, and surface sediment, *Paleoceanography, 11*, 679-699, 1996.

Laj, C., C. Kissel, A. Mazaud, J. E. T. Channell, and J. Beer, North Atlantic Paleointensity Stack since 75 ka (NAPIS-75) and the duration of the Laschamp event, *Philosophical Transaction of the Royal Society, Series A, 358*, 1009-1025, 2000.

Laj, C., A. Mazaud, and J.-C. Duplessy, Geomagnetic intensity and ^{14}C abundance in the atmosphere and ocean during the past 50 kyr., *Geophysical Research Letters, 23*, 2045-2048, 1996.

Lazier, J.R., Temperature and salinity changes in the deep Labrador Sea 1962 – 1986. *Deep-Sea Research, 35*, 1247-1253, 1988.

Lea, D.W., T.A. Mashiotta, and H.J. Spero, Controls on magnesium and strontium uptake in planktonic foraminifera determined by live culturing, *Geochimica Cosmochimica Acta, 63*, 2369-2379, 1999.

Lucotte, M., and C. Hillaire-Marcel, Identification et distribution des grandes masses d'eau dans les mers du Labrador et d'Irminger, *Canadian Journal of Earth Sciences, 31*, 5-13, 1994.

Martinson, D.G., N.G. Pisias, J.D. Hays, et al., Age dating and the orbital theory of the Ice Ages : development of a high-resolution 0 to 300,000-year chronostratigraphy, *Quaternary Research, 27*, 1-29, 1987.

Maslin, M.E., N.J. Shackleton, and U. Pflaumann, Surface water temperature, salinity, and density changes in the northeast Atlantic during the last 45,000 years: Heinrich events, deep water formation, and climatic rebounds, *Paleoceanography, 10*, 527-544, 1995.

Meese, D. A., R.B. Alley, A. J. Gow., P. M. Grootes, P. A. Mayewski, M. Ram, K. C. Taylor, I. E. Waddington, and G. A. Zielinski. *Preliminary depth-age scale for the GISP2 ice core*, Spec. Rep. 94-1, 66 p. Cold Reg. Res. and Eng. Lab., Hanover, N. H., 1994.

NODC (National Oceanographic Data Center), *World Ocean Atlas*. National Oceanic and Atmospheric Administration, Data Sets on CD-Rom, 1994.

Nürnberg, D., A. Müller, and R.R. Schneider, Paleo-sea surface temperature calculations in the equatorial east Atlantic from Mg/Ca ratios in planktic foraminifera: A comparison to sea surface temperature estimates from $U^{K'}_{37}$, oxygen isotopes, and foraminiferal transfer function, *Paleoceanography, 15*, 124-134, 2000.

O'Neil, J.K., K.N. Clayton, and T.K. Mayeda, Oxygen isotope fractionation in divalent metal carbonates, *Journal of Chemical Physics, 51*, 5547-5558, 1969.

Plouffe, D., *Variations morphologiques et composition isotopique* de Neogloboquadrina pachyderma *levogyre dans la mer du Labrador*, B.Sc. dissertation, Université du Québec à Montréal, Montréal, 1994.

Rahmstorf, S., Shifting seas in the Greenhouse ? *Nature, 399*, 523-524, 1999.

Rasmussen T. L., E. Thompsen, T.C.E van Weering, and L. Labeyrie, Rapid changes in surface and deep water conditions at the Faeroe margin during the last 58,000 years, *Paleoceanography, 11*, 757-771, 1996.

Ravelo, A.C. and R.G. Fairbanks, Oxygen isotyopic composition of multiple species of planktonic foraminifera: Recorders of the modern photic zone temperature gradients, *Paleoceanography, 5*, 409-431, 1992.

Rochon, A., A. de Vernal, J.-L. Turon, J. Matthiessen, and M.J. Head, *Distribution of dinoflagellate cyst assemblages in surface sediments from the North Atlantic Ocean and adjacent basins and quantitative reconstruction of sea-surface parameters*, Special Contribution Series, American Association of Stratigraphic Palynologists, Dallas, n° 35, 1999.

Sautter, L.R., and R.C. Thunell, Seasonal succession of planktonic foraminifera: results from a four-year time-series sediment trap experiment in the northeast Pacific, *Journal of Foraminiferal Research, 19*, 253-267, 1989.

Shackleton, N.J., Attainement of isotopic equilibrium between ocean water and the benthic foraminifera genus *Uvigerina*: isotopic changes in the ocean during the last glacial, in *Méthodes quantitatives d'étude des variations du climat au cours du Pléistocène*, edited by J. Labeyrie, pp. 203-209, Editions du C.N.R.S., France, 1974.

Shackleton, N.J., A. Berger, and W.R. Peltier, An alternative astronomical calibration of the lower Pleistocene time scale based on ODP Site 677, *Transactions of the Royal Society of Edimburg: Earth Sciences, 81*, 251-261, 1990.

Spero, H.J., and D.W. Lea, Experimental determination of stable isotope variability in *Globigerina bulloides*: implications for paleoceanographic reconstructions, *Marine Micropaleontology, 28*, 231-246, 1996.

Srinivasan, M.S., and J.P. Kennett, Secondary calcification of the planktonic foraminifer *Neogloboquadrina pachyderma* as a climatic index, *Science, 186*, 630-632, 1974.

Stoner, J.S., J.E.T. Channell, and C. Hillaire-Marcel, Late Pleistocene relative geomagnetic paleointensity from the deep Labrador Sea - Regional and global correlations, *Earth and Planetary Science Letters, 134*, 237-252, 1995.

Stoner, J. S., J.E.T. Channell, and C. Hillaire-Marcel, The mag-

netic signature of rapidly deposited detrital layers from the deep Labrador Sea: relationship to North Atlantic Heinrich layers, *Paleoceanography, 11*, 309-325, 1996.

Stoner, J.S, J.E.T. Channell, and C. Hillaire-Marcel, A 200 kyr geomagnetic stratigraphy for the Labrador Sea: Indirect correlation of the sediment record to SPECMAP. *Earth Planetary Science Letters, 159*, 165-181, 1998.

Stoner, J.S, J.E.T. Channell, C. Hillaire-Marcel, and C. Kissel, Geomagnetic paleointensity and environmental record from Labrador Sea Core MD-2024: global marine sediment and ice core chronostratigraphy for the last 110 kyrs, *Earth Planetary Science Letters, 183*: 161-177, 2000.

Sverdlove, M.S., and A.W.H. Bé, Taxonomic and ecological significance of embryonic and juvenile planktonic foraminifera, *Journal of Foraminiferal Research, 15*, 235-241, 1985.

Sy, A., M. Rhein, J.R.N. Lazier, et al., Surprisingly rapid spreading of newly formed intermediate waters across the North Atlantic Ocean, *Nature, 386*, 675-679, 1997.

Turon, J.-L., C. Hillaire-Marcel, and ship board party, *Cruise Report on the 2nd leg of IMAGES V (Quebec-Reykjavik)*. Open-file Report., 3782, 277 p., Geological Survey of Canada (Atlantic, Dartmouth, NS, Canada), 1999.

Vidal, L., L. Labeyrie, E. Cortijo, M. Arnold, J.C. Duplessy, E. Michel, S.Becqué, and T.C.E van Weering, Evidence for changes in the North Atlantic Deep Water linked to meltwater surges during Heinrich events, *Earth and Planetary Science Letters, 146*, 13-27, 1997.

Wood, R.A., A.B. Keen, J. F. B. Mitchell, and J. M. Gregory, Changing spatial structure of the thermohaline circulation in response to atmospheric CO_2 forcing in a climate model, *Nature, 399*, 572-575, 1999.

Wu, G.-P., and C. Hillaire-Marcel, Oxygen isotope compositions of sinistral *Neogloboquadrina pachyderma* tests in surface sediments: North Atlantic Ocean. *Geochimica Cosmochimica Acta, 58*, 1303-1312, 1994a

Wu, G.-P., and C. Hillaire-Marcel, AMS radiocarbon stratigraphies in deep Labrador Sea cores: paleoceanographic implications. *Canadian Journal of Earth Sciences*, 31, 38-47, 1994b.

Claude Hillaire-Marcel, Anne de Vernal, Laurence Candon and Guy Bilodeau, GEOTOP-UQAM, C.P. 8888, succursale "Centre-Ville", Montreal (Qc) H3C 3P8 Canada

Joseph Stoner, Department of Geology, One Shields Avenue, University of California, Davis, CA 95616

Lower Circumpolar Deep Water Flow Through the SW Pacific Gateway for the Last 190 ky: Evidence From Antarctic Diatoms

Catherine E. Stickley[1], Lionel Carter[2], I. Nick McCave[3], and Phil P.E. Weaver[4]

Endemic Antarctic diatoms are incorporated into downwelling Antarctic Bottom Water (AABW) during its formation in the Weddell and Ross Seas and at the sea-ice interface on the shelves around Antarctica. We infer their subsequent entrainment in Lower Circumpolar Deep Water (LCDW) during upwelling of AABW into Lower LCDW around Antarctica. Antarctic diatoms are very effective tracers of Lower LCDW in the world's oceans. Significant quantities of displaced diatoms are recovered in core-tops along the known flow-path of Lower LCDW in the SW Pacific Gateway, east of New Zealand. In contrast, core-tops bathed in overlying water masses such as core North Atlantic Deep Water (NADW) are devoid of such diatoms. The interface between Lower LCDW and core NADW, at 3735-4097 m depth, is clearly defined in this region by the presence/absence of tracer diatoms, which closely matches the hydrographical observed interface at 3800 m. Tracer diatoms in cores located within the palaeo-LCDW flow-path may provide information on the flow of LCDW through the gateway for the last 190 ky. However, is it difficult to eliminate the effects of diatom productivity and preservation. AABW production rate and its mixing rate with LCDW will also have an effect. However, reworked diatoms suggest that Lower LCDW flow was greater during colder times. Separation of total tracers into open-ocean and sea-ice related taxa reveal variations in AABW source for the last 190 ky. AABW formed in the open-ocean appears to be relatively enhanced during colder times, while that formed over the shelf is relatively enhanced in warmer times.

[1]Environmental Change Research Centre, University College London, London, UK

[2]National Institute of Water and Atmospheric Research (NIWA), Kilbirnie, Wellington, New Zealand

[3]Department of Earth Sciences, Downing Street, Cambridge University, Cambridge, UK

[4]Southampton Oceanography Centre, University of Southampton, Southampton, UK

The Oceans and Rapid Climate Change: Past, Present, and Future
Geophysical Monograph 126

INTRODUCTION

A fundamental controlling factor of world climate is the circulation of cold, deep, oceanic water around the globe [e.g., *Broecker and Denton*, 1990; *Keigwin et al.*, 1994; *Broecker*]. Much emphasis has been placed on the role of North Atlantic Deep Water (NADW) production as a driver of thermohaline circulation [e.g., *Gordon*, 1986; *Gordon et al.*, 1992; *Keigwin et al.*, 1994; *Schmitz*, 1995]. The switching on and off of this watermass at various times in the past is thought to have been implicated in climate changes. However, it has recently been suggested that the Southern Ocean [e.g., *Seidov et al.*, 2000; *Seidov et al.*, 2001] and, in particular, deep-water formed around

Antarctica [e.g., Denton, 2000], may have a greater significance in climate dynamics than previously thought. An understanding of the nature and timing of fluctuations in production of these deep watermasses, and teleconnections, is fundamental for climate modelling and prediction [e.g., *Denton*, 2000].

Debate on Southern Ocean ventilation during glacial-interglacial cycles has focused on the sudden switching off of deep water formation caused by dedensification [*Broecker*, 2000]. The formation of Antarctic Bottom Water (AABW) over Antarctic shelves at the sea-ice interface is effectively halted when ice-sheets ground. This is thought to have occurred during glacial periods [*Denton*, 2000] particularly the last glacial [*Kellogg*, 1987; *Pudsey*, 1992]. Production rates of AABW formed over the shelf are presumed to have been increased during interglacial periods when marginal seas were opened following basal melting and grounding line retreat [*Denton*, 2000]. Denton [2000] suggests that the reintroduction of NADW to the Southern Ocean during interglacials [e.g., *Boyle*, 1995] triggered sea-level rise and Antarctic ice-sheet basal melting (e.g., the West Antarctic Ice Sheet). Conversely, grain size analyses indicate a reduced southern-source deep-water flow into the SW Atlantic during the last interglacial [e.g., *Ledbetter*, 1979], and an enhanced glacial deep western boundary current in the SW Pacific [e.g., *Hall et al.*, 2000]. Changes in flux of NADW or southern-source deep-water at any one time can affect the relative influence of the other in the world's oceans. For instance, reduced NADW influence in glacials in the North Atlantic is matched by an increased influence of 'Antarctic Bottom Water' [*Vidal et al.*, 1997].

This paper addresses the issue of Lower Circumpolar Deep Water (LCDW) flow through the SW Pacific Gateway over the past 190 ky, via an examination of the composition and abundance of a biological tracer (endemic Antarctic diatoms) from cores along its flow-path within Deep Western Boundary Current (DWBC) inflow to the SW Pacific. We speculate on the relative importance of the two main sources of AABW over this time period. So far, there has been very little speculation on possible switches in source during glacial-interglacial cycles. By treating the two components of AABW separately it may be possible to explain some of the discrepancies on ideas of Southern Ocean ventilation, and therefore further improve climate models.

Antarctic Bottom Water (AABW) and Lower Circumpolar Deep Water (LCDW)

In this paper, we use the definitions by Orsi et al. [1999] for two discrete bottom waters of southern origin, Antarctic Bottom Water (AABW) and Lower Circumpolar Deep Water (LCDW). Orsi et al. [1999] define the generic term AABW as the total volume of southern bottom waters being denser than Circumpolar Deep Water (CDW), and which are formed south of the Antarctic Circumpolar Current (ACC). AABW is not circumpolar in its spatial distribution being too dense to flow over the sill of the Drake Passage [*Orsi et al.*, 1999]. Regional names, such as Weddell Sea Deep Water (WSDW), are retained for the several varieties of AABW formed at different areas around Antarctica. LCDW is the lower part of the CDW, a voluminous deep watermass, carried in the ACC around Antarctica. The LCDW is the only watermass of Antarctic origin which is exported to the major ocean basins in the thermohaline flow via the Southern Hemisphere western boundary currents [*Orsi et al.*, 1999]. In the SW Pacific, the LCDW comprises an upper high-salinity (S=34.72-34.73 psu) layer at 2900-3800 m depth. This is the signature of the NADW [*McCave and Carter*, 1997; *Whitworth et al.*, 1999] transported in the ACC. This upper layer is referred to as core NADW herein. The lower layer of the LCDW (herein Lower LCDW) is a cold, lower salinity (S=34.68 psu) zone occupying depths of >3800 m [*McCave and Carter*, 1997]. Its density is attributed to Weddell Sea Deep Water and Ross Sea Deep Water, which is mixed with NADW in the ACC [*Gordon*, 1975].

AABW is formed in two main areas; 1. at the sea-ice interface in coastal and shelf regions, and 2. in the open-ocean [*Gordon*, 1971, 1983, 1988; *Warren*, 1981; *Rintoul*, 1998; *Rintoul and Bullister*, 1999]. Shelf-AABW is formed around much of Antarctica whereas open-ocean AABW production is mainly associated with the gyre systems in the Weddell Sea and Ross Sea [e.g., *Gordon*, 1971, 1975, 1983]. Shelf-AABW is formed within coastal latent heat polynya when katabatic winds push newly formed sea-ice away from the coast. As further sea-ice forms over the exposed area, brine is formed within the surface waters due to salt rejection. Density instability in the water column causes surface water sinking and drainage over the continental slope into the deep ocean. In this way, the process of sea-ice production and its continual removal can generate vast amounts of sea-ice and AABW. Open-ocean sensible heat polynya also generate AABW where surface water is super-cooled via heat loss to the atmosphere. It is thought they are maintained by an overturning convection cell comprising a component of upward-moving warm, deep water and a component of downward-moving cold, surface water [e.g., *Gordon*, 1988]. The Weddell Sea was thought to be the main source of AABW in the Southern Ocean [e.g., *Carmack and Foster*, 1975; *Reid et al.*, 1977], although recent work has shown a very significant and greater production rate along the Wilkes-Adélie coast

(140-150^{0}E) [*Orsi et al.*, 1999; *Rintoul and Bullister*, 1999].

The upper layer of AABW is constantly warmed as it mixes with the overlying Lower LCDW. It is upwelled into Lower LCDW, a mechanism which continuously replenishes the latter [*Orsi et al.*, 1999]. By this mechanism, modified AABW is able reach the major oceanic basins via the thermohaline flow.

Entrainment of Antarctic Diatoms in AABW and LCDW

Diatoms are particularly useful for palaeoceanographic and palaeoclimatic reconstructions in areas of the oceans where carbonate production and/or preservation is minimal. For instance, in the Southern Ocean, diatoms dominate primary production and are the main opal contributor to sediments around Antarctica [e.g., *Zielinski and Gersonde*, 1997]. Some species of Antarctic diatom have a preference for open-ocean Antarctic and Subantarctic waters (e.g., *Fragilariopsis kerguelensis, Thalassiosira lentiginosa*) and others for near-shore, ice, and ice-edge regions (e.g., *F. curta, F. cylindrus*) [e.g., *Burckle et al.*, 1987; *Zielinski and Gersonde*, 1997; *Crosta et al.*, 1998; *Gersonde and Zielinski*, 2000].

The idea that Antarctic diatoms could be downwelled into AABW during its formation in Antarctica [*Walsh*, 1968], was developed by Burckle and Stanton [1974] and Booth and Burckle [1976]. Endemic Antarctic diatoms are entrained in AABW during its formation in the open-ocean (e.g., the Weddell Sea) and at the sea-ice interface on the shelf. The diatoms are then presumably entrained in Lower LCDW during the upwelling of underlying AABW [*Orsi et al.*, 1999], since they are found along its modern flow path, effectively becoming 'tracers' of Lower LCDW (e.g., Vema Channel, SW Atlantic; Valerie Passage, SW Pacific). Booth and Burckle [1976] validated this by comparison with other flow-proxies (e.g., sedimentary bedforms). These tracer diatoms become a permanent record of Lower LCDW flow [*Burckle and Stanton*, 1974; *Booth and Burckle*, 1976], although some resuspension and redeposition along the flow-path is expected during times of high flow.

Downcore fluctuations in tracer diatoms can provide temporal information on palaeo-LCDW flow. However, it is uncertain if diatom flux is linearly proportional to flow. More importantly, it cannot be assumed, that diatom flux is solely related to changes in LCDW flow. Other factors including AABW production rate and subsequent mixing with LCDW around Antarctica are also important, as is diatom productivity and preservation. Effectively, the diatom signal is a combination of all these factors. In this pa-

per, we consider diatom productivity and preservation in the signal to attempt to eliminate their effects. However, it is difficult to speculate with any certainty on the separate effects of all these factors.

Antarctic diatoms may also incorporated into AAIW during its formation at the Antarctic Polar Front (APF) [e.g., *Ettwein et al.*, 2000]. However, this mechanism does not explain the high flux of sea-ice related diatoms in the SW Pacific (see below). Fenner et al. [1992] report on the presence of rare Antarctic diatoms north of the STF on the North Chatham Rise. They attribute this to occasional transport via eddies across the STF. However, the high numbers of such diatoms in the Valerie Passage cores can not be explained by this method.

Since AABW has an origin in surface waters and diatoms dominate primary production in areas of AABW formation, no other biological tracer can contribute as greatly as the diatoms towards monitoring changes in flux of Antarctic-source deep-waters. Moreover, with advances in understanding the ecology and surface-water distribution of Antarctic diatoms [e.g., *Zielinski and Gersonde*, 1997; *Gersonde and Zielinski*, 2000], and with the careful selection of taxa in analysis, diatoms provide the unique opportunity to infer temporal switches of AABW source.

REGIONAL SETTING

This work focuses on an area of the SW Pacific, east of New Zealand called the Eastern New Zealand Oceanic Sedimentary System (ENZOSS) [c.g., *L. Carter et al.*, 1996; *R.M. Carter et al.*, 1996] (Figure 1). The ENZOSS lies close to the obliquely collisional boundary between the Pacific and Australian plates, and defines a system of oceanographic, sedimentary and tectonic connections. A major feature of the ENZOSS is the Deep Western Boundary Current (DWBC), part of the global themohaline flow. South of latitude 49^{0}S, the DWBC interacts with the wind-driven eastward Antarctic Circumpolar Current (ACC) [*Carter and McCave*, 1994; *Carter and Wilkin*, 1999]. At about this latitude, the ACC and associated eddies break away and continue heading east at a volume transport of ~115 x 10^6 m^{3}s^{-1} [e.g., *R.M. Carter et al.*, 1996], while the DWBC heads northwards into the central Pacific Ocean. Today, with a flux of 15.8 ± 9.2 x 10^6 m^{3}s^{-1} [*Whitworth et al.*, 1999], the DWBC transports 40% of the worlds deep water into the Pacific Ocean [*Warren*, 1973, 1981]. In this respect, the SW Pacific Gateway is ideal for studying changes in deep ocean circulation.

Surface hydrography for the region as far south as 60^{0}S is illustrated in Figure 2. Four surface-water masses (Antarctic Water (AAW), Circumpolar Subantarctic Water (CSW), Australasian Subantarctic Water (ASW) and Sub-

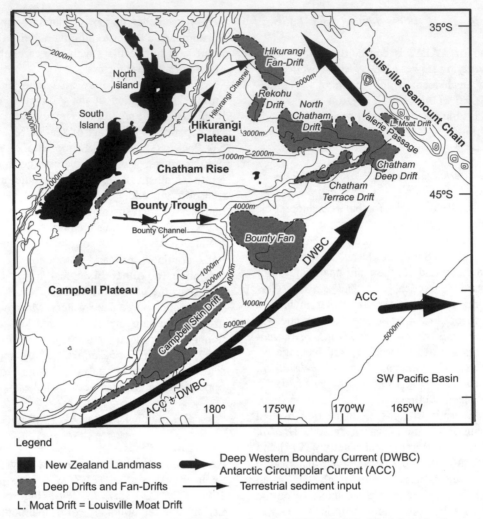

Figure 1. The Eastern New Zealand Oceanic Sedimentary System (ENZOSS). Modified from L. Carter et al. [1996], R.M. Carter et al. [1996] and Carter et al. [1998].

tropical Water (STW)) are separated by three ocean fronts (Antarctic Polar Front (APF), Subantarctic Front (SAF) and Subtropical Front (STF)). Typical summer sea-surface temperatures for the APF, SAF and STF are 5°C and 8°C and 15°C, respectively [*Heath*, 1985]. The STF is located over the crest of the Chatham Rise where it is contained by the East Cape Current in the north and the Southland Current in the south. Although the STF can move by up to 2° latitude, its position over the rise appears to have been stable during the Holocene at least [*Fenner et al.*, 1992; *Nelson et al.*, 1993]. However, it may have shifted slightly northwards at the last glaciation [*Weaver et al.*, 1998].

Subsurface watermasses described above are indicated in Figure 3. The DWBC carries core NADW and Lower LCDW northwards. Intermediate watermasses are the North Pacific Deep Water (NPDW) travelling south and the Antarctic Intermediate Water (AAIW) travelling north.

The DWBC has helped create and shape a number of deep sea drifts and a fan-drift at a variety of depths (Figure 1). This paper concerns two cores positioned directly within the flow path of the DWBC. They are located in the Valerie Passage (Figures 1 and 4), a major corridor for DWBC flow as it swings around the eastern apex of the Chatham Rise (Figure 1). CHAT 3K (4802 m depth) is from the Chatham Deep Drift, a ridge-like drift around the base of the east Chatham Rise, and CHAT 11K (4900 m depth) is from the Louisville Moat Drift.

METHODS

Core and Core-top Material

Two cores and sixteen core-tops from the ENZOSS are used in this study (Figure 4). All sites are located north of

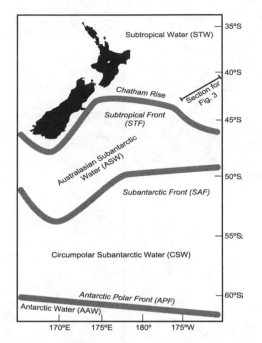

Figure 2. Surface Water Masses and Oceanic Fronts of the ENZOSS. Modified from Nelson et al. [1993] (after Heath [1985]) and Carter et al. [1998].

1993) cruises of the National Institute for Water and Atmospheric Research (NIWA), New Zealand. Core-tops, collected during various NIWA cruises, were chosen to provide a depth transect through the subsurface water masses. All sediments were sampled at NIWA and at core repositories at the Southampton Oceanography Centre and the Department of Earth Sciences, Cambridge University. Cores were sampled at 10 cm intervals and at a higher resolution over important intervals, giving a sampling resolution of ca. 1-7 ky. This relatively low resolution study aims at establishing general trends in order to identify areas for further refinement. All ages are presented in calendar years B.P. The notation ka is used to denote an event 'thousands of years ago' whilst ky refers to an interval of time.

Diatom Analysis

Diatom processing was undertaken on core and core-top material according to the technique outlined in Stickley [1998] based around the waterbath method of Battarbee [1986]. Divinylbenzene (DVB) microspheres [*Battarbee and Kneen*, 1982] were introduced to the cleaned samples prior to slide preparation to determine absolute abundance of diatoms (valves per gram dry weight, gdw^{-1}). All diatoms (Antarctic and others) were analysed, and 350-400 diatom valves, per sample, were counted at a magnification of x1000. Counting procedure followed the scheme of Schrader and Gersonde [1978] with modifications. Abso-

the STF (Figure 2). Water depth data and associated watermasses are provided in Table 1. Kasten cores CHAT 3K and CHAT 11K were recovered during the *Raphuhia* 2050 (October-November 1991) and *Lavrentyev* 3011 (August

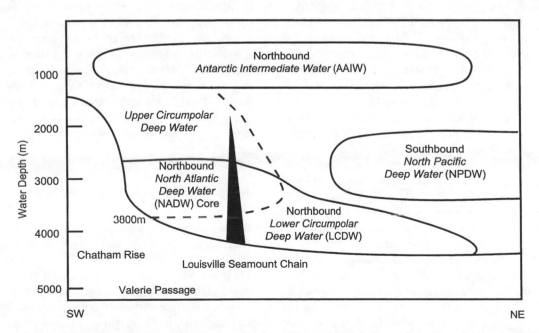

Figure 3. Subsurface Water Masses of the Eastern Chatham Rise and Valerie Passage Region. Section indicated in Figure 2. Modified after R.M. Carter et al. [1996].

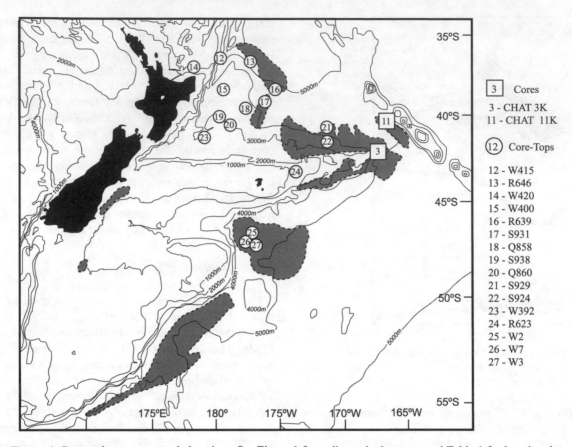

Figure 4. Core and core-top sample locations. See Figure 1 for sediment body names and Table 1 for location data.

lute abundances were converted to diatom accumulation rates (DAR, valves cm^{-2}ky^{-1}) for core samples, to nullify the effects of dilution. DAR is the product of absolute diatom abundance (valves gdw^{-1}), salt corrected dry bulk density (DBD, gcm^{-3}) and linear sedimentation rate (SR, cmky^{-1}) [DBD x SR = dry mass accumulation rate, gcm^{-2}ky^{-1}].

In order to analyse variations in AABW source, Antarctic diatoms were divided into two groups according to their known ecological preference; 1. those with a preference for open-ocean waters, and 2. those with a preference for nearshore, sea-ice or ice-edge environments. The open-ocean group [e.g., *Fenner et al.*, 1976; *Burckle*, 1984a, 1987; *Burckle et al.*, 1987; *Zielinski and Gersonde*, 1997; *Crosta et al.*, 1998] comprises *Fragilariopsis kerguelensis*, *F. ritscheri*, *F. separanda*, *Thalassiosira lentiginosa* and *T. gracilis*. The nearshore and sea-ice related group [e.g., *Burckle*, 1984a; *Burckle et al.*, 1987; *Zielinski and Gersonde*, 1997; *Crosta et al.*, 1998; *Gersonde and Zielinski*, 2000] consists of *Eucampia antarctica*, *F. curta*, bipolar *F. cylindrus* and the *T. antarctica-scotia* group. Zielinski and Gersonde [1997] dispute the association of *E. antarctica* with sea-ice [e.g., Burckle, 1984b] but we retain it in this group since its downcore distribution is very similar to

those of other sea-ice related species (and not those of the open-ocean group). This may point to subtle variations in source water, but its presence is not common enough in the SW Pacific sediments, to make any firm conclusions. *F. curta and F. cylindrus* are the most abundant species of the nearshore, sea-ice and ice-edge group in these sediments. Ratios of one group to the other are used to highlight variations in the AABW source. 'Ice-edge' diatoms (herein) refer to the nearshore, sea-ice and ice-edge group described above.

Stable Oxygen Isotopes

Stable oxygen isotope measurements (δ^{18}O, ‰VPDB) from specimens of benthic (*Uvigerina* spp.) and planktonic (*Globigerina bulloides*) foraminifera were made for CHAT 3K using VG Sira Series II and Prism mass spectrometers. The data are used to provide a chronology. CHAT 11K contains calcium carbonate only at discrete intervals intercalated by intervals of almost no carbonate. Oxygen isotope data from this core are therefore insufficient to construct a reliable isotope stratigraphy and are not presented

Table 1. Location data for cores and core-tops (see Figure 4). Watermasses indicated: AAIW - Antarctic Intermediate Water; LCDW - Lower Circumpolar Deep Water; NADW - North Atlantic Deep Water; NPDW - North Pacific Deep Water.

Core/Core-top	Water Depth (m)	Latitude (Degrees + decimalised minutes)	Longitude (Degrees + decimalised minutes)	Water Mass
CHAT 3K (core)	4802	42.6598 S	167.4962 W	LCDW
CHAT 11K (core)	4900	40.3088 S	166.8917 W	LCDW
W415	2925	36.3928 S	179.7403 W	core NADW
R646	5321	36.6348 S	177.5000 W	LCDW
W420	2177	37.0112 S	178.0945 E	NPDW
W400	3595	38.1353 S	179.8697 W	core NADW
R639	4630	38.4420 S	175.7400 W	LCDW
S931	4097	39.4567 S	176.4183 W	LCDW
Q858	3735	39.8267 S	178.0583 W	core NADW
S938	3003	40.0328 S	179.9957 E	core NADW
Q860	3219	40.2483 S	179.0800 W	core NADW
S929	4240	40.7833 S	171.5490 W	LCDW
S924	3556	41.5833 S	171.5000 W	core NADW
W392	2573	41.5952 S	178.8685 E	NPDW
R623	1128	43.2000 S	174.0000 W	AAIW
W2	4296	46.4242 S	177.8050 W	LCDW
W7	4475	46.6110 S	178.0560 W	LCDW
W3	4683	47.2000 S	177.2330 W	LCDW

in this paper. The diatom data are compared to $\delta^{18}O$ from CHAT 3K in Figure 8.

Chronology

The age-model for CHAT 3K is based on a comparison of the $\delta^{18}O$ curves with the SPECMAP time scale of Martinson et al. [1987]. Further control is provided by dated tephra layers originating from the Central Volcanic Region of the North Island of New Zealand. The age-model for CHAT 11K is based on carbonate and diatom stratigraphy. The lowermost sample taken from CHAT 3K (384 cm) is dated at 193 ka, while that from CHAT 11K (342 cm) is dated at 169 ka. We assume no loss at the top of both cores. CHAT 3K, therefore, extends from oxygen isotope stage (OIS) 1 to latest OIS 7, while CHAT 11K extends from OIS 1 to mid OIS 6. Linear sedimentation rates for both cores average ~1.4-1.6 cmky^{-1} for the period 193-28 ka, increasing to 4.4-5.8 cmky^{-1} for the last 28 ky.

RESULTS AND DISCUSSION

Endemic Antarctic diatoms and those local to the ENZOSS are recovered from the analysed material. Local species include temperate-subtropical taxa as well as *Chaetoceros* resting spores associated with high productivity over the Chatham Rise [e.g., *Fenner et al.*, 1992; *Stickley*, 1998], and associated with the STF [*Heath*, 1985; *Bradford-Grieve et al.*, 1999]. Data for the southern source species only are presented here. The open ocean diatom, *F. kerguelensis*, is by far, the most abundant Antarctic diatom in the core material but also the most abundant overall diatom comprising 41-83% of the entire (local and Antarctic) diatom assemblage in CHAT 11K. This reflects the dominance of this species in Antarctica. It is one of the main opal contributors to sediments of the Southern Ocean comprising 60-90% of total species in core-tops of Antarctica [*Zielinski and Gersonde*, 1997]. Its dominance in the ENZOSS cores and core-tops is something of an enigma since open-ocean southern source water is thought to contribute <10% of the total entering the SW Pacific [*Whitworth et al.*, 1999], see Variations in AABW Source. *Fragilariopsis kerguelensis* in CHAT 3K comprises 15-57% of the entire diatom assemblage in this core. This relative abundance value is lower than for CHAT 11K due to dilution by local diatoms associated with high productivity over the Chatham Rise and STF. Other endemic Antarctic and Subantarctic diatoms recovered include: *Actinocyclus actinochilus, A. exiguus, Azpeitia tabularis, Fragilariopsis rhombica, F. sublinearis, Odontella weissflogii, Porosira pseudodenticulata, Rhizosolenia bergonii, Stellarima microtrias,* and *Thalassiosira oliverana.*

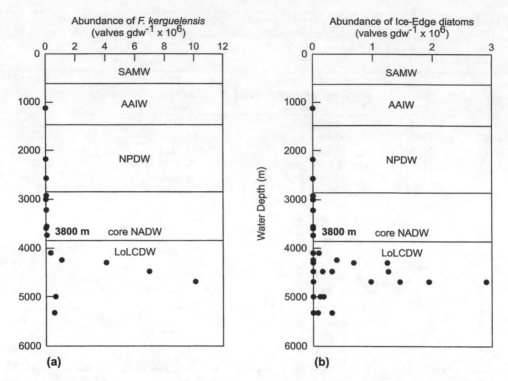

Figure 5. Absolute abundance of Lower LCDW-flow tracer diatoms as a function of water depth (core-tops). Water masses indicated: Subantarctic Mode Water (SAMW, 0-600 m); Antarctic Intermediate Water (AAIW, 600-1400 m); North Pacific Deep Water (NPDW, 1400-2900 m); North Atlantic Deep Water core (NADW, 2900-3800 m); Lower Lower Circumpolar Deep Water (loLCDW, >3800 m). (s) *Fragilariopsis kerguelensis* (open-ocean diatom), (b) Nearshore, sea-ice and ice-edge diatoms (*F. curta, F. cylindrus, Eucampia antarctica, Thalassiosira antarctica-scotia* group).

Core-Tops

Core-top material averages the last few hundred to few thousand years due to bioturbation. Sea-floor sediments, therefore, do not necessarily represent strictly modern day flow. However, our core-top material is still useful in establishing information about flow in the late Holocene. Core-tops (Figure 4; Table 1) were analysed for Antarctic diatoms along a depth-transect through the subsurface watermasses (Figure 3). Unfortunately, Subantarctic Mode Water (SAMW) is not represented and AAIW and NPDW are represented by just one and two core-tops, respectively. However, the main watermasses of interest, the core NADW and Lower LCDW are well-represented.

Displaced Antarctic diatoms are present in significant amounts (e.g., on the order 10^6 valves gdw^{-1}) in core-tops within the flow path of the modern Lower LCDW (Figure 5). Their preservation is good with very little signs of dissolution. Core-tops bathed in overlying water masses (e.g., core NADW) are barren of such diatoms but contain abundant local species. The presence/absence of Antarctic diatoms, between 4097 m (sample S931, tracers common) and

3735 m (sample Q858, tracers absent), defines the interface between Lower LCDW and core NADW and is in excellent agreement with the physical oceanographic observation of 3800 m [e.g., *R.M. Carter et al.*, 1996]. The data for *F. kerguelensis* (Figure 5a) are given as an example of open-ocean conditions in the Weddell and Ross Seas. However, the other species in the open-ocean group are also present in significant quantities in the Lower LCDW-bathed core-tops. Nearshore, sea-ice and ice-edge diatoms also define the Lower LCDW/core NADW interface clearly (Figure 5b). The data indicate that Antarctic diatoms are incorporated into AABW and Lower LCDW as described above and illustrated in Figure 6.

Similar results were found in core-tops from the Vema Channel of the SW Atlantic [*Burckle and Stanton*, 1974; *Jones and Johnson*, 1984]. However, the quantity of tracer diatoms recovered from the SW Pacific [*Booth and Burckle*, 1976; this paper] is about 10^3 times greater than those reported for comparable latitudes in the SW Atlantic [*Burckle and Stanton*, 1974; *Jones and Johnson*, 1984]. This may reflect the greater flux of LCDW entering the SW Pacific than the SW Atlantic. Diatom abundance, however,

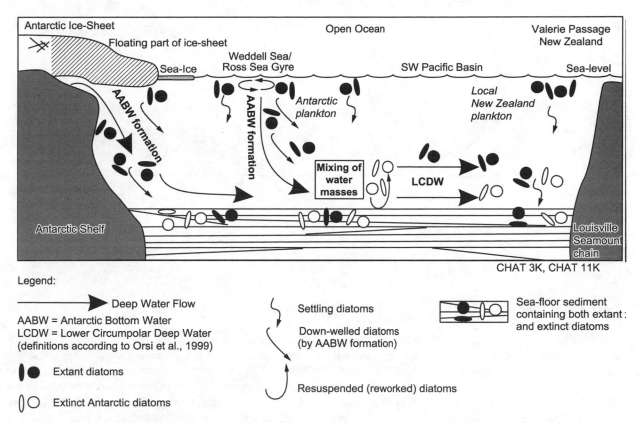

Figure 6. The entrainment of Antarctic diatoms in AABW (via downwelling) and LCDW (via mixing), and their subsequent emplacement in sediments in the Valerie Passage, New Zealand (not to scale).

is not linearly proportional to LCDW transport volume. Transport of CDW into the SW Pacific is $15.8 \pm 9.2 \times 10^6$ m^{3}s^{-1} [*Whitworth et al.*, 1999], while that into the SW Atlantic is 7×10^6 m^{3}s^{-1} [*Schmitz*, 1995]. Also, differences in topography and terrigenous sediment supply between the two areas could also account for the difference in diatom abundance (L. Burckle, pers. comm). In addition, sea-ice related diatoms *F. curta* and *F. cylindrus* are not reported in the SW Atlantic [*Burckle and Stanton*, 1974; *Johnson et al.*, 1977; *Jones and Johnson*, 1984], indicating some variation in source region (and/or a preservational bias) between the two oceanic basins.

Late Quaternary Fluctuations

Figures 7 and 8 illustrate downcore variations in tracer diatom abundance for the last 190 ky. The DAR (flux) of total tracer diatoms (all open-ocean plus sea-ice related taxa) are given in Figure 7. Significant quantities of tracer diatoms (on the order of 10^6-10^7 valves cm^{-2}ky^{-1}) are recovered throughout the time period. Again, absolute abundance values (not shown) are approximately of the order 10^3 greater than those in cores of comparable age-range

from the SW Atlantic [*Johnson et al.*, 1977; *Jones and Johnson*, 1984]. However, diatom abundance does not appear to be linearly proportional to volume of water transport (see above).

The trends for CHAT 3K and CHAT 11K are similar. However, CHAT 11K contains a generally greater flux of tracer diatoms than CHAT 3K. Flow speed is greater at the western boundary (CHAT 3K). Here, during times of increased LCDW flow, the silt-sized diatoms in suspension may not be able to settle out, instead being carried further along the flow-path. CHAT 11K, however, is located further away from the boundary where lower flow speeds [*Carter and Wilkin*, 1999] allow a higher degree of deposition than at CHAT 3K. The difference in DAR between the two cores is greater when DAR values are highest, and reduced when DAR values are lowest (Figure 7).

DAR is enhanced during times of ice build-up *prior* to the penultimate glacial maximum in OIS 6 and the Last Glacial Maximum (LGM) at 21 ka (Figure 7). Broadly, the highest flux of tracer diatoms occur in early and mid glacial OIS 6, late stadial OIS 4, at the OIS 2/3 boundary and in late OIS 1 (CHAT 3K only). Jones and Johnson [1984] also reported high abundances of tracer diatoms prior to

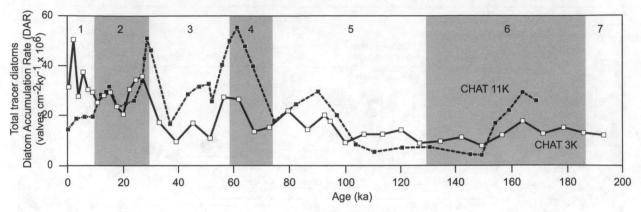

Figure 7. Flux of total diatoms for cores CHAT 3K and CHAT 11K. Oxygen isotope stages (OIS) 1-7 are indicated. Glacials OIS 2,4, and 6 are shaded.

glacial maxima in the Vema Channel, at the OIS 2/3 boundary (although they did not convert the data to DAR). In the SW Pacific, DAR is low during glacial maxima, but lowest during the last interglacial (115-125 ka), interstadial OIS 3 and the present interglacial OIS 1 (CHAT 11K only).

It is difficult to attribute the diatom flux data to a single factor. LCDW-flow rate into the SW Pacific as well as local flow speeds will have an influence on the signal. The rate of AABW production and upwelling into the Lower LCDW around Antarctica [*Orsi et al.*, 1999] must also be considered. Diatom productivity at the source and preservation may also significantly affect the data. Extended sea-ice during glacials suppressed the productivity of open-ocean diatoms [e.g., *Armand*, 2000]. Assuming diatoms are available for entrainment during interglacials, then low DAR values suggest a reduced LCDW flow into the SW Pacific at these times. Abundance values in core-tops support this view, and suggest a relatively reduced modern LCDW compared to glacials. The highest flux of tracer diatoms in CHAT 3K is recorded during OIS 1, however. This may, in part, be caused by the most significant reduction in deep water flow during the last 190 ky, thus allowing more diatoms to settle out at the boundary. However, drifts on the floor of the Valerie Passage show signs of recent erosion [*Carter and Wilkin*, 1999]. Interpretation of the diatom data, therefore, is clearly more complex than this simple model. Post-burial dissolution of diatom valves may, in part, also contribute to the high DAR values in OIS 1 observed in CHAT 3K. Such dissolution, increasing with age/depth, may partly cause the gradual downcore decrease (to 100 ka) in DAR for CHAT 3K (Figure 7), although this trend does not appear to be evident in CHAT 11K.

Diatom Production vs Lower LCDW Flow

Diatom productivity around Antarctica has fluctuated over the last 190 ky. Open-ocean diatoms are dominant north of the maximum winter sea-ice edge [e.g., *Armand* 2000]. Their abundance is suppressed by extended sea-ice during glacial periods [e.g., *Burckle et al.*, 1982; *Armand*, 2000]. Antarctic sea-ice at the LGM was not only more extensive but also more concentrated. Estimations of a 5^0-10^0 extension in latitude to the north for winter LGM sea-ice have been suggested [e.g., *Crosta et al.*, 1998]. Armand [1997, 2000] estimates >40% winter sea-ice cover during the last and penultimate glacials as far north as 56^0S in the SE Indian Ocean, and Gersonde and Zielinski [2000] suggest its extension to the position of the present Polar Frontal Zone (50^0-60^0S in the Atlantic sector). The dominance of open-ocean diatoms throughout CHAT 3K and CHAT 11K, associated with AABW production, indicate the presence of open-ocean polynya sustaining the productivity of such diatoms, or ice-free regions in the summer season [e.g., *Armand*, 1997, 2000; *Crosta et al.*, 1998].

It is difficult to separate the effects of diatom productivity from the Lower LCDW-flow signal. One way may be to analyse material being reworked along the sea-floor since reworked diatoms are not productivity-dependent. Cores CHAT 3K and CHAT 11K contain reworked diatoms endemic to Antarctic and Subantarctic sediments. Relatively lower abundances (up to 1 x 10^6 valves cm^{-2}ky^{-1}) in these cores makes the signal less obvious than that for extant tracers. Nevertheless, these values are significant and make the reworked diatom component fairly meaningful. The taxa encountered, *Actinocyclus ingens*, *Fragilariopsis barronii*, *F. praeinterfrigidaria-interfrigidaria-weaveri* group, *Hemidiscus karstenii*, *Rouxia antarctica*

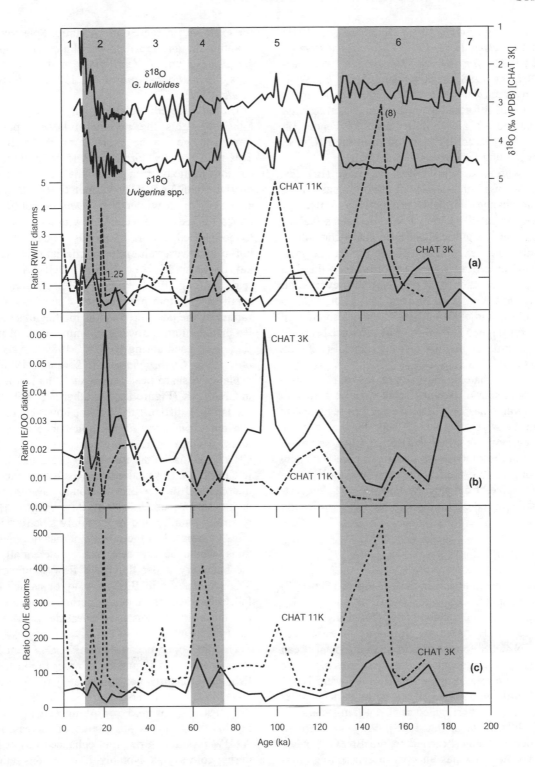

Figure 8. (a) Ratio of reworked (RW) / ice-edge (IE) diatoms for CHAT 3K and CHAT 11K compared to $\delta^{18}O$ *Uvigerina* spp. (‰ VPDB) and $\delta^{18}O$ *Globigerina bulloides* (‰ VPDB) for CHAT 3K. Ratios ≥ 1.25 are considered significant. (b) The ratio ice-edge (IE) / open-ocean (OO) diatoms and its reverse (c) for CHAT 3K and CHAT 11K. (NB: The ratio OO / IE is illustrated to highlight glacial-interglacial variations not immediately obvious from the IE / OO curves). Oxygen isotope stages (OIS) 1-7 are indicated. Glacials OIS 2, 4 and 6 are shaded.

and *R. isopolica*, [e.g., *Fenner*, 1977; *Burckle et al.*, 1978; *Akiba*, 1982; *Ciesielski*, 1983, 1986; *Baldauf and Barron*, 1991; *Fenner*, 1991; *Harwood and Maruyama*, 1992; *Gersonde and Barcena*, 1998] are less common and less well-preserved than extant Antarctic diatoms, but can nevertheless provide useful information on bottom water activity. Reworked diatoms are also reported by J.M. Fenner [in *Carter et al.*, 1999] in cores recovered from the ENZOSS during Ocean Drilling Program (ODP) Leg 181.

The ratio reworked diatoms (RW) / ice-edge (IE) diatoms (Figure 8a) highlights phases of reworking assuming no significant changes in palaeoproductivity. Pulses of reworking are indicated by high RW / IE values (values ≥1.25 are considered significant). These peaks correspond to heavy benthic $\delta^{18}O$ excursions (Figure 8a), not as notable in the planktonic $\delta^{18}O$ signal. Unfortunately, the data are inconclusive for the LGM (21 ka), however, high Lower LCDW-flow is inferred for (broadly) glacials OIS 6 and OIS 2, and for stadial OIS 4. High flow is also inferred during cool substage 5b (~90-95 ka). Reduced Lower LCDW-flow is inferred for the last interglacial (115-125 ka), and interstadial OIS 3.

Of course, extant diatoms may also be reworked. Analysis of valve pore-size of extant tracers may be a way to determine if a valve has been reworked or has remained in suspension in the water colomn (L. Burckle, pers. comm). Sea-water is everywhere undersaturated in silica, so valves showing signs of dissolution (e.g., enlarged pore-size) may indicate reworking. We found little sign of dissolution in extant diatom valves from the SW Pacific cores, suggesting that most displaced extant diatoms are not significantly reworked.

Variations in AABW Source

By separating the total tracer diatom signal (Figure 7) into open-ocean species and nearshore to ice and ice-edge species it is possible to emphasise temporal changes in the two sources for AABW. The vast majority of tracer diatoms in the ENZOSS cores and core-tops are open-ocean diatoms. This raises some interesting questions since the modern production rate of open-ocean source water is < 1 x 10^6 m^3s^{-1} [*Whitworth et al.*, 1999], whereas deep water formed over the shelf is estimated at 13 x 10^6 m^3s^{-1} [*Jacobs et al.*, 1985; *Whitworth et al.*, 1999]. Why are there so many open-ocean diatoms carried to the ENZOSS when <10% of the water flux has an open-ocean source? It is possible that pre-Holocene shifts in the STF [*Weaver et al.*, 1998] may contribute to the high numbers of *F. kerguelensis* and *T. lentiginosa*, at least, since Zielinski and Gersonde [1997] report low abundances (<20% of the total assemblage in core-tops) of these diatoms near the STF,

for example, in the South Atlantic. Selective dissolution resulting in an apparent enrichment of robust taxa *F. kerguelensis* and *T. lentiginosa* in the sediments, is also possible [*DeFelice and Wise*, 1981; *Zielinski and Gersonde*, 1997].

The ratio ice-edge diatoms (IE) / open-ocean (OO) diatoms highlights the periodic relative input of AABW formed at the sea-ice interface on the shelf (Figure 8b). This ratio broadly corresponds to some light isotopic excursions in the benthic $\delta^{18}O$ signal. This implies more productive shelf-AABW at warmer times when AABW was able to form in winter [e.g., *Denton*, 2000]. It may also indicate a lesser sea-ice extent with the ice-edge retreating to the shelf-edge more frequently. The ratio is reduced (but not to its lowest value) during glacial maxima (135 ka; 21 ka) possibly due to inhibited shelf-AABW production [e.g., *Kellogg*, 1987; *Pudsey*, 1992; *Denton*, 2000] and/or suppressed diatom productivity. The signal is complicated further by the possible periodic melting of sea-ice releasing ice-diatoms into the water column, an event which occurs seasonally each spring [*Fryxell*, 1989; *Leventer and Dunbar*, 1996; *Cunningham and Leventer*, 1998]. This may explain the sharp peak values at 18 ka just after the LGM in CHAT 3K (Figure 8b). Another peak at 95 ka in CHAT 3K is more difficult to explain, however. In addition, we can not explain the higher abundance of ice-related diatoms in CHAT 3K compared to CHAT 11K, nor why CHAT 11K lacks the two sharp peaks in IE / OO evident in CHAT 3K. Physical differences between the open-ocean diatoms and the ice-edge diatoms may result in hydrodynamic sorting under a high-flow regime. However, the generally smaller and more thinly silicified ice-edge diatoms (compared to the larger and heavier open-ocean diatoms) would be expected to be preferentially removed at the boundary under the faster flow anticipated at CHAT 3K. Some peaks in IE / OO may be caused by a relative fall in open-ocean diatoms, perhaps due to extended sea-ice. In particular, this is likely at the LGM and cool substage 5d (below).

The ratio open-ocean diatoms (OO) / ice-edge (IE) diatoms is shown in Figure 8c in order to illustrate and highlight some features not immediately obvious from the IE / OO curves (Figure 8b). The OO / IE ratio coincides with the pulses of inferred reworking (Figure 8a) and heavy benthic $\delta^{18}O$ values suggestive of increased open-ocean AABW production rate and enhanced Lower LCDW flow during cold events. Notably, however this ratio is low during the LGM and cool substage 5d (110 ka) when extended sea-ice suppressed diatom productivity. Gersonde and Zielinski [2000] report a permanent extension of sea-ice at both these times as far north as 53°S in the Atlantic sector of the Southern Ocean (and seasonally as far as 49°S).

SUMMARY

Endemic Antarctic diatoms, incorporated into downwelling AABW during its formation and subsequently into Lower LCDW, are used as tracers of the LCDW flow-path in modern times. Core-tops bathed in Lower LCDW contain significant amounts of tracer diatoms, while shallower watermasses are barren of such diatoms. The Lower LCDW/core NADW interface at 3800 m depth is well-defined by the presence/absence of tracer diatoms.

Tracer diatoms were recovered from two kasten cores situated within the palaeo-LCDW flow-path in the SW Pacific. They may be useful for inferring changes in Lower LCDW-flow through the SW Pacific Gateway for the last 190 ky. Diatom accumulation rate (DAR) accounts for dilution of the signal and allows for a direct comparison of the cores. Reworked diatoms suggest that Lower LCDW-flow was greater during colder times. Low DAR values at the LGM may be the result of reduced diatom productivity at the source making inferences about Lower LCDW flow using extant tracers inappropriate for glacial periods. During the present and last interglacials, when tracer diatoms are presumed to have been available for incorporation, reduced DAR is suggestive of a relatively reduced flow. The data suggest that AABW formed in the open-ocean is relatively more important during colder times and that AABW formed over the shelf is relatively more important at warmer times. The majority of tracer diatoms are extant open-ocean species, data which does not corroborate modern oceanographic observations of a largely shelf source for deep water flowing into the SW Pacific. Clearly factors of diatom abundance at the source and preservation are important.

If care is taken in selecting indicative taxa, there is significant potential for the use of displaced Antarctic diatoms as a supplement to geochemical and sedimentological techniques in Antarctic deep-water flow studies. Areas of refinement lie in quantification of data in terms of flow volume and the use of water and sediment trap samples to help refine ideas on diatom productivity vs watermass flow. It is also important to choose core material carefully. During times of high Lower LCDW-flow, the upper limit of the watermass may shallow (L. Burckle, pers. comm). Analysis of core material close to the interface between the NADW core and the Lower LCDW on continental slopes would be useful to test the hypothesis.

Acknowledgements. The first author wishes to thank Lloyd Burckle and John Barron for illuminating discussions on diatoms, and Alan Lord for valued encouragement. This paper was greatly improved by reviews by Lloyd Burckle and Julianne Fenner. Martin Pearce helped to draw and resize some of the figures. This work was carried out as part of Natural Environment Research Council Research Grants GT4/94/212/G and GR3/10133.

REFERENCES

Akiba, F. 1982. Late Quaternary diatom biostratigraphy of the Bellingshausen Sea, Antarctic Ocean. *Report of the Technology Research Center. J.N.O.C.,* 16: 31-74.

Armand, L.K. 1997. The use of diatom transfer functions in estimating sea-surface temperature and sea-ice in cores from the southeast Indian Ocean. *Ph.D. thesis, Australian National University, Canberra,* pp. 392.

Armand, L.K. 2000. A ocean of ice - advances in the estimation of past sea ice in the Southern Ocean. *GSA Today,* 10: 1-7.

Baldauf, J.G. and Barron, J.A. 1991. Diatom biostratigraphy: Kerguelen Plateau and Prydz Bay regions of the Southern Ocean. *In,* Barron, J.A., Larsen, B.L. *et al.* (Eds.). *Proceedings of the Ocean Drilling Program, Scientific Results,* 119: 547-598. College Station, Texas.

Battarbee, R.W. 1986. Diatom analysis. *In,* Berglund, B.E. (Ed.). *Handbook of Holocene palaeoecology and palaeohydrology,* 527-570. John Wiley & Sons Ltd. pp. 869.

Battarbee, R.W. and Kneen, M.J. 1982. The use of electronically counted microspheres in absolute diatom analysis. *Limnology and Oceanography,* 27: 184-188.

Booth, J.D. and Burckle, L.H. 1976. Displaced Antarctic diatoms in the southwestern and central Pacific. *Pacific Geology,* 11: 99-108.

Boyle, E. 1995. Last-glacial-maximum North Atlantic Deep Water: on, off or somewhere in-between? *Philosophical Transactions of the Royal Society of London B,* 348: 243-253.

Bradford-Grieve, J.M., Boyd, P.W., Chang, F.H., Chiswell, S., Hadfield, M., Hall, J.A., James, M.R., Nodder, S.D. and Shushkina, E.A. 1999. Pelagic ecosystem structure and functioning in the Subtropical Front region east of New Zealand in austral winter and spring 1993. *Journal of Plankton Research,* 28: 405-428.

Broecker. W.S. 1998. Paleocean circulation during the last deglaciation: a bipolar seesaw? *Paleoceanography,* 13: 119-121.

Broecker, W.S. 2000. Abrupt climate change: causal constraints provided by the paleoclimate record. *Earth-Science Reviews,* 51: 137-154.

Broecker, W.S. and Denton, G.H. 1990. The role of ocean-atmosphere reorganizations in glacial cycles. *Quaternary Science Reviews,* 53: 305-341.

Burckle, L.H. 1984a. Diatom distribution and paleoceanographic reconstruction in the Southern Ocean - Present and Last Glacial Maximum. *Marine Micropaleontology,* 9: 241-261.

Burckle. L.H. 1984b. Ecology and paleoecology of the marine diatom *Eucampia antarctica* (Castr.) Mangin. *Marine Micropaleontology,* 9: 77-86.

Burckle, L.H., Clarke, D.B. and Shackleton, N.J. 1978. Isochronous last-abundant-appearance datum (LAAD) of the diatom *Hemidiscus karstenii* in the sub-Antarctic. *Geology,* 6: 243-246.

Burckle, L.H., Jacobs, S.S. and McLaughlin, R.B. 1987. Late austral spring diatom distribution between New Zealand and Ross Ice Shelf, Antarctica: Hydrographic and sediment correlations. *Micropaleontology,* 33: 74-81.

Burckle, L.H. Robinson, D. and Cooke, D. 1982. Reappraisal of sea-ice distribution in Atlantic and Pacific sectors of the Southern Ocean at 18,000 yr B.P. *Nature,* 299: 435-437.

Burckle, L.H. and Stanton, D. 1974. Distribution of displaced Antarctic diatoms in the Argentine Basin. *Ehrlich,* 3: 283-292.

Carmack, E.C., and Foster, T.D. 1975. On the flow of water out of the Weddell Sea. *Deep-Sea Research,* 22: 711-724.

Carter, L., Carter, R.M., McCave, I.N. and Gamble, J. 1996. Regional sediment recycling in the abyssal southwest Pacific Ocean. *Geology,* 24: 735-738.

Carter, L., Garlick, R.D., Sutton, P. et al., 1998. Ocean Circulation New Zealand. NIWA Chart Miscellaneous Series 76.

Carter, L. and McCave I.N. 1994. Development of sediment drifts approaching an active plate margin under the SW Pacific Deep Western Boundary Current. *Paleoceanography,* 9: 1061-1085.

Carter, L. and Wilkin, J. 1999. Abyssal circulation around New Zealand - a comparison between observations and a global circulation model. *Marine Geology,* 159: 221-239.

Carter, R.M., Carter, L. and McCave, I.N. 1996. Current controlled sediment deposition from the shelf to the deep ocean: the Cenozoic evolution of circulation through the SW Pacific gateway. *Geologisch Rundschau,* 85: 438-451.

Carter, R.M., McCave, I.N., Richer, C., Carter, L. et al. 1999. *Proceedings of the Ocean Drilling Program, Initial Reports,* 181. [CD ROM]. Available from: Ocean Drilling Program, Texas A&M University, College Station, TX 77845-9547, U.S.A.

Ciesielski, P.F. 1983. The Neogene and Quaternary diatom biostratigraphy of subantarctic sediments, Deep Sea Drilling Project Leg 71. *In,* Ludwig, W. J., Krasheninnikov, V. A., *et al.* (Eds.). *Initial Reports of the Deep Sea Drilling Project* 71 (2): 635-665. Washington (U.S. Govt. Printing Office).

Ciesielski, P.F. 1986. Middle Miocene to Quaternary diatom biostratigraphy of Deep Sea Drilling Project Site 594, Chatham Rise, southwest Pacific. *In,* Kennett, J. P., von der Borch, C. C. *et al* (Eds.). *Initial Reports of the Deep Sea Drilling Project,* 90: 863-885. Washington (U.S. Govt. Printing Office).

Crosta, X., Pichon, J.-J. and Burckle, L.H. 1998. Application of modern analog technique to marine Antarctic diatoms: Reconstruction of maximum sea-ice extent at the Last Glacial Maximum. *Paleoceanography,* 13: 284-297.

Cunningham, W. and Leventer, A. 1998. Diatom assemblages in surface sediments of the Ross Sea: Relationship to present oceanographic conditions. *Antarctic Science,* 10: 134-146.

DeFelice, D.R. and Wise, S.W.J.R. 1981. Surface lithofacies, biofacies, and diatom diversity patterns as models for delineation of climatic changes in the Southeast Atlantic Ocean. *Marine Micropaleontology,* 6: 29-70.

Denton, G. 2000. Does an asymmetric thermohaline-ice-sheet oscillator drive 100,000-yr glacial cycles? *Journal of Quaternary Science,* 15: 301-318.

Ettwein, V.J., Stickley, C.E., Maslin, M.A., Laurie, E.R., Rosell-Melé, A., Vidal, L., and Brownless, M. 2001. Fluctuations in Productivity and Upwelling Intensity at Site 1083, during the Intensification of the Northern Hemisphere Glaciation (2.40-2.65 Ma). *In,* Wefer, G., Berger, W.H., and Richter, C., *et al., Proceedings of the Ocean Drilling Programme, Scientific Results,* 175 [CD-ROM]. Available from: Ocean Drilling Program, Texas A&M University, College Station, TX 77845-9547, U.S.A.

Fenner, J. 1977. Cenozoic diatom biostratigraphy of the Equatorial and southern Atlantic Ocean. *In,* Perch-Nielsen, K., Supko, P. R., et al. (Eds.). *Initial Reports of the Deep Sea Drilling Project,* 39 (supplement): 491-624.

Fenner, J.M. 1991. Late Pliocene-Quaternary quantitative diatom stratigraphy in the Atlantic sector of the Southern Ocean. *In,* Ciesielski, P.F., Kristoffersen, Y., *et al.* (Eds.). *Proceedings of the Ocean Drilling Program, Scientific Results,* 114: 97-121. College Station, TX: Ocean Drilling Program.

Fenner, J., Carter, L. and Stewart, R. 1992. Late Quaternary paleoclimate and paleoceanographic change over northern Chatham Rise, New Zealand. *Marine Geology,* 108: 383-404.

Fenner, J. Schrader, H. and Wienigk, II. 1976. Diatom phytoplankton studies in the southern Pacific Ocean, composition and correlation to the Antarctic Convergence and its palaeoecological significance. *In,* Hollister, C. D., Craddock, C. *et al.* (Eds.). *Initial Reports of the Deep Sea Drilling Project,* 35: 757-813.

Fryxell, G.A. 1989. Marine phytoplankton at the Weddell Sea Ice Edge: Seasonal changes at the specific level. *Polar Biology,* 10: 1-18.

Gersonde, R. and Barcena, M.A. 1998. Revision of the upper Pliocene-Pleistocene diatom biostratigraphy for the northern belt of the Southern Ocean. *Micropaleontology,* 44: 84-98.

Gersonde, R. and Zielinski, U. 2000. The reconstruction of late Quaternary Antarctic sea-ice distribution - the use of diatoms as a proxy for sea-ice. *Palaeogeography, Palaeoclimatology, Palaeoecology,* 162: 263-286.

Gordon, A.L. 1971. Oceanography of Antarctic waters. *In,* Reid, J. L. (Ed.). *Antarctic Oceanology* I. Antarctic Research Series 15: 169-203. Washington (American Geophysical Union).

Gordon, A.L. 1975. An Antarctic oceanography section along 170^0E. *Deep Sea Research,* 22: 357-377.

Gordon, A.L. 1983. Polar Oceanography. *Reviews of Geophysics,* 21: 1124-1131.

Gordon, A.L. 1986. Interocean exchange of thermocline water. *Journal of Geophysical Research,* 91: 5037-5046.

Gordon, A.L. 1988. The Southern Ocean and global climate. *Oceanus,* 31: 39-46.

Gordon, A.L., Zebiak, S.E. and Bryan, K. 1992. Climate variability and the Atlantic Ocean. *EOS, Transactions, AGU,* 79: 161, 164-165.

Hall, I.R., McCave, I.N., Shackleton, N.J., Weedon, G.P. and Harris, S. 2000. Glacial Intensification of the Deep Western Boundary Current inflow to the SW Pacific Ocean over the last 1.2 Ma. *EOS, Transactions, AGU Fall Meeting 2000,* 28: F714 (*abstract*).

Harwood, D.M. and Maruyama, T. 1992. Middle Eocene to Pleistocene diatom biostratigraphy of Southern Ocean sediments from the Kerguelen Plateau, Leg 120. *In*, Schlich, R., Wise, S.W. *et al.* (Eds.). *Proceedings of the Ocean Drilling Program, Scientific Results.* 120, 683-733. College Station, Tx (Ocean Drilling Program).

Heath, R.A. 1985. A review of the physical oceanography of the seas around New Zealand - 1982. *New Zealand Journal of Marine and Freshwater Research*, 19: 79-124.

Jacobs, S.S., Fairbanks, R.G. and Horibe, Y. 1985. Origin and evolution of water masses near the Antarctic continental margin: Evidence from $H_2 {}^{18}O/H_2 {}^{16}O$ ratios in seawater. *In*, Jacobs, S.S. (Ed.), *Oceanology of the Antarctic Continental Shelf.* Antarctic Research Series, 43: 59-85. Washington D.C. American Geophysical Union.

Johnson, D.A., Ledbetter, M. and Burckle, L.H. 1977. Vema Channel paleo-oceanography: Pleistocene dissolution cycles and episodic bottom water flow. *Marine Geology*, 23: 1-33.

Jones, G.A. and Johnson, D.A. 1984. Displaced Antarctic diatoms in Vema Channel sediments: late Pleistocene/Holocene fluctuations in the AABW flow. *Marine Geology*, 58: 165-186.

Keigwin, L., Curry, W.B., Lehman, S.J. and Johnson, S. 1994. The role of the deep ocean in North Atlantic climate change between 70 and 130 kyrs ago. *Nature*, 371: 323-329.

Kellogg, T.B. 1987. Glacial-interglacial changes in global deepwater circulation. *Paleoceanography*, 2: 259-271.

Ledbetter, M. 1979. Fluctuations of Antarctic Bottom Water velocity in the Vema Channel, South Atlantic during the last 160,000 years. *Marine Geology*, 33: 71-89.

Leventer, A. and Dunbar, R.B. 1996. Factors influencing the distribution of diatoms and other algae in the Ross Sea. *Journal of Geophysical Research*, 101: 489-500.

Martinson, D.G., Pisias, N.G., Hays, J.D., Imbrie, J., Moore Jr, T.C. and Shackleton, N.J. 1987. Age dating and the orbital theory of the ice ages: Development of a high-resolution 0 to 300,000-year chronostratigraphy. *Quaternary Research*, 27: 1-29.

McCave, I.N. Carter, L. 1997. Recent sedimentation beneath the Deep Western Boundary Current off northern New Zealand. *Deep-Sea Research I*, 44: 1203-1237.

Nelson, C.S., Cooke, P.J., Hendy, C.H. and Cuthbertson, A.M. 1993. Oceanographic and climatic changes over the past 160,000 years at Deep Sea Drilling Project Site 594 off southeastern New Zealand, Southwest Pacific Ocean. *Paleoceanography*, 8: 435-458.

Orsi, A.H., Johnson, G.C. and Bullister, J.L. 1999. Circulation, mixing, and production of Antarctic Bottom Water. *Progress in Oceanography*, 43: 55-109.

Pudsey, C.J. 1992. Late Quaternary changes in Antarctic Bottom Water velocity inferred from sediment grain size in the northern Weddell Sea. *Marine Geology*, 107: 9-33.

Reid, J.L., Nowlin, W.D. and Patzert, W.C. 1977. On the characteristics and circulation of the southwestern Atlantic Ocean. *Journal of Physical Oceanography*, 7: 62-91.

Rintoul, S.R. 1998. On the origin and influence of Adélie Land Bottom Water. *In,* Jacobs, S.S. and Weiss, R.F. (Eds.), *Interactions at the Antarctic Continental Margins,* (Vol. 75, pp. 151-171), Antarctic Research Series. Washington, DC: American Geophysical Union.

Rintoul, S.R. and Bullister, J.L. 1999. A late winter hydrographic section from Tasmania to Antarctica. *Deep Sea Research I,* 46: 1417-1454.

Schmitz. W.J., Jr. 1995. On the interbasin-scale thermohaline circulation. *Reviews of Geophysics,* 33: 151-173.

Schrader, H-J. and Gersonde, R. 1978. Diatoms and Silicoflagellates. *In*, Zachariasse, W.J., Riedel, W.R., Sanfilippo, A., Schmidt, R.R., Brolsma, M.J., Schrader, H-J., Gersonde, R., Drooger, M.M. and Broekman, J.A. (Eds.). *Micropaleontological counting methods and techniques - an exercise on an eight metres section of the lower Pliocene of Capo Rossello, Sicily. Utrecht Micropaleontological Bulletins,* 129-176.

Seidov, D., Barron, E.J. and Haupt, B.J. 2000. The role of the Southern Ocean in the world ocean deepwater circulation. *EOS, Transactions, AGU Fall Meeting 2000,* 28: F714 (abstract).

Seidov, D., Barron, E.J. and Haupt, B. 2001. Meltwater and the global ocean conveyor: Northern versus southern connections. *Global and Planetary Change (in press).*

Stickley, C.E. 1998. The palaeoceanographical significance of diatoms in late Quaternary sediments from the south-west Pacific. *Ph.D. thesis, University College London,* pp. 495, pl. 9, [with addendum (2000) on CD ROM].

Vidal, L., Labeyrie, L., Cortijo, E., Arnold, M., Duplessy, J.C., Michel, E., Becque, S. and van Weering, T.C.E. 1997. Evidence foe changes in the North Atlantic Deep Water linked to meltwater surges during the Heinrich events. *Earth and Planetary Science Letters,* 146: 13-27.

Walsh, J.J. 1968. The vertical distribution of phytoplankton in the waters around Palmer Peninsula, Antarctica. *M.Sc. Thesis, The University of Miami,* pp. 162.

Warren, B.A. 1973. Transpacific hydrographic sections at Lats. 43^0S and 28^0S: the SCORPIO Expedition - II. Deep water. *Deep Sea Research,* 20: 9-38.

Warren, B.A. 1981. Deep circulation of the world ocean, p.6-41. *In*, Warren, B. A. and Wunsch, C. (Eds.). *Evolution of physical oceanography.* MIT Press, Cambridge, Massachussetts.

Weaver, P.P.E., Neil, H. and Carter, L. 1997. Sea surface temperature estimates from the southwest Pacific based on planktonic foraminifera and oxygen isotopes. *Palaeogeography, Palaeoclimatology, Palaeoecology,* 131: 241-256.

Weaver, P.P.E., Carter, L. and Neil, H.L. 1998. Response of surface masses and circulation to late Quaternary climate change, east of New Zealand. *Paleoceanography,* 13 (1): 70-83.

Whitworth, T., Warren, B.A., Nowlin, W.D., Pillsbury, R.D. and Moore, M.L. 1999. On the deep western-boundary current in the Southwest Pacific Basin. *Progress in Oceanography,* 43: 1-54.

Zielinski, U. and Gersonde, R. 1997. Diatom distribution in Southern Ocean surface sediments (Atlantic sector): Implications for paleoenvironmental reconstructions. *Palaeogeography, Palaeoclimatology, Palaeoecology,* 129: 213-250.

Lionel Carter, National Institute of Water and Atmospheric Research (NIWA), PO Box 14, 901, Kilbirnie, Wellington, New Zealand. (l.carter@niwa.cri.nz)

Nick McCave, Department of Earth Sciences, Downing Street, Cambridge University, Cambridge, CB2 3EQ, UK. (mccave@esc.cam.ac.uk)

Catherine Stickley, Environmental Change Research Centre, Department of Geography, University College London, 26 Bedford Way, London. WC1H 0AP, UK. (c.stickley@ucl.ac.uk)

Phil Weaver, Southampton Oceanography Centre, University of Southampton, Waterfront Campus, European Way, Southampton SO13 3ZH, UK. (ppew@soc.soton.ac.uk)

Modelling Abrupt Climatic Change During the Last Glaciation

Michel Crucifix, Philippe Tulkens [1], and André Berger

Institut d'Astronomie et de Géophysique G. Lemaître,
Université catholique de Louvain, Louvain-la-Neuve, Belgium

A sectorially averaged, ocean–atmosphere–sea-ice model is used to compare the climatic impacts of a freshwater input (FI) in the North Atlantic ocean in pre-industrial and glacial climates. The FI is designed in order to be compatible with paleoceanographic evidence about Heinrich Events (HE) 4 and 5. In agreement with previous studies performed in interglacial conditions, it is shown that the THC turnoff and its resumption along with HE were likely to be responsible for the Antarctic warmings observed during some of them, as well as the subsequent DO abrupt warmings recorded in Greenland ice cores. However, when the FI is applied in glacial climate, the thermohaline circulation recovers in two steps separated in time by about 1470 years. Both steps correspond to abrupt shifts in Greenland temperature by about 4-5 °C induced by a massive release of the heat accumulated at intermediate depth in the Atlantic ocean. This 1470-year time lag, that is here physically linked to the turnover time of the ocean, turns out to be remarkably unsensitive to the duration and the time distribution of the FI, provided that the amplitude of the latter is large enough to induce a complete turnoff of the THC. The implication of these results are discussed by taking into account, in particular, the interpretation limits inherent to the sectorial structure of the model used here.

1. INTRODUCTION

Over the last decade, there has been growing evidence that the Heinrich Events (HE), basically defined as catastrophic inputs of ice-rafted detritus to the North Atlantic ocean induced by iceberg discharges over the last 60 ky (*Heinrich*, 1988; *Grousset et al.*, 1993; *Bond*

and *Lotti*, 1995; *Elliot et al.*, 1998), are connected to large oscillations in sea-surface temperature and salinity in the North Atlantic (*Labeyrie et al.*, 1995), air temperature over Greenland (*Bond et al.*, 1993) and, for some of them, in Antarctica (*Blunier et al.*, 1998; *Blunier and Brook*, 2001) which make them global signals. The reorganization of the thermohaline circulation (THC) induced by the freshwater input appears as a good candidate to explain the link between HE and air temperature variations in polar regions. Indeed, paleonutrient data indicate that HE 1, which occurred between the Last Glacial Maximum (LGM) and the Bølling warm phase, was characterized by a dramatic nutrient enrichment in Atlantic deep water, which suggests a turnoff of the THC (*Sarnthein et al.*, 2000). Similar conclusions

---

[1] Now at Task Force Sustainable Development, Federal Planning Bureau, Brussels, Belgium

The Oceans and Rapid Climate Change: Past, Present, and Future
Geophysical Monograph 126

apply to HE 4 (about 33 ky BP) although local convection sites may have been maintained during the latter (*Vidal et al.*, 1997). Besides, such a cause-effect link between iceberg discharge and abrupt Greenland temperature shifts gets theoretical support from the conceptual model of *Paillard and Labeyrie* (1994). Studies with more comprehensive models show that perturbations of sea-surface forcing in ocean models (*Maier-Reimer et al.*, 1993; *Wright and Stocker*, 1993; *Marchal et al.*, 1998; *Paillard and Cortijo*, 1999) or explicit FIs in ocean-atmosphere general circulation models (*Schiller et al.*, 1997; *Manabe and Stouffer*, 1997) can really induce a turnoff of the THC and have consequences both at regional and global scales. However, the comparison between these model studies and HE is limited by the fact that in the models, the perturbation is applied to a climate in equilibrium with modern or pre-industrial forcing. Indeed, although the circulation in the glacial ocean remains an open subject, it is generally agreed that the glacial Atlantic circulation was characterized by a shallower, and perhaps weaker, North Atlantic outflow towards the Southern ocean and a more pronounced northward penetration of Antarctic Bottom Water (*Duplessy et al.*, 1988; *Boyle*, 1995; *Sarnthein et al.*, 2000). Hence, it can be expected that a same freshwater perturbation will not have the same climatic consequences in peak glacial conditions than in interglacial ones.

The goal of this study is then to compare the simulated climatic impacts of a FI in the North Atlantic under pre-industrial and glacial conditions. For that, we use MoBidiC, a 2.5-dimension ocean–atmosphere–sea-ice model. The response of the model to a perturbation from its present-day equilibrium will be validated through a comparison to earlier published responses of ocean-atmosphere general circulation models (OAGCMs). Then, the same perturbation will be applied on a glacial equilibrium in order to compare the model response in both glacial and interglacial states.

This MoBidiC climate model is briefly described in sections 2 and 3. Section 4 and 5 discuss the interglacial and glacial simulated climates, respectively. Freshwater experiments from glacial and interglacial climates are presented in section 6. Results are discussed and compared to paleodata in section 7.

2. MODEL DESCRIPTION

The model used in this study (called MoBidiC) links the atmosphere, the terrestrial surface, the oceans and sea-ice. The surface is represented on a 5° latitudinal grid and each latitudinal band is divided in two continental sectors (Eurasia-Africa and America) and three oceans (Atlantic, Pacific and Indian). The longitudinal extent of the sectors (ocean and continents) as well as the altitude of the continents are prescribed according to the ETOPO 5 (1986) dataset. Each continental sector can be partly covered by snow and similarly, each oceanic sector can be partly covered by sea ice (with possibly a covering snow layer).

The atmospheric dynamics is represented by a zonally averaged two-level quasi-geostrophic (QG) system of equations written in pressure coordinates (*Sela and Wiin-Nielsen*, 1971) and applied independently over each hemisphere. The numerical implementation follows *Gallée et al.* (1991) but considers the whole earth instead of the northern hemisphere only. The prognostic variables for the atmosphere are the QG potential vorticity (*Holton*, 1979) at 250 hPa and 750 hPa. The model allows to diagnose the temperature at 500 hPa and the vorticity at 250 hPa and 750 hPa, the latter being linearly extrapolated to the surface to compute the zonal surface wind velocity. An additional parameterization of the Hadley cell is implemented to account for the transports of heat and angular momentum by the mean meridional circulation in tropical latitudes (*Crucifix et al.*, 2000). Surface temperature is considered as a prognostic variable, the subsurface heat transfer obeying the relation of *Taylor* (1976) that considers a 12-month oscillation for subsurface heat storage. The radiative transfer is computed by dividing the model atmosphere into 10-15 layers, the exact number depending on the surface pressure over each surface type (*Bertrand*, 1998; *Bertrand et al.*, 1999). The solar radiation scheme is an improved version of the code described by *Fouquart and Bonnel* (1980) and the longwave radiation computations are based on the *Morcrette's* (1984) wide band formulation of the radiation equation. The vertical temperature profile required by this radiative model is determined by assuming that the product of the static stability (prescribed to be 2.5×10^{-6} m^4 s^2 kg^{-2}) and the pressure remains constant along the vertical. Near the surface the temperature lapse rate is modified so that the air has the same temperature as the surface. This lapse rate determinates the convective vertical sensible heat transport from the surface to the atmosphere according to the formulation of *Saltzman and Ashe* (1976). Specific humidity (q) is considered as a diagnostic variable: its surface value is determined from the surface temperature assuming a surface relative humidity of 0.75 and its vertical profile obeys the *Smith* (1966) relation linking the specific humidity at a pressure p to its value at a reference level (pressure p_R) : $q(p) = q(p_R)(p/p_R)^\lambda$, the seasonal and meridional distribution of λ being inferred from the present-day climatology by *Oort* (1983). The reference level is either the surface or either the summit of a possible inversion layer, the latter being then considered of constant specific humidity. Clouds are prescribed. They are represented by a single effective layer, with base and top pressures given by *Ohring and Adler* (1978). The seasonal and meridional distribution of cloudiness is taken from *Warren et al.* (1986).

Evaporation is computed from the bulk formulation depending on surface temperature, vertical sensible heat flux, wind velocity and water availibility developped by *Saltzman* (1980). The zonally averaged precipitation is derived from the zonally averaged evaporation and the vertically integrated water-vapour (WV) meridional transport assuming the steady state approximation for the global WV content. The vertically integrated WV transport is derived from the diagnosed specific humidity assuming contributions by the baroclinic activity (diffusion) and by the mean meridional circulation (*Crucifix et al.*, 2000). The zonally averaged precipitation that is computed in this way for each time step and each zonal band is distributed over the 3 oceanic basins and the 2 continents with ratios derived from the present-day statistics by *Jaeger* (1976). Finally, runoff is implicitly represented by multiplying the precipitation over the oceans such that the freshwater budget over the global ocean is zero at each time step.

Within each individual oceanic basin (Atlantic, Pacific and Indian), the ocean model is based on the sectorially averaged form of the multi-level, primitive equation ocean model of *Bryan* (1969) and *Semtner* (1986). This model is extensively described in *Hovine and Fichefet* (1994) except for some modifications given hereafter. The east–west pressure difference which arises on averaging the momentum equations is taken to be proportional to the meridional pressure gradient as in *Wright and Stocker* (1992). In the northern hemisphere, lateral exchanges of heat and salt between the Atlantic and Pacific basins are permitted between 85 and 90°N. The Bering Strait is represented and is assumed to be 50-m deep. In the southern hemisphere, the three basins are interconnected between 40 and 65°S, where zonal advection and diffusion are represented. The Drake Passage and the Indian-Pacific connection are 4,000 m deep while the Atlantic and Indian basins are interconnected over the whole ocean depth (5,000 m).

The numerical treatment of the scalars (temperature, salinity and possibly tracers) is implicit on the vertical and explicit on the horizontal. The vertical diffusivity is prescribed to be 0.6×10^{-4} m^2 s^{-1}, except when the vertical density profile is diagnosed to be unstable. In this latter case, verical diffusivity is enhanced to 10 m^2 s^{-1} to account for convection (*Goosse and Fichefet*, 2000). The horizontal diffusivity is prescribed to be 1200 m^2 s^{-1}. Besides, the downslopig flow scheme by *Campin and Goosse* (1999) is implemented along Antarctica. Such schemes allow for a better representation of deep-water mass formation and are now used in several ocean general circulation models (e.g., *Beckmann and Döscher*, 1997). A simple thermodynamic-dynamic sea-ice model is coupled to the ocean model. It is based on the 0-layer thermodynamic model of *Semtner* (1976), with modifications introduced by *Harvey* (1988, 1992). A one-dimensional meridional advection scheme is used with ice velocities prescribed as in *Harvey* (1988).

The atmosphere, sea-ice and ocean components are synchronously coupled with a time step of 2 days. The whole system is forced by the seasonal cycle of insolation. Full details about MoBidiC description and parameter values are available in *Crucifix et al.* (2000).

3. SUB-POLAR GYRE PARAMETERIZATION

Without supplementary adaptation to the coupled model, the thermohaline circulation in the Atlantic ocean cannot be sustained under pre-industrial forcing unless the atmosphere freshwater transport is tuned such that the net evaporation over the North Atlantic exceeds about 0.35 Sv. Besides the fact that this value is likely to be overestimated (e.g., *Zaucker et al.*, 1994), such a configuration leads to unsatisfying North Atlantic Deep Water (NADW) properties, i.e., salinity over 35.4 psu and potential temperature around 4 °C instead of 34.9 psu and 3 °C (*Levitus*, 1982). Following A.Ganopolski (*personal communication*) we speculate that this shortcoming might be related to the disability for a sectorially averaged ocean model to account for the the northern North Atlantic horizontal circulation and in particular its East-Greenland branch. Hence, to improve NADW properties, the model uses a sub-polar gyre parameterization between 50 and 75°N. The intensity of this horizontal gyre is fixed to be 1.0 Sv. Details about its numerical implementation are given in the appendix.

4. PRE-INDUSTRIAL EQUILIBRIUM

The control simulation is performed with present-day insolation and a CO_2 concentration of 280 ppmv. The equilibrium — hereafter referred to as IG, for

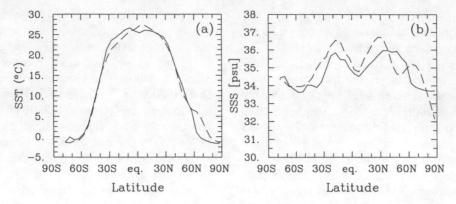

Figure 1. Distribution of Atlantic SST (a) and SSS (b) in latitude. Full lines represent the simulation for pre-industrial climate (IG). Dashed-lines correspond to the observations by *Levitus* (1982) used as restoring boundary conditions for the initial ocean-alone run in *Hovine and Fichefet* (1994).

*inter*glacial — is obtained by starting the model integration from a present-day climatology for the atmospheric component and oceanic temperature and salinity distributions simulated with the ocean-alone model using restoring boundary conditions to *Levitus'* (1982) observations fields (*Hovine and Fichefet*, 1994). The model evolves then towards its own equilibrium characterized by a relatively important cooling in the North

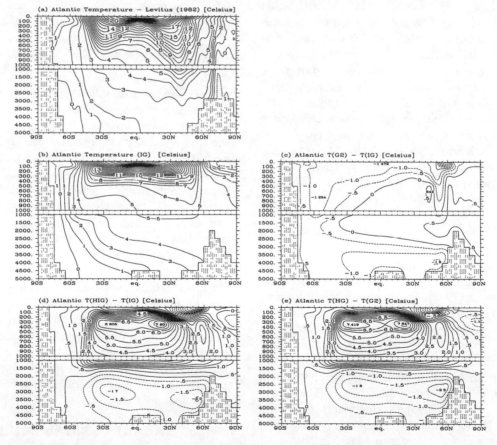

Figure 2. Latitude-depth profile of zonally averaged potential temperature in the North Atlantic [°C]. a) Observations by Levitus (1982) ; b) distribution simulated for pre-industrial equilibrium (IG) ; c) Difference between glacial (G2) and pre-industrial (IG) ; d) difference HIG − IG ; e) difference HG − G2. Values for Heinrich states HIG and HG are averaged over the last 250 years of the freshwater input.

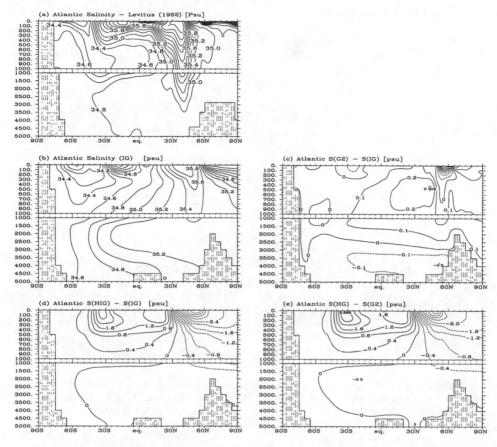

Figure 3. Same as figure 2 for latitude depth profile of the zonally averaged salinity in the North Atlantic [psu].

Atlantic (Figure 1a), associated to an advance of the winter sea-ice margin to 60°N. Besides, the complex structures of sea-surface salinities (SSSs) cannot be maintained (Figure 1b). Nevertheless, the temperature and salinity distributions in intermediate and deep waters remain compatible with *Levitus'* (1982) observations (Figures 2a,b and 3a,b) although the NADW originating from the high latitudes remains slightly too warm (between 3 and 4 °C) and too salty (between 35.2 and 35.4 psu). Cold and fresh, but also slightly too salty, AABW recirculates under NADW up to 40°N and, finally, the low salinity tongue originating from the southern latitudes and recirculating up to the equator in the Atlantic at 500 m depth can be interpreted as a model analogue of Antarctic Intermediate Water (AAIW).

The mean meridional Atlantic flow for this pre-industrial experiment is plotted on Figure 4a. Around 12.5 Sv of NADW reaches the Southern ocean where it is exported to the Pacific and Indian oceans through an Antarctic Circumpolar Current of 148 Sv. The global

stream function (not shown) reveals also the formation of deep water around Antarctica. An amount of 6 Sv is exported northwards as AABW, 1.5 Sv of which recirculate in the Atlantic Ocean. Below 1500 m, deep water penetrates into the Indian basin at a rate of 2 Sv and into the Pacific at 8 Sv. In the latter, 2 Sv recirculate at mid-depth ; the remainder is upwelled over a broad region and finally returns to the Atlantic via the southern convection zone. The global meridional stream function also reveals a 1000-m deep Deacon cell located around 55°S that contributes to the formation of Antarctic Intermediate Water. At last, the tropical Pacific and Indian exhibit a rather symmetrical Ekman circulation in the upper kilometer, with a maximum overturning rate of the order of 15 Sv.

5.   THE GLACIAL CLIMATE

To simulate the glacial climate, the atmospheric CO_2 concentration is fixed to 200 ppmv and, in the northern hemisphere, the topography of the Northern American,

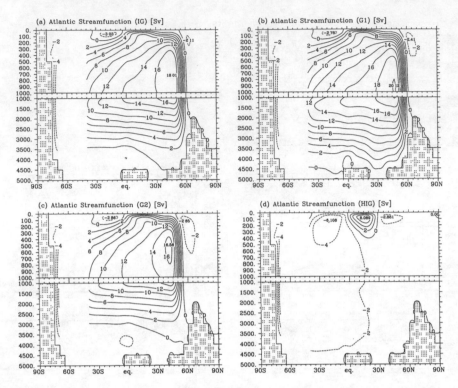

Figure 4. Zonally averaged meridional stream function [Sv] in the Atlantic Ocean. Positive values (full lines) correspond to clockwise circulation. a) Pre-industrial equilibrium (IG); b) Glacial equilibrium G1; c) Glacial equilibrium G2 ; d) Heinrich state (HIG). The flow for Heinrich state is an average over the last 250 years of the discharge.

Eurasian and Greenland ice sheets are fixed according to the simulation by the LLN 2-D NH climate-ice sheet model at 21 ky BP (*Gallée et al.*, 1992). The Antarctic Ice Sheet remains as today. Changes in ocean topography, continental area, and global salinity are not considered here. Insolation at the top of the atmosphere is computed for 21 ky BP following *Berger* (1978).

A first experiment (referred to as G1) consists of a 5000-year integration starting from the IG equilibrium presented in section 4 with the glacial forcing described above. It appears that the glacial forcing generates a strong enhancement of convection in the North Atlantic that is driven by cold sea-surface temperatures (SST) and relatively large SSS. The southward flow of NADW is even stronger than today while recirculation of AABW in the Atlantic vanishes almost completely (Figure 4b). These results, similar to those already obtained with an ocean general circulation model forced by reconstructed SSS (*Winguth et al.*, 1999, experiment GFGNC), conflict with the view of the glacial flow by, e.g., *Duplessy et al.* (1988), *Boyle* (1995) and *Sarnthein et al.* (2000). We retain at least four potential candidates to explain the mismatch between this sim-

ulation and paleodata : (a) The model assumes that precipitation and runoff distribution between the different basins and continents are as today. The reality is certainly different as the large ice sheets that developed in the northern hemisphere have affected both the zonal moisture transport (*Dong and Valdes*, 1995) and drainage basins geometry (*Licciardi et al.*, 1999) ; (b) The East-Greenland outflow, fixed to be 1 Sv for the pre-industrial climate, may have been weaker in glacial peak conditions than today. Indeed, 3-D OGCM simulations (Campin, pers. comm., based on the OGCM described in *Campin and Goosse*, 1999) suggest that the southward barotropic transport along East-Greenland, that presently amounts to 5 Sv, was reduced to 0-1 Sv at the LGM, depending on the salinity forcing ; (c) The model neglects the direct effects of ice-sheet dynamics on thermohaline circulation, both in the southern hemisphere where the progression of the ice-shelf may have influenced AABW formation (*Denton*, 2000) and in the northern hemisphere where the abundant icebergs drifting from Labrador and Greenland likely impacted on SSS up to the west of Ireland (*Ruddiman*, 1977; *Sarnthein et al.*, 2000; *de Vernal and Hillaire-*

Table 1. Main characteristics of the equilibrium experiments. Exp.: name; I.S.: initial state (OA corresponds to equilibrium with ocean alone using restoring boundary conditions). The integration time is indicated between parenthesis [yrs]; SPG: Sub-polar gyre parameterization [Sv]; ψ: NADW outflow towards the Southern ocean at 27.5 S [Sv]; $H_{Atl.,25°N}$: integrated northward heat transport accros 25°N in the Atlantic Ocean (positive northwards) [$10^{15}W$]; $F_{N.Atl}$: net evaporation over the North Atlantic ocean (including run-off) [Sv]; T_{NH}: hemispheric average of surface air temperature over the northern hemisphere [°C]; T_{SH} : idem for southern hemisphere; ACC: barotropic water transport through the Antarctic Circumpolar Current (Eastward positive) [Sv]. Results are averages over the last 250 years of integration

Exp.	I.S.	SPG	ψ	$H_{Atl.,25°N}$	$F_{N.Atl}$	T_{NH}	T_{SH}	ACC
IG	OA (8000)	1.0	12.2	0.936	0.272	15.68	14.56	148.1
G1	IG (5000)	1.0	14.5	1.026	0.286	12.61	12.70	149.4
G2	IG (8500)	0.0	10.1	0.858	0.268	12.64	12.75	148.5
G3	G2 (10000)	1.0	10.3	0.869	0.271	12.65	12.73	148.6

Marcel, 2000) ; (d) Initial conditions might be inappropriate. This last hypothesis gets support from numerous studies both with GCMs as well as simpler models showing the coexistence of multiple equilibrium solutions for the ocean-atmosphere system to a given external forcing and under certain conditions (*Stocker and Wright*, 1991; *Rahmstorf*, 1996; *Manabe and Stouffer*, 1999).

Using the model formulation presented here, it is difficult to determine which of these topics is the most critical to simulate a proper glacial ocean flow. However, reducing the subpolar gyre circulation from 1 Sv to 0 Sv turns out to be an interesting solution. Indeed, our experiment with no subpolar gyre — called G2 — is characterized by a southward shift of the northern boundary of the convection zone in the Atlantic from 60°N to 55°N and by a reduction of the NADW outflow towards the Southern ocean from 12.1 Sv to 10.5 Sv (Figure 4c). Most of convection takes place between 40 and 50°N in agreement with oxygen and carbon data interpretation by *Labeyrie et al.* (1992) and *Sarnthein et al.* (1994). Nevertheless, no convection takes place in the Nordic seas (around 75°N) as this was suggested from density reconstructions by *Sarnthein et al.* (2000). The reorganization of convection patterns impacts on water mass properties in the Atlantic Ocean. The Atlantic is less ventilated by salty surface water and becomes dominated by cold and fresh southern-source water (Figures 2c and 3c). Namely, NADW becomes shallower, warmer (by 0.5 °C) and more salty while the deepest water masses experience a significant cooling and freshening due to the strengthening of AABW recirculation.

As shown in table 1, the northward heat transport at

25°N by the Atlantic circulation is only reduced by 9% with respect to the pre-industrial equilibrium. This is a small reduction when compared to some other model studies, as *Weaver et al.* (1998) and *Winguth et al.* (1999), but on the other hand near-modern ocean transport might have been necessary to explain the enhanced tropical cooling documented in fossil corals and ground water tropical data (*Webb et al.*, 1997). In this model, the global ocean transport in glacial climate is even slightly larger than in pre-industrial climate (from 1.67 to 1.71 PW at 20°N). This increase, already simulated for similar experiments by *Ganopolski et al.* (1998), is induced by a strengthening of winter westerlies in the northern hemisphere by about 20 % that impacts positively on the Ekman transport in the North Pacific ocean. The model also simulates a global surface air temperature (SAT) cooling by 2.4°C. The SAT drop is about 2 to 2.5 °C in tropical areas where it is equally distributed over the oceans and the continents. It is much larger between 60 and 70°N (up 12 °C in zonal average) and is, as expected, maximum over the continents. The global temperature cooling simulated here is in the lower range of the simulation results obtained in the framework of the Paleoclimate Modelling Intercomparison Project (PMIP) (*Joussaume and Taylor*, 2000) predicting a deviation of the 21 ky BP globally averaged SAT by −6 °C to −2 °C from the pre-industrial climate when SSTs are computed with a mixed-layer ocean. Using a model based on a formalism closer to MoBidiC, *Ganopolski et al.* (1998) obtained a value of 6.2 °C. The relatively weak sensitivity of MoBidiC is partly linked to its low CO_2 climate sensitivity, which is of 2.0 °C for CO_2 doubling, but also to the fact that the sea-level change is not taken into account here. Lastly, the ice

Table 2. Main characteristics of the equilibrium experiments. Exp.: name; I.S.: initial state (OA corresponds to equilibrium with ocean alone using restoring boundary conditions). The integration time is indicated between parenthesis [yrs]; SPG: Sub-polar gyre parameterization [Sv]; ψ: NADW outflow towards the Southern ocean at 27.5 S [Sv]; $H_{Atl.,25°N}$: integrated northward heat transport accros 25°N in the Atlantic Ocean (positive northwards) [$10^{15}W$]; $F_{N.Atl}$: net evaporation over the North Atlantic ocean (including run-off) [Sv]; T_{NH}: hemispheric average of surface air temperature over the northern hemisphere [$°C$]; T_{SH} : idem for southern hemisphere; ACC: barotropic water transport through the Antarctic Circumpolar Current (Eastward positive) [Sv]. Results are averages over the last 250 years of integration

Exp.	I.S.	Forc.	SPG	FI	$H_{Atl,eq.}$	$F_{N.Atl}$	T_{NH}	T_{SH}
HIG	IG	IG	1.0	T(1000)	−0.153	0.170	15.06	15.20
HG	G2	G	0.0	T(1000)	−0.212	0.178	12.43	13.46
HG.S1	G2	G	0.0	T(500)	−0.248	0.179	12.30	13.26
HG.S2	G2	G	0.0	T(2000)	0.422	0.237	12.34	12.37
HG.S3	G2	G	0.0	C(500)	−0.230	0.182	12.33	13.27

sheets prescribed here differ from *Peltier's* (1994) reconstructions used for the PMIP experiments.

When compared to the IG equilbrium, the globally averaged precipitation rate decreases by 0.18 mm/day as a direct consequence of the global cooling. In the southern hemisphere, the decrease is rather homogeneous. In the northern hemisphere, the precipitation rate decreases by 1 mm/day over 60-70°N and increases by about 0.1 mm/day over 30-40°N in winter and 40-50°N in summer.

Finally, it must be noted that an active sub-polar gyre is not incompatible with a low glacial circulation, which highlights the importance of initial conditions. Indeed, in an additional sensitivity experiment, called G3, glacial boundary conditions are used and the sub-polar gyre intensity is fixed to be 1.0 Sv but contrarily to G1, the run is started from equilibrium G2, i.e., a weaker circulation than in IG. In this case, the convection between 55 and 60°N is not reactivated and the intensity of NADW as well as the freshwater and heat transport remain very close to values simulated in G2 (table 1).

6. FRESHWATER INPUTS EXPERIMENTS

6.1. *Experimental Setup*

As we are looking towards a model analogue of HE, the duration, amplitude and geographical location of the freshwater inputs (FI) applied to the ocean must be constrained by paleoceanographic estimates. The study of about 20 deep-sea cores by *Grousset et al.* (1993) demonstrates that these events — except HE3 — occurred along the northern boundary of the glacial polar front over most of the North Atlantic, i.e., between 40 and 55°N. This zone corresponds, by and large, to the region defined earlier by *Ruddiman* (1977) to characterize the zone of high accumulation of ice-rafted detritus during the last glacial period. The shifts in $\delta^{18}O$ of planktonic foraminifera established by *Labeyrie et al.* (1995) between 40 and 60°N allow them to estimate that about 1 million km^3 of ice, i.e., a corresponding sea-level change by 2 m, was stripped from the ice sheets during each of these events. However, this should be considered as a first order estimate as it relies on rough hypotheses about the mixing depth and the mixing rate of surface water. At last, according to ^{14}C datings on foraminiferal shells by *Voelker et al.* (1998), the FI input associated with HE4 and HE5 extended over 0.7 and 2.5 kyr respectively, results in good agreement with earlier estimates by *Vidal et al.* (1997).

Consequently, our experiments will be carried in the following way: a total volume of 3×10^6 km^3 of freshwater will be uniformly discharged between 40°N and 60°N in the Atlantic, which corresponds to a 0.06 psu drop in global ocean mean salinity. The FI is assumed to have the same temperature as surface water, so that it does not have a direct effect on SSTs. The time distribution of the FI with time is triangular, i.e., the rate of FI increase equals the rate of its subsequent decrease (as in *Schiller et al.*, 1997). The responses of the glacial and interglacial climate will be compared for a FI duration of 1 kyr. The maximum freshwater flux reaches in this case 0.192 Sv. In a second step, the sensitivity of the glacial climate response to the duration of the FI input will be considered for FI durations growing from 0.5 to 2 kyr and also for a constant FI.

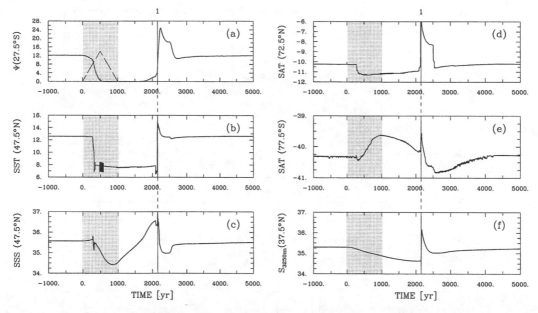

Figure 5. FI simulation in interglacial climate. Time evolution of (a) NADW outflow towards the Southern ocean at 27.5°S; (b) Zonally averaged Atlantic SST at 47.5°N [°C]; (c) Zonally averaged Atlantic SSS at 47.5°N [psu]; (d) Zonally averaged surface air temperature at 72.5°N [°C]; (e) Zonally averaged surface air temperature at 77.5°S [°C]; (f) Zonally averaged Atlantic salinity at 3250 m depth and 37.5°N. The shaded area corresponds to the period of freshwater input and the vertical lines refer to the overshoot. The [-1000,0] time interval corresponds to the end of the equilibrium run IG.

6.2. Freshwater Input in Pre-industrial Climate

The first freshwater experiment is called HIG and starts from the pre-industrial climate equilibrium (IG). The time evolution of the NADW export towards the Southern Ocean is plotted along with the amplitude of the freshwater input on the Figure 5a. During the first 280 years, the FI remains rather low and convection in the North Atlantic is not strongly affected. Both SST and SSS remain close to their initial values (Figure 5b,c). SSS at 47.5°N increases even slightly because the gradual weakening of the northward surface flow induces an accumulation of salty water in temperate latitudes, in spite of the FI. As soon as FI exceeds a threshold value, that is here of 0.124 Sv, the vertical structure of the water column in the North Atlantic becomes stratified such that the convective heat pump can no longer be maintained. The Arctic winter sea-ice southward margin shifts from 60°N to 55°N. There is also an important sea-surface cooling in regions were deep convection was previously active. Namely, SST decrease by 5 °C at 47.5°N while it remains virtually unchanged North of 60°N (Figure 6a). On the contrary, the SSS decrease is gradual and extents all over the northern North Atlantic (Figure 6b). The maximum

SSS anomaly is localized at 87.5°N where it reaches 2 psu. The air temperature experiences the direct effect of the convection turnoff, enforced by the positive albedo feedback related to the advance of winter sea-ice. However, in the polar latitudes, SAT does not decrease by more than 1 °C (Figure 5d). In the ocean, a positive temperature anomaly develops at intermediate depth (between 1000 and 3000 m), because the heat diffusing from the surface in intertropical regions is no longer advected northwards by the THC. Like in the OAGCM simulation by *Manabe and Stouffer* (1997), the maximum temperature anomaly is located in the northern hemisphere, at 20°N and 400 m depth (Figure 2d). In our model, it exceeds 7 °C, which is twice larger than the value obtained by *Manabe and Stouffer* (1997). By the same mechanism, salt accumulates at intermediate depth, mainly in intertropical latitudes (Figure 2d). This positive salinity anomaly is confined between 0 and 1000 m and is of the order of 1 psu. Below that depth, the strengthening of AABW advection induces a water freshening up to 0.5 psu. Also in agreement with the OAGCM study by *Manabe and Stouffer* (1997), the increase in meridional SST gradient produces an intensification of northern westerlies in the atmosphere between 30°N and 60°N up to 0.8 m/s. However, it has

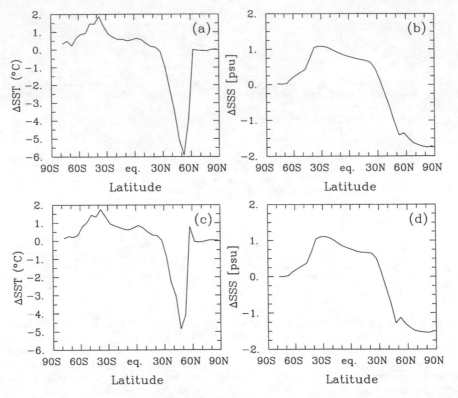

Figure 6.   Meridional distribution of SST and SSS variations induced by the FI in the Atlantic Ocean. (a) and (b) : differences between HIG (after 1000 years) and IG. Bottom: (c) and (d) : difference between HG (after 1000 years) and G2.

little impact on the latent heat transport in our model, the latter increasing by no more than 0.1 PW.

On the other hand, this intermediate-depth warm water pool propagates gradually towards the Southern Ocean, where it induces a gradual warming in Antarctica (SAT at 77.5°S increases by 0.6°C in 800 years, Figure 5e) through the convective activity along the Antarctic shelf. This warming is accompanied by a 50% decrease in Austral sea-ice cover, although the impact of the latter may be underestimated because the convection along the Antarctic shelf associated to the AABW formation prevents the formation of sea ice throughout the whole year in the 72.5°S - 67.5°S latitude band. This shortcomming is inherent to the zonal structure of MoBidiC that is not able to account for the coexistence of convective areas and sea-ice at the same latitude. When the southern temperature reaches its maximum, the North Atlantic overturning cell has completely disappeared and is replaced by a slow inverse circulation where deep water is solely formed near the Antarctic shelf (Figure 4d). In agreement with *Schiller et al.* (1997), the AABW is not strongly intensified but the

circulation is caracterized by a much larger extent of this cell with respect to the initial equilibrium.

Slightly before the end of the FI, salt accumulates again at the surface of the North Atlantic ocean, so that SSS increases gradually between 40°and 60°N and exceeds its equilibrium concentration (Figure 5c). About 650 years after the end of the FI, North Atlantic SSS is so large that the vertical stratification is broken, which turns the convective pump on again (event labelled (1) on the Figure 5). At that time, the heat accumulated at intermediate depth during the THC turnoff-phase is suddenly released to the atmosphere, inducing a sharp increase in SST (by over 6 °C). In the meantime, the salty water that accumulated between 0 and 1000 m depth is massively advected northwards where it contributes along with evaporation in increasing the surface density, which further enhances the convective activity. This process is called overshoot. The large amount of heat realeased in the atmosphere impacts directly on the northern latitude temperatures, the SAT at 72.5°N exhibiting a sharp increase by 5 °C just at the time of the overshoot (Figure 5d). Antarctica SAT experience

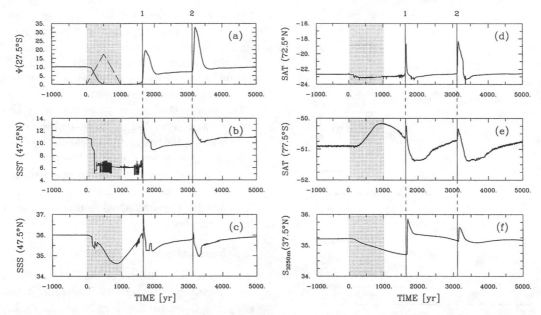

Figure 7. Simulation of FI from glacial climate (HG). Time evolution of (a) NADW outflow towards the Southern ocean at 27.5°S; (b) Zonally averaged Atlantic SST at 47.5°N [°C]; (c) Zonally averaged Atlantic SSS at 47.5°N [°C]; (d) Zonally averaged surface air temperature at 72.5°N [°C]; (e) Zonally averaged surface air temperature at 77.5°S [°C]; (f) Zonally averaged Atlantic salinity at 3250 m depth and 37.5°N. The shaded area corresponds to the period of freshwater input and the vertical lines refer to overshoots 1 and 2. The [−1000,0] time interval corresponds to the end of the equilibrium run G2.

also a temperature spike, but the latter is much smaller than in Greenland. The intense sinking flow of salty water rapidly increase the deep Atlantic salinity above its equilibrium concentration (Figure 5f) while the salty and warm intermediate pool that developed during the FI is replaced by cold and freshwater advected from the Southern ocean, which induces a decrease in North Atlantic SSS damping the THC intensity. This slowdown is first gradual, and then more abrupt when the 60-65°N latitude band becomes vertically stratified. The northern polar SAT decreases as well and finds back its initial level after a brief excursion through lower values.

Both available OAGCM simulations of *Schiller et al.* (1997) and *Manabe and Stouffer* (1997) show no sign of such an overshoot in FI experiments. There is thus here a clear disagreement between the 2-D and 3-D models to which one will have to be conscious in the following. This issue will be discussed more at length in section 7.

6.3. Freshwater Input in Glacial Climate

The second freshwater experiment, called HG, starts from the glacial climate equilibrium G2. The FI induces a turnoff of the THC associated with a sharp and persistent cooling between 35 and 55°N, while SSSs decrease more gradually (Figure 7). The general response of the

glacial ocean circulation to a FI is thus similar to the one simulated in interglacial conditions. However, some quantitative differences appear between HG and HIG. First, the lag between the beginning of the FI and the abrupt SST drop is shorter in HIG than in HG. In HG, there is a first small drop after 144 years and a second drop after 220 years. Besides, the impact of the THC turnoff on polar air temperatures is weaker in HG. As in these experiments, the northward heat transport by the Atlantic ocean is similar for glacial and interglacial conditions, this weaker sensitivity of polar latitudes in glacial climate is attributed to the fact that the albedo feedback is less efficient in the high latitudes in glacial climate because an important fraction of the surface is already covered by snow, continental ice or sea ice before the FI. This is shown by the variation in shortwave radiation reflected at the top of the atmosphere at 72.5°N in response to the freshwater perturbation. The latter increases by 1.4 W/m² in interglacial mode and by only 0.6 W/m² in glacial climate.

In the ocean, the distribution of potential temperature change (Figure 2e) is more symmetric than in the interglacial climate, but its magnitude is the same. Changes in surface salinity (Figure 3e) are virtually the same for both experiments.

After the FI, the THC recovers more quickly in the

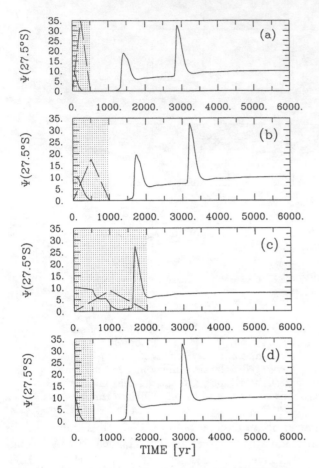

Figure 8. Sensitivity of the time evolution of NADW outflow towards the Southern ocean at 27.5°S [Sv] for different time distributions of the FI, represented by long-dashed curves. (a) Triangular, 500 years (HG.S1); (b) Triangular, 1000 years (HG), (c) Triangular, 2000 years (HG.S2); (d) Constant, 500 years. (HG.S3)

glacial mode (after 650 years instead of 1088 years). Interestingly, the system mode recovers its here initial state by passing through two overshoots, spaced by 1468 years, and both are associated to large shifts in Greenland air temperature labelled (1) and (2) on the Figure 7. Indeed, in glacial conditions, the freshening of the surface and at intermediate depth occuring after the first overshoot is efficient enough to stabilize the circulation with low NADW export rate (around 8 Sv, see Figure 7 between both overshoots), where no convection occurs North of 50°N. During this phase, the salty water pool that formed in the deep ocean during the first overshoot is diffused and advected towards the surface. This is illustrated by the decrease in deep ocean salinity (Figure 7f) and increase in surface salinity (Figure 7c) before overshoot 2. The fresh and cold anomaly

at intermediate depth is then gradually resorbed and a warm pool fed by tropical warm water can develop under the halocline in Atlantic high latitudes. There, water temperature increases gently (up to 2 °C), eroding gradually the vertical stratification until its breakdown, causing the second overshoot.

Therefore, the ocean state between the two overshoots must regarded as meta-stable state. The time necessary for its transition towards the more stable initial state depends on the rate of increase in surface salinity in low and mid latitudes in the Atlantic caused by salt advection and diffusion from the deep ocean, and can thus be considered as a measure of the turnover time of the ocean.

6.4. Sensitivity to FI Duration

The time distribution of the FI as well as its duration are difficult parameters to determine from paleodata. They are consequently highly arbitrary in our experimental setup. Hence, the model response was tested for triangular FI with durations of 500, 1000, and 2000 years (referred to as HG.S1, HG and HG.S2) and for a constant input of 500 years (HG.S3). The total amount of discharged freshwater is in all cases 3×10^6 km^3 and glacial external forcing as described in the section 5 is used. The time dependence of NADW outflow towards the Southern ocean is plotted for each of these sensitivity experiments on Figure 8. The FI experiments fall in two categories: on the one hand we have HG.S1, HG and HG.S3, for which THC disappears completely and is replaced by a slow, inverse circulation; and on the other hand we have HG.S2 for which a residual THC as well as convection can be maintained throughout the FI. In this case, convection is shifted southwards and is confined between 25 and 35°N. The maximum FI input in HG.S2, i.e., 0.092 Sv, appears thus to be too small to induce a complete reorganization of the ocean system and fully active THC can be restored very quickly after the end of the FI. In other words, the ocean system does not reach the 'non-return point' on the system hysteresis earlier described by *Stocker and Wright* (1991) and *Rahmstorf* (1996).

When the FI is large enough to induce a complete turnoff, i.e., for HG.S1, HG and HG.S2, a lag is necessary after the end of the FI to reactivate convection in the high latitudes. Depending on the amplitude of the FI and its time distribution, this lag ranges between 500 and 800 years. Besides, THC recovers always in two steps characterized by two distinct overshoots. This feature is thus recurrent in our model with glacial forcing. Furthermore, the lag between both overshoots is surprisingly unsensitive to the FI duration : 1463 yr for HG.S1, 1468 yr for HG and 1454 yr for HG.S2. At last, the HG.S3 sensitivity experiment reveals that when a strong, constant FI is applied, the ocean surface temperature responds immediately, the density meridional gradient adjusting quickly such that the NADW southern outflow reaches zero in about 250 years.

7. COMPARISON WITH PALEODATA AND DISCUSSION

7.1. Comparison of Heinrich and Glacial Modes

Comparison with SST reconstructions by *Cortijo et al.* (1997) for HE4 shows that the meridional distribution of SST changes in the Atlantic induced by the FI in experiment HG (Figure 6c) is quantitatively supported by the data. These reconstructions suggest a cooling by 3 to 5 °C between 40 and 50°N, and only small changes (about 1 °C) North of this zone. Besides, the reconstructions performed using the modern-analog method by *van Kreveld et al.* (2000) reveal that SST did not drop significantly at the beginning of HE4 and HE5 in the Irminger Sea. This seems to confirm that the effect of the freshwater on SST are mostly important in convective areas, where the convection turn-off induces large local temperature variations. Conversely, the increase in SST occuring after the HE is recorded in all cores, confirming that the subsequent DO event results from a much more global effect.

The simulated SSS variations seem to be in less good agreement with the data. Indeed, the SSS reconstructions by *Cortijo et al.* (1997) suggest that the surface freshening caused by the FI during HE4 was intense (about 3 psu), but confined in the discharge area, i.e., between 40 and 50°N. Our model predicts rather that the freshwarning extended up to the North Pole and reached about 1.5 psu.

7.2. The Simulated Transient Response

In order to compare the simulated response of polar air temperature with polar ice core records, a common time scale between ice-core data and marine records is needed. *van Kreveld et al.* (2000) assumed that isotopic temperature variations recorded in the GISP2 Greenland ice core (*Stuiver and Grootes*, 2000) were synchronous with the Nord Atlantic SST in the Irminger Sea, which is well supported by our simulations. This provides us with datings for HEs consistent with the GISP2 timescale based on annual layer counting (*Meese et al.*, 1994). On the other hand, the GISP2 ice core was synchronized with the BYRD Antarctic ice core record

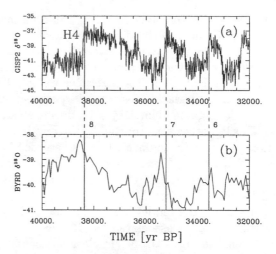

Figure 9. Rapid climatic change records. (a) High resolution $\delta^{18}O$ record from the GISP2 ice core dated on annual layer counting basis (*Grootes and Stuiver*, 1997; *Stuiver and Grootes*, 2000). The timing of Heinrich Event 4 is superimposed according to the synchronisation by *van Kreveld et al.* (2000) based on the correlation between SST reconstructed from core SO82-5, Irminger Sea and GISP2 $\delta^{18}O$. (b) $\delta^{18}O$ record from BYRD, Antarctica (*Johnsen et al.*, 1972). The synchronisation with the GISP2 ice core is based on methane concentration measurements by *Blunier and Brook* (2001). The Dansgaard-Oeschger events 8, 7 and 6 are indicated by vertical lines.

(*Johnsen et al.*, 1972) on the basis of the correlation of atmospheric methane concentration by *Blunier and Brook* (2001).

Figure 9 illustrates the polar temperature data based on these correlations. It turns out that the structure of the GISP2 reconstructed temperatures, mainly characterized by an abrupt increase at the onset of the DO event followed by a gradual decrease, can be recognized in the simulation. However, there are some discrepancies. First, given the relation between $\delta^{18}O$ and temperature calibrated for DO event 19 by *Lang et al.* (1999), the amplitude of the DO8 (which followed HE4) should have been of about 15 °C, which is much more than simulated. Besides, according to the timescale established by *van Kreveld et al.* (2000), DO8 occured a few hundred years after the end of the FI, while the model predicts a lag of 600 years. As this lag is determined by the time necessary to increase the SSS after the FI, it is possible that the amount of discharged freshwater and the drop in surface salinity that it induces are overestimated. At last, model and data do not agree on the length of the simulated interstadial phase, which is much shorter in the simulation. It could be that in the real world the positive feedbacks involved in the overshoot, i.e., mainly northward salt advection and evaporation, are efficient enough to make the convective activity in the high latitudes relatively stable, like it appears in the interglacial freshwater experiment where convection is sustained North of 60°N during about 400 years. This hypothesis, supported by the relatively long duration of the interstadial phases as recorded in the Irminger Sea by *van Kreveld et al.* (2000), was recently developped by *Ganopolski et al.* (2001) on the basis of a model similar to ours.

In Antarctica, the model reproduces successfully the A1 warming that is associated to HE4 in the BYRD record. Consistently with the simulation, this warming is gradual and peaked at the end of the discharge, the onset of the subsequent DO event occuring during the Antarctic cooling phase. After the first overshoot, the time evolution of the simulated Antarctic temperature remains compatible with the BYRD δ^{18}O reconstruction showing first a cooling after DO8, followed by a warming towards DO7.

7.3. Implications

We have seen that the transient response of the model is punctuated by episodes of intense ocean circulation that explain the largest part of the variability in polar temperatures. Such episodes of intense overturning, either called overshoots or flushes, where first described in ocean models by *Marotzke* (1989) and *Weaver and Sarachick* (1991). They appear also as a feature common to most sectorially averaged ocean-atmosphere models (*Sakai and Peltier*, 1997; *Marchal et al.*, 1998; *Wang and Mysak*, 2001; *Ganopolski et al.*, 2001). However, some studies based on 3-D models give another view. First, *Mikolajewicz and Maier-Reimer* (1994) showed that the choice of boundary conditions can strongly affect the overshoot behviour in ocean circulation models. Besides, as earlier stated, both available OAGCM simulations of *Schiller et al.* (1997) and *Manabe and Stouffer* (1997) consistently show no sign of overshoot in their FI experiments. This disagreement between 2-D and 3-D models may result from an overestimation of the advective-salinity feedback strength in zonally averaged formalisms, essentially because the path from equatorial to northern latitudes is then much more direct than in the real world.

Does it mean that the overshoot is pure artefact ? Paleodata give opposite arguments and support, at least partly, the idea that overshoots have occured in the past. In particular, they remain an attractive concept to explain both the abruptness and the amplitude of the DO warmings that followed the Heinrich Events.

Another question is to determine whether the overshoot simulated 1500 years after the first one in the glacial mode can be interpreted as a model analogue for the so-called 'non-Heinrich' DO, as for example the DO7 and DO6 that occured respectively 3000 and 4500 years after the HE4. This hypothesis could be supported by the fact that in the model, the lag between both overshoots is not controlled by the duration and the intensity of the FI, recalling the striking regularity of DO events observed in paleorecords (*Schulz et al.*, 1999; *Wunsch*, 2000; *Sarnthein et al.*, 2000). In this case, such DO events, as well as the important SSS variations that are associated to them, might be due to transitions induced by internal ocean-sea-ice dynamics rather than the direct consequences of a freshwater pulse. In turn, the temperature variations induced by these overshoots can impact on the northern hemisphere ice sheets and induce additional meltwater pulses.

A similar idea was already suggested by *Sakai and Peltier* (1997). However, this tempting interpretation must be considered with caution. First, the transition materialized by the second overshoot concerns here one grid point, which is not sufficient to be conclusive. Secondly, 3-D ocean circulation models generally suggest that the convection in the high latitudes, and in particular in the GIN seas, is more stable than the convection in the Labrador Seas (e.g. *Rahmstorf*, 1995). Therefore, one could expect convection should recover first in the high latitudes instead of the low latitudes as here. In this case, the occurence of a second overshoot would be problematic.

In conclusion, we have extended, with this study, the FI modelling exercice to glacial conditions. It is confirmed that FI having an amplitude and a location compatible with data-based estimates can induce a turnoff of the Atlantic circulation in glacial climate. However, the polar temperatures are little affected by this shutdown. Therefore, despite the important obstacles raised by recent 3-D model simulations, the overshoot remains an attractive mechanism to explain the sharpness of the DO events appearing in polar records. Our model suggest also that a single FI may produce, in glacial conditions, two overshoots spaced by about 1500 years. However, we think that the latter result has still to be considered with reservation, which highlights the importance of a continued effort with ocean-atmosphere models in that field.

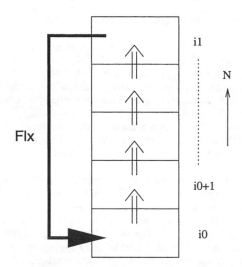

Figure 10. Representation of the North Atlantic, subpolar gyre parameterization in the horizontal plane. A flux Flx is transported from latitude grid point i_1 (corresponding to 72.5°N) to latitude grid point i_0 (52.5°N). The northward return flow is represented by double arrows.

APPENDIX A: PARAMETERIZATION OF THE SUB-POLAR GYRE IN THE ATLANTIC

The subpolar gyre parameterization introduced in the sectorial ocean model attempts to represent the effects of the horizontal circulation in the North Atlantic high latitudes on the distributions of temperature, salinity and any additional tracer. It is intended in general for zonally averaged models where this kind of circulation cannot be explicitly represented through the simulated velocity field.

It is assumed that the East-Greenland current can be represented by a transport (Flx) of water from 75°N (grid i_1) towards 50°N (grid i_0) in the Atlantic ocean. In order to ensure the water volume conservation, a northward return path flow takes place from 50°to 75°N (Figure 10). This transport only affects the scalar properties (temperature, salinity and other tracers, hereafter S) in the first 50 m, i.e., the first depth level in the ocean model. The numerical resolution is achieved through an upwind scheme. Let S_i^n represent the scalar property corresponding to latitude index i at time step n for the first depth level, and V_i the volume of the corresponding grid box. Then the East-Greenland current is represented by:

$$S_{i_0}^n = \tilde{S}_{i_0}^n + \frac{\Delta t}{V_{i_0}} Flx \left[\tilde{S}_{i_1}^n - \tilde{S}_{i_0}^n \right],$$

and the return path flow is represented by the following equation: for $i = i_0 + 1 \to i_1$:

$$S_i^n = \tilde{S}_i^n + \frac{\Delta t}{V_i} Flx \left[\tilde{S}_{i-1}^n - \tilde{S}_i^n \right],$$

where $\tilde{S}_i^n$ represents the scalar property at time step n according to the original advection/diffusion model, and Δt is the time step.

Acknowledgments. We wish to thank M.F. Loutre and F. Lefebre for their comments on the manuscript. Discussions with A. Ganopolski, H. Renssen and J.M. Campin were of great benefit. MC wish to thank the welcome by L.A. Mysak and Z. Wang during his visit at the Department of Atmospheric and Oceanic Sciences, McGill University, Montreal, where part of this work was made. MC is Research Fellow with the Belgian National Fund for Scientific Research.

REFERENCES

Beckmann, A., and R. Döscher, A method for improved representation of dense water spreading over topography in geopotential-coordinate models, *J. Phys. Oceanogr.*, *27*, 581–591, 1997.

Berger, A., Long-term variations of daily insolation and Quaternary climatic changes, *J. Atmos. Sci.*, *35*, 2362–2367, 1978.

Bertrand, C., Climate simulation at the secular time scale, Ph.D. thesis, Institut d'Astronomie et de Géophysique G. Lemaître, Université catholique de Louvain, 208 pp, 1998.

Bertrand, C., J.-P. van Ypersele, and A. Berger, Volcaninc and solar impacts on climate since 1700, *Clim. Dyn.*, *15*, 355–367, 1999.

Blunier, T., and E. Brook, Timing of millnial-scale climate change in Antarctica and Greenland during the last glacial period, *Science*, *291*, 109–112, 2001.

Blunier, T., J. Chappellaz, J. Schwander, A. Dällenbach, B. Stauffer, T. F. Stocker, D. Raynaud, J. Jouzel, H. B. Clausen, C. U. Hammer, and S. J. Johnsen, Asynchrony of Antarctic and Greenland climate change during the last glacial period, *Nature*, *394*, 739–743, 1998.

Bond, G., W. Broecker, S. Johnsen, J. Mc Manus, L. Labeyrie, J. Jouzel, and G. Bonani, Correlations between climate records from North Atlantic sediments and Greenland ice, *Nature*, *365*, 143–147, 1993.

Bond, G. C., and R. Lotti, Iceberg discharges into the North Atlantic on millennial time scales during the last glaciation, *Science*, *267*, 1005–1010, 1995.

Boyle, E., Last-Glacial-Maximum North Atlantic Deep Water: on, off or somewhere in-between?, *Phil. Trans. R. Soc. Lond. B*, *348*, 243–253, 1995.

Bryan, K., A numerical method for the study of the circulation of the world ocean, *J. Comp. Phys.*, *4*, 347–376, 1969.

Campin, J.-M., and H. Goosse, Parameterization of density-driven downsloping flow for a coarse-resolution ocean model in z-coordinate, *Tellus*, *51A*, 412–430, 1999.

Cortijo, E., L. Labeyrie, L. Vidal, M. Vautravers, M. Chapman, J.-C. Duplessy, M. Elliot, M. Arnold, J. L. Turon, and G. Auffret, Changes in sea surface hydrology associated with Heinrich event 4 in the North Atlantic Ocean between 40° and 60°N, *Earth Planet. Sci. Lett.*, *146*, 29–45, 1997.

Crucifix, M., P. Tulkens, M. Loutre, and A. Berger, A reference simulation for the present-day climate with a non-flux corrected atmosphere-ocean-sea ice model of intermediate complexity, Progress Report 2000/1, Institut d'Astronomie et de Géophysique G. Lemaître, Université catholique de Louvain,also available from http://www.astr.ucl.ac.be/publi.php?model=mob, 2000.

de Vernal, A., and C. Hillaire-Marcel, Sea-ice cover, sea-surface salinity and halo-/thermocline structure of the nortwest North Atlantic: modern versus full glacial conditions, *Quat. Sci. Rev.*, pp. 65–85, 2000.

Denton, G. H., Does an asymmetric thermohaline-ice-sheet oscillation drive 100 000-yr glacial cycles?, *Journ. Quat. Sci.*, *15*(4), 301–318, 2000.

Dong, B., and P. J. Valdes, Sensitivity studies of Northern Glaciation using an atmospheric general circulation model, *J. Climate*, *8*, 2471–2496, 1995.

Duplessy, J.-C., N. J. Shackelton, R. G. Fairbanks, L. Labeyrie, D. Oppo, and N. Kallel, Deepwater source variations during the last climatic cycle and their impact on the global deepwater circulation, *Paleoceanogr.*, *3*, 343–360, 1988.

Elliot, M., L. Labeyrie, G. Bond, E. Cortijo, J. L. Turon, N. Tisnerat, and J.-C. Duplessy, Millenial-scale iceberg discharges in the Irminger Basin during the last glacial period: relationship with the Heinrich events and environmental settings, *Paleoceanogr.*, *13*(5), 433–446, 1998.

ETOPO 5, *Global 5′ × 5′ Depth and Elevation*. available from National Geophysical Data Center, NOAA, US Dept of Commerce, Code E/GC3, Boulder, CO 80303, 1986.

Fouquart, Y., and B. Bonnel, Computations of solar heating of the earth's atmosphere: A new parameterization, *Beitr. Phys. Atmos.*, *53*, 35–62, 1980.

Gallée, H., J. P. van Ypersele, T. Fichefet, C. Tricot, and A. Berger, Simulation of the last glacial cycle by a coupled, sectorially averaged climate-ice sheet model. Part I: The climate model., *J. Geophys. Res.*, *96*, 13,139–13,161, 1991.

Gallée, H., J. P. van Ypersele, T. Fichefet, I. Marsiat, C. Tricot, and A. Berger, Simulation of the last glacial cycle by a coupled, sectorially averaged climate-ice sheet model. Part II: Response to insolation and CO_2 variation, *J. Geophys. Res.*, *97*, 15,713–15,740, 1992.

Ganopolski, A., S. Rahmstorf, V. Petoukhov, and M. Claussen, Simulation of modern and glacial climates with a coupled global model of intermediate complexity, *Nature*, *391*, 351–356, 1998.

Ganopolski, A., S. Rahmstorf, Rapid changes of glacial climate simulated in a coupled climate model, *Nature*, 153–158, 2001.

Goosse, H., and T. Fichefet, Open-ocean convection and polynya formation in a large-scale ice-ocean model, *Tellus*, in press, 2000.

Grootes, P. M., and M. Stuiver, Oxygen 18/16 variability in greenland snow and ice with 10−3 to 105-year resolution, *J. Geophys. Res.*, *102*(C12), 26,455–26,470, 1997.

Grousset, F. E., L. Labeyrie, J. A. Sinko, M. Cremer, G. Bond, J. Duprat, E. Cortijo, and S. Huon, Patterns of ice-rafted detritus in the glacial North Atlantic (40-55°N), *Paleoceanogr.*, *8*(2), 175–192, 1993.

Harvey, L. D. D., Development of a sea-ice model for use in zonally averaged energy balance models, *J. Climate*, *1*, 1221–1238, 1988.

Harvey, L. D. D., A two-dimensional ocean model for long-term climatic simulations: stability and coupling to atmospheric and sea-ice models., *J. Geophys. Res.*, *97*, 9435–9453, 1992.

Heinrich, H., Origin and consequences of cyclic ice rafting in the Northeast Atlantic Ocean during the past 130,000 years, *Quat. Res.*, *29*, 142–152, 1988.

Holton, J. R., *An Introduction to Dynamic Meteorology*. Academic Press, San Diego, Calif., 314 pp., 1979.

Hovine, S., and T. Fichefet, A zonally averaged, three-basin ocean circulation model for climate studies, *Clim. Dyn.*, *10*, 313–331, 1994.

Jaeger, L., Monatskarten des nierderschlags für die ganze erde, Tech. Rep. 139, Ber. Dtsch. Wetterdienstes, 38 pp, 1976.

Johnsen, S., W. Dansgaard, H. B. Clausen, and C. C. Langway Jr, Oxygen isotope profiles through the Antarctic and Greenland ice sheets, *Nature*, *235*(5339), 429–434, 1972.

Joussaume, S., and K. E. Taylor, The paleoclimate modelling intercomparison project, in *Paleoclimate modelling intercomparison projects*, edited by P. Braconnot, vol. WCRP-111, WMO/TD No. 1007, pp. 7–24. 2000.

Labeyrie, L., J.-C. Duplessy, J. Duprat, A. Juillet-Leclerc, J. Moyes, E. Michel, N. Kallel, and N. J. Shackelton, Changes in the vertical structure of the North Atlantic ocean between glacial and modern times, *Quat. Sci. Rev.*, *11*, 401–413, 1992.

Labeyrie, L., L. Vidal, E. Cortijo, M. Paterne, M. Arnold, J. C. Duplessy, M. Vautravers, M. Labracherie, J. Duprat, J. Turon, F. Grousset, and T. Van Weering, Surface and deep hydrology of the Northern Atlantic Ocean during the past 150 000 years, *Phil. Trans. R. Soc. Lond. B*, *348*, 255–264, 1995.

Lang, C., M. Leuenberger, and J. Schwander, 16 degree C rapid temperature variation in central Greenland 70 kyr ago, *Science*, *286*(5441), 934–937, 1999.

Levitus, S., *Climatological Atlas of the World Ocean*, NOAA Prof. Paper 13. Nat. Oceanic and Atmos. Admin, U.S. Dep. of Comm., Washington, D. C., 173 pp, 1982.

Licciardi, J. M., J. T. Teller, and P. U. Clark, Freshwater routing by the Laurentide Ice Sheet during the last deglaciation, in *Mechanisms of Global Climate Change at Millenial Time Scles*, edited by P. Clark, R. Webb, and L. Keigwin, vol. 112 of *AGU Monograph Series*, pp. 177–201. American Geophysical Union, 1999.

Maier-Reimer, E., U. Mikolajewicz, and K. Hasselmann, Mean circulation of the Hamburg LSG OGCM and its sensitivity to the thermohaline surface forcing, *J. Phys. Oceanogr.*, *23*, 731–757, 1993.

Manabe, S., and R. Stouffer, Are two modes of thermohaline circulation stable?, *Tellus*, *51A*, 400–411, 1999.

Manabe, S., and R. J. Stouffer, Coupled ocean-atmosphere model response to freshwater input: Comparison to

Younger Dryas event, *Paleoceanogr.*, *12*(2), 321–336, 1997.

Marchal, O., T. Stocker, and F. Joos, Impact of oceanic reorganizations on the ocean carbon cycle and atmospheric carbon dioxide content, *Paleoceanogr.*, *13*(3), 225–244, 1998.

Marotzke, J., Instabilities and multiple steady states of the thermohaline circulation, in *Oceanic Circulation Models: Combining data and dynamics*, edited by D. L. T. Anderson, and J. Willebrand, vol. 284 of *NATO ASI series*, pp. 501–511. Kluwer, 1989.

Meese, D. A., R. B. Alley, A. J. Gow, P. Grootes, P. A. Mayewski, M. Ram, K. C. Taylor, E. D. Waddington, and G. Zielinski, Preliminary depth-age scale of the GISP2 ice core, CRRL Spec. Rep. 94-1, U.S. Army Cold Reg. Res. and Eng. Lab., Hannover, NH, 1994.

Mikolajewicz, U., and E. Maier-Reimer, Mixed boundary conditions in ocean general circulation models and their influence on the stability of the model's conveyor belt, *J. Geophys. Res.*, *99*(C11), 22633–22644, 1994.

Morcrette, J. J., Sur la paramétrisation du rayonnement dans les modèles de la circulation générale atmosphérique, Ph.D. thesis, Univ. des Sci. et Tech. de Lille, Lille, France, 373 pp., 1984.

Ohring, G., and S. Adler, Some experiments with a zonally-averaged climate model, *J. Atmos. Sci.*, *35*, 186–205, 1978.

Oort, H. H., *Global atmospheric circulation statistics 1958-1973*, NOAA Prof. Paper 14. Nat. Oceanic and Atmos. Admin, U.S. Dep. of Comm., Washington, D. C., 180 pp, 1983.

Paillard, D., and E. Cortijo, A simulation of the Atlantic meridional circulation during Heinrich event 4 using reconstructed sea surface temperatures and salinites, *Paleoceanogr.*, *14*(6), 716–724, 1999.

Paillard, D., and L. Labeyrie, Role of the thermohaline circulation in the abrupt warming after Heinrich events, *Nature*, *372*, 162–164, 1994.

Peltier, W. R., Ice age paleotopography, *Science*, *265*, 195–201, 1994.

Rahmstorf, S., Bifurcations of the Atlantic thermohaline circulation in response to changes in the hydrological cycle, *Nature*, *378*, 145–149, 1995.

Rahmstorf, S., On the freshwater forcing and transport of the Atlantic thermohaline circulation, *Clim. Dyn.*, *12*, 799–811, 1996.

Ruddiman, W. F., Late quaternary deposition of ice-rafted sand in the subpolar north atlantic sediments, *Geol. Soc. Am. Bull.*, *83*, 2817–2836, 1977.

Sakai, K., and W. Peltier, Dansgaard-Oeschger oscillations in a coupled atmosphere-ocean climate model, *J. Climate*, *10*(5), 949,970, 1997.

Saltzman, B., Parametrization of the vertical flux of latent heat at the earth's surface for use in statistical-dynamical climate models, *Arch. Met. Geoph. Biokl.*, A *29*, 41–53, 1980.

Saltzman, B., and S. Ashe, Parametrization of the monthly mean vertical heat transfer at the earth's surface, *Tellus*, *28*, 323–331, 1976.

Sarnthein, M., K. Winn, S. J. A. Jung, J.-C. Duplessy, L. Labeyrie, H. Erlenkeuser, and G. Ganssen, Changes in east Atlantic deepwater circulation over the last 30,000 years: Eight time slice reconstructions, *Paleoceanogr.*, *9*, 209–267, 1994.

Sarnthein, M., K. Stattegger, D. Dreger, H. Erlenkeuser, P. Grootes, B. Haupt, S. Jung, T. Kiefer, W. Kuhnt, U. Pflaumann, C. Schäfer-Neth, H. Schulz, M. Schulz, D. Seidov, J. Simstich, S. van Kreveld, E. Vogelsang, A. Völker, and M. Weinelt, Fundamental modes and abrupt changes in North Atlantic circulation and climate over the last 60ky — Concepts, reconstruction, and numerical modelling, in *The Northern North Atlantic: A changing environment*, edited by P. Schäfer, pp. 45–66. Springer-Verlag, 2000.

Schiller, A., U. Mikolajewicz, and R. Voss, The stability of the North Atlantic thermohaline circulation in a coupled ocean-atmosphere general circulation model, *Clim. Dyn.*, *13*, 325–347, 1997.

Schulz, M., W. H. Berger, M. Sarnthein, and P. Grootes, Amplitude variations of 1470-year climate oscillations during the last 100,000 years linked to fluctuations of continental ice mass, *Geophysical Research Letters*, *26*(22), 3385–3388, 1999.

Sela, J., and A. Wiin-Nielsen, Simulation of the atmospheric annual energy cycle, *Mon. Wea. Rev.*, *99*, 460–468, 1971.

Semtner, A. J., A model for the thermodynamic growth of sea ice in numerical investigations of climate, *J. Phys. Oceanogr.*, *6*, 379–389, 1976.

Semtner, A. J., Finite-difference formulation of a world ocean model, in *Advanced Physical Oceanographic Numerical Modelling*, edited by J. J. O'Brien, pp. 187–202. D. Reidel, Dordrecht, 1986.

Smith, W., Note on the relationship between total precipitable water and surface dew point, *J. Appl. Meteor.*, *5*, 726–727, 1966.

Stocker, T. F., and D. G. Wright, Rapid transitions of the ocean's deep circulation induced by changes in surface water fluxes, *Nature*, *351*, 729–732, 1991.

Stuiver, M., and P. M. Grootes, GISP2 oxygen isotope ratios, *Quat. Res.*, *53*(3), 277–284, 2000.

Taylor, K., The influence of subsurface energy storage on seasonal temperature variations, *J. Appl. Meteor.*, *15*, 1129–1138, 1976.

van Kreveld, S., M. Sarnthein, H. Erlenkeuser, P. Grootes, J. S., M. Nadeau, U. Pflaumann, and A. Voelker, Potential links between surging ice sheets, circulation changes and the Dansgaard-Oeschger cycles in the Irminger Sea, 60-18 kyr, *Paleoceanogr.*, pp. 425–442, 2000.

Vidal, L., L. Labeyrie, E. Cortijo, M. Arnold, J. C. Duplessy, E. Michel, S. Becqué, and T. C. E. van Weering, Evidence for changes in the North Atlantic Deep Water linked to meltwater surges during the Heinrich events, *Earth Planet. Sci. Lett.*, *146*, 13–27, 1997.

Voelker, A. H. L., M. Sarnthein, P. M. Grootes, H. Erlenkeuser, C. Laj, A. Mazaud, M.-J. Nadeau, and M. Schleicher, Correlation of marine ^{14}C ages from the nordic seas with the GISP2 isotope record: implications for ^{14}C calibration beyond 25 ka BP, *Radiocarbon*, *40*(1), 517–534, 1998.

Wang, Z., and L. A. Mysak, Ice sheet-thermohaline circulation interactions in a climate model of intermediate complexity, *J. Oceanogr.*, *in press*, 2001.

Warren, S., C. Hahn, J. London, R. Chervin, and R. Jenne, Global distribution of total cloud cover and cloud type amounts over land, Technical note tn-273+str, NCAR, Boulder, CO, 29 pp. + 200 maps, 1986.

Weaver, A. J., and E. S. Sarachick, The role of mixed boundary conditions in numerical models of the Ocean's climate, *J. Phys. Oceanogr.*, *21*, 1470–1493, 1991.

Weaver, A. J., M. Eby, A. F. Fanning, and E. C. Wiebe, Simulated influence of carbon dioxide, orbital forcing and ice sheets on the climate of the last glacial maximum, *Nature*, *394*(6696), 847 – 853, 1998.

Webb, R. S., D. Rind, S. Lehman, R. Healy, and D. Sigman, Influence of ocean heat transport on the climate of the Last Glacial Maximum, *Nature*, *385*, 695–699, 1997.

Winguth, A. M. E., D. Archer, J.-C. Duplessy, E. Maier-Reimer, and U. Mikolajewicz, Sensitivity of paleonutrient tracer distributions and deep-sea circulation to glacial boundary conditions, *Paleoceanogr.*, *14*(3), 304–323, 1999.

Wright, D., and T. F. Stocker, Younger Dryas experiments, in *Ice in the climate system*, edited by W. Peltier, vol. I 12 of *NATO ASI series*, pp. 395–416. Springer-Verlag, 1993.

Wright, D. G., and T. F. Stocker, Sensitivities of a zonally averaged global ocean circulation model, *J. Geophys. Res.*, *97*, 12,707–12,730, 1992.

Wunsch, C., On sharp spectral lines in the climate record and the millennial peak, *Paleoceanogr.*, *15*(4), 417–424, 2000.

Zaucker, F., T. F. Stocker, and W. S. Broecker, Atmospheric freshwater fluxes and their effect on the global thermohaline circulation, *J. Geophys. Res.*, *99*, 12,443–12,457, 1994.

A. Berger and M. Crucifix, Institut d'Astonomie et de Géophysique Georges Lema^, Université catholique de Louvain, chemin du Cyclotron, 2, 1348 Louvain-la-Neuve, Belgium. (e-mail: berger@astr.ucl.ac.be, crucifix@astr.ucl.ac.be)

P. Tulkens, Task Force Sustainable Development, Federal Planning Bureau, Avenue des Arts, 47-49, 1000 Brussels, Belgium. (e-mail: pt@plan.be)

Simulating Climates of the Last Glacial Maximum and of the Mid-Holocene: Wind Changes, Atmosphere-Ocean Interactions, and the Tropical Thermocline.

Andrew B. G. Bush

Department of Earth and Atmospheric Sciences, University of Alberta, Edmonton, Alberta, Canada.

Transition of the Earth's climate from the cold and windy conditions prevailing at the Last Glacial Maximum to the relatively warm and strongly seasonal conditions of the mid-Holocene produced many changes in the climate system. This study compares and contrasts the differences during these two time periods as simulated by a coupled atmosphere-ocean general circulation model.

Compared to today, there is enhanced equatorward flux of easterly momentum in the northern hemisphere during both time periods, but for different reasons. During the mid-Holocene, a majority of the increase is associated with transient eddy activity in the upper troposphere at northern midlatitudes; an additional component arises from changes in the mean meridional circulation. Increased eddy activity is related to increased seasonality associated with mid-Holocene insolation. At the LGM, there are also increases in transient eddy momentum flux near the southern edge of the North American ice sheets, but a majority of the anomalous flux arises from stationary eddies that are also induced by the ice sheets.

These enhanced momentum fluxes increase the strength of the surface equatorial easterlies through intensification of subtropical subsidence and modification of the lower troposphere's meridional pressure gradient. Through atmosphere-ocean interactions, this increases the spatial extent of the tropical Pacific cold tongue in both simulations. Results imply that the mean state of the tropical thermocline may be changed in a similar way either by increasing seasonal radiative forcing or by introducing strong topographic forcing.

1. INTRODUCTION

The continuously changing climate of the late Quaternary has exhibited, over the past 21,000 years, a transition from cold and windy glacial conditions, in which massive continental ice sheets dominated the northern hemisphere, to relatively warm yet highly seasonal conditions during the early-mid Holocene. Global atmospheric winds during these two time periods were likely to have been different according to a variety of proxy data analyses and numerical modeling studies [e.g., *Quinn*, 1972; *Wright et al.*, 1993; *Kutzbach and Otto-Bliesner*, 1982; *Hall et al.*, 1996; *Bush and Philan-*

The Oceans and Rapid Climate Change: Past, Present, and Future
Geophysical Monograph 126
Copyright 2001 by the American Geophysical Union

der, 1999; *Bush*, 1999; *Otto-Bliesner*, 2000]. In particular, evidence exists for stronger equatorial Pacific trade winds at the Last Glacial Maximum (LGM; e.g., *Pedersen*, 1983; *Sarnthein et al.*, 1981; *Farrell et al.*, 1995) whereas the mid-Holocene exhibited stronger seasonal winds, particularly in the region of the south Asian monsoon [e.g., *Prell*, 1984; *Clemens and Prell*, 1990; *Prell and Kutzbach*, 1992].

However, the coupled atmosphere-ocean system can be quite sensitive to changes in wind direction and strength because of atmosphere-ocean feedbacks [e.g., *Philander*, 1985; *Xie*, 1998]. In today's climate, for example, the tropical Pacific Ocean exhibits perhaps the most prominent example of such interactions at play: the El Niño Southern Oscillation. At the heart of this interannual oscillation is the coupling of easterly trade wind strength to thermocline depth, with the El Niño state characterized by weak to nonexistent trade winds, a fairly flat and deep thermocline, and warm SSTs, and with the La Niña state characterized by strong trade winds, a shoaled thermocline in the east, and an extended cold tongue. Climatic consequences of changes in tropical Pacific SST are global in nature and, in many parts of the modern world, can be of sufficient magnitude to warrant the need for long-range forecasts.

Nevertheless, considerable debate continues concerning the history of the tropical Pacific Ocean during the late Quaternary. It appears that at the LGM the thermocline had a steeper east-west tilt than it does today [*Andreasen and Ravelo*, 1997], a result that concurs with evidence for stronger equatorial trade winds [*Sarnthein et al.*, 1981] and increased productivity [*Pedersen*, 1983; *Lyle et al.*, 1992]. However, there is also some evidence for decreased Pacific productivity, evidence which would not be consistent with increased trade wind speeds and so, assuming colder SSTs, is taken to indicate an increase in eastern boundary current temperature advection [*Loubere*, 2000].

In simulations with a coupled atmosphere-ocean general circulation model configured for the LGM and for the mid-Holocene it has been shown that, during both time periods, simulated easterly trade winds were stronger and that the cold tongue was more pronounced compared to today (although much more so at the LGM; *Bush and Philander*, 1998; *Bush*, 1999). The diagnostics presented here detail the dynamical reasons for the wind increases in these simulations and relate them directly to the dominant forcing factors that characterise the two time periods: increased seasonal solar radiation during the mid-Holocene; and the presence of massive continental ice sheets during the LGM.

A brief description of the numerical model is given in Section 2, followed by diagnostics of momentum fluxes and circulation changes in Section 3. A discussion and conclusions are presented in Section 4.

2. THE MODEL

Only a brief overview of the model details will be given here. The interested reader is referred to Bush and Philander (1999) for the LGM simulation and Bush (1999) for the mid-Holocene simulation. The model is a global, fully coupled atmosphere-ocean general circulation model (GCM) whose original atmospheric and oceanic components are described in Gordon and Stern (1982) and Pacanowski *et al.* (1991), respectively. The equivalent resolution in the atmospheric spectral model is 3.75 degrees in longitude and 2.25 degrees in latitude (at the equator); there are 14 unevenly-spaced levels in the vertical extending from near the surface to the 50 millibar level. The finite-difference ocean model has a resolution of $3.62°$ in longitude and $2°$ in latitude. There are 15 unevenly-spaced levels in the vertical, most of which are in the upper ocean in order to maximize resolution for the wind-driven circulation. The thermodynamic sea ice parameterization of Fanning and Weaver (1996) is used with the modification that sea ice albedo depends on ice thickness. The model components exchange daily mean dynamic and thermodynamic boundary condition information at 1 day intervals. Boundary conditions required by the atmospheric model are SST, surface current velocities, and sea ice extent; the ocean model requires daily mean vector wind stress, net heat flux, net freshwater flux, net shortwave radiation, and sea ice extent.

In the mid-Holocene simulation, the only input parameters that are different from the control simulation are the Earth's orbital parameters; values of obliquity, eccentricity, and longitude of perihelion are set to those appropriate for 6,000 B.P. as computed by Berger (1992).

In the LGM simulation, atmospheric CO_2 levels are reduced to 200 ppm, orbital parameters appropriate for 21,000 B.P. are imposed [*Berger*, 1992), glacial height and areal extent are specified according to reconstructions by Peltier (1994), continental margin locations assume a 120 meter drop in sea level [*Fairbanks*, 1989), and LGM bare land surface albedo is imposed [*CLIMAP*, 1981]. Further technical details concerning the simulations may be found in Bush and Philander (1999).

All simulations are decadal in timescale; we are therefore not in a position to discuss changes in the deep

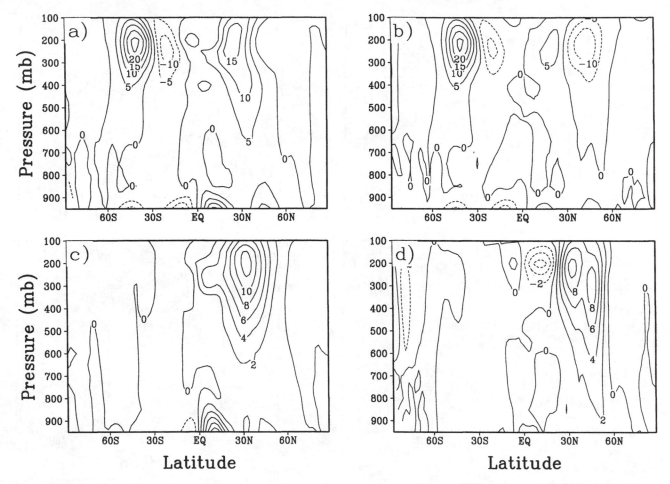

Figure 1. Annual mean, total northward transport of westerly momentum $[\overline{UV}]$ ((a)), and contributions from (b) the mean meridional circulation $[\overline{V}][\overline{U}]$, (c) the transient eddies $[\overline{V'U'}]$, and (d) the stationary eddies $[\overline{V^*U^*}]$ for the control simulation. Units are m^2/s^2.

ocean circulation nor changes in deep water upwelling (for such changes at LGM the reader is referred to *Weaver et al.*, 1998 and references therein). Rather, we are restricted to focus on the radiative balances and wind-driven circulation of the upper ocean, and the annual to decadal timescale atmosphere-ocean feedbacks that act to determine the depth of the tropical Pacific thermocline [*Philander*, 1990; *Gu and Philander*, 1997]. While running this particular coupled model to equilibrium of the meridional overturning circulation in all experiments is impractical, it is necessary to note that if such experiments could be performed then high latitude SSTs could be sufficiently different that zonal mean baroclinicity in the lower atmosphere may be changed. (Since the simulations are initialized from Levitus (1982) data, high latitude SSTs are indicative of modern overturning strength.) We will discuss the implications and limitations of the experiments more in section 4.

3. MODEL DIAGNOSTICS OF MOMENTUM FLUXES

The timestep in the atmospheric model is 216 seconds and, during the course of integration, model fields are collected into daily mean data which are subsequently averaged to produce monthly means. The following diagnostics of energetics were computed using these monthly mean values. As discussed in Peixoto and Oort (1992) this procedure implies that the transient eddy statistics to follow contain fluctuations with timescales of a month or less, so that the following decomposition implies that any interannual variability is therefore included in the stationary eddy statistics.

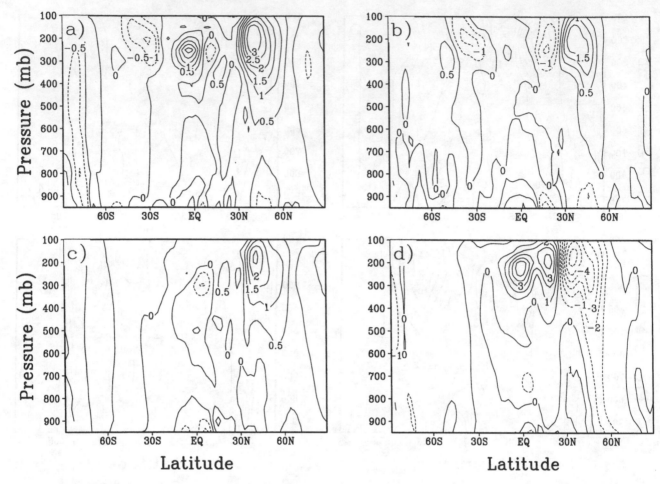

Figure 2. Differences in momentum flux between the mid-Holocene and control simulations (layout as in Figure 1). Units are m^2/s^2.

The total northward angular momentum transport is defined as $[\overline{VU}]$, where an overline denotes a time average and square brackets denote a zonal average. The following decomposition then follows [e.g., *Peixoto and Oort*, 1992]:

$$[\overline{VU}] = [\overline{V}][\overline{U}] + [\overline{V'U'}] + [\overline{V^*U^*}]. \quad (1)$$

In the above expression, a prime indicates a departure from the time mean and a star indicates a departure from the zonal mean. Terms on the right hand side therefore represent contributions to the total momentum flux from, respectively, the mean meridional circulation, the transient eddies, and the stationary eddies.

For reference, a breakdown of momentum transport in the control simulation is shown in Figure 1. In comparison to modern observations [e.g., *Peixoto and Oort*,

1992] the model underestimates the total angular momentum flux in the upper troposphere by a factor of approximately two. It also relies more on the mean meridional circulation and stationary eddies to achieve its total transport (although in the northern hemisphere the transient eddies still generate a majority of the momentum flux). There is also a distinct lack of transient eddy flux in the southern hemisphere. The latter two results are not surprising given that the resolution in the model precludes the observed frequency and intensity of midlatitude baroclinic instabilities which, in their barotropic decay phase, are the primary contributors to the equatorward flux of easterly momentum [e.g., *Simmons and Hoskins*, 1980; *Edmon et al.*, 1980]. Nevertheless, spatial resolution in the model is kept constant between all experiments, so differences between experiments of atmospheric variables should remove these biases. It is also noted that the southern hemisphere, being relatively inactive, does not contribute to the fol-

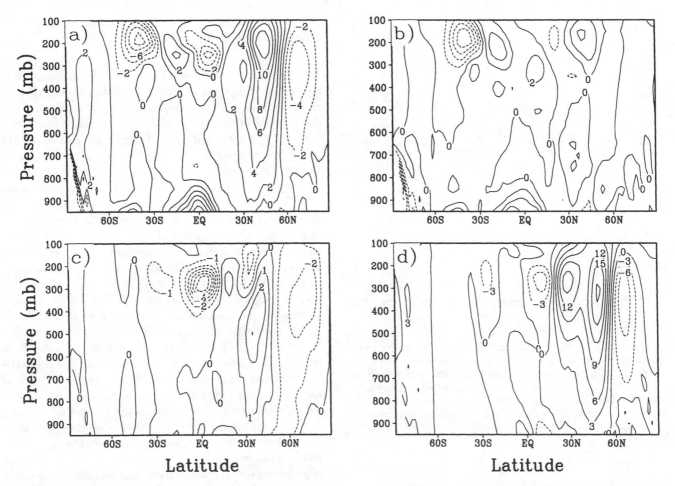

Figure 3. As in Figure 2, except differences are between the LGM and control simulations. Units are m^2/s^2.

lowing changes nearly so much as the northern hemisphere.

3.1. Mid-Holocene Fluxes

The difference (compared to today) of the total, annual mean momentum flux in the mid-Holocene simulation indicates a ~20% strengthening of the net mid-latitude, upper tropospheric flux near 45N (Figure 2a). The dominant positive contribution is from the transient eddies (Figure 2c), although a contribution from the mean meridional circulation is also positive (Figure 2b); their sum is reduced somewhat by negative changes in the stationary eddies (Figure 2d) that arise from a weaker upper level westerly jet downstream of the Himalayas. The net result, however, is a positive contribution to the upper level, northern midlatitude momentum flux with the dominant contribution coming from the transient eddies. Analysis of the spatial

variation of the transient eddy flux at the 200 mb level indicates that the largest increases occur over eastern Asia, over the north-central Pacific, and over the entrance to the North Atlantic storm track.

3.2. LGM Fluxes

The total LGM momentum flux in northern midlatitudes (~45N) is significantly greater than in the control simulation (Figure 3a). Decomposition into its various components indicates that transient eddy fluxes increase in northern midlatitudes (Figure 3c), but not as much as in the mid-Holocene simulation (cf. Figure 2c; and they occur lower in the troposphere). Spatial maps indicate that transient eddy fluxes increase over both the Pacific and Atlantic storm tracks, but also along the southwestern edge of the Cordilleran ice sheet where atmospheric baroclinicity is greatest.

The largest change in momentum flux is caused, how-

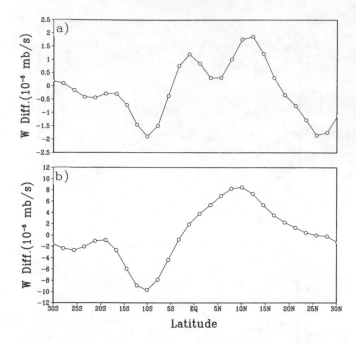

Figure 4. Differences (experiment minus control) in annual, zonal, and vertical mean vertical velocities between (a) the mid-Holocene and (b) the LGM simulations. (The units are 10^{-5} mb/s, so a positive value indicates downwards motion and a negative value indicates upwards motion. Values have been calculated from the terrain following pressure coordinates by using the mean surface pressure from each simulation.)

ever, by a change in stationary eddies in the northern hemisphere (Figure 3d). Changes are of approximately the same magnitude as the total, annual mean momentum flux in the control simulation and therefore represent a significant contribution to the maintenance of the general circulation. Spatial analysis of $[\overline{V^*U^*}]$ indicates that stationary eddy fluxes increase greatly along both east and west margins of the North American ice sheets near 45N, with the largest positive fluxes extending along the east coast into the North Atlantic ocean in the LGM storm track jet (see Figure 8 in *Bush and Philander*, 1999).

The Laurentide and Cordilleran ice sheets therefore impact the momentum fluxes in two ways. First, they impose a topographic barrier which changes the course of the midlatitude jet stream and sets up planetary long waves that have a signature in the stationary eddies both upstream and downstream of the ice sheets. Second, they focus the westerly jet over central North America and over the Atlantic, increasing transient eddy activity in those regions [*Hall et al.*, 1996; *Bush and Philander*, 1999].

3.3. Vertical Velocities, Easterly Trades, and the Tropical Thermocline

A positive northward flux of westerly momentum is equivalent to a positive equatorward flux of easterly momentum and, given the equatorward fluxes during the decay phase of baroclinic eddies (which constitute a majority of the transient eddy fluxes; e.g., *Edmon et al.*, 1980), it is the latter definition that is most appropriate. Positive convergence of easterly momentum into the subtropics in the upper levels during the mid-Holocene and the LGM (Figures 2a and 3a, respectively) implies increased upper level convergence with the poleward flow of the upper branch of the mean Hadley circulation. Increased upper level convergence induces vertical mean tropospheric subsidence (Figure 4), particularly in the northern hemisphere where the anomalous momentum fluxes are the largest. Differences in vertical velocity of pressure surfaces are an order of magnitude larger at the LGM than during the mid-Holocene because of the larger changes in total momentum flux. Maximum differences in mid-Holocene and LGM vertical velocities, as shown in Figure 4, represent changes of approximately 15% and 53% , respectively, of the peak subtropical vertical velocities in the control simulation.

Changes in mean vertical velocities alter surface pressure, particularly where subsidence is greatest, and this changes the meridional pressure gradient in the subtropical latitudes. In particular, the meridional pressure gradient $\partial P/\partial y$ is increased in both simulations from the equator to 10-15N, with differences at the LGM larger than those during the mid-Holocene by approximately an order of magnitude (Figures 5a,b). South of the equator, $\partial P/\partial y$ is more negative in the simulations from $\sim 0 - 10$S.

Mean meridional gradients in surface pressure are crucial to the strength of the surface trade easterlies [e.g., *Grotjahn*, 1993]. Stronger positive meridional gradients north of the equator in both simulations imply stronger easterlies through simple geostrophy, as do the stronger negative gradients south of the equator. Differences in the zonally averaged 950 mb zonal wind speed reflect changes in these lower tropospheric pressure gradients and indicate that trade easterlies are indeed stronger by ~ 0.3 m/s and ~ 2 m/s in the mid-Holocene and LGM simulations, respectively (Figures 5c,d).

Increased zonal mean easterlies, particularly over the Pacific, increase the tilt of the tropical mean thermocline (Figure 6). Consistent with the differences in zonal wind speeds, the tilt of the LGM thermocline is steeper

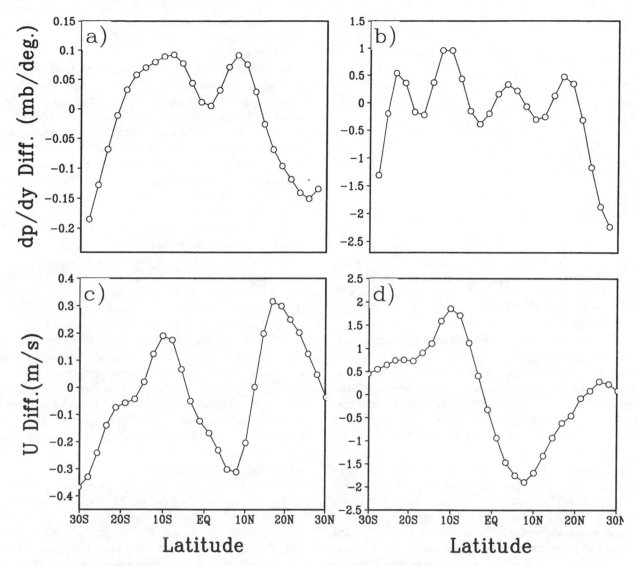

Figure 5. Differences (experiment minus control) between annual mean, zonally averaged surface pressure ((a) and (b)) and 950 mb zonal wind speeds ((c) and (d)). (a) and (c) are the differences between the mid-Holocene and control simulations. (b) and (d) are the differences between the LGM and control simulations. Note that scaling of the ordinate axes is different between experiments. Units in (a) and (b) are mb/degree; units in (c) and (d) are m/s.

than that of the mid-Holocene, and both are steeper than that produced in the control simulation.

4. DISCUSSION AND CONCLUSIONS

The annual mean fields shown in Figure 5 should reflect the dominant, large-scale, dynamical balances near the equator. Since geostrophic balance is maintained quite close to the equator (to within ~ 2°; *Philander*, 1990), differences in zonal mean surface pressure gradients should provide a first-order estimate for changes

in the zonal winds as follows. Compared to the control simulation, the difference in surface pressure between the equator and 10N in the mid-Holocene simulation is ~ 0.15 mb. An order of magnitude estimate for the change in easterly wind strength, assuming geostrophic balance and using a density of ~ 1 kg/m^3, gives zonal easterlies that are ~ 0.7 m/s stronger. An equivalent calculation for the LGM (5 mb over 20 degrees of latitude) gives easterlies ~ 9 m/s stronger. Compared to the 0.3 m/s and 2 m/s wind differences indicated in Figure 5, changes in zonal wind speed are, to first order,

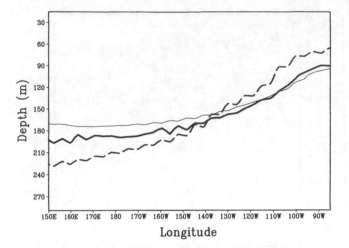

Figure 6. Depth of the 18 degree isotherm (averaged between 5N and 5S) in the tropical Pacific in the control simulation (thin solid line), the mid-Holocene simulation (thick solid line) and the LGM simulation (thick dashed line).

in geostrophic balance with the altered pressure field.

These annual mean fields, however, implicitly include the effects of any atmosphere-ocean interactions that have acted to amplify an initial strengthening of the easterlies. In an ENSO cycle, for example, an increase in trade easterly strength will induce more upwelling of colder water in the east and a steeper thermocline tilt; this will increase the zonal temperature/pressure gradient, which in turn will strengthen the easterlies even more. The net result is a La Niña state, and a reversal of these arguments (i.e., an initial weakening of the easterlies) leads to an El Niño state. Since these atmosphere-ocean feedbacks are positive, it is likely that the direct effect of momentum fluxes on the easterlies is not as large as that shown in Figures 5c,d and that the zonal pressure gradient changes along the equator (implied by Figure 6) contribute in part to the simulated wind changes.

In the LGM simulation, changes in surface pressure are further complicated by the presence of the continental ice sheets, which have sufficient vertical extent that they significantly alter the spatial mass distribution of the atmosphere. For example, typical surface pressures in the tropics are 10 mb greater than in the control simulation because northern hemisphere ice sheets can be greater than 3 kilometers high [*Peltier*, 1994] and therefore displace a large mass of air (as does a larger Antarctic ice sheet in the southern hemisphere). Zonal mean surface pressures are greater in the LGM simulation everywhere between ∼ 70S and ∼ 30N. However, since it is the pressure gradients that are dynamically

important, and not the absolute value of pressure itself, further work is warranted on this particular aspect of the LGM climate.

Simulated changes in the mean state of the tropical Pacific thermocline are therefore caused by changes in the atmosphere's general circulation. In the mid-Holocene simulation, increased northern hemisphere transient eddy flux is ultimately caused by increased seasonality. Mid-Holocene insolation induces colder winters and warmer summers and there is a concomitant increase in midlatitude eddy activity, particularly in wintertime. In the LGM simulation, increased eddy activity is forced primarily by topographic steering of the jet stream by the continental ice sheets. Increased eddy fluxes in both simulations lead to stronger equatorial easterlies through the mechanisms discussed in section 3. It is important to note, however, that the simulated increases occur for different reasons: in the mid-Holocene simulation it is the radiative forcing; in the LGM simulation it is the ice sheet topography.

These results are, of course, limited in the sense that the thermohaline circulation is not in equilibrium with the surface radiative, mechanical, and freshwater forcing that ultimately drive deepwater formation. Changes in this circulation are also believed to have been responsible for a component of late Quaternary climate change [e.g., *Manabe and Stouffer*, 1995; *Ganopolski et al.*, 1998; *Weaver et al.*, 1998; *Broecker*, 1998; *Stocker*, 2000]. While coupling and feedbacks between the deep ocean circulation and the wind-driven circulation remains an outstanding unsolved problem in physical oceanography [e.g., *Pedlosky*, 1996; *Munk and Wunsch*, 1998; *Egbert and Ray*, 2000], global models based on diffusive upwelling parameterizations suggest that surface ocean temperatures do change during the spinup to equilibrium. This implies that the surface baroclinicity of the ocean, particularly in the North Atlantic, may slowly change during the equilibration process and that this process would also modify the transient eddy fluxes presented here.

Nevertheless, we have demonstrated that during the mid-Holocene and at the LGM, simulated changes in thermocline depth in the tropical Pacific Ocean are related in part to changes in transient and stationary eddy momentum fluxes through modification of subtropical surface pressures. While diagnostics of the dynamics that occur in a particular model realization are important to unravel, it is equally essential that we work towards understanding the mechanisms that maintain the mean climatological state of the tropical thermocline (even for the present climate) because, somewhat ironically, we understand variability in the tropical Pacific better than we understand the mean state. It is apparent from this study, however, that atmospheric changes in high latitudes can impact thermocline depth and tilt and that such changes are likely to have played a role in determining the mean state of the tropical Pacific Ocean during the Late Quaternary.

Acknowledgments.   The author thanks the Natural Sciences and Engineering Research Council for support through Research Grant OGP0194151 and the Climate System History and Dynamics Project, as well as UNESCO for sponsoring International Geological Correlation Project 415.

REFERENCES

Berger, A., Orbital variations and insolation database, IGBP PAGES/World Data Center-A for Paleoclimatology Data Contribution Series # 92-007. NOAA/N GDC Paleoclimatology Program, Boulder CO, USA.

Broecker, W.S., Paleocean circulation during the last deglaciation: A bipolar seesaw?, *Paleoceanogr., 13*, 119-121, 1998.

Bush, A.B.G., Assessing the impact of mid-Holocene insolation on the atmosphere-ocean system. *Geophy. Res. Lett., 26*, 99-102, 1999.

Bush, A.B.G. and S.G.H. Philander, The climate of the Last Glacial Maximum: Results from a coupled atmosphere-ocean general circulation model. *J. Geophys. Res., 104*, 24,509-24,525, 1999.

Bush, A.B.G., and S.G.H. Philander, Tropical cooling in a coupled model simulation of the Last Glacial Maximum, *Science, 279*, 1341-1344, 1998.

Clemens, S.C. and W.L. Prell, Late Pleistocene variability of Arabian Sea summer-monsoon winds and dust source-area aridity: A record from the lithogenic component of deep-sea sediments, . *Paleoceanogr., 5*, 109-145, 1990.

Climate: Long-Range Investigation, Mapping, and Prediction (CLIMAP) Project Members, Seasonal reconstructions of the Earth's surface at the last glacial maximum, *Map and Chart Series MC-36*, Geol. Soc. of Am., Boulder, CO, 1981.

Edmon, H.J., B.J. Hoskins, and M.E. McIntyre, Eliassen-Palm cross-sections for the troposphere, *J. Atmos. Sci., 37*, 2600-2616, 1980.

Egbert, G.D. and R.D. Ray, Significant dissipation of tidal energy in the deep ocean inferred from satellite altimeter data, *Nature, 405*, 775-778, 2000.

Fairbanks, R.G., A 17,000-year glacio-eustatic sea level record: Influence of glacial melting rates on Younger Dryas event and deep-ocean circulation, *Nature, 342*, 637-642, 1989.

Fanning, A.F., and A.J. Weaver, An atmospheric energy-moisture balance model: Climatology, interpentadal climate change, and coupling to an ocean general circulation model, *J. Geophys. Res., 101*, 15,111-15,128, 1996.

Farrell, J.W., T.F. Pedersen, S.E. Calvert, and B. Nielsen, Glacial-interglacial changes in nutrient utilization in the equatorial Pacific Ocean, *Nature, 377*, 514-517, 1995.

Ganopolski, A., S. Rahmstorf, V. Petoukhov, and M. Claussen, Simulation of modern and glacial climates with a coupled

global model of intermediate complexity, *Nature, 391*, 351-356, 1998.

Gordon, C.T., and W. Stern, A description of the GFDL global spectral model, *Mon. Weath. Rev., 110*, 625-644, 1982.

Grotjahn, R., *Global atmospheric circulations. Observations and theories*, 430 pp., Oxford University Press, New York, 1993.

Gu, D. and S.G.H. Philander, Interdecadal climate fluctuations that depend on exchanges between the tropics and extratropics, *Science, 275*, 805-807, 1997.

Guilderson, T.P., R.G. Fairbanks, and J.L. Rubenstone, Tropical temperature variations since 20,000 years ago: Modulating interhemispheric climate change, *Science, 263*, 663-665, 1994.

Hall, N.M.J., P.J. Valdes, and B. Dong, The maintenance of the last great ice sheets: A UGAMP GCM study, *J. Clim., 9*, 1004-1019, 1996.

Kutzbach, J.E. and B.L. Otto-Bliesner (1982). The sensitivity of the African-Asian monsoonal climate to orbital parameter changes for 9000 years B.P. in a low-resolution general circulation model. *J. Atmos. Sci., 39*, 1177-1188.

Levitus, S., Climatological atlas of the world ocean, *NOAA Prof. Paper 13*, 173 pp., U.S. Govt. Print. Office, Washington, D. C., 1982.

Loubere, P., Marine control of biological production in the eastern equatorial Pacific Ocean, *Nature, 406*, 497-500.

Lyle, M.W., F.G. Prahl, M.A. Sparrow, Upwelling and productivity changes inferred from a temperature record in the central equatorial Pacific, *Nature, 355*, 812-815, 1992.

Manabe, S. and R.J. Stouffer, Simulation of abrupt climate change induced by freshwater input to the North Atlantic Ocean, *Nature, 378*, 165-167, 1995.

Munk, W. and C. Wunsch, Abyssal recipes II: energetics of tidal and wind mixing, *Deep-Sea Res., 45*, 1977-2010, 1998.

Otto-Bleisner, B.L., El Niño/La Niña and Sahel precipitation during the middle Holocene. *Geophys. Res. Lett., 26*, 87-90, 1999.

Pacanowski, R.C., K. Dixon, and A. Rosati, *The GFDL Modular Ocean Model user guide*, GFDL Ocean Group Tech. Rep. 2, Geophys. Fluid Dyn. Lab., Princeton, N.J., 1991.

Pedersen, T.F., Increased productivity in the eastern equatorial Pacific during the last glacial maximum (19,000 to 14,000 yr B.P.), *Geology, 11*, 16-19, 1983.

Peixoto, J.P. and A.H. Oort, *Physics of Climate*, 520 pp., American Institute of Physics, New York, 1992.

Pedlosky, J., *Ocean Circulation Theory*, 453 pp., Springer, New York, 1996.

Peltier, W.R., Ice age paleotopography, *Science, 265*, 195-201, 1994.

Philander, S.G.H., El Niño and La Niña, *J. Atmos. Sci., 42*, 2652-2662, 1985.

Philander, S.G.H., *El Niño, La Niña, and the Southern Oscillation*, 293 pp., Academic Press, New York, 1990.

Prell, W.L., Monsoonal climate of the Arabian Sea during the late Quaternary: A response to changing solar radiation, in *Milankovitch and Climate*, edited by A. Berger et al., pp. 349-366. Reidel, Dordrecht, 1984.

Prell, W.L. and J.E. Kutzbach, Sensitivity of the Indian monsoon to forcing parameters and implications for its evolution. *Nature, 360*, 647-652, 1992.

Quinn, W.H., Late Quaternary meteorological and oceanographic developments in the equatorial Pacific, *Nature, 229*, 330-331, 1971.

Sarnthein, M., G. Tetzlaff, B. Koopman, K. Wolter, and U. Pflaumann, Glacial and interglacial wind regimes over the eastern subtropical Atlantic and north-west Africa, *Nature, 293*, 193-196, 1981.

Simmons, A.J. and B.J. Hoskins, Barotropic influences on the growth and decay of non-linear baroclinic waves, *J. Atmos. Sci., 37*, 1679-1684.

Stocker, T.F., Past and future reorganizations in the climate system, *Quat. Sci. Rev., 19*, 301-319, 2000.

Weaver, A.J., M. Eby, A.F. Fanning, and E.C. Wiebe, Simulated influence of carbon dioxide, orbital forcing and ice sheets on the climate of the Last Glacial Maximum, *Nature, 394*, 847-853, 1998.

Wright, H.E. Jr., J.E. Kutzbach, T. Webb III, W.F. Ruddiman, F.A. Street-Perrott, and P.J. Bartlein (Eds.), "Global climates since the Last Glacial Maximum", 569 pp, University of Minnesota Press, Minneapolis, 1993.

Xie, S.-P., Ocean-atmosphere interaction in the making of the Walker circulation and the equatorial cold tongue, *J. Clim., 11*, 189-201, 1998.

A. B. G. Bush, 126 Earth Sciences Building Department of Earth and Atmospheric Sciences, University of Alberta, Edmonton, Alberta, Canada T6G 2E3. (email: andrew.bush@ualberta.ca)

Section II

Ocean and Climate Models: Bridges from Past to Future

Ocean Bi–Polar Seesaw and Climate: Southern Versus Northern Meltwater Impacts

Dan Seidov, Bernd J. Haupt, Eric J. Barron, and Mark Maslin

EMS Environment Institute, Pennsylvania State University, University Park, Pennsylvania

Model simulations targeting the ocean circulation response to changes in surface salinity in the high latitudes of both Northern and Southern Hemispheres demonstrate that meltwater impacts in one hemisphere may lead to a strengthening of the thermohaline conveyor driven by the source in the opposite hemisphere. This, in turn, leads to significant changes in poleward heat transport. Further, meltwater events caused largely by sea ice melting can lead to deep–sea warming and thermal expansion of abyssal water, that in turn can cause a substantial sea level change even without a major ice sheet melting. Experiments with a glacial ocean circulation regime prone to northern and southern meltwater events imply that glacial cycles may have been influenced by both northern and southern deepwater sources. Importantly, the experiments suggest that the southern source can be a more powerful modulator of the meridional deep–ocean conveyor that the northern source, which challenges our current vision of the North Atlantic Deep Water as an ultimate driver of deep–ocean circulation. Our experiments show that the southern impact can overpower northern ones. Even in the experiment in which the amplitude of the perturbation in the North Atlantic was as high as -3 psu, and the amplitude in the Southern Ocean was only -1 psu, the deep–water regime was qualitatively the same as in the pure Southern Ocean scenario, with somewhat less deep-ocean warming, yet still global and substantial.

INTRODUCTION

One of the most noticeable attributes of present–day climate is its hemispheric asymmetry, with a warm ocean surface in the northern North Atlantic, a moderately cool northern North Pacific and a much colder Southern Ocean (e.g., *Weyl* [1968]; see also a discussion in *Weaver et al.* [1999]). This prominent hemispheric asymmetry of the thermal state of the ocean surface is determined by three major factors: (i) the freshwater regime of the subtropical Atlantic with evaporated water transferred to the Pacific Ocean over the Panama

Isthmus; (ii) the North Atlantic ocean geometry that facilitates delivery of this warm and saline subtropical water far to the north to form deep convection that drives the global thermohaline circulation (THC), also known as a thermohaline ocean conveyor; and (iii) the Southern Ocean circumcurrent system that causes extreme cooling of the surface water, resulting in the formation of the densest deep–ocean water around Antarctica. Nonlinear interactions of these three factors result in an overturning regime prone to instability and rebounds. The idea of a so–called bi–polar ocean seesaw [*Broecker*, 1998; *Stocker*, 1998; *Broecker*, 2000] (see also a discussion in this volume [*Stocker et al.*, 2001]) serves to explain these rebounds. The bi–polar seesaw is an oscillating meridional overturning regime driven by two deepwater sources – the North Atlantic Deep Water (NADW) in the north, and Antarctic Bottom Water

The Oceans and Rapid Climate Change: Past, Present, and Future
Geophysical Monograph 126

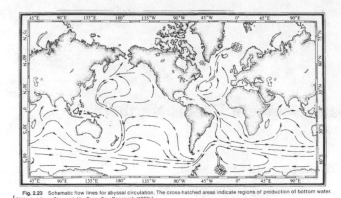

Fig. 2.23 Schematic flow lines for abyssal circulation. The cross-hatched areas indicate regions of production of bottom water.
[Adapted from Stommel, H., *Deep Sea Research* (1958).]

Figure 1. Stommel–Arons scheme of the deep ocean currents.

(AABW) in the south. *Broecker* [2001] argues that the thermohaline circulation flip–flops may be responsible for the 1500 year cycles in ice-rafted debris in the northern North Atlantic, which has been found by *Bond et al.* [1997]. Model results by *Ganopolski and Rahmstorf* [2001] indicate that the climate dynamics in this range of cyclicity is essentially nonlinear, and some caveats are needed in the discussion. However, although it is yet not clear whether the flip-flops might have been somehow linked to major glacial rebounds and whether they follow the same periodicity, as the cycles found by *Bond et al.* [1997], there is a little doubt that the flip-flops can have a substantial impact on the long-term climate dynamics.

The results of numerical experiments presented here are intended not only to better understand the bi–polar seesaw and its role in the ocean climate dynamics on the millennium time–scale, but also to assess the role of the Southern Ocean in driving the THC. This work is an extension of the efforts of *Seidov et al.* [2001b]. The major new element of this cited work and the results described here is the focus on the Southern Ocean response to different meltwater scenarios. Additions to the experiments presented in *Seidov et al.* [2001b] include new meltwater scenarios using present-day sea surface climatology and new scenarios for the post-glacial meltwater episodes.

This paper is structured as follows: The section "Meltwater events and THC" which follows gives an (incomplete) overview of the meltwater episodes that can be used as prototype for numerical simulations targeting the THC changes induced by such episodes, as well as some ideas that inspired our work. This section also contains an overview of numerical modeling efforts by many researchers that provide a starting point for our study. In the section "Scenarios of meltwater events" we describe how the basic knowledge of meltwater episodes in the past are translated into specific boundary conditions employed in the numerical experiments. The section entitled "Numerical experiments" deals

with the specifics of the completed experiments including the numerical model and data utilized, and explains how to interpret the boundary conditions and the model output. In the next section "Present-day experiments", the results of the experiments with perturbations of the present-day surface climatology are discussed and illustrated (including sealevel change caused by thermal expansion). The results in the post-glacial meltwater scenarios are analyzed in the section "LGM experiments". These latter two sections have subsections representing the northern meltwater impacts, and various scenarios of southern impacts. The concluding section, "Discussion and Conclusions", contains our interpretation of the results as a whole, with some caveats added to emphasize the limitations associated with the preliminary character of our conclusions, and which serve to direct future efforts with more complete models.

1. MELTWATER EVENTS AND THC

During the last two decades, many researchers have contributed substantially to understanding the conveyor dynamics on decadal, centennial and millennial time scales (many studies are cited below). However, the role of the deepwater source in the Southern Ocean is still unclear, especially if compared to the far better understood role of the North Atlantic Deep Water formation. The goal of our work is to investigate further this issue of the competitive role of these two deepwater sources. To narrow our goals we focus on the high-latitudinal freshwater impacts as one of the most potent modulators of the THC dynamics.

The concept of meridional deep–ocean circulation controlled by high–latitudinal deepwater sources in two hemispheres is the foundation of the Stommel–Arons theory [*Stommel et al.*, 1958; *Stommel and Arons*, 1960]. Figure 1 reviews their theory emphasizing that the deep ocean equatorward flows are western boundary currents driven by deepwater sources in the high latitudes of the two hemispheres. The present–day deepwater regime largely depends on the strong NADW source, which was not necessarily always as today. High–latitudinal cooling causes convection and isopycnal outcropping in winter seasons. The strength of the deep western boundary currents (and therefore THC as a whole; see Figure 1) decisively depends on the volume of water either descending along these outcropping density surfaces, or mixing in the "convective chimneys", or both. The geologic record indicates that these volumes might have varied widely in the past climates, which in turn might have affected the climate on the centennial to millennial time scale. For example, these variations might have had a substantial impact on the glacial cycles of the late Quaternary (e.g., *Sarnthein et al.* [1995]; *Broecker* [1998]; *Broecker* [2000]).

A new twist of the Stommel–Arons vision of meridional overturning is given by the above–mentioned idea of the bi–polar seesaw [*Broecker*, 1998; *Broecker*, 2000] combined with the data from two ice cores, one in Greenland and one in Antarctica (see also *Blunier et al.* [1998] and *Stocker* [1998]). These two cores suggest [*Blunier et al.*, 1998] that some of the millennial glacial cycles of the Pleistocene in the Northern Hemisphere might have been out of phase with and preceded by those in the Southern Hemisphere.

Broecker et al. [1999] and *Broecker* [2000] argue that the deepwater production in the Southern Ocean has reduced from 15 Sv (1 Sv = 10^6 m^3/s) to 5 Sv over the last 800 years. They also suggest that the Little Ice Age (approximately 1350 – 1850 AD) might have been caused by far stronger deep ocean ventilation in the Southern Ocean. One of the reasons for the speed up of the Southern Ocean ventilation could be an increase in Atlantic Ocean salinity [*Broecker et al.*, 1999], whereas the slowdown could be caused by reduced surface salinity associated with warming and sea ice, icebergs, or ice sheet melting in the Southern Ocean after the Little Ice Age [*Broecker*, 2000]. *Seidov and Maslin* [2001] argue that the "heat piracy" of the North Atlantic might have been replaced by heat piracy of the Southern Hemisphere. Heat piracy means that the cross–equatorial heat transport is non–zero and can cause warming of one hemisphere by cooling another. Changing the sign of the cross–equatorial heat transport causes the hemispheres to trade places. For example, positive northward cross–equatorial oceanic heat transport, characteristic of the present–day Atlantic overturning regime, is thought to be replaced by southward cross–equatorial heat transport, characteristic of meltwater events in the North Atlantic [*Seidov and Maslin*, 2001]. Thus, meltwater events might have caused cooling of the North Atlantic and might have contributed to further cooling of the northern hemisphere because of the conveyor reduction or reversal.

Millennium time–scale climatic variations in recent geological history may also give a clue to possible future changes. The most obvious future example is a "green–house" climate that would be accompanied by major ice melting in either, or in both hemispheres. One of the most important implications of recent modeling and proxy data analyses is that ice melting may cause cooling of the hemisphere where the melting occurs, and warming of the opposite hemisphere (e.g., *Schiller et al.* [1997]; *Broecker* [2000]). Therefore our study targets both past and future THC change caused by meltwater (or equivalent thermal changes that would cause similar de–densification of the sea surface) events in either, or both hemispheres. Moreover, we question whether melting of the ice around Antarctica, or the ice in the Antarctic Circumpolar Current (ACC) is the crucial factor. We also explore sea level change caused by ocean density restructuring in the course of meltwater impacts.

Substantial evidence exists for variations in THC and in freshwater fluxes that can modify the character of deepwater currents. For example, the thermohaline meridional circulation of the present–day type was reduced during glacial periods due to alterations in atmospheric circulation as well as due to input of freshwater from melting icebergs in the North Atlantic (e.g., *Duplessy et al.* [1988, 1991]; *Sarnthein et al.* [1994, 1995]; *Seidov et al.* [1996]), and collapsed during meltwater episodes (e.g., *Manabe and Stouffer* [1988, 1995]; *Maslin et al.* [1995]; *Rahmstorf* [1995a]; *Rosell-Melé et al.* [1997]; *Zahn et al.* [1997]). However, further complicating the problem, *Wang and Mysak* [2000] show in their climate model that the THC could have been stronger prior to massive iceberg melting, especially during major ice sheet build-up (~114 ka BP).

A combination of paleoceanographic proxy data and ocean general circulation models can help to assess the circulation effects and origins of the quasi–periodic ice rafting pulses called Heinrich events (e.g., *Heinrich* [1988]; *Bond et al.* [1992]; *Andrews* [1998]; *Bradley* [1999]). Heinrich events occur every 7 to 13 k.y. and have duration of between 100 and 500 yr [*Dowdeswell et al.*, 1995a] and are thought to be an essential element of millennial-scale climate variability (see a review in this volume [*Maslin et al.*, 2001]). The ice–rafted debris found in deep–sea sediment during the Heinrich events may have originated from either the Laurentide or the European ice sheets (e.g., *Grousset et al.* [1993]; *Robinson et al.* [1995]; *Gwiazda et al.* [1996]; *Rasmussen et al.* [1997]; see Figure 2). At present it is debated whether the Heinrich events are caused by internal ice–sheet dynamics (e.g., *MacAyeal* [1992]), external climate changes [*Broecker*, 1994a; *Hulbe*, 1997], or ice sheet – THC interactions [*Wang and Mysak*, 2001]. The above-mentioned flip-flops of THC in the Atlantic Ocean might have been an important element in climatic changes caused by these meltwater events.

Substantial concern has been recently expressed about the stability of the West Antarctic Ice Sheet (WAIS) (see review ref. *Oppenheimer* [1998]). Uncertainties about Antarctic ice sheet mass balance and its contribution to global sea level rise is a major issue of debate [*Vaughan et al.*, 1999] even without a large–scale collapse of the WAIS. Hence, the potential for changes in freshwater fluxes or salinity variations to influence the Southern Ocean is also clearly evident. Moreover, *Schmittner and Stocker* [1999] give another reason for dilution of the sea surface during global warming, invoked by an increased equator–to–pole freshwater transport in a warmer atmosphere. The increased poleward moisture flux results in increased precipitation in the high latitudes, which, in concert with cryosphere melting due to a

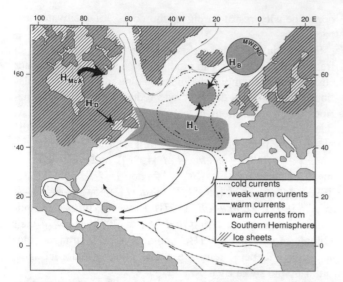

Figure 2. Reconstruction of the last glacial maximum circulation based on paleoceanographic proxy data (from *Seidov and Maslin*, 1999). Two possible source regions of icebergs, Laurentide ice sheet (H$_L$) and the Barents shelf (HB), are indicated; (MWENS = meltwater event in Nordic Seas).

warmer atmosphere, could enhance the meltwater impact on the ocean circulation.

Although sea surface salinity controls both NADW and AABW, the source and character of the meltwater in the Southern Hemisphere is noticeably different from the Northern Hemisphere. The meltwater episodes of the Pleistocene are usually associated with meltdown of iceberg flotillas that surged from the major glacial ice sheets or shelves (Laurentide ice sheet, or Barents ice shelf; see, e.g., *Sarnthein et al.* [1994]; *Dowdeswell et al.* [1995a]; *Maslin et al.* [1995]) in deglaciation cycles. As many believe, the Nordic Seas were at least seasonally ice–free even during the LGM [*Sarnthein et al.*, 1995; *Seidov and Maslin*, 1996]. In any case, cryosphere resources in the north are smaller than those surrounding Antarctica (e.g., *Cronin* [1999]). Although much less is known about meltwater events in the south, the shear volume of sea ice suggests that the Southern Hemisphere has a far greater potential.

Early studies [*Toggweiler and Samuels*, 1980] and more recent modeling results from *Goosse and Fichefet* [1999] confirm the importance of brine rejection during the formation of the sea ice, or equivalently, of increasing sea surface salinity caused by trapping the freshwater part of any freezing water volume in the sea ice. Moreover, AABW is not the only water mass that may be affected by diluting the surface water. Sea surface salinity controls both AABW and Antarctic Intermediate Water [*England*, 1992; *Stocker et al.*, 1992], which makes the southern impact via surface water freezing/melting cycles even more plausible. Brine rejection

alone could have a substantial impact over the global thermohaline circulation, if it decreases due to reduced freezing around Antarctica caused by global warming or some other factors.

The key is that the northern North Atlantic and Nordic Seas as well as the Southern Ocean are marginally stable, and therefore are prone to destabilizing effects of even moderate de–densification of the sea surface, either due to its warming, or dilution. However, the brine rejection, ice melting, and water freezing all happen without a sea surface temperature rise because of the latent heat of melting, i.e. surface water temperature stays at the freezing/melting point as the phase change continues. (Only the near–surface temperature of water in direct contact with ice will not rise; the water below the ice can be warmed by warmer and saltier underlying water).

New research [*Seidov et al.*, 2001] substantially extends a preliminary study [*Seidov and Maslin*, 2001] by adding new northern and southern meltwater simulations. *Seidov and Maslin* [2001], following a large number of authors (e.g., *Gordon* [1986]; *Manabe and Stouffer* [1988]; *Broecker and Denton* [1989]; *Broecker* [1991]; *Gordon et al.* [1992]; *Maier-Reimer et al.* [1993]; *Broecker* [1994b]; *Weaver and Hughes* [1994]; *Manabe and Stouffer* [1995]; *Rahmstorf* [1995a]; *Sarnthein et al.* [1995]; *Schmitz* [1995]; *Weaver* [1995]; *Seidov and Haupt* [1997, 1999a]), believed that the ocean conveyor is most affected by the NADW production. *Seidov and Maslin* [2001], following *Broecker* [1998, 2000], debate the role of the southern source, but it was not directly modeled. *Seidov et al.* [2001] simulated both the southern and northern meltwater events and indicate that the southern source can be far more powerful in controlling the ocean thermohaline circulation than was previously thought. This new vision is in agreement with the arguments of [*Broecker*, 2000]. The work by *Seidov et al.* [2001] is also in line with model study of *Goosse and Fichefet* [1999], although substantial additional simulations and new views on the implications for sea level change are added. These recent works challenge the current paradigm of the global thermohaline conveyor being controlled largely by NADW formation (e.g., *Gordon et al.* [1992]).

Seidov et al. [2001] emphasize the role of the Southern Ocean and favor the southern source as having a more powerful influence in future changes if the global warming continues. This is in line with a growing modeling effort designed to investigate the climatic role of the Southern Hemisphere [*Goodman*, 1998; *Stoessel et al.*, 1998; *Hirschi et al.*, 1999; *Scott et al.*, 1999; *Wang et al.*, 1999a, 1999b; *Kamenkovich and Goodman*, 2001]. These studies describe the potential importance of feedbacks between northern and southern sources of deepwater. They suggest that freshwater forcing in the Southern Hemisphere may influence the

NADW formation and examine the importance of brine re-lease rates on AABW formation.

To conclude the overview of THC response to meltwater events, a sketch in Figure 3 illustrates the idea of the bi–polar seesaw (see also *Broecker* [2000]) in the light of recent numerical experiments by *Seidov et al.* [2001] and *Seidov and Maslin* [2001]. The upper panel (Figure 3a) represents the present–day, or interglacial mode of the conveyor, whereas two other panels show the northern meltwater mode with AABW dominance (Figure 3b) and the warm–house mode of the conveyor with complete NADW dominance (Figure 3c) in the Atlantic Ocean. The cross–equatorial heat transport follows the conveyor rebounds. The role of the cross–equatorial heat transport seems to be a very powerful climate feedback element and throughout the text we revisit its possible role in climate change caused by meltwater events. However, although our experiments give some clue to how the flip-flops of the conveyor and corresponding flip-flops of the cross-equatorial heat transport can affect the climate, a coupled cryosphere-ocean-atmosphere model is the only alternative that can fully address the climate system dynamics. Therefore, we consider the results and conclu-sions outlined in the next sections as a preliminary estimate of a possible behavior of THC if southern meltwater events occur, and of the differences between northern and southern meltwater events based on the same model.

2. SCENARIOS OF MELTWATER EVENTS

The numerical experiments discussed in the next sections describe idealized meltwater impacts that model either southern, northern, or combined southern and northern meltwater events. Scenarios of changes of the sea surface climatology are simple and straightforward. There are two large groups of sensitivity experiments that simulate de–densification of sea surface in the high–latitudes: (1) scenar-ios of possible future meltwater events that might happen due to major climate change (e.g., due to ongoing global warming, or yet unknown large–scale climate fluctuations on century time scales), and (2) idealized scenarios of melt-water events derived from the glacial cycles of the Pleisto-cene, with the LGM as a model for a glacial state. Two con-trol runs in each group have been completed. Two major present–day sub–groups and one LGM sub–group are shown in Figure 4 (see detailed description of the experiments be-low and in Tables 1 and 2).

In the scenarios of possible future change of the THC, the upper layer temperature and salinity are restored to the pre-sent–day annual mean sea surface temperature (SST) and sea surface salinity (SSS) [*Levitus and Boyer*, 1994; *Levitus et al.*, 1994]. The *Hellerman and Rosenstein* [1983] annual mean wind stress is used in present–day simulations.

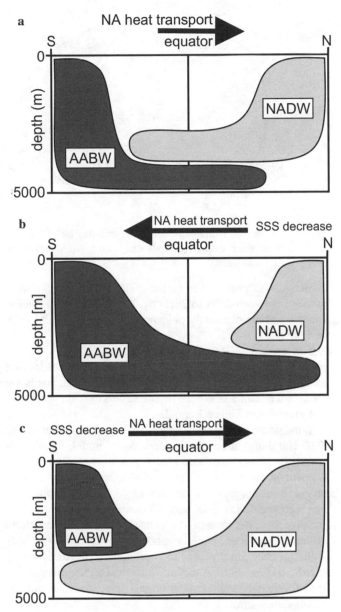

Figure 3. Schemes of water mass layering and overturning struc-ture: (a) present–day; (b) present–day northern meltwater event; (c) present–day southern meltwater event. Direction of cross–equatorial oceanic heat transport is shown by arrows above each scheme.

In idealized scenarios of impacts of the meltwater events of the Pleistocene, basically falling in to either Heinrich Events, or Dansgaard–Oeschger Events categories (see above), the upper layer temperature was restored to recon-structed annual mean sea surface temperature (SST) during the Last Glacial Maximum (LGM). The LGM SST are modified *CLIMAP* [1981] SST, which are systematically reduced in the tropics to reflect new results indicating a

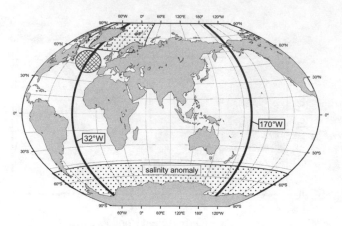

Figure 4. Idealized meltwater events. Present–day and LGM NA and SO meltwater events (dotted) and LGM NNA meltwater event (double–hatched).

cooler glacial tropical sea surface. The CLIMAP SST were reduced between 20°S to 20°N by 4°C and this perturbation was zeroed exponentially poleward between 20°S and 60°S and 20°N and 60°N.

The annual LGM see surface salinity (SSS) was constructed as described in *Seidov et al.* [1996] and *Seidov and Haupt* [1997]. Surface salinity in meltwater scenarios is the present-day and LGM SSS modified in key areas as shown in Figure 4 and Tables 1 and 2.

Wind stress fields were extracted from the output of the T42 Hamburg atmospheric circulation model, which was forced by present–day and glacial sea–surface climatologies [*Lorenz et al.*, 1996]. In the Southern Ocean, a low–salinity signal is superimposed on the LGM SSS, retaining unchanged the LGM SST and wind stress, in the very same manner as in the present–day runs (see above). In the North Atlantic, however, experiments that model a low–salinity signal in the central part of the northern North Atlantic were carried out in addition to those identical to the present–day runs (double–hatched area in Figure 4). As the runs with the low–salinity impact confined in the Nordic Seas simulate Dansgaard–Oeschger Events (e.g., *Oeschger* [1984],

Dowdeswell et al. [1995b]; see also discussion in *Seidov and Maslin* [1999]), the scenarios with the southward shifted low–salinity signal simulate impacts of the Heinrich Events. Heinrich events are thought to be caused by melting of iceberg armadas from the Laurentide Ice Sheet [*Ruddiman and McIntyre*, 1981] (and possibly from the Barents Shelf, e.g., *Sarnthein et al.* [1995]). The essential difference from the Dansgaard–Oeschger events is that meltwater in Heinrich events caps the southward–shifted convection and therefore are associated with complete shut–off or even reversal of the Atlantic thermohaline conveyor.

The rebound of ocean conveyor depends crucially on the SSS anomalies. The perturbations we apply in our idealized scenarios do not exceed those found in paleoreconstructions (e.g., *Duplessy et al.* [1991]; *Sarnthein et al.* [1995]). However, there may be significant uncertainties in the SSS reconstructions. To test the effect of these uncertainties on the global circulation, *Seidov and Haupt* [1999a] have performed a number of numerical experiments in which the SSS in the meltwater pools was altered by as much as 1 psu. In none of these runs did the SSS modification prevent the capping of convection and the depression of the conveyor (see also *Seidov and Maslin* [1999]). Thus, despite the uncertainties, the SSS reconstructions by *Duplessy et al.* [1991] and *Sarnthein et al.* [1994] give a very solid foundation for numerical simulations based on these data [*Seidov et al.*, 1996]. The thermohaline circulation collapse due to freshwater impact in *Seidov et al.* [1996] compares well with the paleoreconstructions in *Sarnthein et al.* [1995]. It has been confirmed that the results of the regional NA model by *Seidov et al.* [1996] are still valid in global circulation experiments [*Seidov and Haupt*, 1997, 1999b; *Seidov and Maslin*, 1999]. Moreover, it has been shown in these studies that the meltwater signal in the Nordic Seas during the MWE is so strong that the circulation response is a robust feature in all numerical experiments. These results from previous realistic and idealized simulations, as well as the implications based on proxy analyses, provide more confidence in the results of the idealized study presented here.

Table 1. Amplitudes of sea surface salinity anomalies (in psu).

Exp.	NA	SO	WED	ANT	ACC	
CRPDC (#1)	–	–	–	–	–	CRPDC – Control Case (annual mean present–day sea surface climatology); NA – North Atlantic; SO – Southern Ocean; WED – the Weddell Sea; ANT– Antarctica coastline; ACC– ACC bound signal.
#2	-2.0	–	–	–	–	Salinity anomalies are added to the present–day annual mean sea surface salinity in the bands between 60°N and 80°N in NA, and/or 50°S and the coast of the Antarctica (SO).
#3	–	-1.0	–	–	–	The modified salinity was merged using a cosine filter to the unchanged field within two
#4	-3.0	-1.0	–	–	–	latitudinal grid points (8°). In the SO the anomalies are circumglobal. The WED low
#5	–	–	-3.0	–	–	salinity is confined to the Weddell Sea only. The ANT signal is in the band of 4° thickness
#6	–	–	–	-1.0	–	around Antarctica; ACC is the signal in the band between around 50°S, approximate
#7	–	–	–	–	-1.0	position of the ACC axes.

Table 2. Meridional overturning in the Atlantic Ocean (north of 30°S) in Sv (1 Sv = 10^6 m^3/s).

Exp.	NADW production	Convection depth in NA (km)	NADW outflow at 30°S	AABW inflow at 30°S
#1 (CRPDC)	16	3–4	10	6
#2 (NA-2 psu)	10	2	4	4
#3 (SO-1 psu)	25	bottom (> 4 km)	20	0
#4 (NA-3 +SO-1)	20	bottom (> 4 km)	14	4
#5 (WED-3 psu)	15	bottom (> 4 km)	10	4
#6 (ANT-1 psu)	20	bottom (> 4 km)	12	4
#7 (ACC-1 psu)	20	bottom (> 4 km)	12	4

3. NUMERICAL EXPERIMENTS

All experiments are completed using the GFDL MOM version 2, a well documented and extensively utilized ocean circulation model [*Pacanowski*, 1996]. A newer version 3 is now the current version. However, to our knowledge, no changes that are critical to a scenario-type coarse resolution modeling have been reported so far. A rather coarse resolution of 6°x4° with 12 levels is employed, as is appropriate for a pilot comparison between several low–salinity regimes in which the focal point is the large–scale thermohaline circulation. This resolution is comparable with other coarse resolution studies [*Toggweiler et al.*, 1989; *Weaver et al.*, 1994; *Manabe and Stouffer*, 1995; *Rahmstorf*, 1995b; *Seidov and Haupt*, 1999a; *Seidov and Maslin*, 1999] addressing similar problems, and has proven to be sufficient for studying the response of ocean meridional overturning to freshwater signals. For example, *Seidov and Haupt* [1999a] demonstrate that water transports, convection depths, and inter–basin water exchanges are reasonably well–simulated in a study with a similar spatial resolution using similar boundary conditions.

The world ocean is bounded by Antarctica in the south and 80°N in the north. Barents shelf area is included, and the eastern boundary in the North Atlantic Ocean sector is at 40°E. The Mediterranean Sea and the Arctic region are excluded.

In our experiments we use isopycnal mixing [*Gent and McWilliams*, 1990] as it is implemented in *MOM-2* [1996]. Inclusion of isopycnal mixing is crucial because, as it has been shown (see discussion in *McWilliams* [1998]), ocean circulation models without isopycnal mixing suffer from exaggerated convection, especially in the Southern Ocean region. *Hirst et al.* [2000] indicate that using of Gent-McWilliams mixing scheme substantially improved performance of a coupled ocean atmosphere model as well.

A great number of idealized numerical experiments have been completed in our study of northern and southern melt-water events. There is no way to arrange them all in one observable table. However, they can be classified into two major groups, with 5 subsets in the first group comprising the present–day runs, and 3 subgroups in the second group, comprising the LGM runs.

First subsets in each group consists of just one experiment, that is, the control run: The control run in the present–day climate (PDC) scenario (CRPDC), and the control run in the LGM scenario (CRLGM). Other runs in any group represent different meltwater scenarios. In the experiments, either the present–day or LGM sea–surface climatology are used for the surface boundary conditions (see above). Every run is compared to the corresponding control runs within its main groups, i.e., either to CRPDC, or CRLGM. All runs are 2000 model years long, with 5–fold acceleration in the deep layers (in MOM, this means that the deep ocean is effectively run for 10000 years). A complete steady state is reached in all numerical experiments (with global temperature and salinity trends in the end of experiments smaller than 10^{-4} °C and 10^{-5} psu per century respectively).

Within the PDC main group of 7 experiments, there are following subsets (Tables 1 and 2): (1) Three experiments (Exp. 2-4) that target the northern versus southern meltwater impacts; this is the core of our study; (2) two experiments (Exp. 5 and 6) that quantify the difference between local sources in the Southern Ocean around Antarctica (for example, the role of the Weddell Sea, Ross Sea and the Antarctica continental shelf; and (3) one experiment (Exp. 7) exploring whether melting of the ice around Antarctica, or the ice in the Antarctic Circumpolar Current (ACC) is the crucial factor. Exp. 1 is the control run, CRPDC (see above), with unchanged present-day climatology.

In the LGM group there are three major sub–groups: (1) LGM runs with southern and northern meltwater impacts that have the same character as in the subgroups 1 of the present–day runs, and (2) experiments with the northern impact shifted southward, as in *Seidov and Maslin* [1999, 2001].

All steady states of THC in sensitivity runs with perturbed surface salinity are compared to the steady states in the control runs. Solving this system simulates evolution of the ocean from an initial state to a steady state of the ocean currents and thermohaline structure that are fully adjusted to the sea–surface boundary conditions.

In the cold climates of the Pleistocene low salinity signals in the high latitudes are mainly due to melting of sea ice or icebergs, with poleward water vapor transport from the tropics and THC feedbacks being important but as a secondary factor (e.g., *Bradley* [1999]. High salinity signals are due to brine rejection process that accompanies seawater freezing (e.g., *Gill* [1982]). In warm ice–free climates, poleward water vapor transport or river run–off can be the only cause of a low salinity signal, whereas increased evaporation (unlikely in high latitudes) might be a cause of an increased surface salinity elsewhere. However, in our approach, temperature changes could be modeled by changing salinity, as our concern is sea surface density rather than temperature or salinity specifically. A useful rule–of–thumb is that the same density increases can be achieved by either increase of salinity by approximately 1 psu, or by decrease of temperature by about -5°C [*Pond and Pickard*, 1986]. Hence, meltwater is a convenient means by which to control density changes that actually drive the convection.

Regarding the strength of the low–salinity signal, even 1 psu is a rather moderate estimate of the possible dilution of sea surface water during a southern post-LGM meltwater event. For instance, *Goosse and Fichefet* [1999] argue that even the reduction of brine rejection alone can cause a 1 psu decrease in sea surface salinity. *Duplessy et al.* [1996] show that the low salinity anomalies in the Southern Ocean could be up to -1.8 psu during the LGM. Thus, much stronger anomalies might be expected in a southern meltwater episode. *Labeyrie et al.* [1986] argue that the periphery of the Antarctic ice sheet was eroded during some of the glacial cycles of Pleistocene, and that only 10,000 km^3 of meltwater could have reduced the sea surface δ^{18}O by 1 ‰ (which translates to approximately 2 psu in sea surface salinity [*Duplessy et al.*, 1996]). *Anderson and Andrews* [1999] revisited the problem of the late Quaternary Antarctic meltdown, and argue that significant deglaciation of the Weddell Sea continental shelf could have taken place prior to the last glaciation. *Birchfield and Broecker* [1990] point out that a relatively small freshwater flux converted to a low–salinity signal will hamper the conveyor operation. For instance, they show that a freshwater flux of 0.1 to 0.3 Sv in the North Atlantic can cause 0.3 to almost 1 psu reduction of salinity in 1000 years. The 0.3 Sv flux during a thousand years would convert to 10,000 km^3 a year, a value that is only about 4 times greater than the present–day annual meltwater

production in Antarctica of about 2,500 km^3/year (e.g., *Vaughan et al.* [1999]).

In contrast to many studies aimed at sensitivity of the circulation to freshwater fluxes (e.g., *Maier-Reimer et al.* [1993; *Weaver and Hughes* [1994]; *Rahmstorf* [1995a]; *Manabe and Stouffer* [1997]) we use a restoring boundary condition on salinity (e.g., *Bryan* [1987]). Normally, a restoring boundary condition on salinity is considered as inferior to a flux formulation. A compromise could be the use of implied freshwater flax (e.g., see a discussion in *Weaver* [1999] and *Ezer* [2001] in this volume). However, in a numerical model, it is difficult to control a salinity change to match observations using a flux formulation. In order to assess compatibility issue, we have calculated apparent freshwater volumes that would be needed to dilute the surface layer to achieve a respective salinity change in the imposition domain. These volumes can be thought of as virtual freshwater volumes that would have to be added to a thin surface layer, where the SSS is specified, to dilute the water in this layer to a prescribed SSS reduction. (In the runs with increased SSS, this would be the freshwater to be removed to achieve the respective increase in SSS caused by brine rejection).

Following *Manabe and Stouffer* [1995], we estimate the rates by which these virtual freshwater fluxes would have to be added within 10 years. The total amount of freshwater added at those rates is fairly realistic. For instance, to dilute a 10 m layer of water with salinity of 35 psu by 1 psu in 10 years in one of the Southern Ocean experiments (Exp. 8 in Table 1), a freshwater (or equivalent sea ice) layer of about only 0.3 m thickness would be needed. The freshwater fluxes to maintain the low salinity signals employed in this study are comparable to those in *Manabe and Stouffer* [1995]. For example, to maintain – 1 psu anomaly in the SO experiment (Table 1), a freshwater flux of 0.06 Sv is needed.

3.1 Present–day Experiments

Table 1 includes the control run and examples from the first of the two major groups (see also Figure 4): Exp.1 is the CRPDC. Exp. 2 (NA) has a low–salinity perturbation in the high–latitudinal North Atlantic; Exp. 3 (SO) has a low–salinity signal in the Southern Ocean, Exp. 4 (NA and SO combined) has low–salinity signals in both these two regions, Exp. 5 (WED) has a low–salinity signal confined to the Weddell Sea only (we have also run an experiment with low–salinity signals in the Ross Sea; as the results do not substantially differ from those in WED, these runs are not present here), Exp. 6 (ANT) has only the area within one grid step thickness around Antarctica (repeating the shoreline) affected by meltwater, and Exp.7 (ACC) has a freshwater band between 40°S and 50°S, i.e. in a water band approximating the ACC. The experiments shown in Table 1 are the runs that have been done with varying the amplitude of the salinity perturbations.

The low–salinity signal is applied as a negative salinity anomaly with different amplitudes, from 0.5 to 3 psu in the North Atlantic, from 0.2 to 1 psu in the Southern Ocean, and from 1 to 3 in the combined cases. The low salinity in the Weddell Sea is lower than in CRPDC, with maximum differences from 0.5 to 3 psu. In the experiments exploring the relative importance of Antarctic coastal meltwater impacts versus ACC impacts the low–salinity signals are 1 psu.

We discuss the results of experiments in the same sequence as they are shown in Table 1. Six major sub–groups of events are selected, as listed above, in Exp. 2 through 7. However, only three major cases of NA, SO and the combined NA+SO cases (see notations in Table 1) are illustrated, with others only briefly discussed in the text.

North Atlantic events. An important characteristic of the thermohaline circulation is the meridional overturning streamfunction. Notably, the most important is the overturning in the Atlantic Ocean because this ocean is the most active element of the THC. Moreover, the largest component of meridional heat transport is determined by the meridional overturning.

The results of the experiments simulating low–salinity impact in the North Atlantic conform to what is already well known from previous work (see "Meltwater events and THC" section above). The conveyor is weaker and shallower and convection is shifted southward. For comparison with all sensitivity tests, the overturning in the CRPDC is shown in Figure 5a and Figure 6a. Figure 5a depicts overturning in the Atlantic Ocean, whereas Figure 6a shows the World Ocean. Figure 7a illustrates present–day model temperature sections in the Atlantic and Figure 8a illustrates the section in the Pacific Oceans in CRPDC, along meridians 32°E and 170°W respectively. Figure 9 depicts northward oceanic heat transport in the Atlantic Ocean, whereas Figure 10 shows global northward oceanic heat transport. Similarly, each of the above fields is given in Figures 5–10, part b, for a low–salinity signal in the North Atlantic. Figures 5–10, part c, showing the results with a low salinity signal in the Southern Ocean, represents southern impacts.

Northward cross–equatorial heat transport in the North Atlantic, which is a characteristic of the present–day climate, is dramatically reduced in the scenario with a strong northern freshwater impact (Figures 9 and 10). This result conforms to *Manabe and Stouffer* [1995, 1997] results showing the possibility of a northern cold episode following a northern meltwater event. However, the impact of freshwater on the conveyor depends crucially on the location of the freshwater lid. In our simulations even an excessive northern low–salinity signal of -2 psu did not cause a complete termination

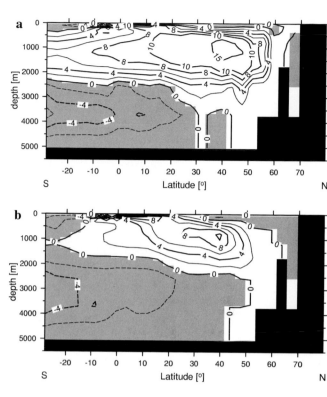

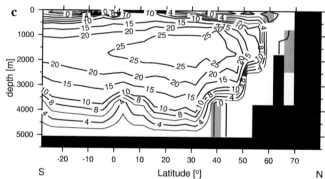

Figure 5. Meridional overturning in the Atlantic Ocean; (a) CRPDC; (b) NA event; (c) SO event (see Table 1). Streamfunction is shown in Sv (1 Sv = 10^6 m³/s).

of the conveyor (in contrast to the runs in *Seidov and Haupt* [1997] and *Seidov and Maslin* [1999], where more southward spread of meltwater caused complete cessation or even reversal of the conveyor). Importantly, during meltwater events in the North Atlantic, only the North Atlantic is affected by southward cross-equatorial heat transport, whereas during meltwater events in the Southern Ocean the whole Northern Hemisphere gains heat at expense of heat loss in the Southern Hemisphere (Figure 9).

Temperature differences between this low–salinity scenario (Exp. 2) and the control case CRPDC (Exp. 1) at 3000

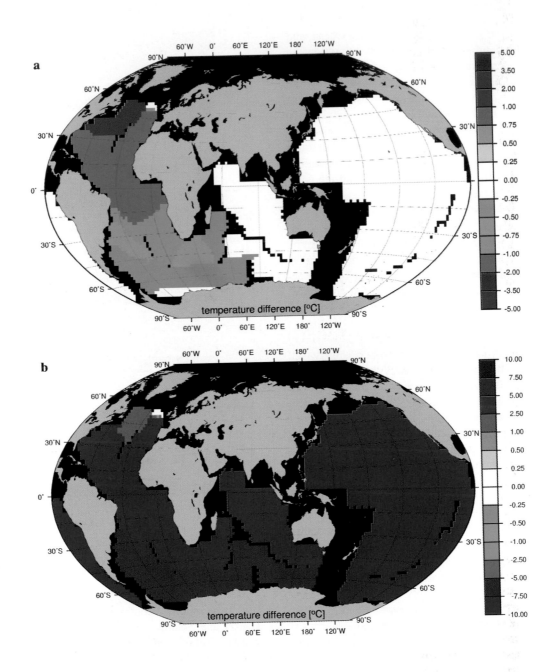

Plate 1. Temperature differences at 3000 m depth between (a) NA and CRPDC experiments and (b) SO and CRPDC experiments (see Table 1).

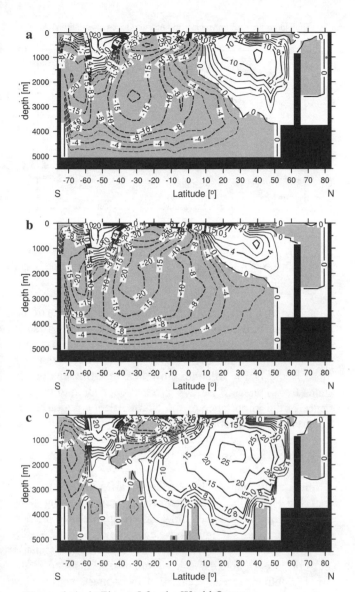

Figure 6. As in Figure 5 for the World Ocean.

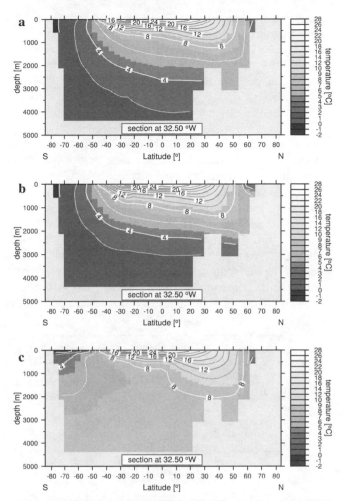

Figure 7. Temperature sections in the Atlantic Ocean at 32°W: (a) CRPDC; (b) NA event; (c) SO event.

m depth are shown in the color plate in Plate 1a. Northern events indicate cooling in high latitudes of the Atlantic Ocean. This occurs because the reduced NADW production led to a shallow conveyor, with deep water characterized by cooler and fresher water than today in these latitudes. The reduced NADW outflow, coincident with reduced replacement water crossing the equator in the North Atlantic, has an evident imprint on the oceanic heat transport.

Southern Ocean events. In contrast to the predictable and understandable results of the northern low–salinity impact, the results of the Southern Ocean surface freshening are less intuitive. Two aspects are particularly noteworthy. First, the circulation changes driven by the low–salinity signal were much stronger, and second, they led to a very strong warm-

ing of the deep ocean. Figures 5–10, part c, shows results of sea surface de–densification in the Southern Ocean in Exp. 3. Increased overturning and northward heat transport are the signatures of the southern low–salinity impacts. Temperature sections in the Atlantic Ocean (Figure 7c) and in the western Pacific Ocean (Figure 8c) and temperature differences between the SO and CRPDC cases at 3000 m depth (Plate 1b) show the dramatic worldwide warming caused by the retreat of AABW and increase of NADW.

It has long been recognized that the increase of NADW production can cause cooling of the upper waters of the Southern Atlantic as the poleward heat flux increases (e.g., *Manabe and Stouffer* [1988]; *Crowley* [1992]). However, we emphasize that the deep ocean thermal trend in the southern meltwater impact scenario can be of opposite sign to those in the upper layers. In contrast to the North Atlantic meltwater scenario, there is a substantial warming of the deep ocean everywhere. The warming takes place over the entire deep ocean and its maximum shifts to the southern edges. This deep–sea warming is caused not only by a substantially increased (by 40 to 60 %) NADW production, but also largely because the meridional overturning takes over the entire deep ocean, pushing away the lessened AABW. In the North Atlantic scenario, the meltwater impact on the conveyor caused thermal effects only in the deep Atlantic Ocean, whereas in the Southern Ocean, the meltwater scenario impact is global. The increased NADW outflow in the deep layers leads to increased, compensating northward surface water flow, which might further increase NADW production until the atmosphere warms up to reduce the cooling of the sea surface and subsequently reduce deep convection.

North Atlantic versus Southern Ocean. Although in Exp. 4 (Table 1; not shown in figures) the amplitude in the North Atlantic (-3 psu) perturbation was three times the Southern Ocean (-1 psu) event, the deep–water regime is qualitatively similar to the experiment with a Southern Ocean–only perturbation. There is somewhat less deep–ocean warming, but

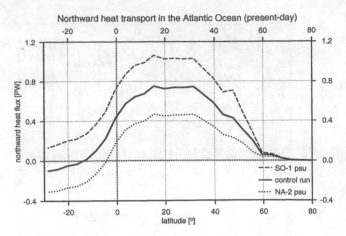

Figure 9. Northward heat transport (in PW; 1 PW = 10^{15} W) in the Atlantic Ocean in CRPDC (solid line), NA (dotted line) and SO (dash line) runs.

it remains global and substantial. Basically, the results of the runs with perturbations to two sources, in the North Atlantic and Southern Ocean, demonstrates a more powerful response to a meltwater event in the Southern Ocean than for those in the North Atlantic. However, much of this power stems from increased NADW production, adding to the evidence of the importance of the North Atlantic region. Most importantly, the problem of deep–ocean teleconnections is now seen from a very different angle.

Weddell Sea, Antarctica–bound, and ACC–bound scenarios.

Importantly, neither the Weddell Sea ice melting (Exp. 5 in Table 1), nor the Antarctica near–shore ice melting alone (Exp. 6), or the ACC–bound alone (Exp. 7) have impacts that are nearly as strong as those caused by the whole Southern Ocean events. Surprisingly, the Weddell Sea scenario did not give as noticeable a warming as was found in the whole Southern Ocean scenario. This contrasts with the belief that the Weddell Sea is the key point for the THC. Even with the amplitude of the freshwater signal in the Weddell Sea of -3 psu, the impact was far less than in the Southern Ocean scenario with only -0.2 psu. This model result implies that AABW formed around Antarctica may be more important for the conveyor dynamics than the major portion originating in the Weddell Sea. It is not clear, however, whether this would be the case in a coupled ocean–atmosphere model, with the low–salinity signal spreading from the Weddell Sea circumpolarly. In this case, the results would conform more to the SO cases, rather than to experiments with the Weddell Sea as the center of action. Our grid resolution is not sufficient to resolve the Weddell Sea in detail; therefore, we can indicate only that a local impact tied to this area would be far less powerful than a distributed impact of the whole Southern Ocean, or a large part of it.

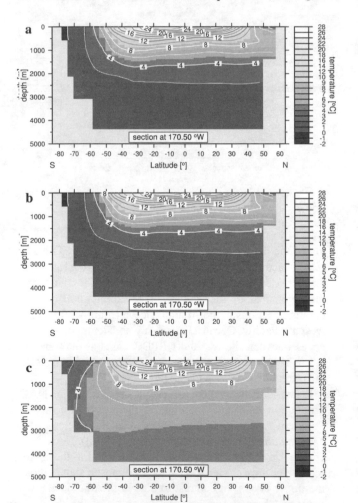

Figure 8. As in Figure 7 for the Pacific Ocean at 170°W.

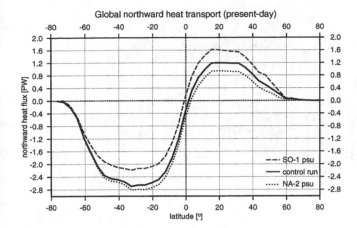

Global northward heat transport (present-day)

Figure 10. As in Figure 9 for the World Ocean.

In the case of a meltwater confined to either the Antarctica coastline, or the ACC bound scenario, far weaker impact is recorded in the numerical runs (Exp. 6 and 7 in Table 1).

Table 2 shows NADW production rates, the depths to which the convection reaches at the NADW convection sites, the meridional overturning at the critical latitude of 30°S, and the outflow of NADW and inflow of AABW. The balance of these two flows determines the state and intensity of the conveyor.

Implications for changes in sea level. Sea level rise caused by melting of major ice sheets is a central issue of global warming forecasts (see references in *Karl* [1993]; *Warrick et al.* [1993]; *Houghton* [1997]). However, there is also an indirect sea level effect of meltwater events caused by thermal restructuring of the world ocean. Therefore, the impact of de–densification of the sea surface is not limited to changes in oceanic circulation. As the deep ocean warms up, the sea elevation will change because of the thermal expansion of sea water. Historic hydrographic data suggest that thermal expansion of the ocean can contribute tens of centimeters to the observed sea level rise over the last century [*Godfrey and Love*, 1992]. Ocean circulation models predict ocean level rise caused by thermal expansion due to THC changes (e.g., *Church et al.* [1991]; *Weaver and Wiebe* [1999]; *Jackett et al.* [2000]; *Knutti and Stocker* [2000]). Some simulations (e.g., *Church et al.* [1991]) indicate that the thermal expansion of the ocean associated with a global warming of 3°C temperature rise by the year 2050 will result in up to a 30 cm sea level rise. On the other hand, cooling of large segments of the world ocean would compensate for the land ice sheet melting and reduce the sea level rise caused by such melting.

The difference of the sea elevation relative to the sea floor was calculated for each of the sensitivity experiments. Sea level change in Exp. 3 relative to CRPDC is shown in Figure

11. These differences in sea surface height are due to the differences in the 3–D density field caused by different T and S distribution in the world ocean. Significant sea level rise (up to 2–3 m in the SO case, Figure 11b) is evident. At the same time, the only region where sea level rise was significant in the NA case (Exp. 2; Figure 11a) is the Nordic Seas. Lowering of sea level was found in NA experiments (Figure 11a); interestingly, this lowering is far away from the northern source. Strong uprise of sea level in the SO or combined cases are largely due to deep–ocean warming. As the deep ocean is warmed up in response to the southern meltwater scenario, the water column expands and sea level rises. Notably, in many sensitive coastal areas the sea level rise can be over 1 m. Hence, it is possible that a meltwater episode, especially in the Southern Ocean (with a substantial global deep ocean warming and salinity redistribution), could strongly impact island nations and coastal regions through a noticeable sea–level change. Importantly, this sea–level rise could occur without significant melting of the ice sheets, including WAIS, which is considered the most vulnerable to climate change. If some melting of WAIS, or any other ice sheet were to happen, the effect would lead to even more dramatic changes than those shown in Figure 11.

Evolution of sea level in Exp. 2 and Exp. 3 relative to sea level in CRPDC is shown in Figure 12. The thermal response of the deep ocean and related sea level change was remarkably fast. In an integration of the model for 1000 years (tracer time without deep-ocean acceleration) with the low–salinity signal superimposed on the steady state of the control run, the first 50 % of total sea level change and warming of deep ocean was reached within first 150 years of the 2000 years of the spin up, with a much slower increase followed. Note, however, that sea level rise in Exp. 2 is almost an order of magnitude smaller than in Exp. 3. In fact, a negative global sea level change could have been expected in Exp. 2, with the deep ocean cooling. However, the deep ocean cooling in Exp. 2 is counterbalanced by substantially more warming of intermediate depths in the northern meltwater scenario (caused by shoaled NADW outflow). Therefore, there is a net rise, rather than lowering of the global sea level. Figure 12 implies that a southern meltwater episode can be a more dangerous environmental challenge than a northern meltwater episode, at least within the range considered for de–densification of the present–day sea surface.

3.2 LGM Experiments

The second large group of experiments are the runs with low salinity perturbations applied to the LGM, rather than to the present–day sea surface conditions. Experiments from this group are shown in Table 3, with all notations, except for NNA (northern North Atlantic) as in Table 1. LGM scenarios that we discuss in this paper are reduced to only four

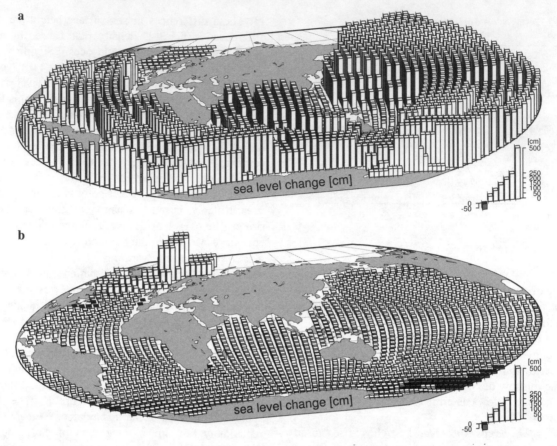

Figure 11. Sea level change in present–day NA and SO meltwater scenario (Exp. 2 and 3 in Table 1). Heights of the bars show the level change relative to the sea level in CRPDC run (Exp. 1 in Table 1).

runs (Table 3). These runs are: Exp. 1L (NA), which is the control run for LGM (CRLGM) with undisturbed LGM surface climatology as was described in the scenario section (LGM surface climatology was corrected to incorporate colder tropics; see above); setups of Exp. 2L and 4L are similar to analogues NO and SO present–day runs (Exp. 2 and 3 in Table 1). Exp. 3L is also similar to NA present–day run, except for the low–salinity signal is in the central part of the northern North Atlantic, south of Iceland, rather than in the Nordic Seas (see Figure 4). Exp. 5L combines the signals in Exp. 3L and Exp. 4L to produce an analogue to the pre-

sent–day run in Exp. 4. We also refer to the results in *Seidov and Maslin* [1999, 2001].

Northern LGM events. The modeled LGM conveyor (Figure 13a), though weaker and shallower, is not as different from the modern conveyor as one might expect based on strong the surface condition changes relative to the present-day climatology. As in earlier results, the rate of overturning is up to half of the modern rate and the shape of the overturning resembles those of the present–day, though convection sites and NADW outflow was shifted southward (to reflect the shift of the convection sites) and is noticeably

Table 3. Amplitudes of sea surface salinity anomalies (in psu).

Exp.	NA	NNA	SO	
CRLGM (#1L)	–	–	–	CRLGM – Control Case (annual mean LGM sea surface climatology); NA – North Atlantic; NNA – (central) northern North Atlantic; SO – Southern Ocean
#2L	-2.0	–	–	Salinity anomalies are added to the present–day annual mean sea surface salinity in the bands between 60°N and 80°N in NA, in the lens at about 50°N in NNA, and band
#3L	–	-2.0	–	between 50°S and the coast of the Antarctica (SO). The modified salinity was merged
#4L	–	–	-1.0	using a cosine filter to the unchanged field within two latitudinal grid points (8°). In the
#5L	–	-2.0	-1.0	SO the anomalies are circumglobal.

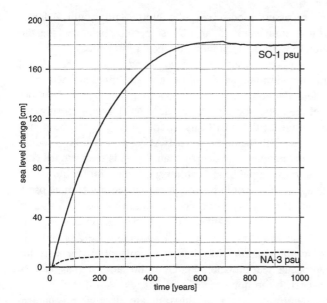

Figure 12. Evolution of the sea level change relative to the CRPDC sea level in time: solid line in SO event, and dash line in NA event.

shallower. This occurs despite the shift of NADW production to the middle of the northern North Atlantic and strongly reduced convection in the Nordic Seas (e.g., *Seidov and Haupt* [1999a]; *Seidov and Maslin* [2001]). This led *Seidov and Maslin* [2001] to conclude that convection in the northern North Atlantic is a key factor in maintaining the

"normal" mode of the conveyor during glacial periods. However, it is important that although the conveyor operates in its "normal" or "conveyor-on" mode, there is very little cross–equatorial heat transport in the Atlantic Ocean (Figure 14) indicating an approximate balance between southern and northern heat budgets in this basin. Nevertheless, the global southward cross–equatorial heat transport does occur and could be instrumental in maintaining the glaciation in the north (see Figure 15).

In contrast, the meridional overturning stream function of northern meltwater events shows a reduction of the deep–water conveyor at Dansgaard – Oeschger (D-O) events (NA, Exp. 2L; Figure 13b), and a complete collapse at a Heinrich event (NA, Exp. 3L; Figure 13c). *Seidov and Maslin* [1999, 2001] indicate independence of the collapse event on the source of the meltwater based on different sources of meltwater capping convection in the central northern North Atlantic. In all scenarios of meltwater events the South Atlantic gains oceanic heat from the north due to a reversed conveyor and a change in the sign of the cross-equatorial heat transport in the Atlantic Ocean (see Figure 14 and a sketch in Figure 3b).

Southern LGM events. Stronger southern impacts, similar to those in the present–day runs, are registered in the LGM runs with low–salinity signals in the Southern Ocean. The overturning streamfunction in Exp. 4L confirms that a meltwater event in the Southern hemisphere can reverse the conveyor and heat piracy signs, and therefore reverse cooling to

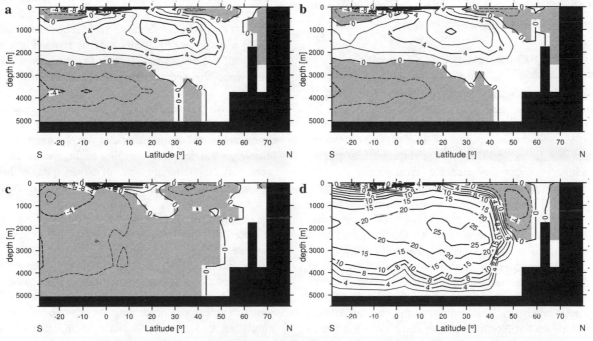

Figure 13. Meridional overturning in CRLGM (a), Exp. 2L (b), Exp. 3L (c), and Exp4L (d) (see Table 3).

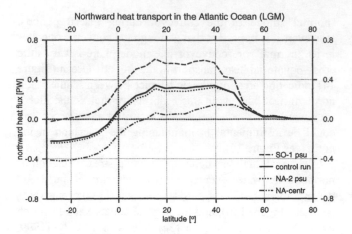

Figure 14. Northward heat transport in the Atlantic Ocean during LGM. As in Figure 9.

warming in the Northern Hemisphere (Figure 13d and Figure 14). Therefore, the rebound of the conveyor, or bi–polar seesaw, could be the cause of glacial age terminations in the Northern Hemisphere.

The combined northern and southern impacts favors the idea of stronger AABW control of the conveyor, and confirms that as soon as AABW production is curtailed, the increase of NADW production converts a cold trend to warm trend and can reverse the cooling in the north, despite the ongoing northern meltwater event. Alternatively, an increase of AABW production would curtail NADW production and lead to cooling in the Northern Hemisphere, in a harmony with Broecker's idea (e.g., *Broecker* [2000]) of the ocean bi–polar seesaw.

4. DISCUSSION AND CONCLUSIONS

Based on their numerical experiments assessing Heinrich and Dansgaard–Oeschger type events, *Seidov and Maslin* [2001] presented an explanation of hemispheric asymmetry by variations in relative amount of deepwater formation in the two hemispheres and thus by the resulting heat piracy. For example, present–day oceanic cross–equatorial heat transport is northward, whereas during a Heinrich event the cross–equatorial transport was southward. Thus, South Atlantic Ocean post–glacial heat piracy, replacing the present–day North Atlantic Ocean piracy, causes the Southern Hemisphere oceans to warm, while the Northern Hemisphere oceans cool. When the iceberg armada ceases, the freshwater cap on NADW formation is removed and northward cross–equatorial heat transport kicks in to restore a North Atlantic heat piracy condition. Accordingly, the North Atlantic Ocean warms up the Northern Hemisphere while the Southern Hemisphere cools down. If the glacial boundary conditions re–assert themselves, the North and South Atlantic Oceans come back to an almost perfect balance (see the sketch in Figure 3 explaining the heat piracy idea).

An important result of the *Seidov and Maslin* [1999, 2001] computer simulations is that a high–latitudinal meltwater (or de–densification, in general) scenario, although it does not reject the upper–ocean impact, does NOT require a substantial change in the heat transported by the upper ocean currents. The imbalance between NADW and AABW productions, i.e. between the deep–ocean flows, is the primary control on cross–equatorial heat transport, and thus could be the sole agent responsible for the observed seesaw climate oscillations. This seems to be counter–intuitive, as it is usually assumed that only changes of the upper–ocean currents can affect the heat transport to high latitudes (see Appendix in *Seidov and Maslin* [2001] for an explanation for why the deep ocean heat transport can be of the same sign as the heat transport in the upper ocean). It is suggested that the deep–ocean currents could be the ultimate internal mechanism capable of reversing cross–equatorial heat transport within the required time scale of the oscillations (hundreds of years or longer).

Support for this deep–ocean driving force hypothesis comes from the relative timing of the north–south lead–lag observed in the ice cores [*Blunier et al.*, 1998]. For example, the substantial weakening of the conveyor would happen within a matter of years, as soon as a sufficient number of icebergs travel to the convection sites in the northern North Atlantic to provide the meltwater capping of the convection. However, the actual warming of high latitudinal waters, due to change of overturning and the related change of the heat transport may take hundreds of years. For instance, *Broecker* [2000] argues that 200 years may be needed to reduce the density of the sea surface waters in one hemisphere to the point when the hemispheres change the role and the conveyor flip–flops.

This is why, when the GRIP and Byrd ice cores are compared, the peak warming in Antarctica is delayed, occurring almost at the end of the Heinrich event [*Blunier et al.*, 1998]. Moreover, the dramatic warming at the end of the Heinrich event and the switching "on" of North Atlantic heat transport regime would take hundreds of years to steal enough heat to cool down the Southern Hemisphere. Hence it can explain the delay seen in the ice core data (*Broecker* [2000] estimated that the time needed to rebound of the conveyor may as long as several hundred years).

Our simulations demonstrate the stronger sensitivity of the global ocean circulation to variations in the southern than to northern deepwater source. A new vision of the role of the Southern Ocean role in driving the ocean conveyor becomes significant in view of possible consequences of the global warming process. In this scheme, the seesaw oscillating mechanism, prone to an AABW regime, follows a simple rule. The rule is that when the NADW subsides, the AABW picks up and affects the heat transport regime in the opposite way. Alternatively, when AABW subsides, NADW spins up and warms up the deep ocean.

Although it is believed that there is less AABW variation than NADW variability during the glacial cycles of the Pleistocene, the reduction of NADW implies an increased impact of AABW even if the latter source does not change at all. Therefore, the perturbation of the conveyor only at the NADW sites could be a sufficient mechanism for sustaining the suggested climatic oscillations. If, however, sufficient evidence were found that glacial AABW production had varied substantially (e.g., *Stephens and Keeling* [2000]), theoretically it alone could drive the climate seesaw. *Broecker* [2000] suggests that AABW increase drove the Little Ice Age that ended after AABW production subsided. Hence we should anticipate a true bi–polar seesaw driven by two hemispheric sources. The key element is that the seesaw rebounds are driven by changes of deep–ocean currents rather than by sea–surface circulation alterations. However, although a southern meltwater might be a powerful near–future climate element, the role of the southern source of AABW in modulating variability has received much less attention then northern impacts, limiting the development of a complete understanding of decadal to millennial time–scale climate change. Therefore, our major task was to compare the roles of the hemispheric de–densification sources.

The source and character of the salinity perturbations in the Southern Hemisphere is substantially different from the Northern Hemisphere, and involves brine rejection in the formation of sea ice, freshwater fluxes from ablation and calving of the Antarctic Ice Sheet and the stability of the WAIS. Hence, the potential for changes in freshwater fluxes or salinity variations to influence the Southern Ocean is clearly evident. Our results demonstrate that changes in surface salinity of the Southern Ocean can significantly alter deepwater structure and temperatures. In addition, the experiments demonstrate that the high latitude sea surface need not be very warm, or very salty, or both to produce significant deep ocean warming.

In principle, an imbalance of meltwater impacts in high latitudes may lead to warming up of the deep ocean even if the ocean in high latitudes stays relatively cool in one of the hemispheres. In this case, the warming is in response to increased overturning in one of the hemispheres, whereas the other hemispheric source of deepwater might become stagnated. In other words, a strong meltwater impact in one of the two hemispheres may lead to a strengthening of the thermohaline conveyor driven by the source in another hemisphere. This, in turn, may lead to an increased surface poleward compensating flow in the active hemisphere. Fur-

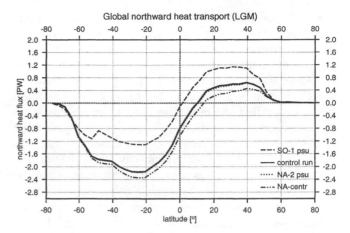

Figure 15. Northward heat transport in the Atlantic Ocean during LGM. As in Figure 9.

ther, this warming can have a substantial sea level impact even without ice sheet melting.

We have found that the warming of deep–ocean is caused by NADW intensification that accompanies the reduction of AABW, rather than by reduction of the southern source itself. The study presented here emphasizes the competitive nature of the northern and southern sources and indicates the role of the southern source as a strong modifier of NADW on centennial and millennial time scales. Further, this study calls attention to the important implications of changes in the THC, both in terms of heat transport and sea level rise.

Our results on the amplitude of deep–ocean warming during hypothesized near–future meltwater impacts can be compared with observations. As shown by *Levitus et al.* [2000], the global volume warming is 0.06°C per 40 years. Our global warming rates vary between 4°C to 7°C per 1000 years which is about two to three times higher than the observed present–day trend [*Levitus et al.*, 2000], would it continue for a thousand years. These rates could be dependent on the fact that there are no atmospheric feedbacks that could slow down the ocean warming induced by the southern meltdown (see below). Alternatively, the warming could speed up as the AABW weakens with the addition of meltdown (if is the major cause of the warming trend). Note that the ocean warming is not linear in time (see above), even if the low–salinity signal is kept constant.

In discussing the role of the Southern Ocean, we emphasize that neither the Weddell or Ross Sea deepwater sources, nor the meltwater around Antarctica are powerful enough to overwhelm the direct impact over the NADW source. Only the entire Southern Ocean impact has the power to control the world ocean deep circulation (albeit not directly but via modulating the NADW operation). Hence, the southern ice sheet meltdown and sea ice melting work synergetically to counterbalance the northern hemisphere influence.

Finally, we employed the last glacial maximum sea surface climatology to test our conclusions about the bi–polar seesaw sensitivity to the southern de–densification impact. All our above–formulated conclusions, as well as those presented in *Seidov and Maslin* [2001], stay intact in these paleoceanographic experiments. Regardless of the amplitude, origin or location of a meltwater impact in the North Atlantic, a counterbalancing de–densification of sea surface water in the Southern Ocean has a potential for again reversing the conveyor and restoring the present–day North Atlantic heat piracy pattern. As a restored northward cross–equatorial oceanic heat transport might cause a recurrence of northern cooling, the warming of the northern North Atlantic might revoke melting. Hence the ocean seesaw can rebound again, and the whole cycle will repeat multiple times until the northern meltwater resources are nearly exhausted (like at present or milder climates of the Holocene). Thus, in our simulations, we end up with a simple scheme of millennial scale climate oscillations driven by deep–ocean circulation, which is controlled by two hemispheric deepwater sources, in the North Atlantic and in the Southern Ocean. This scheme can also explain the apparent hemispheric asymmetry of the glacial record. And most importantly, it implies that the Southern ice melting impact can be a real threat in the climate change trends.

Although a stand–alone ocean model helps to outline the nature of the problem and enables a wealth of sensitivity experiments it has inherent limitations as a true climate change study. Basically, our approach is largely a first–order analysis. As such, the more extreme scenarios provide a higher level of sensitivity that clearly illustrate the potential response of the ocean without atmospheric and cryospheric feedbacks. The response of a coupled model may be different than a stand–alone ocean model because of the potential importance of feedbacks associated with the atmospheric response to an altered poleward ocean heat transport, or the impact of wind stress changes on global thermohaline overturning. Hence, our work is only a first step in assessing the climate response to changes in freshwater inputs at high latitudes. A use of a coupled ocean-atmosphere model may call for some corrections of the ocean seesaw dynamics as it is seen here. However, despite all the advances in getting supercomputers faster, scenario-type simulations of past climates are still hardly affordable. Palliative solution can be sought in using either "enhanced" ocean models, with atmospheric part downsized to energy balance models (e.g., *Bjornsson et al.* [1997]; *Weaver et al.* [1998]; *Weaver and Wiebe* [1999]; *Bjornsson and Toggweiler* [2001] (this volume), or so-called models of "intermediate complexity", with ocean-atmosphere and other climatic feedback included using simplified components of the climate systems (e.g.,

Ganopolski et al. [1998]; *Wang and Mysak* [2000, 2001]; *Ganopolski and Rahmstorf* [2001] (this volume)).

There are many other caveats that may be added to underline the limited nature of our "ocean-only" modeling results. Most of them are obvious and are due to absence of feedbacks between ice, ocean and atmosphere components of the climate system. However, we believe that the main result of importance of the southern freshwater impacts in seesaw behavior is valid despite of all the limitations of our approach. The Southern Ocean has the potential to overpower the Northern Hemisphere oceans and become a major climatic player in long–term climate change in some combination of climatic factors. Yet it is the NADW that is a universal driver of the conveyor. Amplified by southern meltwater episodes or reduced by the northern meltwater impacts, it remains the strongest player in bi–polar thermohaline conveyor variability. The two hemisphere sources, if reduced, lead to principally different consequences: If the NADW is reduced, the deep ocean cools down, whereas reduction of AABW leads to warmer abyssal waters. Therefore, a change of sea–ice or WAIS state tending toward less saline surface waters in the Southern Ocean can cause unfavorable sea-level changes, whereas the collapse of the northern source might cause cooling of the northern climate. Both scenarios may pose a serious threat to climate–sensitive environments.

Acknowledgments. We are very grateful to Lawrence Mysak, Ron Stouffer, Andrew Weaver, Andreas Schmittner, and Katrin Meissner for their useful and extended comments, which helped to substantially improve the manuscript. This study was partly supported by NSF (NSF project #9975107 and ATM 00-00545).

REFERENCES

Anderson, J.B. and J.T. Andrews, 1999: Radiocarbon constraints on ice sheet advance and retreat in Weddell Sea, Antarctica, *Geology*(27): 179-182.

Andrews, J.T., 1998: Abrupt changes (Heinrich events) in late Quaternary North Atlantic marine environments, *Journal of Quaternary Science*, 13: 3-16.

Birchfield, G.E. and W.S. Broecker, 1990: A salt oscillator in the glacial Atlantic? 2. A "scale" analysis model, *Paleoceanography*, 5: 835-843.

Bjornsson, H., L.A. Mysak and G.A. Schmidt, 1997: Mixed boundary conditions versus coupling with an energy-moisture balance model for a zonally averaged ocean climate model, *Journal of Climate*, 10: 2412-2430.

Bjornsson, H. and J.R. Toggweiler, 2001: The climatic influence of Drake Passage, *Geophysical Monograph (This Volume)*, D. Seidov, B.J. Haupt and M. Maslin (Editors), American Geophysical Union, Washington, D.C.

Blunier, T., J. Chappellaz, J. Schwander, A. Daellenbach, B. Stauffer, T.F. Stocker, D. Raynaud, J. Jouzel, H.B. Clausen,

C.U. Hammer and S.J. Johnsen, 1998: Asynchrony of Antarctic and Greenland climate change during the last glacial period, *Nature*, 394: 739-743.

Bond, G., H. Heinrich, W. Broecker, L. Labeyrie, J. McManus, J. Andrews, S. Huon, R. Jantschik, S. Clasen, C. Simet, K. Tedesco, M. Klas, G. Bonani and S. Ivy, 1992: Evidence for massive discharges of icebergs into the North Atlantic ocean during the last glacial period, *Nature*, 360: 245-249.

Bond, G., W. Showers, M. Cheseby, R. Lotti, P. Almasi, P. deMenocal, P. Priore, H. Cullen, I. Hajdas and G. Bonani, 1997: A pervasive millennial-scale cycle in North Atlantic and glacial climates, *Science*, 278: 1257-1366.

Bradley, R.S., 1999, Paleoclimatology: Reconstructing climates of the Quaternary, Academic Press, San Diego, 613 pp.

Broecker, W., 1991: The great ocean conveyor, *Oceanography*, 1: 79-89.

Broecker, W., 2001: The big climate amplifier ocean circulation-sea ice-storminess-dustiness-albedo, *Geophysical Monograph (This Volume)*, D. Seidov, B.J. Haupt and M. Maslin (Editors), American Geophysical Union, Washington, D.C.

Broecker, W.S., 1994a: Massive iceberg discharges as triggers for global climate change, *Nature*, 372: 421-424.

Broecker, W.S., 1994b: An unstable superconveyor, *Nature*, 367: 414-415.

Broecker, W.S., 1998: Paleocean circulation during the last deglaciation: A bipolar seesaw? *Paleoceanography*, 13: 119-121.

Broecker, W.S., 2000: Was a change in thermohaline circulation responsible for the Little Ice Age? *Proc. Nat. Acad. Sci.*, 97(4): 1339-1342.

Broecker, W.S. and G.H. Denton, 1989: The role of ocean atmosphere reorganizations in glacial cycles, *Geochimica Cosmochimica Acta*, 53: 2465-2501.

Broecker, W.S., S. Sutherland and T.-H. Peng, 1999: A possible 20th century slowdown of Southern Ocean deep water formation, *Science*, 286: 1132-1135.

Bryan, F., 1987: Parameter sensitivity of primitive equation ocean general circulation models, *Journal of Physical Oceanography*, 17: 970-985.

Church, J.A., J.S. Godfrey, D.R. Jackett and T.J. McDougall, 1991: A model of sealevel rise caused by ocean thermal expansion, *Journal of Climate*, 4: 438-456.

CLIMAP, 1981: Climate: Long-Range Investigation, Mapping, and Prediction (CLIMAP) Project Members, Seasonal reconstructions of the Earth's surface at the Last Glacial Maximum. *Map and Chart Ser. MC-36*. Geological Society of America, Boulder, Colorado, pp. 1-18.

Cronin, T.M., 1999, Principles of Paleoclimatology, Perspectives in Paleobiology and Earth History. Columbia University Press, New York, 560 pp.

Crowley, T.J., 1992: North Atlantic deep water cools the southern hemisphere, *Paleoceanography*, 7: 489-497.

Dowdeswell, J.A., M.M. A., A.J. T. and M.I. N., 1995a: Iceberg production, debris rafting, and the extent and thickness of Heinrich layers (H1, H2) in North Atlantic sediments., *Geology*, 23: 301-304.

Dowdeswell, J.A., M.A. Maslin, J.T. Andrews and N.I. McCave, 1995b: Iceberg production, debries rafting, and the extent and thickness of Heinrich layers (H1,H2) in North Atlantic sediments, *Geology*, 23: 301-304.

Duplessy, J.-C., L. Labeyrie, A. Julliet-Lerclerc, J. Duprat and M. Sarnthein, 1991: Surface salinity reconstruction of the North Atlantic Ocean during the Last Glacial Maximum, *Oceanologica Acta*, 14: 311-324.

Duplessy, J.-C., L. Labeyrie, M. Paterne, S. Hovine, T. Fichefet, J. Duprat and M. Labracherie, 1996: High latitude deep water sources during the Last Glacial Maximum and the intensity of the global oceanic circulation, In: G. Wefer, W.H. Berger, G. Siedler and D. Webb (Editors), *The South Atlantic*. Springer, NY, pp. 445-460.

Duplessy, J.C., N.J. Shackleton, R.G. Fairbanks, L. Labeyrie, D. Oppo and N. Kallel, 1988: Deepwater source variations during the last climatic cycle and their impact on the global deepwater circulation, *Paleoceanography*, 3: 343-360.

England, M.H., 1992: On the formation of Atlantic intermediate and bottom water in ocean general circulation models, *Journal of Physical Oceanography*, 22: 918-926.

Ezer, T., 2001: On the response of the Atlantic Ocean to climatic changes in high latitudes: Sensitivity studies with a sigma coordinate ocean model, *Geophysical Monograph (This Volume)*, D. Seidov, B.J. Haupt and M. Maslin (Editors), American Geophysical Union, Washington, D.C.

Ganopolski, A. and S. Rahmstorf, 2001: Stability and variability of the thermohaline circulation in the past and future: a study with a coupled model of intermediate complexity, *Geophysical Monograph (This Volume)*, D. Seidov, B.J. Haupt and M. Maslin (Editors), American Geophysical Union, Washington, D.C.

Ganopolski, A., S. Rahmstorf, V. Petoukhov and M. Claussen, 1998: Simulation of modern and glacial climates with a coupled global model of intermediate complexity, *Nature*, 391: 351-356.

Gent, P.R. and J.C. McWilliams, 1990: Isopycnal mixing in ocean circulation models, *Journal of Physical Oceanography*, 20: 150-155.

Gill, E.G., 1982, Atmosphere-ocean dynamics, International Geophysical Series, 30. Academic Press, San Diego, 666 pp.

Godfrey, J.S. and G. Love, 1992: Assessment of sealevel rise, specific to the South Asian and Australian situations In: Seal Level Changes: Determination and Effects, *Seal Level Changes: Determination and Effects. Geophys. Monograph 69*. AGU, Washington, D.C., pp. 87-94.

Goodman, P.J., 1998: The role of North Atlantic Deep Water formation in an OGCM's ventilation and thermohaline circulation, *J. Phys. Oceanogr.*, 28: 1759-1785.

Goosse, H. and T. Fichefet, 1999: Importance of ice-ocean interactions for the global ocean circulation: A model study, *Journal of Geophysical Research*, 104(C10): 23337-23355.

Gordon, A., 1986: Interocean exchange of thermocline water, *Journal of Geophysical Research*, 91: 5037-5046.

Gordon, A.H., S.E. Zebiak and K. Bryan, 1992: Climate variability and the Atlantic Ocean, *Eos, Transactions, American Geophysical Union*, 79: 161,164-165.

Grousset, F.E., L. Labeyrie, J.A. Sinko, M. Cremer, G. Bond, J. Duprat, E. Cortijo and S. Huon, 1993: Patterns of ice-rafted detritus in the glacial North-Atlantic (40-55°N), *Paleoceanography*, 8: 175-192.

Gwiazda, R.H., S.H. Hemming and W.S. Broecker, 1996: Tracking the sources of icebergs with lead isotopes: The provenance of ice rafted debris in Heinrich event 2., *Paleoceanography*, 11: 77-93.

Heinrich, H., 1988: Origin and consequences of cyclic ice rafting in the northeast Atlantic Ocean during the past 130,000 years., *Quaternary Research*, 29: 142-152.

Hellerman, S. and M. Rosenstein, 1983: Normal monthly wind stress over the world ocean with error estimates, *Journal of Physical Oceanography*, 13: 1093-1104.

Hirschi, J., J. Sander and T.F. Stocker, 1999: Intermittent convection, mixed boundary conditions and the stability of the thermohaline circulation, *Climate Dynamics*, 15(4): 277-291.

Hirst, A.C., S.P. O'Farrell and H.B. Gordon, 2000: Comparison of a coupled ocean-atmosphere model with and without oceanic eddy-induced advection. Part I: Ocean spinup and control integrations, *Journal of Climate*, 13: 139-163.

Houghton, J., 1997, Global warming. Cambridge University Press, N.Y., 251 pp.

Hulbe, C., 1997: An ice shelf mechanism for Heinrich layer production, *Paleoceanography*, 12: 711-717.

Jackett, D.R., T.J. McDougall, M.H. England and A.C. Hirst, 2000: Thermal expansion in ocean and coupled general circulation models, *Journal of Climate*, 13: 1384-1405.

Kamenkovich, I.V. and P.G. Goodman, 2001: The effects of vertical mixing on the circulation of the AABW in the Atlantic, *Geophysical Monograph (This Volume)*, D. Seidov, B.J. Haupt and M. Maslin (Editors), American Geophysical Union, Washington, D.C.

Karl, T.R., 1993: A new perspective on global warming, *Eos, Transactions, American Geophysical Union*, 74: 25.

Knutti, R. and T.F. Stocker, 2000: Influence of the thermohaline circulation on projected sea level rise, *Journal of Climate*, 13(12): 1997-2001.

Labeyrie, L.D., J.J. Pichon, M. Labracherie, P. Ippolito, J. Dupart and J.-C. Duplessy, 1986: Melting history of Antarctica during the past 60,000 years, *Nature*, 322: 701-706.

Levitus, S., J.I. Antonov, T.P. Boyer and C. Stephens, 2000: Warming of the World Ocean, *Science*, 287: 2225-2229.

Levitus, S. and T.P. Boyer, 1994: World Ocean Atlas, vol. 3, Salinity, 99 pp., Natl. Ocean and Atmos. Admin., Washington, D. C.

Levitus, S., R. Burgett and T.P. Boyer, 1994: World Ocean Atlas, vol. 4, Temperature, 117 pp., Natl. Ocean and Atmos. Admin., Washington, D. C.

Lorenz, S., B. Grieger, H. P. and K. Herterich, 1996: Investigating the sensivity of the atmospheric general circulation Model ECHAM 3 to paleoclimate boundary conditions., *Geol. Rundsch.*, 85: 513-524.

MacAyeal, D.R., 1992: Irregular oscillations of the West Antarctic ice sheet, *Nature*, 359: 29-32.

Maier-Reimer, E., U. Mikolajewicz and K. Hasselmann, 1993: Mean circulation of the Hamburg LSG OGCM and its sensitivity to the thermohaline surface forcing, *Journal of Physical Oceanography*, 23: 731-757.

Manabe, S. and R. Stouffer, 1997: Coupled ocean-atmosphere model response to freshwater input: Comparison to Younger Dryas event, *Paleoceanography*, 12: 321-336.

Manabe, S. and R.J. Stouffer, 1988: Two stable equilibria of a coupled ocean-atmosphere model, *Journal of Climate*, 1: 841-866.

Manabe, S. and R.J. Stouffer, 1995: Simulation of abrupt change induced by freshwater input to the North Atlantic Ocean, *Nature*, 378: 165-167.

Maslin, M., D. Seidov and J. Lowe, 2001: Synthesis of the nature and causes of rapid climate transitions during the Quaternary, *Geophysical Monograph (This Volume)*, D. Seidov, B.J. Haupt and M. Maslin (Editors), American Geophysical Union, Washington, D.C.

Maslin, M.A., N.J. Shackleton and U. Pflaumann, 1995: Surface water temperature, salinity, and density changes in the northeast Atlantic during the last 45,000 years: Heinrich events, deep water formation, and climatic rebounds, *Paleoceanography*, 10: 527-544.

McWilliams, J.C., 1998: Oceanic géneral circulation models, In: E.P. Chassignet and J. Verron (Editors), *Ocean Modeling and Parameterization*. Kluwer Academic Publishers, Boston, pp. 1-44.

MOM-2, 1996: Documentation, User's Guide and Reference Manual (edited by R. C. Pacanowski), GFDL Ocean Technical Report No. 3.2, Geophysical Fluid Dynamics Laboratory/NOAA, Princeton, N.J.

Oeschger, H., J. Beer, U. Siegenthaler, B. Stauffer, W. Dansgaard and C.C. Langway, 1984: Late glacial climate history from ice cores, In: J.E. Hansen and T. Takahashi (Editors), *Climate Processes and Climate Sensitivity. Geophys. Monogr. Ser.* American Geophysical Union, Washington D.C., pp. 299-306.

Oppenheimer, M., 1998: Global warming and the stability of the west Antarctic ice sheet, *Nature*, 393: 325-332.

Seidov, D. and B.J. Haupt, 1999b: Numerical study of glacial and meltwater global ocean, In: J. Harff, J. Lemke and K. Stattegger (Editors), *Computerized Modeling of Sedimentatry Systems*. Springer, New York, pp. 79-113.

Seidov, D. and M. Maslin, 1996: Seasonally ice freee glacial Nordic Seas without deep water ventilation, *Terra Nova*, 8: 245-254.

Seidov, D. and M. Maslin, 1999: North Atlantic Deep Water circulation collapse during the Heinrich events, *Geology*, 27: 23-26.

Seidov, D. and M. Maslin, 2001: Atlantic Ocean heat piracy and the bi-polar climate sea-saw during Heinrich and Dansgaard-Oeschger events, *Journal of Quaternary Science*, 16: in press.

Seidov, D., M. Sarnthein, K. Stattegger, R. Prien and M. Weinelt, 1996: North Atlantic ocean circulation during the Last Glacial

Maximum and subsequent meltwater event: A numerical model, *Journal of Geophysical Research*, 101: 16305-16332.

Stephens, B. and R. Keeling, 2000: The influence of Antarctic sea ice on glacial-interglacial CO_2 variations, *Nature*, 404,: 171-174.

Stocker, T.F., 1998: The seesaw effect, *Science*, 282: 61-62.

Stocker, T.F., R. Knutti and G.-K. Plattner, 2001: The future of the thermohaline circulation - a perspective, *Geophysical Monograph (This Volume)*, D. Seidov, B.J. Haupt and M. Maslin (Editors), American Geophysical Union, Washington, D.C.

Stocker, T.F., D.G. Wright and W.S. Broecker, 1992: The influence of high-latitude surface forcing on the global thermohaline circulation, *Paleoceanography*, 7: 529-541.

Stoessel, A., S.-J. Kim and S.S. Drijfhout, 1998: The impact of Southern Ocean sea ice in global ocean model, *J. Phys. Oceanogr.*, 28: 1999-2018.

Stommel, H. and A.B. Arons, 1960: On the abyssal circulation of the world ocean, I, Stationary planetary flow patterns on a sphere, *Deep Sea Research*, 6: 140-154.

Stommel, H., A.B. Arons and A.J. Faller, 1958: Some examples of stationary planetary flow patterns in bounded basins, *Tellus*, 10: 179-187.

Toggweiler, J.R., K. Dixon and K. Bryan, 1989: Simulations of radiocarbon in a coarse-resolution world ocean circulation model, 1, Steady state prebomb distribution, Journal *of Geophysical Research*, 94: 8217-8242.

Toggweiler, J.R. and B. Samuels, 1980: Effect of sea ice on the salinity of Antarctic Bottom Water, *J. Phys. Oceanogr.*, 25: 1980-1997.

Vaughan, D.G., J.L. Bamber, M. Giovinetto, J. Russel and A.P.R. Cooper, 1999: Reassessment of net surface mass balance in Antarctica, *Journal of Climate*, 29(4): 933-946.

Wang, X., P. Stone and J. Marotzke, 1999a: Global thermohaline circulation. Part I. Sensitivity to atmospheric moisture transport, *J. Climate*, 12: 71-82.

Wang, X., P. Stone and J. Marotzke, 1999b: Global thermohaline circulation. Part II. Sensitivity with interactive atmospheric transports, *J. Climate*, 12: 83-91.

Wang, Z. and L.A. Mysak, 2000: A simple coupled atmosphere-ocean-sea-ice-land surface model for climate and paleoclimate studies, *Journal of Climate*, 13: 1150-1172.

Wang, Z. and L.A. Mysak, 2001: Ice sheet-thermohaline circulation interactions in a climate model of intermediate complexity, Journal of Oceanography: in press.

Warrick, R.A., E.M. Barrow and T.M.L. Wigley (Editors), 1993. Climate and Sea Level Change: Observations, Projections and Implications. Cambridge University Press, N.Y.

Weaver, A., 1999: Millennial timescale variability in ocean/climate models, In: P.U. Clark, S.R. Webb and L.D. Keigwin (Editors), *Mechanisms of global climate change*. AGU, Washington, DC, pp. 285-300.

Weaver, A.J., 1995: Driving the ocean conveyor, *Nature*, 378: 135-136.

Weaver, A.J., S.M. Aura and P.G. Myers, 1994: Interdecadal Variability in an Idealized Model of the North Atlantic, *Journal of Geophysical Research - Oceans*, C99: 12423-12441.

Weaver, A.J., C.M. Bitz, A.F. Fanning and M.M. Holland, 1999: Thermohaline circulation: high latitude phenomena and the difference bewteen the Pacific and Atlantic, *Annual Review of Earth and Planetary Sciences*, 27: 231-285.

Weaver, A.J., M. Eby, A.F. Fanning and E.C. Wiebe, 1998: Simulated influence of carbon dioxide, orbital forcing and ice-sheets on the climate of the last glacial maximum., *Nature*, 394: 847-853.

Weaver, A.J. and T.M.C. Hughes, 1994: Rapid interglacial climate fluctuations driven by North Atlantic ocean circulation, *Nature*, 367: 447-450.

Weaver, A.J. and E.C. Wiebe, 1999: On the sensitivity of projected oceanic thermal expansion to the parameterisation of sub-grid scale ocean mixing, *Geophysical Research Letters*, 26: 3461-3464.

Weyl, P.K., 1968: The role of the oceans in climatic change: a theory of ice ages, *Meteorological Monographs*, 8: 37-62.

Zahn, R., J. Schonfeld, H.-R. Kudrass, M.-H. Park, H. Erlenkeuser and P. Grootes, 1997: Thermohaline instability in the North Atlantic during meltwater events: Stable isotopes and ice-rafted detritus from core SO75-26KL, Portuguese margin, *Paleoceanography*, 12: 696-710.

E.J. Barron, EMS Environment Institute, Pennsylvania State University, University Park, PA 16802. (eric@essc.psu.edu)

B.J. Haupt, EMS Environment Institute, Pennsylvania State University, University Park, PA 16802. (bjhaupt@essc.psu.edu)

M. Maslin, Environmental Change Research Centre, Department of Geography, University College London, 26 Bedford Way, London. WC1H 0AP, UK. (mmaslin@ucl.ac.uk)

D.Seidov, EMS Environment Institute, Pennsylvania State University, University Park, PA 16802. (dseidov@essc.psu.edu)

Glacial–to–Interglacial Changes of the Ocean Circulation and Eolian Sediment Transport

Bernd J. Haupt, Dan Seidov, and Eric J. Barron

EMS Environment Institute, Pennsylvania State University, University Park, Pennsylvania

Glacial–to–interglacial climate transitions are characterized by distinct basin–wide sediment accumulation patterns that can reveal accompanying ocean circulation changes that occur during these transitions. A combination of an ocean global circulation model (OGCM) and a large–scale 3–D sediment transport model is used to model the global ocean thermohaline conveyor (THC) and distribution of the sedimentation rates for different time periods. Two different OGCM numerical mixing schemes are used to test the sensitivity of the sedimentation pattern to changes in model parameterizations of the thermohaline circulation, and especially the mixing technique. A sediment transport model is employed to identify glacial–to–interglacial changes of the deep–ocean currents. This model is suitable to identify the regions of the world ocean that are most responsive to the strong changes in glacial and interglacial circulation patterns and therefore are suitable for comparison of the computer simulations and geologic record. Two different eolian dust sources were prescribed at the sea surface: (1) A spatially homogeneous inorganic atmospheric dust source is used to depict sedimentation patterns that result only from the circulation changes. (2) A realistic present–day eolian dust pattern was specified in order to assess a more realistic sedimentation pattern. The results show that the modeled THC changes are traceable in the sediment accumulation patterns. These simulations indicate that the changes in sediment deposition rates are often associated with the changes in inter–basin water exchange, enabling verification of the simulated deep circulation using the geologic record.

INTRODUCTION

Decades of ocean sediment collection combined with major advances in ocean general circulation modeling provide a growing opportunity to address quantitatively a number of fascinating paleoceanographic problems. However, most of the existing sedimentation models are two–

dimensional and are designed for small basins, targeting specific features such as alluvial or deltaic basin fill (cf. *Ericksen et al.* [1989]; *Tetzlaff and Harbaugh* [1989]; *Bitzer and Pflug* [1990]; *Paola et al.* [1992]; *Syvitski and Daughney* [1992]; *Cao and Lerche* [1994]; *Slingerland et al.* [1994]). In such models, sediment transport is gravity–driven, with the sediment load proportional to the basin slope, water discharge, and diffusion coefficient (e.g, *Granjeon and Joseph* [1999]). Usually, these models are not coupled with models that simulate the ocean general circulation. As such they are not capable of assessing the impact of global climate change on the ocean sediment

The Oceans and Rapid Climate Change: Past, Present, and Future
Geophysical Monograph 126

drifts on millennial and longer time scales. For these problems, a three–dimensional coupled ocean circulation–ocean sedimentation model is required. A coupled model is key to addressing the extent of the linkages between sedimentation and the deep–ocean, and whether modeled ocean paleocirculation can be unambiguously verified by observed changes in inorganic sedimentation rates and sediment drift dynamics. Our objective is to begin to assess whether it is possible to trace the changes in the total terrigenous sedimentation rates induced by millennium–scale changes of the ocean currents in a combined circulation–sedimentation model.

The following goals are pursued: (i) to simulate glacial–to–interglacial changes of the global ocean thermohaline conveyor; (ii) to test whether the deep–sea circulation signal is overwhelmed by changes in sediment input by simulating terrigenous (eolian) sedimentation rates for each conveyor regime with different sedimentation input rates and patterns; (iii) to test the method of using inorganic sediment as a tracer that can constrain past ocean circulation modeling. This study expands previous work by *Seidov and Haupt* [1997b] and *Seidov and Haupt* [1999] by addressing three additional questions: (1) Can we adequately specify an eolian sediment input at the sea surface in the sedimentation model? (2) Does the sediment transport by deep-ocean currents substantially modify the sedimentation pattern, or is the latter essentially determined by the input at the sea surface? (3) If the deep-ocean circulation can substantially influence the sedimentation rates, are there sufficient sediment data to verify the results of the simulations? Answering these questions is the main focus of this publication.

The global oceanic thermohaline circulation (THC), also known as the "salinity conveyor belt" [*Stommel*, 1958; *Broecker and Denton*, 1989; *Broecker*, 1991] is driven by deepwater production in the high-latitudes, especially by the production of North Atlantic Deep Water (NADW). The variability of deepwater formation in the North Atlantic has been addressed in a number of model studies during the last decade (e.g., *Maier-Reimer et al.* [1991]; *Manabe and Stouffer* [1988, 1995, 1997]; *Seidov and Haupt* [1999]; *Seidov and Maslin* [1999]). These studies show that changes in freshwater flux in the high–latitudinal North Atlantic (NA) can substantially modify the deep ocean circulation. Recent studies show that the NADW is not the only driver of the global conveyor belt (e.g., *Goodman* [1998]; *Stoessel et al.* [1998]; *Goosse and Fichefet* [1999]; *Scott et al.* [1999]; *Wang et al.* [1999]; *Seidov et al.* [2001a]). For example, a recent study by *Seidov et al.* [2001a] demonstrates that a relatively small disturbance in the freshwater flux in the present–day climate asymmetry, e.g. caused by melting of icebergs around Antarctica, could

drastically change the THC. The study presented here is a companion to *Seidov et al.* [2001a] and *Seidov et al.* [2001b] (this volume) in that these contributions provide the circulation characteristics for the sediment transport simulations described here.

The effects of changes in glacial–to–interglacial interbasin water transport on sedimentation are evaluated by examining present–day and glacial ocean circulation experiments influenced by freshwater perturbations in the high latitudes of both hemispheres. The Late Quaternary and, more specifically, a scenario for the last glacial maximum (LGM) (22–18 ka BP) and for idealized subsequent meltwater events (MWE), are chosen as examples. The meltwater event scenarios in the Northern Hemisphere are designed to mimic real meltwater events caused by iceberg flotillas from either the Laurentide Ice Sheet (e.g., *Ruddiman and McIntyre* [1981]) or the Barents Ice Shelf (e.g., *Sarnthein et al.* [1994]). As these events substantially altered the deep–ocean circulation, we anticipate that they led to abrupt and noticeable changes in sedimentation patterns.

MODEL DESCRIPTIONS

A well–known ocean general circulation model (OGCM), the modular ocean model (MOM version 2.2, hereafter MOM) [*Bryan*, 1969; *Cox*, 1984; *Pacanowski et al.*, 1993; *MOM-2*, 1996], is utilized to compute the ocean circulation in order to drive sediment transport. The model domain is global for the ocean with a horizontal grid spacing of $6° \times 4°$ and 12 unequally spaced vertical layers. From top to bottom, the layer thicknesses are 100, 100, 100, 100, 100, 100, 200, 360, 584, 848, 1107, and 1301 m. The bathymetry with this resolution is derived from the *ETOPO5* [1986] data set and represents a smoothed version of the real World Ocean bottom topography (Figure 1). A rather coarse resolution is utilized, following other recent modeling efforts that target long–term climate change (e.g., *Toggweiler et al.* [1989]; *Manabe and Stouffer* [1995]; *Rahmstorf* [1995b]; *Seidov and Haupt* [1999]). A coarse resolution study is also justified by the limited amount of data for boundary conditions and for model verification [*Lessenger and Lerche*, 1999].

There are, however, many drawbacks to coarse resolution. For example, Iceland is represented by a seamount in the model bottom topography. Thus, velocity vectors, water mass transports, and non–zero sedimentation rates are artifacts at the exact position of this island. Even so, as we address scenario–type problems, we believe coarse resolution is sufficient in preliminary experiments. An important issue is whether the resolution of the models is appropriate to resolve 'sediment drifts', or if the results of

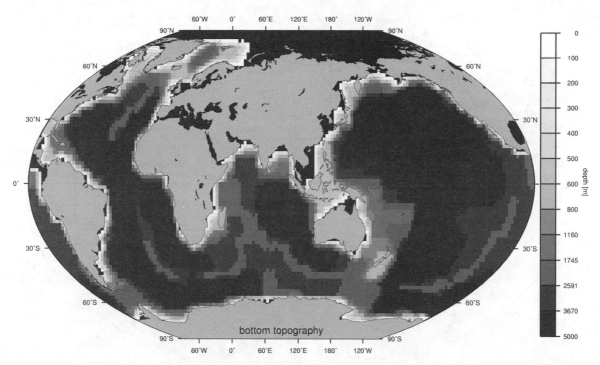

Figure 1. Model bathymetry with a horizontal grid spacing of 6°x4° and 12 unequally spaced vertical layers. The bathymetry is derived from the [*ETOPO5*, 1986] data set and represents a smoothed version of the real World Ocean bottom topography.

the sedimentation models merely indicate regions where they might occur. Clearly, such a coarse resolution does not allow identification of single known sediment drifts and waves, respectively, that are in close proximity. However, in previous studies *Haupt et al.* [1994], *Haupt* [1995], *Haupt et al.* [1995], *Stattegger et al.* [1997], *Haupt and Stattegger* [1999], *Haupt et al.* [1999], and *Seidov and Haupt* [1997b] showed with earlier versions of both OGCM and sediment transport models at much higher horizontal resolutions, 0.5°x0.5° and 2°x2° respectively, that the sediment transport model is capable of reproducing and distinguishing between 'small' drifts in the NA and around Iceland. Therefore, we mainly focus on the impact of glacial–to–interglacial changes in deep–ocean currents on the large–scale sedimentation patterns, especially in the areas of higher sediment accumulation, rather than on identification of single sediment drift bodies.

The off–line large–scale dynamic sediment transport model, SEDLOB, can be utilized to simulate: (1) sediment distribution patterns on the sea floor, especially accumulation and erosion of sediments integrated over time intervals long enough to represent the stratigraphic architecture; and (2) transport paths of water volumes and defined sediment particles from prescribed sources. SEDLOB consists of two components: a 3–D sediment transport model in the ocean interior and a 2–D sediment transport model in a

near–bottom layer following smoothed bottom topography. The 3–D and 2–D sub-models are coupled by vertical exchange of suspended sediment. This exchange is crucial for redistribution of suspended sediment. Sediment suspended in the 2–D bottom layer (e.g., due to erosion) can re–enter the near-bottom 3–D currents. These currents are capable of transporting the suspended sediment to another deposition site (e.g., *Haupt et al.* [1994, 1995, 1999]; *Haupt and Stattegger* [1999]; see also discussion in *Beckmann* [1998]). This approach enables redistribution of sediment by the reduced bottom currents [*Sündermann and Klöcker*, 1983].

SEDLOB's 3–D component is used to simulate the lateral sediment transport and inorganic sediment material entry at the sea surface. This entry can include any kind of inorganic sediment including dust, ice–rafted debris and riverine sediment input. However, to simplify our task at this stage of the research, we include only eolian dust particles. The mass of sediment covering the seafloor depends only on the balance of sources and sinks, whereas the spatial variation of the sediment rate depends on the circulation pattern and particle grain size.

The sediment content of the 2–D bottom layer is updated at every time step by exchanging sediment between the main water body (near bottom water layer of the 3-D sub-model) and the ocean floor (effectively the 2-D bottom

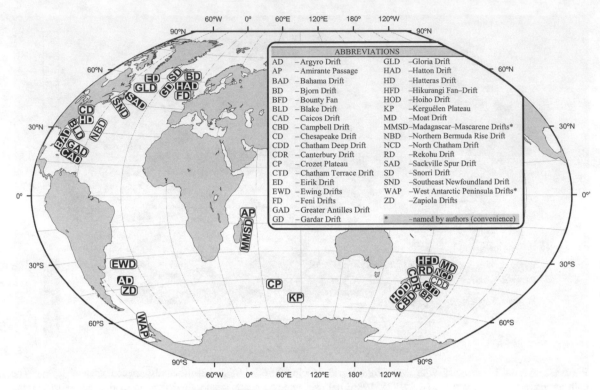

Figure 2. Locations of sediment drifts.

layer). The sediment in the bottom layer is transported by a corrected benthic flow, which is largely a projection of the OGCM velocity field onto the smoothed 1 cm thick bottom layer. Additionally, the near–bottom velocities are reduced as appropriate to take bottom friction into account [*Miller et al.*, 1977; *Sündermann and Klöcker*, 1983].

The equation for sediment concentration in the model is similar (except for the added settling term) to the OGCM equations for the 3-D advection–diffusion of heat and salt. This equation is solved numerically in the same way as in the OGCM, with horizontal and vertical mixing of sediment concentration and an added term describing sediment–settling rate.

Annual mean sea surface conditions are used in all experiments. The OGCM is run to steady state (see details in section 'Input Data and Setup of Numerical Experiments'). Thus, the OGCM provides velocity, temperature, salinity, and convection depths (in the locations where convection occurs due to hydrostatic instability) at steady state, which enter SEDLOB as external parameters. Internal variable parameters that are specified in SEDLOB are the sediment properties. Sediment source properties are characterized by their physical properties, sinking velocity of 0.05 cm s^{-1} = 43.2 m day^{-1} [*Shanks and Trent*, 1980], density (ρ_s = 2.6 g cm^{-3}) and porosity (γ = 0.75) of sediment, grain size and sedimentological grain diameter, and form factor of sediment particles (FF = 0.7) [*Zanke*, 1982]. SEDLOB can be coupled to any 3–D OGCM as long as adequate input fields (velocity, temperature, salinity, and convection depth) are provided (e.g., *Haupt et al.* [1994, 1995]; the most recent and detailed description of this model can be found in *Haupt and Stattegger* [1999] and *Haupt et al.* [1999]).

INPUT DATA AND SETUP OF NUMERICAL EXPERIMENTS

Near–bottom currents over the sediment layer consist of three major components: the water flows controlled by the sea surface conditions (i.e., by wind and heat, and freshwater fluxes across the sea surface), periodical tidal flows, and sporadic benthic flows caused by mesoscale eddies and other short–term processes (e.g., benthic storms). Our numerical experiments are based on mean currents (see model description). Consideration of other components of deep–ocean currents is beyond reasonable present–day computation capabilities. Moreover, the simulation of short–term components in a global ocean model is difficult because many events, such as turbidity currents and benthic storms, are often triggered by short-term atmospheric variability. *Flood and Shor* [1988] argue that if the effects of tidal and benthic storm flows are small, sediment drifts can be modeled using time–averaged deep–ocean currents with short–term variability filtered out. Furthermore, *Flood and Shor* [1988] state that if benthic storms are common in a region, the impact of stationary currents cannot be traced in the sediment patterns because continuous, persistent near–bottom currents are needed to produce strong sediment drifts [*Flood and Shor*, 1988].

Three major glacial–to–interglacial modes of the conveyor are simulated. The output fields are used to model sedimentation patterns based on two different inorganic eolian atmospheric dust sources. The goal is to trace changes of THC in the transported, re–suspended, and re–deposited sediment over time intervals that are long enough to represent the stratigraphic changes in the sediment covering the sea floor. One group of experiments consists of the runs with highly idealized, spatially homogeneous eolian dust sources prescribed at the sea surface. These experiments have the goal of depicting only the circulation imprint in transporting eolian sediment. This homogeneous flux is equal 0.0864 mg m^{-2} day^{-1} [*Miller et al.*, 1977; *Honjo*, 1990]. Sediment deposition rates in these experiments are not real world rates. They are hypothetical sedimentation patterns calculated to explore the potential of the ocean circulation to transport the suspended material over the world ocean. In particular, these sedimentation rates illustrate differences between the near–bottom currents during the three major glacial–to–interglacial modes. Another set of experiments consists of runs with realistic present–day eolian dust patterns [*Rea*, 1994] with the objective of predicting a more realistic sedimentation pattern. Figure 3 shows the present–day eolian dust pattern (after *Rea* [1994]).

In the absence of the ocean currents (i.e., when sediment particles only settle downward and are not transported by the ocean flow), the pattern of sediment deposition rates repeats the pattern of the dust input at the surface. With currents, the spatial structure of deposition is entirely determined by the transportation, erosion, and re–deposition of sediment by the three-dimensional ocean currents. Although the model has the potential to include various biological organic sediment sources, such as dying plankton and fecal pellet production that can act as sediment sources, our first focus is only on the non–biogenic sediment because it behaves as a passive tracer and can be simulated without the complication of biochemical processes. For slow sinking particles like dust the issue is how they are transported from the ocean surface to the sea floor. For example, coccolith distribution at first appears to be puzzling because they project their distribution at the sea surface waters onto the sea floor without reflecting any features of the ocean currents. According to *Honjo* [1996] they take the fast track to the sea floor as fecal pellets or as aggregates instead of sinking as settling particles. Such biologically mediated aggregations of upper–ocean–

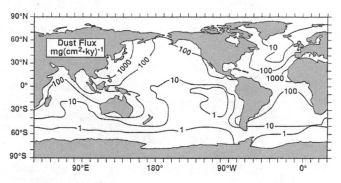

Figure 3. Estimate of the rate of deposition of mineral dust based on consideration of atmospheric transport (reproduced from [*Rea*, 1994], after [*Duce et al.*, 1991]).

generated particles settle much faster than individual fine particles. The bulk speed is up to 200 m d^{-1}. Here, we focus on non–biogenic and non–dissolvable particles that are slow settling and therefore can be transported by ocean currents.

Plate tectonic processes like continental displacement, subsidence, subduction, and sea–floor spreading, as well as sediment compaction are all neglected because they occur on much longer time scales than millennial–scale glacial cycles that we can simulate with reasonable computer resources. The glacial 100 m sea level lowering during the LGM [*Fairbanks*, 1989] including glaciation of shelf areas down to 200 m [*CLIMAP*, 1981; *Lehman*, 1991; *Mienert et al.*, 1992] is also disregarded because its effect on the global thermohaline circulation is minor.

Additionally, we investigate the effect of two different mixing schemes used in the OGCM on the sedimentation patterns as calculated with the sediment transport model SEDLOB (SEDimentation in Large Ocean Basins). Both models are briefly described below. Detailed descriptions of the SEDLOB model can be found in *Haupt et al.* [1999].

Previous numerical experiments using a combination of MOM and SEDLOB address three issues: (1) different numerical schemes used in MOM, (2) different climatic sea surface boundary conditions for different glacial–to–interglacial modes with variations in the sea surface salinity to test the impact of meltwater on the sedimentation pattern, and (3) different eolian sediment fluxes at the sea surface. Only those experiments that are shown in Table 1 (control runs for all three climatic modes) and Table 2 (sensitivity experiments) are discussed in the text. However, the conclusions are based on the large ensemble of runs that are represented in Tables 1 and 2. All OGCM runs are 2000 model years long, with a 5–fold acceleration in the deep layers (which means that the deep ocean is effectively run for 10,000 years). All have reached a complete steady state. For example, the global temperature and

Table 1. Control runs for three different climate scenarios, two different mixing schemes used in the OGCM, and two different eolian dust sources for SEDLOB. Note that experiments 1a and 1b as well as 2a and 2b have identical OGCM runs; they differ in the SEDLOB runs.

Exp.	time slice	OGCM mixing scheme	eolian dust source	
				OGCM – ocean general circulation model
				GMMS – Gent–McWilliams [1990] isopycnal mixing scheme
				VAMS – 'vertical adjustment' mixing scheme
#1a	PD	VAMS	H	LGM – last glacial maximum
#2a	PD	GMMS	H	MWE – meltwater event
#1b	PD	VAMS	PD–ADT	PD – present–day
#2b	PD	GMMS	PD–ADT	H – homogeneous inorganic eolian dust source instead of 'real' present–day distribution
#3	LGM	GMMS	PD–ADT	PD–ADT – present–day atmospheric dust transport (reproduced from *Rea* [1994], after
#4	MWE	GMMS	PD–ADT	*Duce et al.* [1991]).

salinity changes for the last 100 model years are less than 10^{-4} °C and 10^{-5} psu, respectively. The tracer time step length is 86400s (1 day) and the time step length for velocity is 250 s. A more detailed view of the OGCM results is presented in another paper in this volume [*Seidov et al.*, 2001b].

The SEDLOB runs are 1000 years long to compute sedimentation rates for a particular climate state and circulation regime. Unlike the OGCM, it is not possible to run SEDLOB to a steady state condition. Forward time integration leads in every time step to continuous changes in the bottom slope, and therefore the critical velocities for initiating bed and suspension loads change as well, altering both bed and suspension load transports. These changes may influence maximum possible sediment concentrations and transport in the fluid because they are dependent. In fact, it may be viewed as an equivalent of changing sediment availability.

In the OGCM present–day control run, the upper layer of the ocean temperature and salinity are restored to the present–day sea surface climatology (PDC), that is to sea surface temperature (SST) and sea surface salinity (SSS). In the LGM and MWE control runs, the upper ocean tem-

Exp.	NA	SO	
#5	-3	–	NA – North Atlantic
#6	-3	+1	SO – Southern Ocean
#7	–	-1	
#8	–	+1	

Table 2. Present–day sensitivity experiments and their anomalies in psu. Salinity anomalies are added to the present–day annual mean sea surface salinity in the bands between 60°N and 80°N in NA, and a band between 50°S and the coast of the Antarctica (SO). The modified salinity was merged using a cosine filter to the unchanged field within two latitudinal grid points (8°). In the SO the anomalies are circumglobal. All OGCM runs were performed with the Gent–McWilliams [1990] isopycnal mixing scheme.

perature and salinity are restored to the sea surface climatology of the LGM and MWE (see above). Sea surface boundary conditions utilizing prescribed SST and SSS are outlined in detail in Table 1 in *Seidov and Haupt* [1999] (see also *Seidov et al.* [1996]; *Seidov and Haupt*, [1997a]). Additional discussion can be found in *Seidov et al.* [2001b] (this volume). Wind stress fields are from the Hamburg atmospheric circulation model, which was forced with present–day and glacial sea surface climatologies [*Lorenz et al.*, 1996]. For the LGM and MWE the same wind stress is used, although it has been shown that the alteration of wind stress in the Southern Ocean (SO) can by itself cause changes in the Atlantic deep–water circulation (e.g. *Toggweiler and Samuels* [1995]; *Rahmstorf and England* [1997]). We focus on the meltwater control, which provides a distinctive signature of glacial–to interglacial changes. Each of the OGCM control runs for a given climate state was performed twice: one set was done with the classical/conventional 'vertical adjustment' mixing scheme (VAMS) (convective adjustment) that mixes vertical thermohaline instabilities by averaging pairs of layers from top to bottom until all instabilities are removed [*MOM-2*, 1996]; the other set was done with the more sophisticated isopycnal mixing scheme developed by *Gent and McWilliams* [1990] and recently incorporated into MOM–2. Each of the runs with the OGCM found in Table 1 is accompanied by a corresponding SEDLOB run.

Another set of experiments combines sensitivity experiments that model freshwater surface impacts in either the Northern Hemisphere, Southern Hemisphere (SH), or in both hemispheres (Table 2; only the experiments using the Gent–McWilliams mixing scheme (GMMS) are presented). Furthermore, we use realistic present–day eolian dust patterns [*Rea*, 1994] (Figure 3) instead of the idealized homogenous source used in previous runs. Although the atmospheric dust load and its distribution are likely to have varied during glacial cycles [*Ram and Koenig*, 1997], we use the same present–day distribution for all three time slices (present–day, LGM, and MWE). The only justification for such an approach is that we do not have reconstructed dust loads of the same quality for past times as we have for the present–day. In our first–order approach we only address the problem of whether ocean circulation changes can be seen in sedimentation patterns, provided that the dust load remains unchanged.

During the Quaternary, glaciation created an abundant source of eolian material. Deglaciation in regions of strong thermal contrasts and intensive winds created very large areas of unconsolidated glacial deposits [*Lisitzin*, 1996]. In contrast, terrigenous sediments are largely controlled by climatic zonality. As the distribution of aridity shifted with glaciation, the intensity of eolian deposits changed accordingly [*Lisitzin*, 1996; *Allen*, 1997]. Thus, eolian deposition rate is expected to increase during glaciation due to an increase in aridity. The atmosphere of the Northern Hemisphere has a higher concentration of eolian material than the SH today, and this asymmetry apparently also existed in the past. In the past, increases in wind speed had a strong impact on the dust transport and on oceanic sedimentation in the North Pacific (the largest in area), in the equatorial Atlantic and Indian Oceans, and in the regions adjacent to Australia [*Lisitzin*, 1996]. Although the aridity, wind speed and direction, and eolian sediment availability are known qualitatively for the last 20–30 thousand years, there is a great deal of uncertainty about the spatial distribution of eolian dust during LGM and MWE. Given this uncertainty the use of the same eolian sea surface sediment source for all three time slices is considered more useful for our study. However, the fact that these uncertainties could overwhelm the circulation differences in governing sediment patterns also introduces uncertainty in this study.

RESULTS AND DISCUSSION

The results of the experiments in Table 1 are discussed in the following sequence: (1) the results of changes in thermohaline circulation caused by different numerical mixing schemes used in the OGCM, and (2) the response of the thermohaline circulation to different sea surface boundary conditions for different geological time slices. In each case, the OGCM experiments are accompanied by calculation of sediment transport for two sea–surface sediment sources: (i) homogeneous eolian dust (constant everywhere), and (ii) a more realistic present–day eolian dust distribution over the sea surface [*Rea*, 1994].

The Effect of the Mixing Scheme on the Thermohaline Circulation

The key–element of the present–day meridional circulation is the southward North Atlantic Deepwater (NADW)

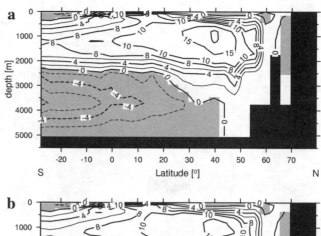

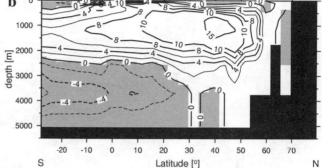

Figure 4. Meridional overturning in the Atlantic Ocean for the present–day experiments (in Sv; 1 Sv = 10^6 m³s⁻¹): (a) Exp. 1a and (b) Exp. 1b. The positive values depict clockwise motion while negative values (shaded) depict counterclockwise motion. The Atlantic overturning is valid only within this ocean's geographical boundary (with meridional walls at both sides; therefore the area south of 30°S is not shown).

outflow from convection sites in the northern North Atlantic (NNA). Figure 4 shows the Atlantic meridional overturning stream function (in Sverdrups (Sv); 1 Sv = 10^6 m³s⁻¹), a representation of the zonally averaged and depth integrated meridional ocean circulation, for the VAMS (Figure 4a) and the GMMS (Figure 4b). Except for the different mixing schemes, all parameters were kept identical in both experiments. Although the overturning looks reasonable in both of the runs, there are important differences that reflect differences in the numerical schemes. The NADW penetrates to the same depths in both runs, but the production rate is 20 Sv in the VAMS run and only 15 Sv in GMMS experiment. Both are within the estimated NADW production rates [*Schmitz*, 1995]. AABW production rates are 6 Sv (GMMS) and 8 Sv (VAMS), which is again in good agreement with climatic averages [*Schmitz*, 1995].

Another way to illustrate the magnitude of the three–dimensional horizontal circulation is to show the water volume transport across vertical diagonal sections of the model's grid cells in Sv. For this purpose we multiplied the

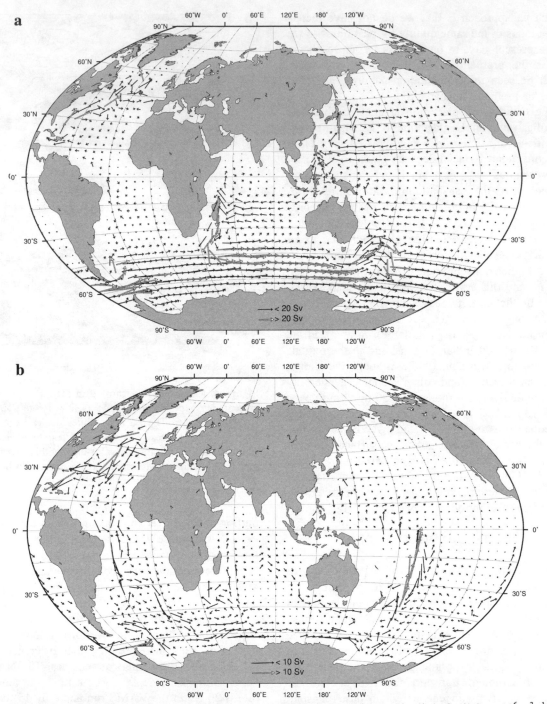

Figure 5. Total water mass transport from Exp. 1a across vertical sections of the grid cells in Sv ($1 \, Sv = 10^6 \, m^3 s^{-1}$): a) upper (above 1.5 km) ocean transport; b) deep (below 1.5 km) ocean transport. Note different scales of vectors for upper and deep ocean.

horizontal velocity components by the area of the side surfaces of the grid cells, and summed vertically from the sea surface to 1.5 km (upper ocean) and from 1.5 km to the bottom (deep ocean). Figure 5 depicts the present–day conveyor mode (Exp.1a) with a strong deep ocean flow of NADW that penetrates the Antarctic Circumpolar Current (ACC). Then, carried by the ACC in the deep ocean, NADW spreads into the Indian and Pacific oceans.

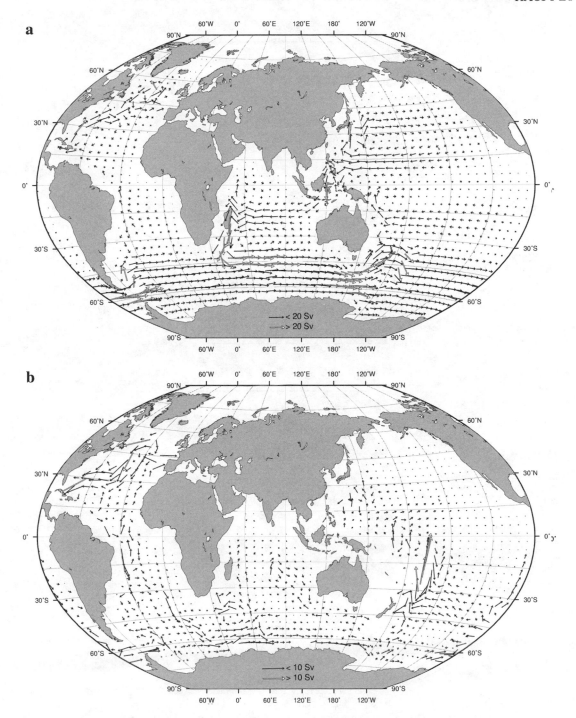

Figure 6. As Figure 5 for Exp. 2a.

Horizontal water mass transport in the upper and deep ocean layers is shown for Exp. 2a in Figure 6. The three–dimensional currents are far more complex than the vertically integrated flows shown in Figures 5 and 6. For example, the deep inflow of the Antarctic Bottom Water (AABW) into the central and NA is masked in these fig-ures. However, integrated volume transports do give a good schematic view of the global thermohaline conveyor.

Deep–ocean currents show stronger sensitivity to millennial–scale climate changes than the surface currents (e.g. *Seidov and Haupt* [1997a]). Moreover in case of re–suspension, deep–ocean flow can effectively influence the

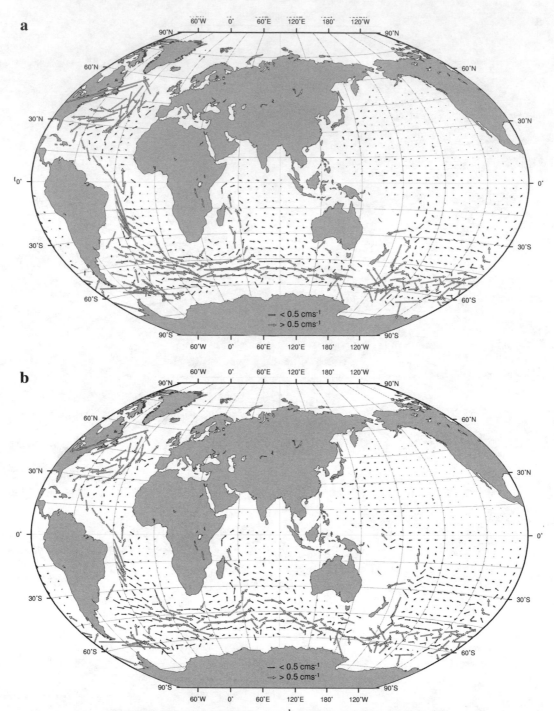

Figure 7. Velocity vectors (in cm s^{-1}) at 2.5 km: a) Exp. 1a; Exp. 2a.

sediment drifts. Figure 7 depicts the present–day circulation at 2.5 km depth for the VAMS (Figure 7a) and GMMS (Figure 7b). It appears that deep ocean currents are not very different in the VAMS and GMMS runs. Figures 5–7 show weaker western boundary currents and a weaker Antarctic Circumpolar Current in the GMMS experiment.

The water mass transport in Sv across selected meridional and zonal vertical sections for different depths in different oceans is depicted in Figure 8a (Exp. 1a) and 8b (Exp. 2a) illustrating water volume exchanges between different parts of the world ocean. In some cases, volume transports in three, rather than in two layers are shown to differentiate

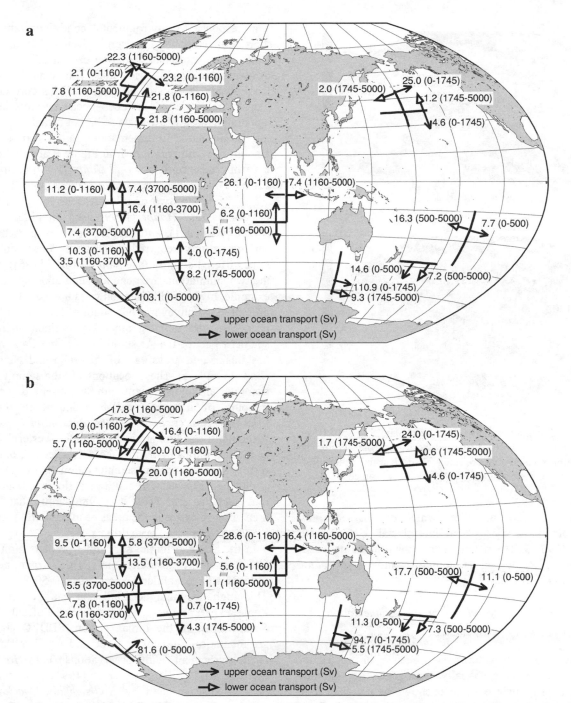

Figure 8. The water mass transport in Sv (1 Sverdrup (Sv) = 10⁶ m³s⁻¹) across chosen meridional and zonal vertical sections for different depths in different oceans: a) Exp. 1a and b) Exp. 2a. Primarily, the transports in the upper and deep ocean are shown; in cases when upper and intermediate water movement essentially differs, the transports in three layers are shown to differentiate between the upper, intermediate–to–deep, and deep–to abyssal flows.

between the flows in the upper, intermediate–to–deep, and deep–to–abyssal flows.

As *Seidov and Haupt* [1999] used VAMS, the results of Exp. 1a shown in Figure 4a, 5, 6a, and 7a match their re-

sults better than those obtained with GMMS. However, there are substantial differences between the two numerical schemes used in the OGCM. Because GMMS is accepted as the most advanced mixing scheme [*Pacanowski*, 1996],

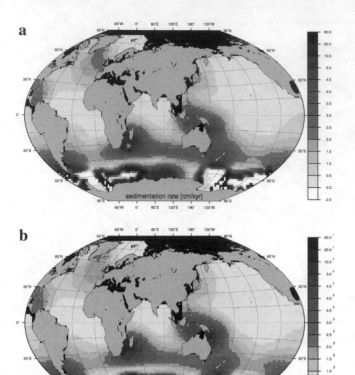

Figure 9. Sedimentation rates from Exp. 1a and (b) Exp. 2a. The color bar gives the sedimentation rates in cm/1000 years. The sedimentation rates shown are not realistic because of an idealized character of the sea surface eolian sediment source.

all of the sections that follow which discuss sensitivity experiments targeting glacial–interglacial changes are based on GMMS. The new results, shown in the next sections, can be considered as an improvement over *Seidov and Haupt* [1999].

The Effect of the Mixing Scheme on the Sedimentation Pattern

The sedimentation patterns have been reconstructed for the global ocean using spatially homogeneous eolian dust supply at the sea surface. Plate 2a shows the results of Exp. 1a whereas Plate 2b shows the computed sediment distribution of Exp. 2a. Using the homogeneous sediment source, the sedimentation rates are not realistic, but they do act as a passive tracer of the circulation. Even though observed sediment distributions cannot be used for a direct model–observation comparison, there are a number of known sediment drifts that reflect development times of tens to hundreds of thousands of years. These drifts, invariably formed in fine–grained sediments, reflect a long–

term response to environmental conditions rather than a short–term response to discrete events [*Flood and Shor*, 1988]. Their presence has often been used as an indicator of the environment associated with sediment deposited from steady, sediment–laden flows. This correlation may help to trace the role of deep–ocean currents in creating and maintaining large–scale sediment drifts.

The simulated sedimentation patterns (not the total amount of sediment) shown in Figure 9 are slightly different from those in *Seidov and Haupt* [1999] (compare with Plate 1a in *Seidov and Haupt* [1999]; note the different scale). There are two possible reasons for the differences: (a) *Seidov and Haupt* [1999] used an older version of MOM (version 1), and (b) the tuning of present–day sea surface salinity required to produce "observed" NADW rates differed slightly (see *Seidov et al.* [1996]). However, the two simulations are generally in a good agreement. For example, higher sedimentation rates are associated with the present–day conveyor mode south of Iceland, in the Caribbean Sea, and in most other key regions, including the southern part of the Argentine Basin, the Cape and Agulhas Basin, and areas east of Australia and New Zealand (Exp. 1a and 2a). These locations of high sediment deposition rates are in agreement with the present–day concept of trapping of inorganic eolian sediment in these areas [*McCave and Tucholke*, 1986; *Bohrmann et al.*, 1990]. Although the sedimentation rates in the western NA are not as high as indicated by the geological record, our modeled results show increased sedimentation in the western NA along the deep western southward flowing boundary current. *McCave and Tucholke* [1986] argue that high sedimentation rates are in the areas of maximum kinetic energy in the western boundary currents.

In the South Atlantic, our model predicts high sedimentation rates in the southern part of the Argentine Basin in the area of the Falkland Escarpment (~50°S, ~55–40°W). This high sediment accumulation can be linked to the Zapiola Drift produced by the Falkland Current [*Flood and Shor*, 1988]. In the Indian Ocean, SEDLOB gives high sedimentation rates in the Cape and Agulhas Basins. This is in agreement with observations (e.g. *Hollister and McCave* [1984]; *Faugeres et al.* [1993]). The map of suspended material load by *Hollister and McCave* [1984] corresponds to the high kinetic energy of surface currents and the spread of cold deep–ocean water flowing northward. SEDLOB also gives high sedimentation rates southeast of New Zealand. These drifts may be under control of the East Australian Current.

Model sediment thicknesses maxima (higher values than in the surrounding basins) match the geologic record in the Yellow Sea and East China Sea [*Nittrouer and Wright*, 1994]. In the real world, the sediment accumulation is, in general, much higher than in the model. If added, the riverine inputs, e.g., from the Huanghe (Yellow River), running into the Gulf of Bohai could be substantial regional source [*Nittrouer and Wright*, 1994]. However, for large–scale open–ocean patterns this source is not essential to tracing the deep ocean currents (see above).

The model's low Pacific sedimentation rate in the central North Pacific is also consistent with measurements [*Rea et al.*, 1985] (see also e.g., *Lisitzin* [1996]). Sedimentation rates derived from their data range from a minimum of 0.02 cm/ka (30°N) to a maximum of about 1.4 cm/ka (40°N).

The model sediment deposition rates around Antarctica are too high. This may be caused by overestimated sediment supply at the sea surface in this region, as represented by our idealized homogeneous boundary conditions. *Rea* [1994] indicates that Antarctica does not substantially contribute eolian dust to the world ocean (see also e.g., *Lisitzin* [1996]) (compare Figure 3). There are mainly two reasons for this: (1) the wind around Antarctica blows predominantly parallel to or even towards the shoreline, (2) Antarctica is covered by ice and delivers very little dust, and (3) part of the SO is covered by sea ice shielding the ocean from eolian source.

An overall comparison of the two experiments, Exp. 1a and 2a, shows that the sedimentation rates in Exp. 2a, especially near the western boundary currents, are slightly smaller than in Exp. 1a. As we use the same initial sea surface boundary conditions in both OGCM runs, this discrepancy represents the difference between the GMMS and the VAMS. However, except for smaller–scale details, these two numerical mixing schemes do not produce significantly different large–scale deep–sea sediment distributions. The experiments which follow are based only on the results obtained using MOM 2 with the more advanced GMMS mixing scheme.

The Effect of Climate Variability on Deep–Sea Sedimentation

In this section, we focus on the linkage between changes in the sedimentation rates in key areas and changes in the water transports during glacial–to–interglacial cycles. Plate 1 shows sedimentation rates in Exp. 2b, Exp. 3, and Exp. 4. In contrast with the idealized simulations described in the previous section, a realistic present–day surface dust distribution from *Rea* [1994] (Figure 3) is employed. Thus, we now focus on changes in deep–sea sediment deposition caused by changes in deep–sea circulation and inter–basin water exchange as a result of climatic variability that can be more directly compared to present–day observations.

Present–day sediment distribution. The simulated present–day sediment distribution is shown in Plate 1a. The

sedimentation pattern in Plate 1a differs from that in Figure 9b. There are many regions where the differences are substantial. For example, the sediment accumulation is higher in the Sea of Japan, the Coral Sea northeast of Australia, the North American Basin, in the Caribbean Sea, under the Gulf Stream System [*Hollister and McCave*, 1984], the southern part of the Argentine Basin (approximately 50°S), the Falkland Escarpment, and in particular, in the Zapiola Drift. The high sedimentation rates in the Zapiola Drift are consistent with observations (e.g., *Flood and Shor* [1988]). Also, we find good agreement between modeled and observed (e.g., *Johnson and Damuth* [1979]) patterns east and northeast of Madagascar and in the Mozambique Channel. *Johnson and Damuth* [1979] state that these drifts were produced by the deep western boundary current in the Indian Ocean (IDWBC), which flows northward along the western margins of the Madagascar Basin through the Mascarene Basin and then through the Amirante Passage into the Somali Basin [*Kolla et al.*, 1980]. The thickness of the drift deposits in the passage suggests that the IDWBC has been active for several to tens of million years. Therefore, the model should predict high sedimentation rates in the IDWBC region for all three time slices in these simulations (compare Plate 1b, c) [*Johnson and Damuth*, 1979; *Kolla et al.*, 1980].

In the eastern part of the Argentine Basin the model predicts a high accumulation area seaward from the edge of the continental shelf. There are other separate sediment drifts in this area, for example the Argyro Drifts and the Ewing Drifts [*Faugeres et al.*, 1993]. Although our coarse horizontal grid resolution does not distinguish between different drift bodies, it does predict that major sediment drifts should be found within the Argentine Basin. This result is in good agreement with the modeled and observed undirected flow pattern in this basin, a basin where major sediment drifts are found mostly out outside the area of benthic storms [*Flood and Shor*, 1988]. *Flood and Shor* [1988] mention that the resolution of their 3.5 kHz profiles is not sufficient to determine Holocene sedimentation rates because the post–glacial sediment is not thick enough. Therefore, it is difficult to compare the modeled sedimentation rates with observations.

Smaller areas of accumulation that also match observations (e.g., *Rebesco et al.* [1994]) were found along the continental rise west of the Antarctic Peninsula between 63 and 69°S. In this region, a group of eight mud waves has been detected [*Rebesco et al.*, 1994]. *Rebesco et al.* [1994] identified the northern four mounds as sediment drifts in the area where the currents weaken and turn northeastward. Again, in coarse resolution simulations one cannot resolve small drift bodies, yet the pattern of large–scale sedimentation in the area is considered an acceptable match.

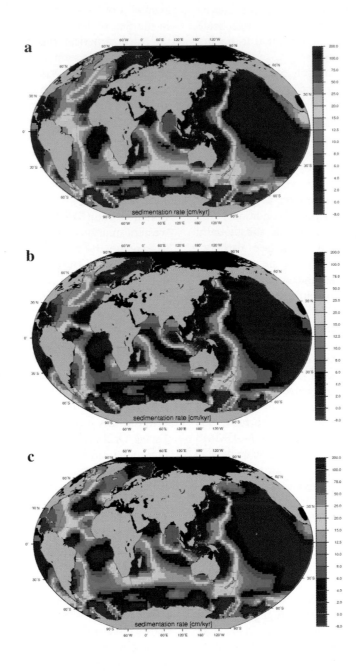

Plate 1. Sedimentation rates from Exp. 2b (a), Exp. 3 (b), and Exp. 4 (c). The color bar gives the sedimentation rates in cm/1000 years. All three climate scenarios were forced with the present–day sea surface eolian sediment distribution [*Rea*, 1994]. The goal is to link the changes of sediment drifts to changes of the ocean circulation patterns and inter–basin water exchange (see text).

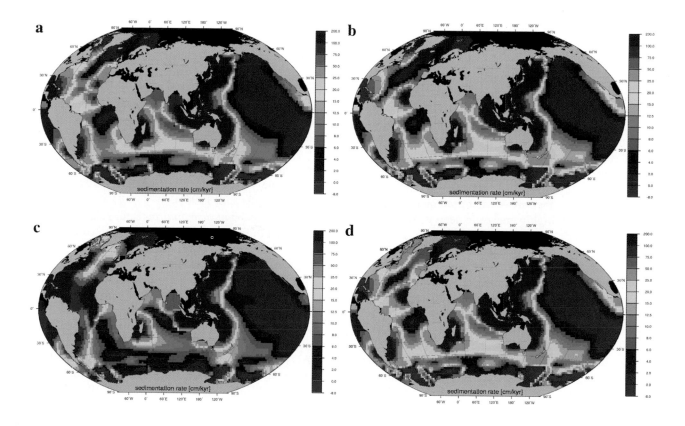

Plate 2. Sedimentation rates from Exp. 5 (a), Exp. 6 (b) Exp. 7 (a), and Exp. 8 (d). The color bar gives the scale of the thickness (in cm) of sediment accumulated during 1000 years, or sedimentation rates in cm/1000 years. All three climate scenarios were forced with the present–day sea surface eolian sediment distribution [*Rea*, 1994]. The goal is to link the changes of sediment drifts to changes of the ocean circulation patterns and inter–basin water exchange (see text).

Despite substantial differences in surface dust sources – between the idealized homogeneous, and observed present–day fluxes –, many high accumulation areas (e.g., the Hikurangi Fan–drift, Moat Drift, North Chathman Drift, Chathman Deep Drift, Chathman Terrace Drift, Canterbury Drift, Bounty Fan, Hoiho Drift, and Campbell Drift) east, southeast, and south of New Zealand [*Carter and Mitchell*, 1987] have similar shape (compare results from Exp. 1b and 2a) albeit higher rates. This result reflects the fact that the deep–ocean currents are not very different. For example, forty percent of the Earth's cold deep water enters the Southwest Pacific as the world largest deep western boundary current (SWP–DWBC), the so–called western intensification, flowing northward at bathyal to abyssal depths, east of New Zealand. South of 50°S, the SWP–DWBC is intimately linked with the ACC [*Carter et al.*, 1996]. *Carter et al.* [1996] argue that the eastern New Zealand oceanic sedimentary system is on a margin where sediment supply was steady for at least 10 My (compare sedimentation pattern in this region for LGM (Plate 1b) and MWE (Plate 1c)).

There are several other areas where sediment accumulation can be linked to highly energetic deep western boundary currents (DWBC) [*Hollister and McCave*, 1984]. For example, the high sedimentation area in the Tasman Sea is beneath the East Australian Current and, over large areas, is beneath the Antarctic Circumpolar Current – southeast of the Crozet Plateau near the Kerguelen Plateau.

Exp. 2b shows a large sediment buildup in the subtropical North Atlantic in comparison to Exp. 2a, since eolian material is more abundant in the air of the Atlantic Ocean of the northern hemisphere than anywhere else [*Lisitzin*, 1996] (Figure 3). This may be caused by increased concentration of eolian dust off the Sahara [*Lisitzin*, 1996; *Allen*, 1997]. Satellite images show that during much of the year several hundred million tons of African dust is transported primarily toward the Caribbean [*Prospero*, 1996; *Prospero et al.*, 1996; *Shinn et al.*, 2000]. With seasonally shifting winds and ocean currents, the dust is also deposited in the Gulf of Mexico and equatorial regions of South America [*Shinn et al.*, 2000]. There are other large regions influenced by the higher concentration of dust in the atmosphere of the Northern Hemisphere in comparison with the Southern Hemisphere. For example, the equatorial Indian Ocean and the western part of the North Pacific receive large amounts of dust from Asia [*Lisitzin*, 1996; *Allen*, 1997]. In the Southern Hemisphere the continental shelf and regions of the deep ocean adjacent to Australia receive their eolian sediment as the dust coming from Australian deserts [*Lisitzin*, 1996; *Allen*, 1997]. Our model results reflect these features in the runs with the realistic dust

supply at the surface. Yet, ocean circulation imposes significant impact on the fate of eolian sediment.

For instance, before applying the realistic eolian dust source to SEDLOB, we expected high accumulation everywhere in the deep sea where the dust concentration is high at the surface. Surprisingly, this is not always true in our simulations. Although the atmospheric dust concentration is high in the Arabian Sea, the model predicts relatively low sediment–settling rates for this region when compared to the adjoining basins due to the role of currents.

A lack of correlation between the surface supply and bottom accumulation is also found in the Bay of Bengal and several other regions. In the Arabian Sea, for example, simulated sedimentation rates are, by an order of magnitude higher due to a higher eolian dust supply. However, the model sedimentation rates in the northern part of the Indian Ocean are much lower than observed. The reason for this discrepancy is the lack of strong riverine sediment input in our simulations (riverine sediment discharge is responsible at least for approximately 90% of sediment deposition [*Lisitzin*, 1996]).

Glacial sediment distribution. Both paleoreconstructions and ocean modeling indicate that the global ocean thermohaline conveyor worked differently in the past than it works now (e.g., *Boyle and Keigwin* [1987]; *Broecker* [1991]; *Crowley and Kim* [1992]; *Maier-Reimer et al.* [1993]; *Weaver et al.* [1993]; *Fichefet et al.* [1994]; *Rahmstorf* [1994, 1995a]; *Manabe and Stouffer* [1997]; *Ganopolski et al.* [1998]; *Seidov and Haupt* [1999]). Our simulations are in good agreement with these publications. NADW production was much weaker during the LGM than today. Figure 10 depicts the LGM conveyor mode (Exp. 3) with its weakened flow of NADW penetrating the ACC (Figure 11a). Figure 12a shows glacial meridional overturning in the Atlantic Ocean with a maximum of approximately 8 Sv, about 30% less than today. Northward heat transport in the Atlantic is weaker because of the reduced meridional overturning (Figure 13). The present–day North Atlantic current brings about 6 Sv into the Nordic Seas (Exp. 2b). During the LGM there is almost no inflow (0.2 Sv). Here we show that the differences in the circulation regimes are traceable in the pattern of sedimentation (see also *Seidov and Haupt* [1999]). Plate 1b displays sediment patterns in a LGM simulation (Table 1, Exp. 3). Reduction of the North Atlantic current carrying suspended sediment northward leads to lower sedimentation rates around Iceland (Plate 1b). Along with this reduction, a concurrent predicted increase in sedimentation rates in the eastern NA is strongly supported by reconstructions indicating stronger glacial sedimentation along the eastern

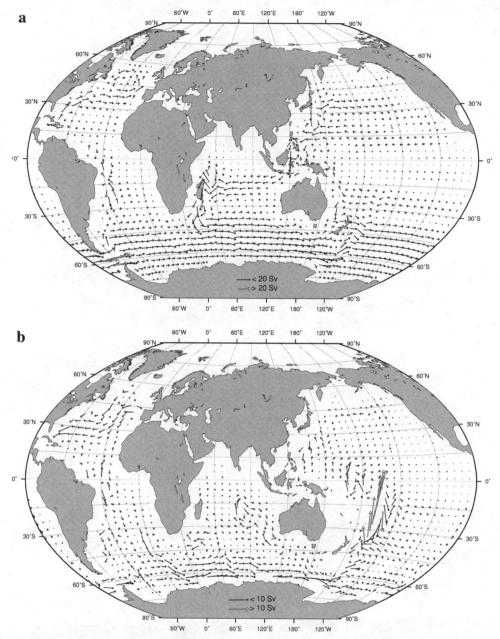

Figure 10. Total water mass transport for the LGM (Exp. 3) across vertical sections of the grid cells in Sv (1 Sverdrup (Sv) = 10^6 m^{3}s^{-1}): a) upper (above 1.5 km) ocean transport;   b) deep (below 1.5 km) ocean transport. Note different scales of vectors for upper and deep ocean.

flank of the Mid–Atlantic Ridge (e.g., *Dowling and McCave* [1993]; *Robinson and McCave* [1994]). High glacial sedimentation rates are found west of the North African coast. Dust from the Sahara and Sahel is found in sediment thousands of kilometers away, as far away as in the Caribbean Basin [*Allen*, 1997]. As we use present–day rather than genuine glacial dust patterns (Saharan dust input in the Atlantic is thought to have been twice as high

as today [*Kolla et al.*, 1980], during glacial times), the model tends to underestimate sediment accumulation in these areas.

In the LGM experiment, the sedimentation rates in the Nordic Seas around Iceland are also lower than observed. Although the winds during glacial times were not generally stronger [*Rea*, 1994], continental dust concentrations in Greenland ice older than 10750 years indicate an increased

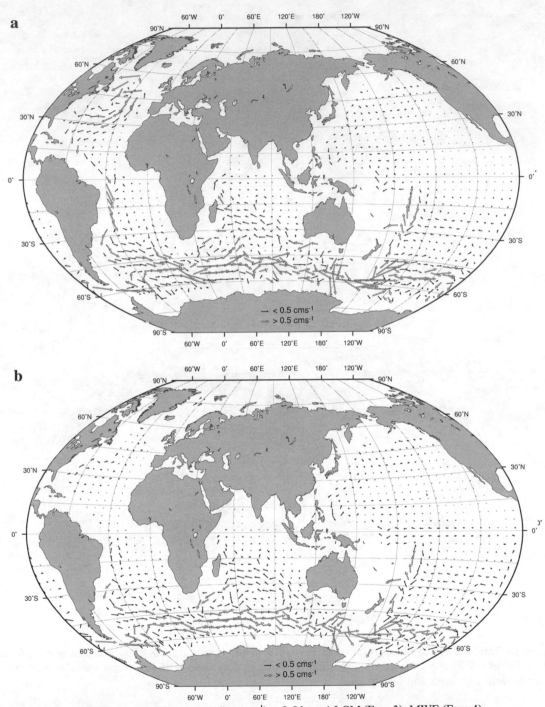

Figure 11. Velocity vectors (in cm s⁻¹) at 2.5 km: a) LGM (Exp. 3); MWE (Exp. 4).

eolian transport in the Arctic [*Pfirmann et al.*, 1989].

Western boundary currents off the American east coast are about 20 to 25% weaker in the LGM experiment than in present–day simulation. As a result, sedimentation rates differ by the same ratio in the western South Atlantic (Plate 1a, b).

The reduced sediment build–up along the Java Trench for the LGM simulation (in comparison to the PDC simulation) is the result of a 10–20% reduction of water transport from the Pacific Ocean via the Indonesian through flow into the Indian Ocean, as well as to a weakening of the South Indian Current [*Seidov and Haupt*, 1999].

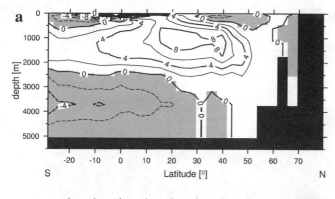

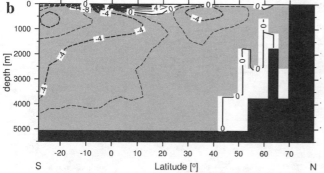

Figure 12. Meridional overturning in the Atlantic Ocean (in Sv; 1 Sverdrup (Sv) = 10^6 m³s⁻¹): (a) LGM (Exp. 3) and (b) MWE (Exp. 4). The positive values depict clockwise motion while negative values (shaded) depict counterclockwise motion. The Atlantic overturning is valid only within this ocean's geographical boundary (with meridional walls at both sides; therefore the area south of 30°S is not shown).

Hollister and McCave [1984] describe the relation of deep-sea sediment accumulation to deep ocean storm intensity. Deep ocean storms are proportional to the intensity of the large-scale currents and mesoscale eddies. The change of the ACC strength in the past, as compared to today, gives a good example of the impact of a change in the intensity of the deep ocean circulation on sediment accumulation. The ACC is about 25% weaker than in the PDC experiment in our LGM run. The sediment accumulation over large areas beneath the ACC – south of the Crozet Plateau, east of the Kerguelen Plateau, in the Weddell Sea, on the southwestern side of the New Zealand Platcau, and in the Tasman Sea – is reduced by approximately the same percentage.

Interglacial sediment distribution. Northern meltwater events (Table 1) are easily traceable in sedimentation patterns. The freshwater impact of Heinrich events in the Nordic Seas and NNA includes the collapse of the conveyor (Figure 12b), leading to a totally different circulation in the NA in comparison to the PDC circulation regime.

This applies to the upper as well to the deep ocean circulation, as shown in Figure 11b and Figure 14. Figure 12b implies that the thermohaline conveyor collapsed completely during the meltwater event (Figure 13). As a result, the cross–equatorial heat transport changed its sign. Almost the same amount of heat that was transported northward in the PDC experiment is now allocated to the SH. The dominance of NADW is replaced by dominance of AABW, and the southbound deep-ocean western boundary currents in the Atlantic Ocean were much weaker. Simulated sedimentation rates in the NA are now determined more closely by the eolian dust source, rather than by ocean currents (Plate 1c). In the absence of the southbound DWBC, the velocity vectors show a slow reversed northward water motion (Figure 11b and 14b). The collapse of the conveyor leads to a higher sedimentation rate around Iceland since sediment is not removed from southward flowing water masses (Plate 1c).

Meltwater sedimentation patterns around Antarctica look similar to those in the PDC and LGM runs. However, meltwater patterns show a further reduction of the LGM sedimentation rates, which, in turn, are already lower than the present–day rates. The sedimentation rate ratio MWE/PDC is close to the ratio of the water flows through the Drake Passage in these two runs. The same reduction ratio is found between the sedimentation rates and water transports across a vertical section between Australia and Antarctica. Sedimentation patterns in the Indian and Pacific Oceans, and in the seas around Australia and around New Zealand, are similar to those from our LGM simulation since the conveyor/circulation in those areas is much less affected by the collapse of the conveyor in the Atlantic Ocean (compare Figure 15a and b).

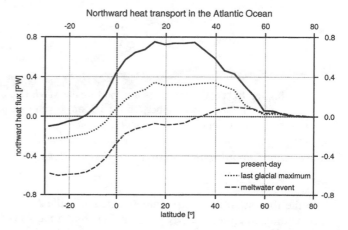

Figure 13. Northward heat transport in the Atlantic Ocean in present–day run (Exp. 2b), LGM run (Exp. 3), and MWE run (Exp. 4).

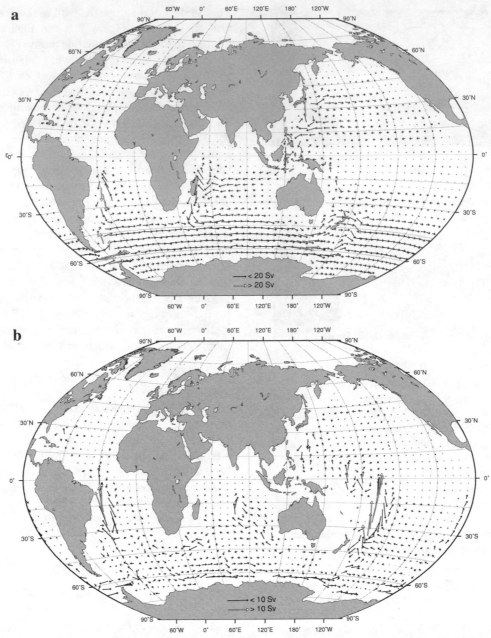

Figure 14. Total water mass transport for the MWE (Exp. 4) across vertical sections of the grid cells in Sv (1 Sverdrup (Sv) = 10^6 m³s⁻¹): a) upper (above 1.5 km) ocean transport; b) deep (below 1.5 km) ocean transport. Note different scales of vectors for upper and deep ocean.

Sensitivity of global sedimentation to high–latitudinal surface density variations

In the four sensitivity experiments listed in Table 2, surface salinity in either the NNA, in the SO around the Antarctica, or in both locations was altered to simulate a decrease in sea surface density (Table 2; Figure 16). A more detailed description of the OGCM experiments is given in

Seidov et al. [2001a] and *Seidov et al.* [2001b] (this volume). Note that salinity variation is just a convenient means of changing sea surface density in an OGCM. A 1 psu change of salinity would require about 5°C change of SST [*Pond and Pickard*, 1986]. Yet, an even smaller high–latitudinal warming may cause cryosphere melting, with a resulting freshwater input that would cause a far greater dilution of surface water. Therefore, we selected surface

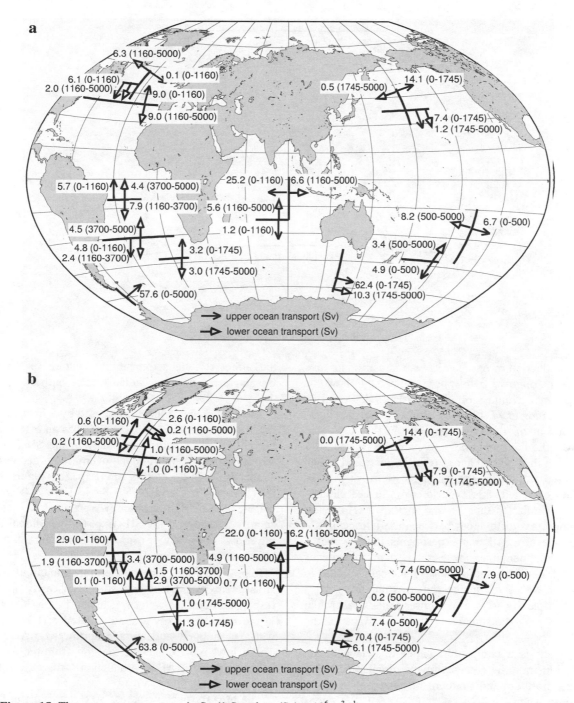

Figure 15. The water mass transport in Sv (1 Sverdrup (Sv) = 10^6 m³s⁻¹) across chosen meridional and zonal vertical sections for different depths in different oceans: a) LGM (Exp. 3) and b) MWE (Exp. 4). Primarily, the transports in the upper and deep ocean are shown; in cases when upper and intermediate water movement essentially differs, the transports in three layers are shown to differentiate between the upper, intermediate–to–deep, and deep–to abyssal flows.

salinity as a far stronger and more realistic influence on surface water density.

The major implication of the sensitivity experiments is that a northern meltwater event results in a decreased NADW production and therefore an increased inflow of AABW into the NA, whereas during a southern meltwater event the NADW production is increased, as the AABW production weakens. In other words, an increase in the

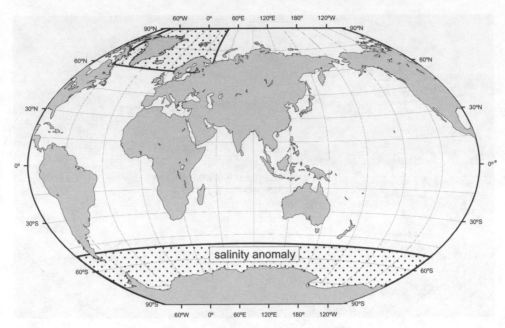

Figure 16. Idealized present–day meltwater events.

deepwater production in one hemisphere is equivalent to a reduction in the opposite hemisphere. A sketch of how the changed conveyor would look in the Atlantic is shown in Figure 3 in *Seidov et al.* [2001b] (this volume). To illustrate the concept of a seesaw character in the deepwater controls over the thermohaline conveyor, we reversed the signs of the salinity disturbances (that is, increased rather then decreased salinity in the same locations) in the SH (Exp. 8). As expected, this experiment produced circulation changes similar to those found in Exp. 5. However, there are important differences between the impacts in the SO and in the NA. All our experiments indicate that a smaller salinity change in the SO has a much stronger effect on the circulation than a stronger salinity change applied to the NA. Even in the run in which the salinity disturbance in the SO (-1 psu) was three times weaker than in the NA (-3 psu), the southern change overwhelmed the northern one.

As this work is a companion paper to Seidov et al., 2001b, we illustrate the changes of the circulation during the idealized northern and southern meltwater episodes by showing only water transports through specified sections (Figure 17). Meridional overturning for the Atlantic and global ocean can be found in *Seidov et al.* [2001b] (their Figures 5 and 6; this volume). All experiments (Table 2) demonstrate that the North Pacific is practically unaffected by salinity changes in either one or both hemispheres. This can be seen in both the OGCM calculations (Figure 17) and in sedimentation patterns from SEDLOB (Plate 2). In contrast, slowdown or speedup of the THC has a substan-

tial effect on the sedimentation pattern in the Atlantic Ocean, in the western Indian Ocean, and around Antarctica.

The slowdown of the THC caused by a freshening of the surface water in the NNA (Exp. 5) has an effect comparable to a positive salinity anomaly in the SH (Exp. 8). In Exp. 5, the NA meridional overturning decreases from almost 20 Sv in the control run (Exp. 2b) (Figure 8b) to 6.6 Sv (Figure 17a), and to 5.9 Sv in Exp. 8 (Figure 17d). During a northern meltwater event intensified by a positive salinity anomaly in the SH (Exp. 6), meridional overturning shows only 2 Sv of deepwater production in the central North Atlantic (Figure 17b).

Exp. 7 illustrates that the thermohaline circulation in the Atlantic in its present–day mode can easily speed up in response to surface water freshening in the SO. Reduction in the surface density around Antarctica suppresses deepwater production and curtails northward AABW inflow. This allows more deepwater outflow from the Atlantic and hence stronger NADW production (~30 Sv) with a very strong DWBC (Figure 17c).

The NADW outflow across 30°S, transport in the Drake Passage, the East–Australian DWBC, cross-equatorial heat transport and maximum northward heat transport in the PDC for the four experiments from Table 2 are shown in Table 3. All overturning related parameters for the NA are proportional to the NADW production, whereas total transport through the Drake Passage and the global cross–equatorial heat transport are inversely proportional to the NADW production.

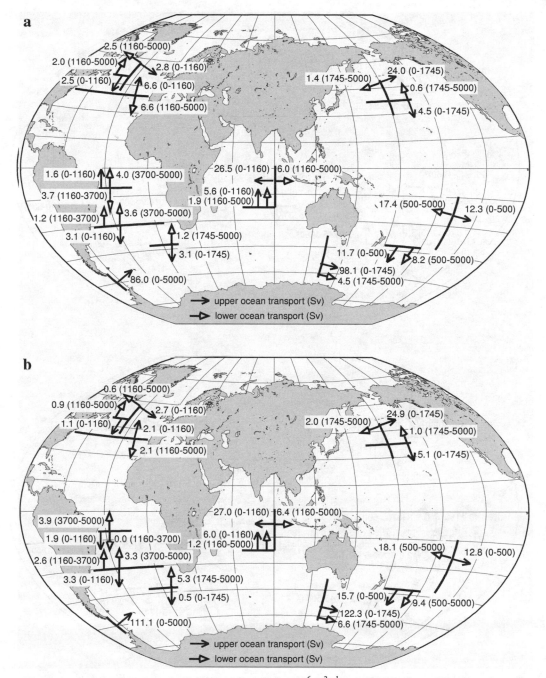

Figure 17. The water mass transport in Sv (1 Sverdrup (Sv) = 10^6 m^{3}s^{-1}) across chosen meridional and zonal vertical sections for different depths in different oceans: a) Exp. 5, b) Exp. 6, c) Exp. 7, and d) Exp. 8. Primarily, the transports in the upper and deep ocean are shown; in cases when upper and intermediate water movement essentially differs, the transports in three layers are shown to differentiate between the upper, intermediate–to–deep, and deep–to abyssal flows.

Sedimentation rate maps for these experiments (Plate 2b, 2a, 2d, 1a, and 2c; same order as the previously listed experiments) reveal that the strength of the DWBC in the western North Atlantic, especially in the Caribbean Sea, is the key governing factor for sedimentation rates under the DWBC and in adjacent areas. For other regions in the world ocean, it is not obvious that the circulation controls the accumulation. For instance, even substantial changes in

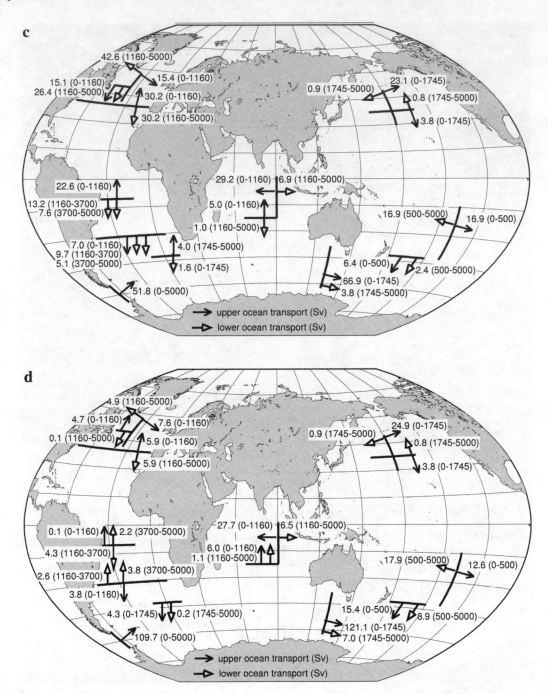

the THC do not noticeably affect the sedimentation pattern in the Pacific Ocean.

SUMMARY AND CONCLUSIONS

As ocean currents at least partially control sediment drifts, these drifts may be indicative of major changes in circulation patterns. We have performed numerical experiments designed to evaluate this idea. In these experi-

ments, we pursued three goals: (1) The first group of the experiments is designed to identify the effect of different ocean model numerical mixing schemes on sedimentation patterns and sedimentation rates. To achieve this goal, we used two different surface sources of eolian dust, a highly idealized and a more realistic present–day global distribution. (2) The second set of experiments simulates the impact of different sea surface boundary conditions for different glacial–to–interglacial modes. This goal has been

Table 3. Results of present–day sensitivity experiments with different amplitudes of sea surface salinity anomalies in psu. Salinity anomalies are added to the present–day annual mean sea surface salinity in the bands between 60°N and 80°N in NA, in the lens at about 50°N in NNA, and band between 50°S and the coast of the Antarctica (SO). The modified salinity was merged using a cosine filter to the unchanged field within two latitudinal grid points (8°). In the SO the anomalies are circumglobal. All OGCM runs were performed with the Gent–McWilliams [1990] isopycnal mixing scheme.

Exp.	NA	SO	NADW production north of 30°S (Sv)	maximal northward heat transport (PW)	NA cross–equatorial heat transport (PW)	global cross–equatorial heat transport (PW)	water transport through Drake Passage (Sv)	
#6	-3	+1	2.1	0.25	-0.1	-0.88	111.1	NA – North Atlantic
#8	–	+1	5.8	0.35	0.1	-0.74	109.7	SO – Southern Ocean
#5	-3	–	6.5	0.4	0.15	-0.66	86.0	1 Sv = 10^6 m³/s
#2b	–	–	20.0	0.75	0.4	-0.38	81.6	1 PW = 10^{15} W
7#	–	-1	30.2	1.0	0.75	0.01	51.8	

achieved through applying sea surface conditions from paleo–climatic reconstructions using the more realistic present–day eolian dust load in the lower atmosphere. (3) The third set of experiments is designed to test sensitivity of the global thermohaline conveyor and deep–sea sediment accumulation to changes of the sea–surface density in high latitudes, introduced as idealized meltwater events in the Northern and Southern Hemispheres. These runs assess the key issue of large–scale climatic changes and whether they can be unambiguously traced in coupled OGCM–sediment transport models. Thus, the results may tell us whether the sedimentologic record can be used to evaluate the impact of global warming through comparison with past meltwater and glacial cycles.

The link between millennial–scale changes in ocean currents and sedimentation is evident in our experiments and may enable model-data comparison, especially if sea surface sediment sources are better known. As a first step in this direction, we have modeled the LGM and a subsequent meltwater event characterized by substantial changes in the thermohaline ocean conveyor. The simulations suggest stronger dependence of the sediment drift patterns on the deep-ocean circulation regime rather than on currents in the upper and mid–depth ocean. Importantly, these experiments indicate that the geologic record of eolian dust accumulation can be used to trace and verify modeled circulation patterns, provided that the deep–ocean current system was substantially distinctive from other circulation modes (for example, LGM mode is different from the present–day mode, and MWE is strikingly different from both these two modes).

However, there are many limitations to our first–order, coarse resolution, modeling study. The coarseness of our resolution limits the results in many ways. All modeled currents except for the ACC are weaker than the observed currents, so our ability to model realistic sedimentation rates is limited as they are strongly linked to the strength of sea floor currents. However, we are able to reconstruct

schematic patterns of changes in sediment accumulation in the regions known for high sediment deposition. Despite significant circulation changes, the results obtained from the sedimentation model for all three time slices, present–day, LGM, and MWE, show low sedimentation rates for the central Pacific Ocean [Rea et al., 1985; Lisitzin, 1996]. The consistency of this pattern may enable the modeled paleocirculation to be verified against geologic record. This result could dramatically increase our confidence in modeling of past ocean climates.

For this first–order study, we were unable to compile global sea surface sources of eolian dust for the past, neither for the LGM nor the MWE. This must be viewed as a major task for future studies.

Importantly, several changes in the ocean model, including a slight imbalance in the northern–southern hemispheric asymmetry driven by warm ocean surface waters in the NNA, a moderately cool northern North Pacific, and much colder SO, appear to be traceable in the sediment transport model. The decrease in sea surface density in the SO and in the NNA lead to a "seesaw" effect, as envisioned by Broecker [1998, 2000] and Stocker [1998]. This seesaw effect was an issue in the recent numerical simulations done by Seidov and Maslin [2001] and Seidov et al. [2001a] that confirmed a flip–flop operation of the global conveyor. The seesaw effect implies that changes in the ocean conveyor mirror freshwater impacts, i.e., a response of the deep–ocean currents to an increase in the deep–water production in one hemisphere is equivalent to a decrease in the opposite hemisphere. In other words, simulations suggest that a decreased deep–water production caused by a de–densification in one hemisphere can be achieved by a densification of the high–latitudinal sea surface water in the opposite hemisphere. These changes of the thermohaline circulation appeared to be easily traceable in the alteration of sediment drifts, justifying the attempts to use sediment as a tracer of change of the ocean circulation through ages.

APPENDIX

Abbreviations and symbols	
AABW	=Antarctic Bottom Water
ACC	=Antarctic Circumpolar Current
DWBC	=deep western boundary current
FF	=form factor
GMMS	=Gent–McWilliams mixing scheme
IDWBC	=Indian Ocean deep western boundary current
LGM	=last glacial maximum
NA	=North Atlantic
NADW	=North Atlantic Deep Water
NNA	=northern North Atlantic
MOM	=modular ocean model
MWE	=meltwater event
OGCM	=ocean global circulation model
PDC	=present–day sea surface climatology
SEDLOB	=SEDimentation in Large Ocean Basins (sediment transport model)
SH	=Southern Hemisphere
SO	=Southern Ocean
SSS	=sea surface salinity
SST	=sea surface temperature
Sv	=Sverdrup (1 Sv = 10^6 m^{3}s^{-1})
SWP–DWBC	=Southwest Pacific deep western boundary current
THC	=thermohaline circulation
VAMS	=‘vertical adjustment’ mixing scheme
γ	=porosity of sediment
ρ_s	=density of sediment

Acknowledgments. We thank the reviewers Jan Harff, Bill Hay, and John Tipper for their critical remarks that essentially improved the manuscript. This study is partly supported by National Science Foundation (grant #9975107 and ATM 00-00545).

REFERENCES

Allen, P.A., 1997, Earth surface processes. Blackwell Science, London, 404 pp.

Beckmann, A., 1998: The representation of bottom boundary layer processes in numerical ocean circulation models, In: E.P. Chassignet and J. Verron (Editors), *Ocean Modeling and Parameterization.* Kluwer, Boston, pp. 135-154.

Bitzer, K. and R. Pflug, 1990: DEPOD: A three-dimensional model for simulating clastic sedimentation and isostatic compensation in sedimentary basin, In: T.A. Cross (Editor), *Quantitative Dynamics Stratigraphy.* Prentice Hall, N.Y., pp. 335-348.

Bohrmann, G., R. Henrich and J. Thiede, 1990: Miocene to Quaternary paleoceanography in the northern North Atlantic: Variability in carbonate and biogenic opal accumulation, In: U. Bleil and J. Thiede (Editors), *Geological History of the Polar Oceans: Arctic versus Antarctic.* Kluwer Academic Publications, Norwell, Massachusetts, pp. 647-675.

Boyle, E.A. and L.D. Keigwin, 1987: North Atlantic thermohaline circulation during the past 20,000 years linked to high-latitude surface temperature, *Nature,* 330: 35-40.

Broecker, W., 1991: The great ocean conveyor, *Oceanography,* 1: 79-89.

Broecker, W.S., 1998: Paleocean circulation during the last deglaciation: A bipolar seesaw?, *Paleoceanography,* 13: 119-121.

Broecker, W.S., 2000: Was a change in thermohaline circulation responsible for the Little Ice Age?, *Proc. Nat. Acad. Sci.,* 97(4): 1339-1342.

Broecker, W.S. and G.H. Denton, 1989: The role of ocean atmosphere reorganizations in glacial cycles, *Geochimica Cosmochimica Acta,* 53: 2465-2501.

Bryan, K., 1969: A numerical method for the study of the circulation of the world ocean, *Journal of Computational Physics,* 4: 347-376.

Cao, S. and I. Lerche, 1994: A quantitative model of dynamical sediment deposition and erosion in three dimensions, *Computers & Geosciences,* 20: 635-663.

Carter, L. and J.S. Mitchell, 1987: Late Quaternary sediment pathways through the deep ocean, east of New Zealand, *Paleoceanography,* 2(4): 409.

Carter, R.M., L. Carter and I.N. McCave, 1996: Current controlled sediment deposition from the shelf to the deep ocean: The Cenozoic evolution of circulation through the SW Pacific gateway, *Geologische Rundschau,* 85: 438-451.

CLIMAP, 1981: Climate: Long-Range Investigation, Mapping, and Prediction (CLIMAP) Project Members, Seasonal reconstructions of the Earth's surface at the Last Glacial Maximum. *Map and Chart Ser. MC-36.* Geological Society of America, Boulder, Colorado, pp. 1-18.

Cox, M., 1984: A primitive equation, 3-dimensional model of the ocean, 1, Technical Report No. 1, 250 pp. Ocean Group, Geophys. Fluid Dyn. Lab., Princeton Univ., Princeton University, New Jersey.

Crowley, T. and K.-Y. Kim, 1992: Complementary roles of orbital insolation and north Atlantic deep water during late Pleistocene interglacials, *Paleoceanography,* 7: 521-528.

Dowling, L.M. and I.N. McCave, 1993: Sedimentation on the Feni drift and late glacial bottom water production in the northern Rockall Trough, *Sedimentary Geology,* 82: 79-87.

Ericksen, M.C., D.S. Masson, R. Slingerland and D. Swetland, 1989: Numerical simulation of circulation and sediment transport in the late Devonian Catskill Sea, In: T.A. Cross (Editor), *Quantitative Dynamic Stratigraphy.* Prentice Hall, Englewood Cliffs, pp. 295-305.

ETOPO5, 1986: Digital Relief of the Surface of the Earth,, National Geophysical Data Center, Boulder, Colorado.

Fairbanks, R.G., 1989: A 17,000-year glacio-eustatic sea level record: Influence of glacial melting rates on the Younger Dryas event and deep-ocean circulation, *Nature,* 342: 637-642.

Faugeres, J.C., M.L. Mezerais and D.A.V. Stow, 1993: Contourite drift types and their distribution in the North and South Atlantic Ocean basins, *Sedimentary Geology,* 82: 189-203.

Fichefet, T., S. Hovine and J.-C. Duplessy, 1994: A model study of the Atlantic thermohaline circulation during the Last Glacial Maximum, *Nature,* 372: 252-255.

Flood, R.D. and A.H. Shor, 1988: Mud waves in the Argentine Basin and their relationship to bottom circulation pattern, Part A, *Deep Sea Research*, 35: 943-971.

Ganopolski, A., S. Rahmstorf, V. Petoukhov and M. Claussen, 1998: Simulation of modern and glacial climates with a coupled global model of intermediate complexity, *Nature*, 391: 351-356.

Gent, P.R. and J.C. McWilliams, 1990: Isopycnal mixing in ocean circulation models, *Journal of Physical Oceanography*, 20: 150-155.

Goodman, P.J., 1998: The role of North Atlantic Deep Water formation in an OGCM's ventilation and thermohaline circulation, *J. Phys. Oceanogr.*, 28: 1759-1785.

Goosse, H. and T. Fichefet, 1999: Importance of ice-ocean interactions for the global ocean circulation: A model study, *Journal of Geophysical Research*, 104(C10): 23337-23355.

Granjeon, D. and P. Joseph, 1999: Concepts and applications of a 3D multiple lithology, diffusive model in stratigraphic modeling, In: J. Harbaugh, W.L. Watney, E.C. Rankey, R. Slingerland, R.H. Goldstein and E.K. Franseen (Editors), *Numerical Experiments in Stratigraphy: Recent Advances in Stratigraphy/Sedimentologic Computer Simulations. Special publication Series #62.* SEPM (Society for Sedimentary Geology)/Special Publication Series #62, Kansas, pp. 197-210.

Haupt, B.J., 1995: Numerische Modellierung der Sedimentation im nördlichen Nordatlantik, 54, Sonderforschungsbereich 313, Universität Kiel, Kiel.

Haupt, B.J., C. Schäfer-Neth and K. Stattegger, 1994: Modeling sediment drifts: A coupled oceanic circulation-sedimentation model of the northern North Atlantic, *Paleoceanography*, 9(6): 897-916.

Haupt, B.J., C. Schäfer-Neth and K. Stattegger, 1995: Three-dimensional numerical modeling of late Quaternary paleoceanography and sedimentation in the northern North Atlantic, *Geologische Rundschau*, 84: 137-150.

Haupt, B.J. and K. Stattegger, 1999: The ocean-sediment system and stratigraphic modeling in large basins, In: J. Harbaugh, W.L. Watney, E.C. Rankey, R. Slingerland, R.H. Goldstein and E.K. Franseen (Editors), *Numerical Experiments in Stratigraphy: Recent Advances in Stratigraphy/Sedimentologic Computer Simulations. Special publication Series #62.* SEPM (Society for Sedimentary Geology)/Special publication Series #62, Kansas, pp. 313-321.

Haupt, B.J., K. Stattegger and D. Seidov, 1999: SEDLOB and PATLOB: Two numerical tools for modeling climatically-forced sediment and water volume transport in large ocean basins, In: J. Harff, J. Lemke and K. Stattegger (Editors), *Computerized Modeling of Sedimentary Systems.* Springer, New York, pp. 115-147.

Hollister, C.D. and I.N. McCave, 1984: Sedimentation under deep-sea storms, *Nature*, 309: 220-225.

Honjo, S., 1990: Particle fluxes and modern sedimentation in the Polar Oceans, In: W.O. Smith (Editor), *Polar Oceanography, Part B.* Acad. Press, Boston, Massachusetts, pp. 687-739.

Honjo, S., 1996: Fluxes of particles to the interior of the open oceans, In: V. Ittekkot, P. Schäfer, S. Honjo and P.J. Depetris (Editors), *Particle Flux in the Ocean. SCOPE Report 57.* Chichester, Chichester, pp. 91-154.

Johnson, D.A. and J.E. Damuth, 1979: Deep thermohaline flow and current-controlled sedimentation in the Amirante Passage: Western Indian Ocean, *Marine Geology*, 33: 1-44.

Kolla, V., S. Eittreim, L. Sullivan, J.A. Kostecki and L.H. Burckle, 1980: Current-controlled, abyssal microtopography and sedimentation in Mozambique Basin, southwest Indian Ocean, *Marine Geology*, 34: 171-206.

Lehman, S.J.e.a., 1991: Initiation of Fennoscandina ice-sheet retreat during the last glaciation, *Nature*, 349: 513-516.

Lessenger, M. and I. Lerche, 1999: Inverse modeling, In: J. Harbaugh, W.L. Watney, E.C. Rankey, R. Slingerland, R.H. Goldstein and E.K. Franseen (Editors), *Numerical Experiments in Stratigraphy: Recent Advances in Stratigraphy/Sedimentologic Computer Simulations. Special publication Series #62.* SEPM (Society for Sedimentary Geology)/Special publication Series #62, Kansas, pp. 29-31.

Lisitzin, A.P., 1996, Oceanic sedimentation: lithology and geochemistry. American Geophysical Union, Washington, D.C., 400 pp.

Lorenz, S., B. Grieger, H. P. and K. Herterich, 1996: Investigating the sensivity of the atmospheric general circulation Model ECHAM 3 to paleoclimate boundary conditions., *Geol. Rundsch.*, 85: 513-524.

Maier-Reimer, E., U. Mikolajewicz and K. Hasselmann, 1991: On the sensitivity of the global ocean circulation to changes in the surface heat flux forcing, *Max-Plank-Institut für Meteorologie, Hamburg, Report No. 68*: 67.

Maier-Reimer, E., U. Mikolajewicz and K. Hasselmann, 1993: Mean circulation of the Hamburg LSG OGCM and its sensitivity to the thermohaline surface forcing, *Journal of Physical Oceanography*, 23: 731-757.

Manabe, S. and R. Stouffer, 1997: Coupled ocean-atmosphere model response to freshwater input: Comparison to Younger Dryas event, *Paleoceanography*, 12: 321-336.

Manabe, S. and R.J. Stouffer, 1988: Two stable equilibria of a coupled ocean-atmosphere model, *Journal of Climate*, 1: 841-866.

Manabe, S. and R.J. Stouffer, 1995: Simulation of abrupt change induced by freshwater input to the North Atlantic Ocean, *Nature*, 378: 165-167.

McCave, I.N. and B.E. Tucholke, 1986: Deep current controlled sedimentation in the western North Atlantic, In: P.R. Vogt and B.E. Tucholke (Editors), *The Geology of North America.* The Geological Society of America, Boulder, Colorado, pp. 451-468.

Mienert, J., J.T. Andrews and J.D. Milliman, 1992: The East Greenland continental margin (65°N) since the last deglaciation: Changes in seafloor properties and ocean circulation, *Marine Geology*, 106: 217-238.

Miller, M.C., I.N. McCave and P.D. Komar, 1977: Threshold of sedimentation under unidirectional currents, *Sedimentology*, 24: 507-528.

MOM-2, 1996: Documentation, User's Guide and Reference Manual (edited by R. C. Pacanowski), GFDL Ocean Technical

Report No. 3.2, Geophysical Fluid Dynamics Laboratory/NOAA, Princeton, N.J.

Nittrouer, C.A. and L.D. Wright, 1994: Transport of particles across continental shelves, *Review of Geophysics*, 32(1): 85-113.

Pacanowski, R., K. Dixon and A. Rosati, 1993: The GFDL Modular Ocean Users Guide, Geophys. Fluid Dyn. Lab., Princeton Univ., Princeton, N. J.

Pacanowski, R.C., 1996: MOM 2. Documentation, User's Guide and Reference Manual, GFDL Ocean Technical Report #3.2 edited by R. C. Pacanowski, Geophysical Fluid Dynamics Laboratory/NOAA, Princeton, N.J.

Paola, C., P.L. Heller and C.L. Angevine, 1992: The large scale dynamics of grain–size variation in alluvial basins, 1: Theory, *Basin Research*, 4: 73–90.

Pfirmann, S., I. Wollenburg, J. Thiede and M.A. Lange, 1989: Lithogenic sediment on Arctic pack ice: Potential aeolian flux and contribution to deep sea sediments, In: M. Leinen and M. Sarnthein (Editors), *Paleoclimatology and Paleometeorology: Modern and Past Patterns of Global Atmospheric Transport*. Kluwer Academic Publishers, New York, pp. 463-493.

Pond, S. and G.L. Pickard, 1986, Introductory Dynamical Oceanography. Pergamon Press, Oxford, UK, 329 pp.

Prospero, J.M., 1996: The atmospheric transport of particles to the ocean, In: V. Ittekkot, P. Schäfer, S. Honjo and P.J. Depetris (Editors), *Particle Flux in the Ocean. SCOPE Report 57*. John Wiley & Sons, Chichester, pp. 19-52.

Prospero, J.M., K. Barrett, T. Church, F. Dentener, R.A. Duce, H. Galloway, I.I. Lewy, J. Moody and P. Quinn, 1996: Atmospheric deposition of nutrients to the North Atlantic Basin, *Biogeochemistry*, 35: 27-37.

Rahmstorf, S., 1994: Rapid climate transitions in a coupled ocean-atmosphere model, *Nature*, 372: 82-85.

Rahmstorf, S., 1995a: Bifurcations of the Atlantic thermohaline circulation in response to changes in the hydrological cycle, *Nature*, 378: 145-149.

Rahmstorf, S., 1995b: Climate drift in an ocean model coupled to a simple, perfectly matched atmosphere, *Climate Dynamics*, 11: 447-458.

Rahmstorf, S. and M.H. England, 1997: Influence of Southern Hemisphere winds on North Atlantic Deep Water flow, *Journal of Physical Oceanography*, 27: 2040-2054.

Ram, M. and G. Koenig, 1997: Continous dust concentration profile of pre-Holocene ice from the Greenland Ice Sheet Project 2 ice core: Dust stadials, interstadials, and the Emian, *Journal of Geophysical Research*, 102(C12): 26641-26648.

Rea, D.K., 1994: The paloclimatic record provided by eolian deposition in the deep sea: The geologic history of wind, *Review of Geophysics*, 32(1): 159-195.

Rea, D.K., M. Leinen and T.R. Janecek, 1985: Geologic approach to the long-term history of atmospheric circulation, *Science*, 227: 721-725.

Rebesco, M., R.D. Larter, P.F. Barker, A. Camerlenghi and L.E. Vanneste, 1994: The history of sedimentation on the continental rise west of the Antarctic Peninsula, *Terra Antarctica*, 1(2): 277-279.

Robinson, S.G. and I.N. McCave, 1994: Orbital forcing of bottom-current enhanced sedimentation on Feni Drift, NE Atlantic, during the mid-Pleistocene, *Paleoceanography*, 9(6): 943-972.

Ruddiman, W.F. and A. McIntyre, 1981: The mode and mechanism of the last deglaciation: Oceanic evidence., *Quaternary Research*, 16: 125-134.

Sarnthein, M., K. Winn, S.J.A. Jung, J.C. Duplessy, L. Labeyrie, H. Erlenkeuser and G. Ganssen, 1994: Changes in east Atlantic deepwater circulation over the last 30,000 years: Eight Time Slice Reconstructions, *Paleoceanography*, 9: 209-267.

Schmitz, W.J., Jr., 1995: On the interbasin-scale thermohaline circulation, *Reviews of Geophysics*, 33: 151-173.

Scott, J.R., J. Marotzke and P.E. Stone, 1999: Interhemispheric thermohaline circulation in a coupled box model, *Journal of Physical Oceanography*, 29: 351-365.

Seidov, D., E.J. Barron and B.J. Haupt, 2001a: Meltwater and the global ocean conveyor: Northern versus southern connections, *Global and Planetary Change*, in press.

Seidov, D. and B.J. Haupt, 1997a: Global ocean thermohaline conveyor at present and in the late Quaternary, *Geophysical Research Letters*, 24: 2817-2820.

Seidov, D. and B.J. Haupt, 1997b: Simulated ocean circulation and sediment transport in the North Atlantic during the Last Glacial Maximum and today, *Paleoceanography*, 12: 281-306.

Seidov, D. and B.J. Haupt, 1999: Last glacial and meltwater interbasin water exchanges and sedimentation in the world ocean, *Paleoceanography*, 14: 760-769.

Seidov, D., B.J. Haupt, E.J. Barron and M. Maslin, 2001b: Ocean bi-polar seesaw and climate: Southern versus northern meltwater impacts, *Geophysical Monograph (This Volume)*, D. Seidov, B.J. Haupt and M. Maslin (Editors), American Geophysical Union, Washington, D.C.

Seidov, D. and M. Maslin, 1999: North Atlantic Deep Water circulation collapse during the Heinrich events, *Geology*, 27: 23-26.

Seidov, D. and M. Maslin, 2001: Atlantic Ocean heat piracy and the bi-polar climate sea-saw during Heinrich and Dansgaard-Oeschger events, *Journal of Quaternary Science*, 16: in press.

Seidov, D., M. Sarnthein, K. Stattegger, R. Prien and M. Weinelt, 1996: North Atlantic ocean circulation during the Last Glacial Maximum and subsequent meltwater event: A numerical model, *Journal of Geophysical Research*, 101: 16305-16332.

Shanks, A.L. and J.D. Trent, 1980: Marine snow: Sinking rates and potential role in vertical flux, *Deep Sea Research*, 27, Part A: 137-143.

Shinn, E.A., G.W. Smith and J.M. Prospero, 2000: African dust and the demise of Caribbean coral reefs, *Geophysical Research Letters*, 27(19): 3029-3032.

Slingerland, R., J.W. Harbaugh and K.P. Furlong, 1994, Simulating Clastic Sedimentary Basins. PTR Prentice Hall, Englewood Cliffs, 220 pp.

Stattegger, K., B.J. Haupt, C. Schäfer-Neth and D. Seidov, 1997: Numerische Modellierung des Ozean–Sediment-Systems in großen Meeresbecken: Das Spätquartär im nördlichen Nordatlantik, *Geowissenschaften*, 15(1): 10-15.

Stocker, T.F., 1998: The seesaw effect, *Science*, 282: 61-62.

Stoessel, A., S.-J. Kim and S.S. Drijfhout, 1998: The impact of Southern Ocean sea ice in global ocean model, *J. Phys. Oceanogr.*, 28: 1999-2018.

Stommel, H., 1958: The abyssal circulation, *Deep-Sea Research*, 5: 80-82.

Sündermann, J. and R. Klöcker, 1983: Sediment transport modeling with applications to the North Sea, In: J. Sünderman and W. Lenz (Editors), *North Sea Dynamics*. Springer-Verlag, New York, pp. 453-471.

Syvitski, J.P.M. and S. Daughney, 1992: Delta2: Delta progradation and basin filling, *Computer and Geosciences*, 18(7): 839-897.

Tetzlaff, D.N. and J.W. Harbaugh, 1989, Simulating Clastic Sedimentation. Van Nostrand Reinhold, New York, 202 pp.

Toggweiler, J.R., K. Dixon and K. Bryan, 1989: Simulations of radiocarbon in a coarse-resolution world ocean circulation model, 1, Steady state prebomb distribution, *Journal of Geophysical Research*, 94: 8217-8242.

Toggweiler, J.R. and B. Samuels, 1995: Effect of Drake Passage on the global thermohaline circulation, *Deep Sea Research*, 42: 477-500.

Wang, X., P. Stone and J. Marotzke, 1999: Global thermohaline circulation. Part I. Sensitivity to atmospheric moisture transport, *J. Climate*, 12: 71-82.

Weaver, A.J., J. Marotzke, P.F. Cummins and E.S. Sarachik, 1993: Stability and variability of the thermohaline circulation, *Journal of Physical Oceanography*, 23: 39-60.

Zanke, U., 1982, Grundlagen der Sedimentbewegung. Springer–Verlag, Berlin, Heidelberg, New York, 402 pp.

E.J. Barron, EMS Environment Institute, Pennsylvania State University, University Park, PA 16802. (eric@essc.psu.edu)

B.J. Haupt, EMS Environment Institute, Pennsylvania State University, University Park, PA 16802. (bjhaupt@essc.psu.edu)

D.Seidov, EMS Environment Institute, Pennsylvania State University, University Park, PA 16802. (dseidov@essc.psu.edu)

On the Response of the Atlantic Ocean to Climatic Changes in High Latitudes: Sensitivity Studies With a Sigma Coordinate Ocean Model

Tal Ezer

Program in Atmospheric and Oceanic Sciences, Princeton University, Princeton, New Jersey

The oceanic adjustment processes during the transition from present climate conditions to atmospheric forcing conditions that are significantly different than present climate have been simulated with a sigma coordinate Atlantic Ocean model. An idealized "global warming" scenario has been simulated by imposing surface flux anomalies, representing warming and freshening in the high latitudes of the North Atlantic. The results show a relatively short oceanic adjustment process that takes place within a period of a few decades in which the thermohaline overturning circulation (THC) is adapted to a new state with a transport smaller by 4-5 Sv than that obtained with a control run without the surface anomalies. While the total change in the intensity of the THC in this ocean model is consistent with that obtained by coarse resolution coupled ocean-atmosphere climate models, this study shows a shorter adjustment time scale and more pronounced spatial changes than climate models do. Interesting results include a weakening and cooling of the Gulf Stream, reversal of the Labrador Sea circulation and considerable weakening in the deep western boundary current and in downslope near-bottom flows in high latitudes. Sensitivity experiments explore how the parameterization of horizontal diffusion in this sigma coordinate ocean model affects long-term climate simulations. These experiments show that horizontal diffusion in the model affects the transition process and local gyres, but the climate change in the THC and in the meridional heat flux are quite robust and insensitive to the way horizontal diffusion is parameterized in the model.

1. INTRODUCTION

The world oceans, and in particular the Atlantic Ocean where most deep-water formation occurs, play an important role in long-term climatic changes. For example, changes in freshwater fluxes associated with changes in ice coverage

The Oceans and Rapid Climate Change: Past, Present, and Future
Geophysical Monograph 126
Copyright 2001 by the American Geophysical Union

influence the thermohaline ocean circulation (THC), meridional heat transport and eventually global overturning circulation patterns through changes in the so-called "ocean conveyor belt" [*Broecker*, 1991]. Some numerical models suggest that the THC may even have multiple stable states [*Manabe and Stouffer*, 1995, 1997; *Marotzke and Willebrand*, 1991; *Delworth et al.*, 1993] associated with different climates. Paleoclimate records such as the Greenland ice cores indicate several abrupt changes in the North Atlantic climate, such as those occurred during the last cold Younger Dryas period, about 11,000 years BP. The last glacial maximum (LGM), about 18,000 years BP, was also the subject of considerable research by observational and model studies [*Imbrie et al.*, 1992; *Prange et al.*, 1997; *Seidov and Haupt*, 1997, 1999]. Radiocarbon analyses suggest that during the LGM the THC was considerably weaker than today, with weaker and shallower North Atlantic Deep-Water (NADW) transport, but possibly larger Antarctic Bottom Water (AABW) transport.

While past climatic changes are documented in paleoclimate records, possible future climatic changes, such as those associated with increasing greenhouse gas concentrations, are based mostly on the prediction of coupled ocean-atmosphere climate models [e.g., *Manabe et al.*, 1991; *Manabe and Stouffer*, 1988, 1997; *Haywood et al.*, 1997]. Some of these climate models, when integrated for hundreds of years into the future, assuming different scenarios for greenhouse gas concentrations, indicate the possibility that the THC may collapse due to increasing surface temperatures and freshwater fluxes in high latitudes. In such a case, ocean circulation and consequently the world climate may be very different than current climate, and may resemble, in some aspects, past climates with weaker THC. In order to run global climate models for hundreds or even thousands of years, some simplifications must be done. For example, these models have low horizontal resolution (in the range of 3-4 degrees), they are highly diffusive, and they do not accurately represent many oceanic processes such as vertical mixing and the interaction between flow and bottom topography. Western boundary currents are only poorly resolved in most climate models. On the other hand, more realistic, high-resolution ocean models require too-much computations to allow long integrations of hundreds of years. However, if the ocean is adjusted to climatic changes within a period of time that is relatively short compared with the long time scale of the climate system, then a relatively short simulations with realistic models may be useful in order to study the adjustment process. This approach has been done before [e.g., *Seidov and Haupt*, 1997; 1999], but it has not been tried yet with a model of the typed used here. *Gerdes and Koberle* [1995] and *Döscher et al.* [1994] have shown that the anomalous temperature signal in high latitudes propagates by coastal waves along the western boundary and affects the Gulf Stream transport and the thermohaline circulation of the North Atlantic within a time scale of 10-20 years.

Since future climate prediction depends to large extent on model configurations and parameterizations, it is important to try variety of models with different levels of complexities. Most of the earliest climate ocean models were based on different versions of the z-level Bryan-Cox model [*Bryan* 1969] and its successor, the Modular Ocean Model,

MOM, however, more recently isopycnal models become an attractive alternative for climate simulations [*Oberhuber*, 1993; *New and Bleck*, 1995; *Halliwell*, 1998]. A different class of models, such as the Princeton Ocean Model, POM [*Blumberg and Mellor*, 1987], originally developed for coastal and regional applications, are now being used for climate studies as well [*Ezer and Mellor*, 1994, 1997; *Häkkinen*, 1995, 1999; *Ezer et al.*, 1995; *Ezer*, 1999, 2001]. Sigma coordinate ocean models may be attractive for simulating processes involved in bottom boundary layers (BBL) and deep water formation along slopes; these processes are sometimes difficult to simulate with coarse resolution z-level models [*Gerdes*, 1993; *Beckmann and Döscher*, 1997; *Winton et al.*, 1998; *Pacanowski and Gnanadesikan*, 1998]. A recent comparison study between z-level and sigma coordinate ocean models using otherwise identical numerics, shows that sigma coordinate models are able to handle much lower horizontal diffusivities than z-level models do [*Mellor et al.*, 2001]. This model attribute may be important for simulations of interdecadal variabilities, which can be sensitive to the horizontal diffusivity of ocean models [*Huck et al.*, 1999]. On the other hand, a disadvantage of sigma coordinate models is that bottom topography smoothing and other remedies are needed in order to reduce pressure gradient and along-sigma diffusion errors near steep topographies [*Mellor et al.* 1994, 1998; *Ezer and Mellor*, 2000]. Of particular concern for climate studies using POM is the use of along-sigma diffusion. A common procedure to minimize diapycnal mixing along sloping bottoms and improve the representation of bottom boundary layers in the model is to subtract climatological temperature and salinity data from the diffusion terms [*Mellor and Blumberg*, 1985]. This technique may affect the model ability to simulate climatic changes, since it involves a weak relaxation of the model climatology to the observed (i.e., today's) climatology. Advanced advection schemes that do not require implicit diffusion [*Häkkinen*, 1999], or isopycnally oriented diffusion [*Barnier et al.*, 1998] are other alternative approaches. However, it is important also to evaluate the implications for climate simulations of the standard along-sigma diffusion as has been used by the majority of POM users and in previous climate simulations with this model [*Ezer and Mellor*, 1994, 1997; *Ezer et al.*, 1995; *Ezer*, 1999]. Here, a modification of the along-sigma diffusion scheme that eliminates the climatological relaxation in POM, is being tested.

The approach in this study is to look at various oceanic indicators, such as meridional heat fluxes, THC and temperature fields, and to evaluate how they change over time

(A) MODEL GRID

(B) BOTTOM TOPOGRAPHY

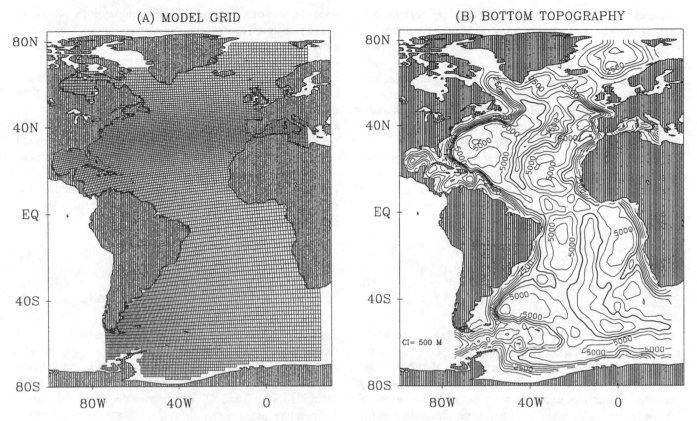

Figure 1. (a) Curvilinear orthogonal model grid and (b) bottom topography. Contour interval in (b) is 500m; a heavy line indicates the 4000m-isobath.

when atmospheric forcing conditions are significantly different than current conditions. These "climate change" simulations will be compared with control simulations representing today's climate. The experiments will be repeated with different model parameterizations, in order to evaluate the effect of model diffusion on current and future climate simulations. This study is a natural follow-up on the climatological simulations of *Ezer and Mellor* [1997] and the decadal variability simulations of *Ezer* [1999], using a similar model domain and configuration, but extending the calculations to resolve longer time scales.

Two main goals are included in this study. First, to study the processes involved in the adjustment of the ocean to extreme climatic changes in high latitudes, and second, to test the sensitivity of climate simulations with a sigma coordinate ocean model to diffusion parameterization.

The paper is organized as follows: First, in section 2, the ocean model and the experiments are described, then in sections 3, the results are discussed. Discussion and conclusions are offered in section 4.

2. THE OCEAN MODEL AND THE EXPERIMENTS

The Princeton Ocean Model, POM [*Blumberg and Mellor*, 1987] is a bottom-following sigma coordinate ocean model, which employs the Mellor-Yamada turbulence scheme [*Mellor and Yamada*, 1982] for vertical mixing parameterization. The model configuration, lateral boundary conditions and the Atlantic Ocean grid, between 80°S and 80°N (Figure 1), are similar to those used by *Ezer and Mellor* [1997] and *Ezer* [1999]. The horizontal curvilinear orthogonal grid has variable resolution with a grid size of 50-100 km; the vertical grid has 16 sigma layers with a higher resolution near the surface. The lateral boundary conditions include three open boundaries, two in the south for the inflow and outflow of the Antarctic Circumpolar Current, and one in the north for the connection with the Arctic Ocean. The distribution of vertically integrated transports along each open boundary is constant in time and specified according to estimates based on observations and models, but internal velocities at each level are controlled by radiation conditions [see *Ezer and Mellor*, 1997 for details]. The possible effect of the lateral boundary conditions on the calculations will be discussed later. There are three main differences between this study and the previous studies with this model. First, surface heat and salt fluxes are used as boundary conditions instead of surface temperature and salinity, second, the along-sigma diffusion has been modified, and third, longer simulations with forcing representing climate changes are performed.

The horizontal mixing coefficients for momentum (viscosity) and tracers (diffusivity) are calculated by a Smagorinsky-type formulation [*Smagorinsky et al.*, 1965] such that

$$(A_M, A_H) =$$
$$(C_{vis}, C_{dif}) \Delta x \Delta y \left[\left(\frac{\partial u}{\partial x} \right)^2 + \frac{1}{2} \left(\frac{\partial v}{\partial x} + \frac{\partial u}{\partial y} \right)^2 + \left(\frac{\partial v}{\partial y} \right)^2 \right]^{1/2} \quad (1)$$

where u and v are the horizontal velocity components in the x and y direction, respectively (x and y are along the curvilinear model grid lines). Therefore, horizontal viscosity and diffusion are reduced with decreasing grid size and velocity gradients. Sensitivity experiments using different values of the coefficients C_{vis} and C_{dif}, have been the subject of the study by *Ezer and Mellor* [2000]; here these coefficients are set to a constant value of 0.2. The formulation of horizontal diffusion along sigma layers in POM, following *Mellor and Blumberg* [1985], has shown some advantages in maintaining bottom boundary layer structures; this approach, however, may cause undesired diapycnal mixing over steep bottom slopes, so steps must be taken to minimize this numerical problem as discuss below. The changes in temperature due to the along-sigma diffusion in the model can be expressed as

$$\frac{\partial T}{\partial t} = \frac{\partial}{\partial x} \left(A_H \frac{\partial (T - T_{clim})}{\partial x} \right)$$
$$+ \frac{\partial}{\partial y} \left(A_H \frac{\partial (T - T_{clim})}{\partial y} \right) \quad (2)$$

where $T_{clim}(x,y,z)$ represents climatological data. A similar formulation applies also to salinity. The subtraction of climatological data removes most of the undesired diapycnal mixing, since only (small) anomalies are involved in the horizontal diffusion terms, and in fact, the results resemble calculations with zero external diffusion [*Ezer and Mellor*, 2000]. This formulation may affect climate simulations since the use of say, *Levitus* [1982] climatology for T_{clim}, adds a relaxation term to the model temperature, which may prevent the model from developing a climate state that is significantly different than the imposed T_{clim}. This hypothesis will be tested here by replacing the constant Levitus T_{clim} with a time varying model climatology, thus eliminating any relaxation of the model fields to the observed climate.

The surface forcing includes the monthly climatological surface wind stress of the Comprehensive Ocean-Atmosphere Data Set (COADS) analyzed by *da Silva et al.* [1994]. Surface heat flux (Q_T) and surface salinity flux (Q_S) from the atmosphere to the ocean are calculated according to

$$Q_T = Q_c + \gamma_T \left(T_m^0 - T_c^0 - \Delta T \right) \quad (3a)$$

$$Q_S = (E - P)_c S_c^0 + \gamma_S \left(S_m^0 - S_c^0 - \Delta S \right). \quad (3b)$$

Subscripts "c" and "m" represent fields derived from the COADS monthly climatology and from the model, respectively. T^0 and S^0 are the surface temperature and salinity, respectively, and E-P is the observed evaporation minus precipitation. ΔT and ΔS are imposed departures of surface temperature and salinity from the observed climatologies; γ_T and γ_S are coupling coefficients. The coupling coefficient γ_T has been chosen as -50 W m^{-2} K^{-1}, similar to the value of $\partial Q / \partial T$ calculated from COADS observations. The chosen coefficient $\gamma_S = 5 \times 10^{-6}$ m s^{-1} is based on sensitivity studies when simulating the current climate (i.e., with $\Delta T = \Delta S = 0$). While surface flux boundary conditions as in (3) are common in ocean modeling, they do have considerable deficiencies as they do not represent realistic atmospheric feedback processes, which are better represented in coupled ocean-atmosphere models. In the previous studies of *Ezer and Mellor* [1997] and *Ezer* [1999] surface temperature and surface salinity were used as boundary conditions (the equivalence of setting the coupling coefficients in (3) to infinity). A boundary condition with an imposed surface property may constrain model variability more than a flux boundary condition does [*Greatbatch et al.*, 1995; *Seager et al.*, 1995; *Huck et al.*, 1999], so there is some advantage in the approach taken here.

To test the model response to surface forcing and diffusion parameterization, four simulations are performed, each one executed for 60 years:

(a) Experiment 1 is a control experiment with standard diffusion (labeled "CS"). Surface boundary conditions in (3) include monthly climatological values obtained from COADS and zero surface temperature and salinity anomalies ($\Delta T=\Delta S=0$); T_{clim} (and S_{clim}) in (2) were obtained from the *Levitus* [1982] climatology (interpolated into the model grid).

(b) Experiment 2 is another control experiment but with modified diffusion (labeled "CM"), i.e., it is similar to CS, but T_{clim} and S_{clim} were obtained from the model climatology of the previous 5 years, eliminating any relaxation to the observed climatology. The actual horizontal diffusion in this experiment (which varies temporally and spatially) is generally smaller than that used in CS and it is also reduces with time as the model approaches an equilibrium state.

(c) Experiment 3 represents a warm-climate experiment with standard diffusion (labeled "WS"). In (3) ΔT and ΔS are functions of latitude, with values of zero for latitudes south of 40°N and linearly varying values between 40°N and 80°N such that at 80°N $\Delta T=3$°C and $\Delta S=-1$psu. These warming and refreshing of the surface of the ocean in high latitudes resemble the climate changes under increase greenhouse gas concentrations as predicted by coupled ocean-atmosphere models [*Haywood et al.*, 1997; *Manabe et al.*, 1991; *Manabe and Stouffer*, 1995, 1997].

(d) Experiment 4 represents another warm-climate experiment, but with a modified diffusion (labeled "WM"), i.e., forcing is similar to that in WS, but diffusion is as in CM.

This set of experiments can isolate the effects of high latitude surface anomalies and model diffusion parameterization, since they show how horizontal diffusion affects both, today and future climate simulations. By comparing CS with WS and CM with WM any model drift other than that associated with the surface conditions is eliminated. In order to isolate buoyant-driven from wind-driven effects, the same COADS monthly climatological wind stresses are used in all the experiments. In reality, or in coupled ocean-atmosphere models, the wind is affected by the climatic changes of surface temperature [*Kushnir*, 1994; *Latif*, 1998]. The possible consequence of neglecting wind effects will be discussed later. The experiments start from the end of the 30-years integration of *Ezer and Mellor* [1997]. The control experiments will thus include the adjustment of the ocean model from surface boundary conditions of imposed temperature and salinity to surface flux boundary conditions, an interesting test by itself.

3. MODEL RESULTS

3.1 The Transient Adjustment Process

We first look at the transient process and how the ocean model adjusts itself to a new climatic state under the imposed surface anomalies. Of particular interest is the question whether or not the relaxation of the model to climatology in (2) will prevent experiment WS from reaching a new oceanic state. Figure 2 shows the zonally averaged temperature difference between the two calculations with standard diffusion formulation (i.e., WS minus CS). The plume of relatively warmer water under the surface anomaly conditions propagated southward to about 40°N and downward, reaching almost a steady state after about 30 years. Around the tropics, and seemingly disconnected from the high latitude warming, subsurface warming grows in amplitude with time during the entire 60-years integration, we will explain this later. Despite the surface warming conditions, two regions show cooling trends, the upper layers around 40°N and the near-bottom deep layers; the former and the later relate to changes in the Gulf Stream and in the Antarctic Bottom Water cell, respectively, as discussed later. The spatial structure in the thermal field seen in Figure 2 is quite different than that obtained by course resolution climate models [e.g., *Manabe et al.* 1991], where warming of the upper layers in the entire North Atlantic is more homogeneous. The downslope penetration of the anomaly due to the near bottom advection and turbulent mixing is more pronounced in the sigma model than it is in the climate (z-level) models.

The temperature change at 500m in Figure 3 (which focuses only on the North Atlantic portion of the model domain) shows that the high latitude warm anomaly propagated along the western boundary and reached the tropics within a period of about 20 years. Our results are quite similar to the response of the Atlantic Ocean to high latitude anomalies, as described by previous studies. For example, *Gerdes and Koberle* [1995] describe adjustment process that involves propagation of anomalies by coastal topographic waves along the North America continent and advective response in the western boundary undercurrent when cooling of surface water near Iceland was imposed; the adjustment time scale was about 10 years. *Döscher et al.* [1994] describe an adjustment process of the Atlantic Ocean to high latitude anomalies that takes place within about 20 years. Despite the zonally averaged imposed surface anomalies, considerable zonal spatial variations in the subsurface temperature changes are apparent in Figure 3. Regions with relatively larger warming trend include the Greenland Sea, the Labrador Sea and the Gulf of Mexico/Caribbean Sea, where warm pools of water accumulate around local recirculation gyres. Cooling occurs downstream of the Gulf Stream and the North Atlantic currents; this cooling will be explained later in relation with weakening of the Gulf Stream and the reduction in the amount of warm subtropical water masses advected by the Gulf Stream into higher latitudes.

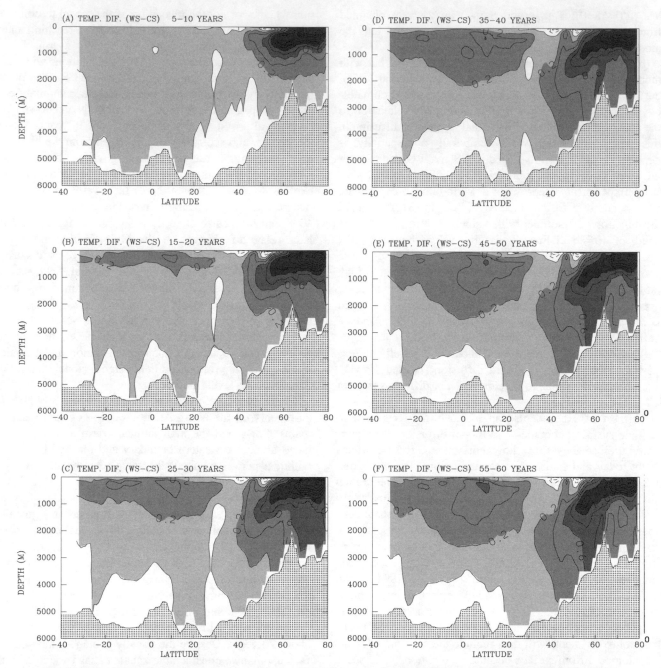

Figure 2. Zonally and 5-year averaged temperature differences (experiment WS minus CS) for years: (a) 5-10, (b) 15-20, (c) 25-30, (d) 35-40, (e) 45-50 and (f) 55-60. Contour interval is 0.2°C; white regions represent negative values and shaded regions represent positive values.

A useful diagnostic parameter in analysis of climate simulations is the thermohaline circulation (THC) index, defined here as the maximum value of the meridional stream function between 50°N and 70°N and expressed in Sv (1 Sverdrup = 10^6 m^3 s^{-1}). A comparison of the THC index in high latitudes between all four experiments is shown

in Figure 4. The experiments with modified (and very low) diffusivities (Figure 4b) show more transient changes during the first stages of the adjustment process and larger variations in the seasonal cycle. These experiments seem to be close to an equilibrium state during the last 30 years of the integration (longer simulations are probably needed to

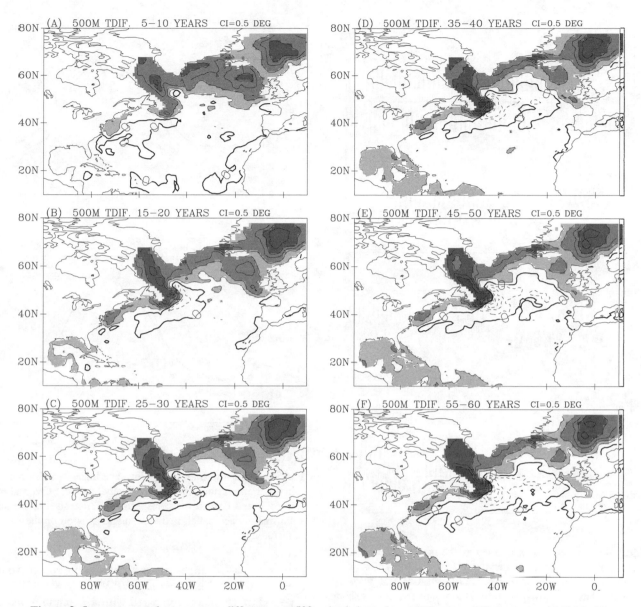

Figure 3. 5-year averaged temperature differences at 500m depth (experiment WS minus CS) for years: (a) 5-10, (b) 15-20, (c) 25-30, (d) 35-40, (e) 45-50 and (f) 55-60. Contour interval is 0.5°C; heavy lines indicate the zero contour, dashed lines negative values and shaded regions represent positive values above 0.5°C.

show if such a state exists at all). While one anticipates that the two control simulations (CS and CM) be very similar to each other and the two warm-climate experiments (WS and WM) be very different from each other due to the relaxation of CS and WS to today's climatology, surprisingly, the results are quite different than expected. In fact, both warm-climate experiments reached almost identical THC intensity after 60 years of integration, pointing to the conclusion that the relaxation effect in (2) is small enough so that it does not prevent the model from

reaching a THC state that is different than current climate. Moreover, the results indicate that the changes in the intensity of the THC are primarily driven by surface fluxes rather than by diffusion. The increase in THC intensity in the two control runs during the first 10 years represents the adjustment from surface temperature and salinity boundary conditions, as used in *Ezer and Mellor* [1997], to flux surface boundary condition; similar results obtained by the box model experiments of *Huck et al.* [1999]. The THC index in the control run with modified diffusion reached

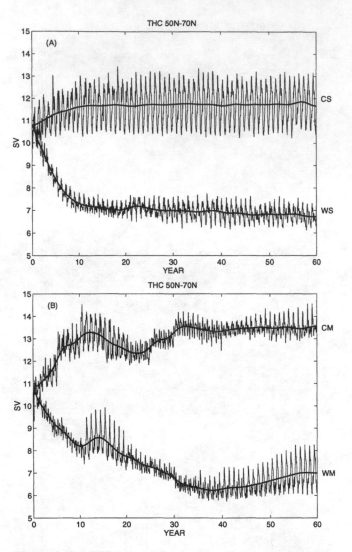

THC 50N-70N

THC 50N-70N

Figure 4. The THC index defined as the maximum value of the meridional stream function between 50°N and 70°N, for (a) experiments CS and WS and (b) experiments CM and WM. The heavy line has been low-pass filtered to remove variations shorter than 2 years.

higher values at the end of the integration than the standard run did. When T_{clim} in (2) is obtained from the model itself in run CM, the actual horizontal diffusion is smaller than that in CS, resulting in more intense circulation patterns, larger changes in convection and subduction, and a larger interannual variability.

3.2. Meridional Heat Flux and Overturning Circulation

The poleward heat transport by the ocean plays an important role in maintaining the earth climate and may change as climate changes. For example, diagnostic calcu-

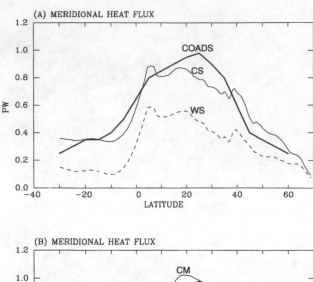

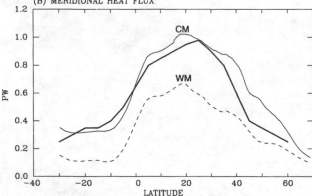

Figure 5. The average meridional heat flux calculated from the last 5 years of the integrations for experiments (a) CS and WS, thin solid and dashed lines, respectively, and (b) CM and WM, thin solid and dashed lines, respectively. Also shown, the heavy solid line in each panel, is the calculations based on the COADS climatology.

lations of past changes suggest changes in North Atlantic meridional heat flux of about 0.2 PW (1 PW= 10^{15}W) over a period of 15 years, compared with a maximum value of about 1 PW. Considerable latitudinal variations due to changes in circulation were also obtained by these calculations [*Greatbatch and Xu*, 1993; *Ezer et al.*, 1995]. The meridional heat flux (MHF) as integrated zonally and vertically over the Atlantic basin is calculated for the four experiments and compared with the value obtained from COADS (Figure 5). Despite the fact that only latitudes north of 40°N were directly affected by the surface anomalies, considerable reduction in the MHF, up to 0.4 PW, are shown in both WS and WM experiments for all latitudes. The different diffusion parameterization in CS and WS compared with CM and WM experiments seems to affect the latitudinal variations in the MHF but had little effect on the climatic changes (i.e., the difference between the

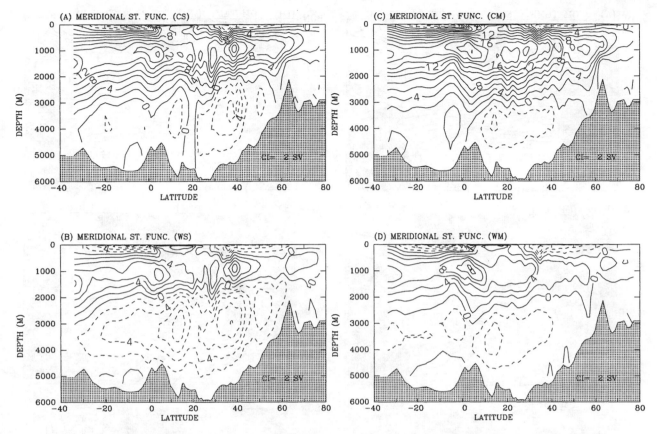

Figure 6. The zonally integrated meridional stream function calculated from the last five years of each experiment: (a) CS, (b) WS, (c) CM and (d) WM. Contour interval is 2 Sv and negative contours are indicated by dashed lines.

dashed and solid-thin lines in Figure 5a and 5b are similar). The experiments with modified diffusion (Figure 5b) have MHF that more closer resembles the observed values than the experiments with a standard diffusion (Figure 5a) showing that the modified diffusion formulation may be advantageous over the standard formulation.

The MHF is closely related to the THC, which is shown in terms of the meridional stream function in Figure 6. Both, the standard and the modified diffusion calculations, show considerable climatic weakening in the intensity of the THC (Figures 6b and 6d). These changes include a reduced North Atlantic Deep Waters transport between 1000 and 2500m depth and an almost complete shutoff in the transport of deep waters across the sills between the Greenland Sea and the North Atlantic Ocean. On the other hand, near bottom circulation, associated with northward transport of Antarctic Bottom Water (see also Figure 2), increased. The suggested climatic changes here due to climate warming resemble possible past changes such as the climate during the Last Glacial Maximum period as obtained from paleoclimate records [*Broecker*, 1991;

Imbrie et al., 1992] and models [*Seidov and Haupt*, 1999]. The results are generally consistent with previous modeling studies [*Döscher et al.*, 1994; *Gerdes and Koberle*, 1995], who show that cooling/warming in high latitudes causes intensifying/weakening of the THC and increasing/decreasing in MHF. It is also interesting to note some changes in the upper ocean transport and in particular the deepening of the shallow circulation cells (indicated by dashed lines in Figure 6) in both hemispheres. This change implies that while the majority of the poleward subsurface transport around 500m depth subducts at middle latitudes into deeper layers in the control experiments, in the warm-climate experiments larger portion of this flow joins the near surface equatorward transport instead. Decadal-scale changes in subduction and the ventilated thermocline [*Luyten et al.*, 1983] may thus affect the upper ocean connection between subtropical and tropical regions [*Liu et al.*, 1994]. Note also the development of a shallow circulation cell around 10°N, near the latitude of the local maximum warming seen in Figure 2. The development of this cell implies that under warm-climate conditions, larger

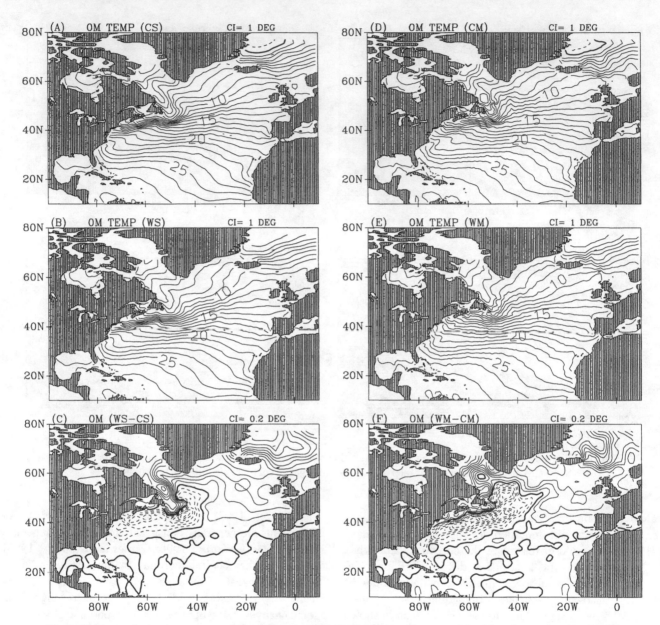

Figure 7. Sea surface temperature and the difference between the warm climate and the control experiments, (a) CS, (b) WS, (c) WS-CS, (d) CM, (e) WM and (f) WM-CM, calculated from the last 5 years of the integrations. Contour interval is 1°C, except (c) and (f) were it is 0.2°C; dashed lines represent negative values.

portion of the northward surface transport in the tropical North Atlantic recirculates back equatorward below the surface instead of transporting tropical waters into higher latitudes.

3.3 Spatial Changes

The imposed surface anomalies in (3), ΔT and ΔS, are zonally uniform, thus any spatial departure from a zonally averaged change (e.g., Figures 2 and 3) must be the result

of changes in ocean circulation. In fact, the use of surface fluxes instead of surface temperature and salinity [as was the case in *Ezer and Mellor*, 1997 and *Ezer*, 1999] allows the model to develop its own surface temperature change (Figure 7) which is very different than the imposed zonal anomaly. Note that surface temperature gradients across the Gulf Stream are reduced in WM (and to lesser extent in WS), indicating a weakening of the geostrophic component of the Gulf Stream. The weaker Gulf Stream reduces the northward transport of warm subtropical waters, resulting

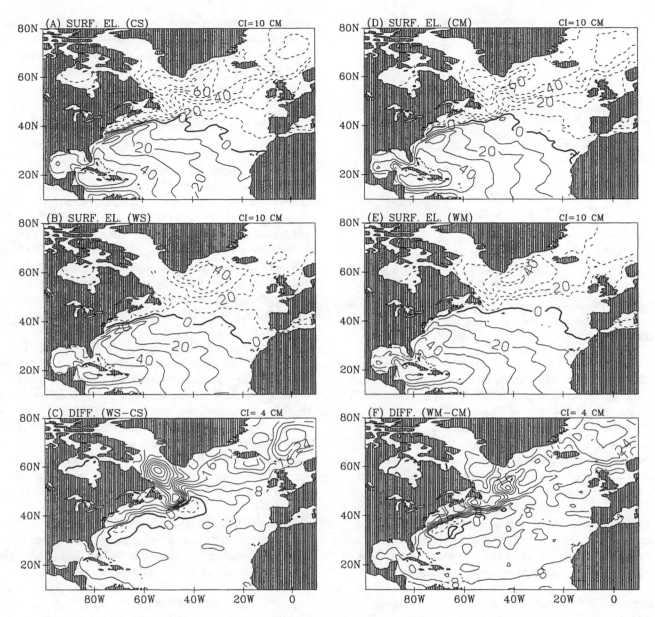

Figure 8. Same as Figure 7, but for the sea surface elevation. Contour interval is 10 cm, except (c) and (f) were it is 4 cm.

in relative cooling along its path. This upper ocean cooling occurs despite the fact that the imposed surface heat flux anomaly implies an additional surface warming north of 40°N in the warm-climate calculations. Figure 8 shows the corresponding surface elevation (and thus surface geostrophic velocity) at the end of the integrations. In the warm-climate experiments, WS and WM, the circulation of the subpolar gyre and the Labrador Sea weakened considerably, and the gradients of surface elevation across the Gulf Stream smaller. Although quantitative spatial differences exists between the experiments with different diffusions (comparing the right and left panels), qualitatively the climatic changes show similar trends. The process of spatial climatic changes can be explained as follows. The warmer than normal surface water masses in high latitudes are entrained with subsurface layers and mixed downward (Figure 2) while the signal also propagates along the western American coast (Figure 3 and Figure 7) around Newfoundland and into the northern recirculation gyre north of the Gulf Stream. The warming of the cold side of the Gulf Stream results in smaller thermal gradients across the Stream and in its intensity. As

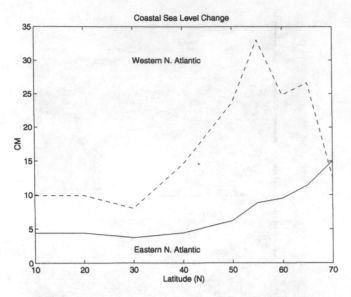

Figure 9. The coastal sea level change (WS-CS) along the eastern coast (solid line) and the western coast (dashed line) of the North Atlantic.

a result, the northeastward transport of warm waters by the Gulf Stream decreases, which further decreases the thermal gradients, creating a positive feedback mechanism. The process does not continue forever. A semi-equilibrium is reached due to the negative feedback of the surface heat flux (i.e., if surface model properties depart from the imposed values, the last term in 3a and 3b produces a flux correction). It should be noted that the freshening effect due to the imposed salinity anomaly works in the same manner (not shown) as the temperature does. However, the spatial change in the salinity field is not as pronounced as in the temperature field, because the imposed heat flux anomaly is at the coldest regions while the maximum imposed salinity anomaly is at regions that are already relatively fresh.

The spatial changes described above may have important implications for the detection of climatic changes from coastal sea level observations. Sea level records reflect the thermal expansion of the global ocean, but local changes in circulation due to the oceanic adjustment to spatial thermal changes can be significant [*Mellor and Ezer*, 1995]. Climatic changes detected at coastal sea level stations vary significantly from one station to another even when only short distance separates between them, however, ocean models can simulate these variations quite well, from climatic changes in the open ocean [*Ezer et al.*, 1995]. The coastal sea level change (WS-CS in Figure 8) shows for example, different responses for the western and for the eastern coasts (Figure 9). The sea level rise along the eastern coast resembles that expected from the imposed zonal surface temperature anomaly, but the sea level rise along the

western boundary is much larger and spatially varying due to the ocean circulation changes described before. Therefore, these calculations indicate the difficulty in predicting local sea level change without taking into account changes in ocean dynamics.

3.4 Changes in Near-Bottom Flows and Deep Water Formation

Figure 10 shows the vertically integrated stream function and the climatic changes. Note first the differences between the two control runs (Figure 10a and Figure 10d), in particular, the differences in the recirculation gyres in the Gulf of Mexico and south of the Gulf Stream. The relaxation of the diffusion terms to the observed climatology in CS and WS may have caused inconsistencies between the model and the observed climatologies that affected those gyres. On the other hand, the control case with modified diffusion, CM, shows more realistic, though somewhat weak, northern recirculation gyre north of the Gulf Stream; it has been demonstrated by *Ezer and Mellor* [1992] that the intensity of this gyre influences the position of the Gulf Stream. The climatic changes under warm-climate conditions include weakening (and even a possible reversal in WS) of the cyclonic circulation in the Labrador Sea, owing to the warm pool created there (Figure 3). The weakening of the Gulf Stream is more pronounced in the case with no relaxation to climatology (Figure 10e) than in the standard case (Figure 10b), as the relaxation helps to maintain the Gulf Stream signature despite the climatic changes.

Since the model uses a bottom-following sigma coordinate system and has a prognostic turbulence closure scheme, dynamics of bottom boundary layers are represented in the model quite well; one can also easily analyze the near-bottom flow, looking at the lowest active layer in the model. *Ezer and Mellor* [1997] shows that even coarse resolution sigma models generate strong bottom flows, such as the deep western boundary current (DWBC), while z-level models may need much higher horizontal and vertical resolutions to generate intense bottom currents [*Winton et al.*, 1998]. The mean kinetic energy at the bottom layer (Figure 11) shows that the DWBC, at about 1500m depth, dominates the near bottom flows. The DWBC plays a major role in the mechanism of transporting cold water masses to the deep North Atlantic from the formation zone near the straits between the Greenland Sea and the Atlantic Ocean. In the warm-climate cases (WS and WM) the DWBC intensity is reduced significantly, and thus contributing to the weakening of the THC (Figure 6) and to the reduction in northward MHF (Figure 5). This result is

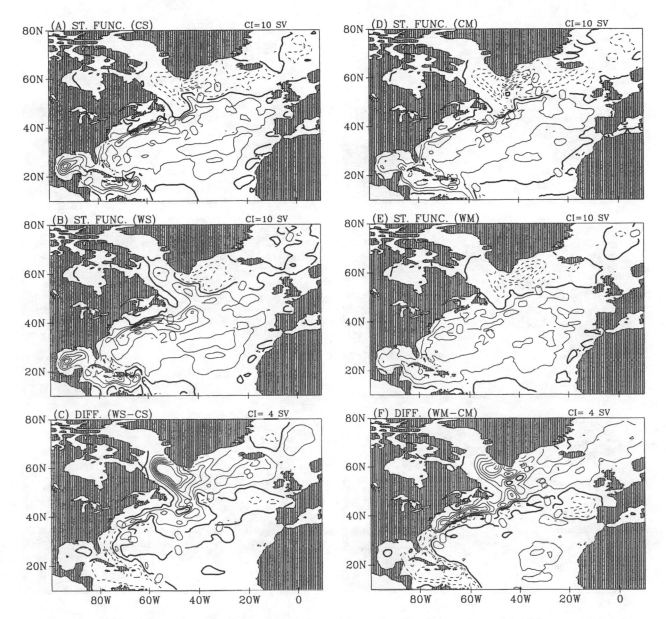

Figure 10. Same as Figure 7, but for the vertically integrated stream function. Contour interval is 10 Sv, except (c) and (f) were it is 4 Sv.

consistent with previous studies using other models [*Döscher et al.*, 1994; *Gerdes and Koberle*, 1995]. However, it is also demonstrated here that horizontal diffusion in a sigma coordinate ocean model may have a significant effect on the bottom flow, which could explain the changes seen in the deep gyres of the THC (Figure 6).

Changes occurred at the region of deep-water formation in high latitudes are of particular importance, thus the vertical component of the bottom layer flow is shown in Figure 12 (only the case with modified diffusion is shown here since the standard diffusion case is almost identical in those

calculations). In the control run (Figure 12a), downslope flows (shaded areas) are evident in the Denmark Straits, the Faroe Island-Iceland Straits and along the Greenland coast, with particular strong downslope component south of Iceland and south of Greenland. This downslope flow is consistent with bottom boundary layer dynamics, where Ekman veering of the near-bottom flow to the left of the mean current is expected. Under warm-climate conditions, the stratification of the upper layers in high latitudes is more stable (Figure 2) so downward mixing of cold surface waters is reduced, and the boundary current is warmer and

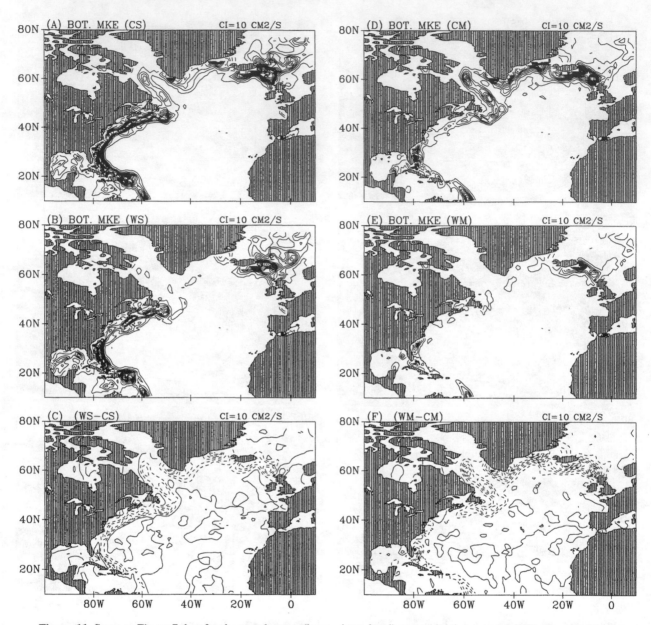

Figure 11. Same as Figure 7, but for the near-bottom (lower sigma level) mean kinetic energy (MKE). Contour interval is 10 cm^2 s^{-1}.

weaker (Figure 11). The result is a significant reduction of downslope flows in regions most important for deep-water formation. The offshore upwelling southeast of Greenland is also reduced. The reduction of surface mixing and the change in circulation also affect another important area of water formation, the Labrador Sea.

4. DISCUSSION AND CONCLUSIONS

This study follows closely on the footsteps of previous studies [*Ezer and Mellor*, 1994, 1997; *Ezer et al.*, 1995;

Ezer, 1999, 2001] which used the same sigma coordinate coastal ocean model (POM) as a tool to study basin-scale climate processes previously studied mostly with other types of ocean models. The model, with its low horizontal diffusivities, turbulence mixing scheme and sensitivity to coastline features and bottom topography, provides a different perspective on some processes that may not be well-resolved in standard coarse resolution climate models.

The ocean model response to an idealized "warm-climate" forcing is consistent with those obtained by coupled ocean-atmosphere climate models [*Manabe et al.*, 1991;

Manabe and Stouffer, 1995, 1997; *Haywood et al.*, 1997] to some extent, for example, in the net change of the intensity of the THC. However, significant differences between the sigma coordinate ocean model calculations and previous studies require further explanations. The adjustment time-scale to abrupt climate changes in the ocean-only model, 10-20 years, is generally shorter than the adjustment time-scale in coupled ocean-atmosphere climate models, 50-100 years. This difference can be partly attributed to the lack of realistic atmospheric feedback in the surface forcing. However, it is also quite likely that the more intense deep flows and more realistic mixing processes in the higher resolution ocean model contribute to a faster transfer of the climate change signal to the ocean interior. The horizontal and overturning circulation in this model have more spatial gyre structure than coarse resolution climate models do and thus climate changes results in spatial differences and local circulation changes that are absent from coarse resolution models. Flows over sills and in particular the exchange of water masses between the Greenland and the Atlantic Ocean are represented in the sigma ocean model, while in many climate models most of the deep water formation occurs unrealistically south of the sills. An interesting result was the relative cooling of middle latitudes (despite the imposed surface heat flux) owing to the weakening of the Gulf Stream current, which transports warm subtropical waters to higher latitudes. Reduction in deep water formation near the Greenland and Iceland coasts have been clearly identified, a result of weakening of the western boundary current and consequently a reduction in its downslope Ekman component.

The study also examine the sensitivity of climate simulations to the along-sigma diffusion in the Princeton Ocean Model, and in particular, the effect of the relaxation of model temperature and salinity fields to the observed climatology. An alternative approach, using the model own climatology (with some time-delay), has been tested successfully. An unexpected result was the fact that the diffusion relaxation term affects the control runs more than the warm-climate runs (Figure 4), implying that the climatic changes are primarily controlled by surface fluxes rather than by horizontal diffusion; near bottom flows were affected by the diffusion parameterization to some extent. Changes in thermohaline circulation and meridional heat flux from current climate to a warm climate conditions are quite robust independent of the particular diffusion used, and the relaxation term does not seem to prevent the model from simulating climatic oceanic changes, as originally feared.

While idealized studies of the type presented her are useful in studying the adjustment process that takes place in

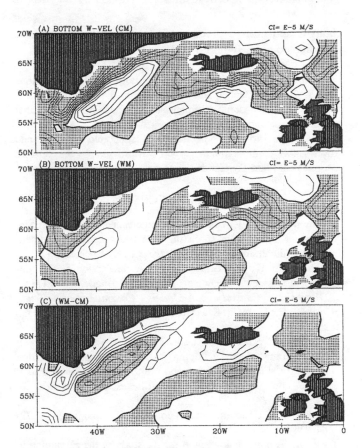

Figure 12. The near-bottom vertical velocity component calculated from the lower sigma level in the model averaged over the last 5 years of the model calculations: (a) CM, (b) WM and (c) WM-CM. Shaded regions indicate negative values, i.e., downslope flows. Contour interval is 10^{-5} m s^{-1}.

the ocean in response to abrupt climate changes, one should also mention the possible limitations of the study compared with studies with global coupled climate models. For example, the atmospheric feedback would have modified the heat, salt and momentum (wind stress) fluxes, but was ignored here (climate models indicate, though, that the THC is more affected by surface heat and salt fluxes than by wind stress, so neglecting wind changes may not significantly alter the results). The use of a basin model requires specification of open boundary conditions; climatic changes in the exchange of heat and momentum between the Greenland Sea and the Arctic Ocean were ignored, as well as effects associated with sea-ice. The lateral boundary conditions in the south should not have a significant effect on the results for the time scales involved here, but the boundary conditions in the north may have, since anomalies have predominantly propagated from the north southward. In any case, the implications of unresolved oceanic spatial changes and mixing processes in coupled climate models, as indicated here, should be further investigated.

Acknowledgments. The study was motivated by discussions with S. G. H. Philander and G. L. Mellor. Two anonymous reviewers provided useful comments and suggestions. The original model was developed with support from NOAA's Atlantic Climate Change Program. The author is supported by ONR and MMS grants. Computational support was provided by the NOAA's Geophysical Fluid Dynamics Laboratory.

REFERENCES

Beckmann, A., and R. Döscher, A method for improved representation of dense water spreading over topography in geopotential-coordinate models, *J. Phys. Ocean.*, *27*, 581-591, 1997.

Blumberg, A. F., and G. L. Mellor, A description of a three-dimensional coastal ocean circulation model, In *Three-Dimensional Coastal ocean Models , Coastal Estuarine Stud.*, vol. 4, edited by N. S. Heaps, 1-16, AGU, Washington, D.C., 1987.

Barnier, B., Marchesiello, P., De Miranda, A. P., Molines, J.-M. and Coulibaly, M., A sigma coordinate primitive equation model for studying the circulation in the South Atlantic. Part I: Model configuration with error estimates, *Deep-Sea Res. 45*, 543-572, 1998.

Broecker, W. S., The great ocean conveyor, *Oceanography*, *4*, 79-89, 1991.

Bryan, K., A numerical method for the study of the circulation of the world ocean, *J. Comput. Phys.*, *4*, 437-376, 1969.

Delworth, T., S. Manabe, and R. J. Stouffer, Interdecadal variations of the thermohaline Circulation in a coupled ocean-atmosphere model, *J. Climate*, *6*, 1993-2011, 1993.

da Silva, A. M., C. C. Young, and S. Levitus, *Atlas of surface marine data 1994, vol. 3, Anomalies of heat and momentum fluxes. NOAA Atlas NESDIS 8*, vol. 8, 413 pp., Natl. Environ. Satell. Data Inf. Serv., Washington, D.C., 1994.

Döscher, R., C. W. Böning and P. Herrmann, Response of circulation and heat transport in the North Atlantic to changes in thermohaline forcing in northern latitudes: A model study, *J. Phys. Oceanogr.*, *24*, 2306-2320, 1994.

Ezer, T., Decadal variabilities of the upper layers of the subtropical North Atlantic: An ocean model study, *J. Phys. Oceanogr.*, *29*, 3111-3124, 1999.

Ezer, T., Can long-term variability in the Gulf Stream transport be inferred from sea level?, *Geophys. Res. Let.*, *28*, 1031-1034, 2001.

Ezer, T., and G. L. Mellor, A numerical study of the variability and the separation of the Gulf Stream induced by surface atmospheric forcing and lateral boundary flows, *J. Phys. Oceanogr.*, *22*, 660-682, 1992.

Ezer, T., and G. L. Mellor, Diagnostic and prognostic calculations of the North Atlantic circulation and sea level using a sigma coordinate ocean model, *J. Geophys. Res.*, *99*, 14,159-14,171, 1994.

Ezer, T., and G. L. Mellor, Simulations of the Atlantic Ocean with a free surface sigma coordinate ocean model, *J. Geophys. Res.*, *102*, 15,647-15,657, 1997.

Ezer, T., and G. L. Mellor, Sensitivity studies with the North Atlantic sigma coordinate Princeton Ocean Model, in *Ocean Circulation Model Evaluation Experiments for the North Atlantic Basin, Dyn. Atmos. Ocean*, vol. 32, edited by E. P. Chassignet and P. Malanotte-Rizzoli, pp. 155-208, 2000.

Ezer, T., G. L. Mellor, and R. J. Greatbatch, On the interpentadal variability of the North Atlantic ocean: Model simulated changes in transport, meridional heat flux and coastal sea level between 1955-1959 and 1970-1974, *J. Geophys. Res.*, *100*, 10,559-10,566, 1995.

Gerdes, R., A primitive equation ocean circulation model using a general vertical coordinate transformation. 1. Description and testing of the model, *J. Geophys. Res.*, *98*, 14,683- 14,701, 1993.

Gerdes, R., and C. Koberle, On the influence of DSOW in numerical model of the North Atlantic general circulation, *J. Phys. Oceanogr.*, *25*, 2624-2642, 1995.

Greatbatch, R. J., and J. Xu, On the transport of volume and heat through sections across the North Atlantic: Climatology and the pentads 1955-1059, 1970-1974, *J. Geophys. Res.*, *98*, 10,125 10,143.1993

Greatbatch, R. J., G. Li, and S. Zhang, Hindcasting ocean climate variability using time- dependent surface data to drive a model: An idealized study, *J. Phys. Oceanogr.*, *25*, 2715- 2725, 1995.

Häkkinen, S., Simulated interannual variability of the Greenland Sea deep water formation and its connection to surface forcing, *J. Geophys. Res.*, *100*, 4761-4770, 1995.

Halliwell, G. R, Simulation of North Atlantic decadal/multidecadal winter SST anomalies driven by basin-scale atmospheric circulation anomalies, *J. Phys. Oceanogr.*, *28*, 5-21, 1998.

Haywood, J. M., R. J. Stouffer, R. T. Wetherald, S. Manabe and V. Ramaswamy, Transient response of a coupled model to estimated changes in greenhouse gas and sulfate concentrations, *Geophys. Res. Lett.*, *24*, 1335-1338, 1997.

Huck, T., A. C. De Verdiere, and A. J. weaver, Interdecadal variability of the thermohaline circulation in box-ocean models forced by fixed surface fluxes. *J. Phys. Oceanogr.*, *29*, 865-892, 1999.

Imbrie, J., E. A. Boyle, S. C. Clemens, A. Duffy, W. R. Howard, G. Kukla, J. Kutzbach, C. G. Martinson, A. McIntyre, A. C. Mix, B. Molfino, J. J. Morley, L. C. Peterson, N. G. Pisias, W. L. Prell, M. E. Raymo, N. J. Shackleton, and J. R. Toggweiler, On the structure and origin of major glaciation cycles. 1. Linear responses to Milankovitch forcing, *Paleoceanogr.*, *7*, 701-738, 1992.

Kushnir, Y. Interdecadal variations in North Atlantic sea surface temperature and associated atmospheric conditions, *J. Clim.*, *7*, 142-157, 1994.

Latif, M., Dynamics of interdecadal variability in coupled ocean-atmosphere models, *J. Clim*, *11*, 602-624, 1998.

Levitus, S., *Climatological atlas of the world ocean, NOAA Prof.*

Pap., *13*, NOAA, 173 pp., U.S. Gov. Print. Off., Washington, D. C., 1982.

Liu, Z., S. G. H. Philander and R. Pacanowski, A GCM study of tropical- subtropical upper- ocean water exchange, *J. Phys. Oceanogr.*, *24*, 2606-2623, 1994.

Luyten, J. R., J. Pedlosky, and H. Stommel, The ventilated thermocline, *J. Phys. Oceanogr.*, *13*, 292-309, 1993.

Manabe, S., R. J. Stouffer, M. J. Spelman, and K. Bryan, Transient responses of a coupled ocean-atmosphere model to gradual changes of atmospheric CO_2. Part I: Annual mean response, *J. Clim.*, *4*, 785-818, 1991.

Manabe, S., and R. J. Stouffer, Simulation of abrupt climate change induced by freshwater input to the North Atlantic ocean. *Nature*, *378*, 165-167, 1995.

Manabe, S., and R. J. Stouffer, Climate variability of a coupled ocean-atmosphere-land surface model: Implication for the detection of global warming, *Bull. Amer. Met. Soc.*, *78*, 1177-1185. 1997.

Marotzke, J., and J. Willebrand, Multiple equilibria of the global thermohaline circulation, *J. Phys. Oceanogr.*, *21*, 1372-1385, 1991.

Mellor, G. L., and A. F. Blumberg, Modeling vertical and horizontal diffusivities with the sigma coordinate system, *Monthly Weather Review*, *113*, 1380-1383, 1985.

Mellor, G. L., and T. Yamada, Development of a turbulent closure model for geophysical fluid problems. *Rev. Geophys.*, *20*, 851-875, 1982.

Mellor, G. L., and T. Ezer, Sea level variations induced by heating and cooling: An evaluation of the Boussinesq approximation in ocean model, *J. Geophys. Res.*, *100*, 20,565-20,577, 1995.

Mellor, G. L., S. Häkkinen, T. Ezer, and R. Patchen, A generalization of a sigma coordinate ocean model and an intercomparison of model vertical grids, *in Ocean Forecasting, Conceptual Basis and Applications*, edited by N. pinardi and J. D. Woods, pp. 55-72, Springer, 2001.

New, A. L., and R. Bleck, An isopycnic model study of the North Atlantic. Part II: Interdecadal variability of the subtropical gyre, *J. Phys. Oceanogr.*, *25*, 2700-2714, 1995.

Oberhuber, J. M., Simulation of the Atlantic circulation with a coupled sea ice-mixed layer isopycnal general circulation model. Part II: Model experiment, *J. Phys. Oceanogr.*, *23*, 830-845, 1993.

Pacanowski, R. C., and A. Gnanadesikan, Transient response in a z-level ocean model that resolves topography with partial cells, *Mon. Weather Rev.*, *126*, 3248-3270, 1998.

Prange, M., G. Lohmann, and R. Gerdes, Sensitivity of the thermohaline circulation for different climates- investigations with a simple atmosphere-ocean model, *Paleoclim.*, *2*, 71-99, 1997.

Seager, R., Y. Kushnir, and M. Cane, On heat flux boundary conditions for ocean models, *J. Phys. Oceanogr.*, *25*, 3219-3230, 1995.

Seidov, D., and B. J. Haupt, Simulated ocean circulation and sediment transport in the North Atlantic during the last glacial maximum and today, *Paleoceanogr.*, *12*, 281-305, 1997.

Seidov, D. and B. J. Haupt, Last glacial and meltwater interbasin water exchanges and sedimentation in the world ocean, *Paleoceanogr.*, *14*, 760-769, 1999.

Smagorinsky, J., S. Manabe, and J. L. Holloway, Numerical results from a nine-level general circulation model of the atmosphere, *Monthly Weather Review*, *93*, 727-768, 1965.

Winton, M., R. Hallberg, and A. Gnanadesikan, Simulation of density-driven frictional downslope flow in z-coordinate ocean models, *J. Phys. Oceanogr.*, *28*, 2163-2174, 1998.

Tal Ezer, Program in Atmospheric and Oceanic Sciences, P.O.Box CN710, Sayre Hall, Princeton University, Princeton, NJ 08544-0710.

The Effects of Vertical Mixing on the Circulation of the AABW in the Atlantic[*]

Igor V. Kamenkovich and Paul J. Goodman[1]

University of Washington, Seattle, Washington

We study the dependence of the volume and transport of the Antarctic Bottom Water (AABW) in the Atlantic Ocean on vertical mixing. Numerical results from a set of OGCM runs show that, while the vertical extent of the AABW cell decreases with intensifying vertical mixing, the cell's transport increases. An analytical model of the deep boundary layers is then used to interpret the results. The decrease in the AABW thickness is explained by the downward expansion of the upper, North Atlantic Deep Water cell. The intensification of the AABW transport is attributed to the increase in the deep meridional pressure gradient, which drives the flow. An estimate of the AABW transport is then derived from the density balance in the deep western boundary layer and compared with the OGCM results.

1. INTRODUCTION.

The ocean's deep circulation and its variability strongly influence the climate system. The thermohaline circulation is believed to have played a major role in dramatic climate reorganizations in the past and has a potential of causing major changes in the future climate of the Earth [e.g. *Broecker*, 1997; *Stocker*, 1999]. Our ability to successfully model both past and future climate changes therefore depends on understanding of mechanisms controlling different branches of the global thermohaline circulation.

There are two main sources of deep water in the World Ocean. North Atlantic Deep Water (NADW) forms in the northern North Atlantic where surface waters lose buoyancy due to cooling, sink to great depth and then flow southward. The resulting large meridional heat transport makes NADW a key player in climate variability on a variety of time scales [*Manabe* and *Stouffer*, 1988; *Maier-Reimer* et al., 1991; *Stocker*, 1999]. This water mass has been the focus of many studies addressing its sensitivity to a variety of factors in numerical general circulation models (GCMs), including vertical mixing [*Bryan*, 1987; *Marotzke*, 1997], surface fluxes of moisture [*Stocker, at al.*, 1992; *Rahmstorf*, 1995; *Manabe and Stouffer*, 1995; *Seidov and Maslin*, 1999; *Seidov and Haupt*, 1999], and momentum [*Toggweiler and Samuels*, 1998].

The important role of the vertical diffusivity in the ocean circulation has been long known to the oceanographic community [*Stommel*, 1958; *Munk*, 1966]. Numerical models show that water mass properties and their formation rates are very sensitive to the value of this parameter. The main known effects of increasing vertical diffusion are the deepening of the thermocline, the

[*]Joint Institute for the Study of Atmosphere and Ocean Contribution Number 785.

[1]Now at Lamont Doherty Earth Observatory Oceanography, Palisades, New York.

intensification of upwelling at low latitudes, and the enhancement of North Atlantic Deep Water (NADW) formation. In his sensitivity study, Bryan [1987] demonstrated a cube root dependence of the both the thermocline depth and the strength of the North Atlantic (NA) overturning on the vertical diffusion coefficient, k_v. His model, however, only included the Northern Hemisphere and therefore excluded effects of the Southern Ocean. In a recent study, Gnanadesikan [1999; G99 hereafter] proposed a simple theoretical model of the oceanic pycnocline, relating the pycnocline depth, the NA overturning (T_n), and the Southern Ocean winds and eddies. He assumed that the oceanic pycnocline is maintained by the balance between low latitude heating, Southern Ocean freshening and upwelling of cold water from below the pycnocline. Toggweiler and Samuels [1998] emphasize the role of the southern winds in driving thermohaline circulation and demonstrate that a wind-powered overturning circulation can exist without much vertical mixing.

Antarctic Bottom Water (AABW) forms in the Weddell and Ross seas at the Antarctic coast, where the cold surface waters become salty due to salt rejection during the local winter and sink to the bottom of the Southern Ocean. This water then propagates northward, mixes with Circumpolar Deep Water and forms a northward-moving tongue of very dense bottom water. There is growing evidence of a significant role for the Southern Ocean in the past glacial cycles [Broecker, 1994; Stocker, 1998; Seidov et al., 2001]. AABW can affect global circulation in two major ways. Firstly, AABW determines the deep density structure in the World Ocean and sets the deep vertical stratification [England, 1993; Seidov et al., 2001]. Secondly, AABW can influence the overlaying NADW cell, controlling the penetration depth and the strength of the latter.

The relationship between the AABW and the NADW circulation is rather complex and has been addressed in a number of studies. Intensification of NADW formation generally leads to a reduction of AABW in the deep ocean [Cox, 1989], AABW recedes southward and is confined to greater depths. England [1993], on the other hand, suggests that the density of AABW in turn controls the inter-ocean exchange of NADW. A decrease in characteristic density of AABW allows more NADW to leave the Atlantic basin in his model. The decrease in the southern deepwater formation leads to stronger and deeper NADW circulation and warmer abyssal waters [Seidov et al., 2000].

As in the case of NADW, the formation rate of AABW strongly depends on the surface exchanges of moisture and heat. England [1992], [1993] and Stocker, et al. [1992] demonstrated the importance of surface salinity at the Antarctic coast in setting the density properties and ultimately the formation rate of this water mass. Goose and Fichefet [1999] show that heat and freshwater exchanges between sea ice and the water are crucial for AABW production. In addition to the importance of the brine rejection during ice formation, they point to the fact that contact with sea ice lowers the seawater temperature to the freezing point, which corresponds to the maximum density of seawater at the given salinity. Seidov et al. [2000] demonstrate that a meltwater event in the Southern Ocean can weaken the AABW production by as much as two-thirds. In fact, their study demonstrates a more powerful response of the thermohaline circulation to a meltwater event in the Southern Ocean than to one in the North Atlantic.

Diapycnal vertical mixing is an important term in the density balance of AABW, due to the contact of the flow with the seafloor. Whitehead and Worthington [1982] have described density changes in the AABW in the North Atlantic north of 4°N as due to the downward fluxes of heat and salt across isopycnal surfaces. McDougall and Whitehead [1984] conclude that the dominant mixing process is diapycnal and estimate a diffusion coefficient to be near 3.9×10^{-4} m^2s^{-1}. A decrease in the horizontal mixing intensity in the Southern Ocean reduces the erosion of AABW during its sinking to the ocean floor, and enhances its density and northward transport [England, 1993].

Previous studies of the formation and circulation of AABW mainly focused on the sensitivity to the processes in the Southern Ocean. Our study aims to understand how the mixing processes in the Atlantic basin can affect the northward propagation of AABW. The changing vertical diffusivity in our study has a dual effect on the bottom water transport: the diffusivity controls the deep diffusion-induced upwelling in the low latitudes, and also alters the NADW overturning, which, in turn, affects the distribution of AABW in the Atlantic basin. In what follows we will attempt to estimate the effects of both of these processes.

2. NUMERICAL RESULTS.

We carried out a series of experiments in a GCM (MOM 1.0) within an idealized global geometry, which consists of three rectangular idealized basins to represent Atlantic, Pacific and Indian oceans (Figure 1). The model's domain and forcing profiles are identical to those described in Goodman [1998]. The grid spacing is coarse: 3.75 degrees in the zonal and 4 degrees in the meridional directions. There is no bottom topography except in the Drake Passage where the sill depth is 2500m and in the

Indonesian Passage where there is a plateau at 1551m. Restoring surface boundary conditions are applied for both temperature and salinity. The restoring timescale is 50 days over 50m and zonal basin averages of values from Levitus [1982] are used. Following England [1993], the restoring salinity off the Antarctic coast is increased to simulate effects of wintertime salt rejection. The model is forced by zonally uniform zonal wind stress from Hellerman and Rosenstein [1983].

Two groups of numerical experiments are carried out. In the first group, we use vertical diffusion with a constant coefficient k_v. We therefore do not account for the fact that vertical mixing is distributed unevenly in the World Ocean with higher values concentrated near rough topography [*Polzin et al.*, 1997, etc.]. The sensitivity experiments described here are identical to one another except for the value of k_v, which varies from 10^{-6} m^{2}s^{-1} to 2×10^{-4} m^{2}s^{-1}. The horizontal diffusion coefficient is 10^3 m^{2}s^{-1}.

We also conducted a second group of experiments with different settings. In the first one (BL), the vertical diffusivity is not constant but rather increases with depth from 3×10^{-5} m^{2}s^{-1} to 1.05×10^{-4} m^{2}s^{-1} at the bottom [*Bryan and Lewis*, 1979]. The profile is flipped in the second run (IBL), with vertical diffusivity decreasing with depth from to 1.05×10^{-4} m^{2}s^{-1} to 3×10^{-5} m^{2}s^{-1}. The third run (DW) has constant diffusivity of 5×10^{-5} m^{2}s^{-1} but the winds between 58°S and 38°S are doubled. Other parameters are the same as in other runs.

As k_v increases, we observe both deepening and strengthening of the NA overturning cell (Figures 2a,c,e). The values of T_n from the series of GCM experiments versus the corresponding values of k_v are shown in Figure 3. As one can see, the experimental points closely follow the curve $T_o + Bk_v^{2/3}$. The 2/3-power dependence on the vertical diffusivity contradicts findings by Bryan [1987], who reports a 1/3-power dependence in his model with a *single* cell, but is in general agreement with theoretical scaling [*Marotzke*, 1997; G99]. It is interesting to note that values for NADW overturning strength in the two runs with variable vertical diffusion (BL and IBL) also fall on the same curve if the corresponding values of k_v are computed as the average from 500m to the 1100m depth. This depth range was chosen to correspond to the maximum upwelling through the base of the pycnocline; the result therefore emphasizes that NADW overturning is controlled by the vertical diffusivity in the upper part of the ocean. The overturning is stronger in the experiment with doubled winds over the Southern Ocean than in the case with standard winds and the same vertical diffusivity, in agreement with previous studies [*McDermott*, 1996; *Toggweiler and Samuels*, 1998].

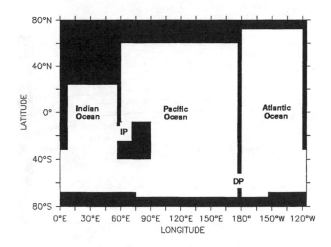

Figure 1. The model domain. DP refers to the location of Drake Passage, IP: the Indonesian Passage.

The bottom meridional circulation cell in the Atlantic, which exists due to the northward propagation of AABW, responds differently to the intensification of the vertical mixing. The northward transport of AABW *increases*, while the thickness of the cell *decreases* (Figures 2a,c,e). Since the northward flow is confined to a boundary current of invariant width, set by the momentum viscosity in our model (see the next section), the decreasing thickness also implies a reduced volume of AABW present at any moment. The increasing transport, T_a, in conjunction with the decreasing vertical extent of the cell, may seem counterintuitive. In the next section, we will provide an explanation for this phenomenon.

3. ANALYTICAL INTERPRETATION OF THE RESULTS.

In what follows we attempt to interpret the numerical results by analyzing a simple analytical model of the Atlantic basin (Figure 4). Our analysis is confined to the processes in the Atlantic basin only, and does not explicitly take into account the role of other basins and connections between them. The model includes two overturning cells: the upper, NADW cell with the thickness $2H$ with the transport T_n and the lower, AABW cell with the thickness $2H_a$ and total transport T_a. Other notations in Figure 4 are explained in the section 3.2.

3.1 The Transport and Thickness of the NADW Cell.

We start by deriving a relation between transport in the NADW cell and its thickness. The derivation will help to illustrate the differences in the dynamics of the NADW

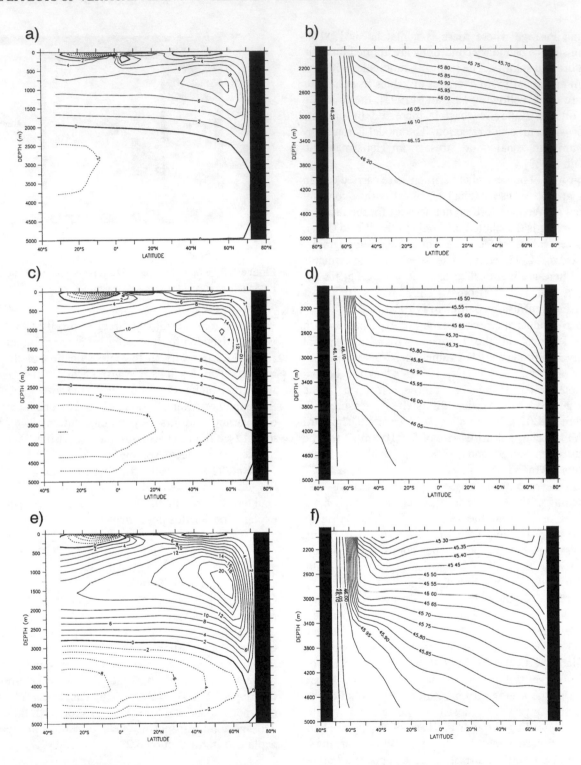

Figure 2. Left column: Meridional overturning streamfunction (Sv) the model Atlantic. Right column: potential density (referenced to 4000m) zonally averaged from 180E to 169W to in the bottom 3000m of the Atlantic basin. Three values of the vertical diffusivity k_v: **(a,b)** 10^{-5} m^2 sec^{-1}, **(c,d)** 5×10^{-5} m^2 sec^{-1}, and **(e,f)** 1×10^{-4} m^2 sec^{-1}.

and AABW cells. In this subsection, we closely follow the argument presented in G99 to relate the transport of NA cell T_n and its thickness H.

We now analyze the momentum balance in the deep western boundary current. We stress that we consider only a northward-flowing upper portion of the current within an upper half of the NADW cell. The total *depth-averaged* transport of the western boundary current in a model is set by the Sverdrup relation and is determined by the mid-latitude winds, which are kept constant in all runs. Western boundary layer theory implies that a balance exists between the horizontal diffusion of momentum and the meridional pressure gradient:

$$A_h \frac{V}{L_m^2} \sim \frac{1}{\rho_0}\left[\frac{\partial p}{\partial y}\right]$$

Where V is the meridional velocity in a western boundary layer of the width $L_m = (A_h/\beta)^{1/3}$. A_h is a horizontal momentum viscosity and ρ_0 is the reference density. The velocity then scales as:

$$V \sim L_m^2 \frac{1}{A_h \rho_0}\left[\frac{\partial p}{\partial y}\right] = \frac{1}{\beta L_m \rho_0}\left[\frac{\partial p}{\partial y}\right]$$

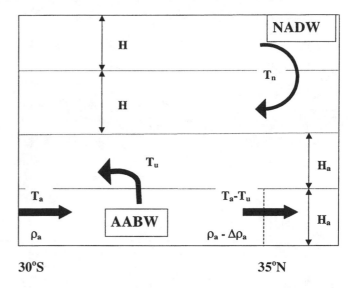

30°S   35°N

Figure 4. Schematics of the zonal-mean mass circulation in the Atlantic basin.

We assume that most of the northward transport takes place within the western boundary current; the NADW transport T_n can then be estimated as

$$T_n \sim V \times H \times L_m = \frac{H}{\rho_0 \beta}\left[\frac{\partial p}{\partial y}\right] \qquad (1)$$

Next, we scale the meridional pressure gradient $[\partial p/\partial y]$ as $gH\Delta\rho/L_u$, where L_u is the meridional extent of the density gradient in the upper layers and $\Delta\rho$ is the meridional density contrast in the upper layers. We finally obtain the expression for T_n and cell's thickness H:

$$T_n = C\frac{g\Delta\rho}{\rho_0 \beta L_u}H^2 \qquad (2a)$$

$$H = \left\{\frac{T_n \rho_0 \beta L_u}{Cg\Delta\rho}\right\}^{1/2} \qquad (2b)$$

where the constant C incorporates the effects of geometry and boundary layer structure.

Note that according to equation (2a), NADW transport T_n depends on both density contrast $\Delta\rho$ and the cell thickness H. In this study, we assume that $\Delta\rho$ is fixed in the upper layers, mainly because of the unchanging surface boundary conditions. In general, this term is affected by changes in the winds and surface buoyancy fluxes. The transport of the NADW cell in our study thus increases only due to the cell's deepening and there is a direct relationship between cell's thickness H and T_n. Other scenarios are possible in which the transport of a cell

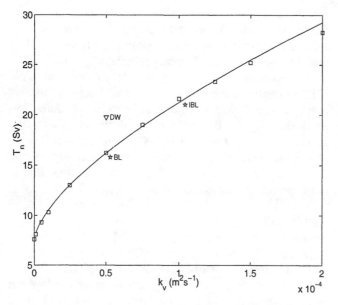

Figure 3. NA sinking rate T_n (Sv) vs. vertical diffusivity k_v. The solid line shows $T_o + Bk_v^{2/3}$. The squares show values from the runs with constant diffusivity. The stars show the values from the two additional runs with depth-dependent vertical diffusivity (BL and IBL) plotted against k_v averaged between 500m and 1100m depth. The triangle shows the value for DW experiment.

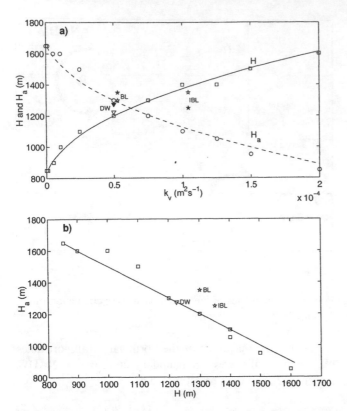

Figure 5. a) Cells' thickness. The squares show H from the OGCM runs; the circles show H_a from OGCM, the solid line shows H given by (2b); the dashed line shows $2500m$-H with H given by (2b). The open stars show the H values for the runs with depth-dependent vertical diffusivity (BL and IBL) plotted against k_v averaged between 500m and 1100m depth; the filled stars show the corresponding H_a values. The empty and filled triangles show the H and H_a for the DW run. **b)** Values of H_a plotted against H. The squares show constant diffusivity cases, the stars show depth-dependent diffusivity cases, and the triangle shows DW case. The solid line is $2500m$ – H.

intensifies mainly due to the increase in $\Delta\rho$, as for example in the AABW cell in this study (see the next subsection). Another example is a single-hemisphere thermohaline overturning, with a cell reaching the bottom [e.g. *Bryan*, 1987]. The thickness of the cell is thus not allowed to change; its transport can however vary due to changes in the meridional density gradients in the upper half of the cell.

3.2 The Thickness and Ttransport of the AABW Cell.

In our study with no bottom topography, AABW has the form of a viscous boundary current, which flows right under the NADW deep western boundary current. AABW thus occupies the volume left available by the upper

overturning cell and the decrease in H_a is explained by the downward expansion of the upper overturning cell. The presence of bottom topography would make the relation between cells' thicknesses more complicated, but a decrease in the AABW volume with intensification of the NADW cell can still be expected. On the other hand, it would be incorrect to conclude, that AABW simply passively responds to the changes in the upper cell and does not in turn influence the NADW circulation. The intensity of the AABW cell has been shown to affect the vertical extent of the NADW cell [*England*, 1993; *Seidov* et al., 2000].

In our analytical model, we assume that $2(H+Ha)=5000m$ (Figure 4). Strictly speaking, this is true only if the level of no motion is exactly in the middle of each cell, and each cell is therefore symmetrical in the vertical direction. Although it is not exactly the case (Figures 2a,c,e), we make this assumption for simplicity and determine H_a as equal to 2500m – H with H given by equation (2b). Doing this will also allow us to check the validity of the assumptions of section 3.1. We use constant values of L_u = 1,500km and $\Delta\rho$=0.1 kgm^{-3}. The value of the constant C is obtained by matching (2a) to GCM result for k_v=5×10^{-5} m^{2}s^{-1}. The resulting values of H and H_a agree well with those from the numerical experiments (Figure 5a).

Our additional runs help to confirm that thickness of the AABW cell is indeed controlled by the vertical extent of the NADW cell. For example, since H in the run with smaller diffusivity in the upper ocean (BL) is smaller than that in the run with larger upper-layer k_v (IBL), the thickness of the AABW cell in BL is larger than that in the IBL. In contrast, the diffusivity in the deep ocean has a much smaller effect on the AABW cell's thickness. In particular, H_a in BL, which has deep diffusivity (the average between 4500m and 3900m) of 1.03×10^{-4} m^{2}s^{-1} and upper-ocean diffusivity (the average from 500m to the 1100m depth) of 5.3×10^{-5} m^{2}s^{-1}, is much larger than that in the experiment with the constant k_v = 1×10^{-4} m^{2}s^{-1}, but is rather close in value to H_a in the run with constant k_v of 5×10^{-5} m^{2}s^{-1}. Figure 5b further stresses that H_a is determined by H. Values of H_a in all of the conducted experiments, including those with doubled winds and the depth-dependent diffusivity, follow the $2500m$-H curve rather closely. In particular, the agreement in the DW run is as good as that in the standard case with unchanged winds and same diffusivity, although the NADW cell intensifies and deepens as a result of increased southern winds (Figure 3).

We now estimate the transport T_a of the AABW cell. As in the upper ocean (see the preceding section), we assume that the northward flow of AABW is concentrated within the boundary current of width L_m and is driven by the meridional gradient in pressure. We estimate only a northward transport of the AABW within the boundary layer. In a numerical model with meridional lateral walls and no bottom topography, the AABW flow is a northward flowing western boundary current. In more realistic configurations, the bottom current will be steered by the bottom topography. Nevertheless, it will still take the form of the viscous boundary layer flowing immediately to the east from either the seamounts or the continental line in the western Atlantic [see for example, *Gent et al.*, 1998]. The current's width in that case is set by the friction and topography, but still is not determined by mixing and density structure. Our derivation below is valid as long as the width of the current can be assumed constant; then only a geometrical constant C_a (see equation 3 below) has to be changed for each particular model. We then repeat the above derivation of equation (1) from section 3.1 to obtain

$$T_a \sim \frac{H_a}{\rho_0 \beta} \left[\frac{\partial p}{\partial y} \right]$$

Our next step is to estimate the deep meridional pressure gradient in the above expression. Since the meridional flow changes direction at $z = -2H - H_a$ (Figure 4), the meridional pressure gradient in the western boundary current has to be nearly zero at that depth. Therefore, in order to estimate the pressure gradient near the bottom, we need to integrate meridional density contrast over the AABW thickness H_a only. $[\partial p / \partial y]$ is then scaled as $g H_a \Delta \rho_a / L_a$, where L_a is the meridional extent of the density gradient in the AABW layer and $\Delta \rho_a$ is the meridional density contrast (Figure 4). We obtain

$$T_a = C_a \frac{g \Delta \rho_a}{\rho_0 \beta L_a} H_a^2 \qquad (3)$$

The constant C_a depends on geometry of the boundary current. As we know, the thickness H_a of the AABW cell decreases in response to intensifying NA overturning. According to equation (3), the decrease in H_a will act to decrease the transport of AABW. On the other hand, the intensified diffusion and the increased downward advection of lighter waters by NADW cell decrease density in the low latitudes (ρ_a-$\Delta \rho_a$ in Figure 4); see Figures 2b,d,f. The lightening of this deep low-latitude water enhances its density difference $\Delta \rho_a$ with dense bottom water entering Atlantic from the Southern Ocean (ρ_a in Figure 4). Increasing $\Delta \rho_a$ will drive more AABW into the Atlantic and will act to increase T_a.

Our next step is to quantify this idea and estimate $\Delta \rho_a$. For simplicity, we assume here a linear dependence of

density on temperature and salinity; density then satisfies the same equations as temperature and salinity. Consider then density balance in the AABW region in the Atlantic sketched in Figure 4. The water with density ρ_a enters from the south at the rate T_a; a part of it (T_u) then upwells through the upper surface A_a of the boundary current, the reminder leaves in the north with density ρ_a-$\Delta \rho_a$. The resulting convergence of the horizontal advective density flux is balanced by the upwelling of dense waters (with the rate of T_u) and density loss due to the vertical diffusion:

$$T_a \rho_a - (T_a - T_u)(\rho_a - \Delta \rho_a) = k_v A_a \frac{\delta \rho}{H_a} + T_u \rho_a,$$

from which we find

$$\Delta \rho_a (T_a - T_u) = k_v A_a \frac{\delta \rho}{H_a} \qquad (4)$$

$\delta \rho / H_a$ is our estimate of the vertical density gradient at the top of the AABW layer. We take $\delta \rho$ to be a constant equal to 0.25 kg m^{-3}; the vertical gradient then increases as the AABW layer becomes thinner. Our simple estimate therefore attempts to account for the fact that the deepening NADW cell presses the isopycnals near the bottom closer together (Figures 2b,d,f and 6a). In general, the vertical density gradient should also be sensitive to the processes in the Southern Ocean. In particular, increase in the salinity at the Antarctic coast will enhance the density of AABW and its vertical contrast with overlaying waters, which will act to increase the transport in agreement with previous numerical studies [*England*, 1992]. Our simple analytical model of the Atlantic basin, however, cannot easily include processes of the Southern Ocean. Nevertheless, Figure 6a demonstrates that our assumption works reasonably well for larger diffusivities; the agreement is worse for smaller diffusivities, which suggests that other mechanisms may be at work in that regime. The vertical gradient is larger in the experiment with the doubled southern winds (DW run) illustrating dependence of the deep density field on the processes in the Southern Ocean.

Note that we neglect the horizontal diffusion in our density balance (eq.4). Our estimates show that the magnitude of lateral diffusion term is small in the bottom layers. In addition, diapycnal diffusion associated with lateral diffusion should in principle be even smaller since tracers are widely believed to mix mainly along isopycnal surfaces. In a model with the mixing in the horizontal direction, however, the deep version of the Veronis effect can alter the density balance. The neglect of the horizontal diffusion term may be an additional reason for

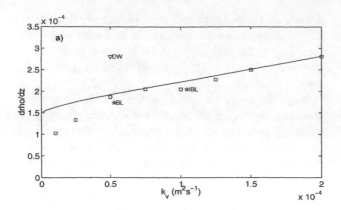

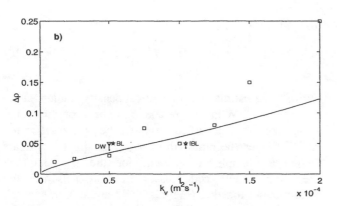

Figure 6. a) The values of the vertical density gradient in the middle of the AABW cell from the numerical runs with: constant diffusivity (the squares), depth-dependent diffusivity (the stars), and doubled southern winds (the triangle). The solid line shows $0.25/(2500\text{-}H)$, with H given by (2b). Units are kg m^{-4}. **b)** The values of the deep meridional density contrast from the numerical runs (marked as in the upper panel). The dots connected with the corresponding experimental points show values given by (5) where H_a is estimated by $2500m\text{-}H$ with H given by (2b) and k_v averaged between 4500m and 2900m. Units are kg m^{-3}.

disagreement between our theory and the results from the OGCM that employs horizontal diffusion.

From the horizontal velocity maps, which are not shown here, we estimate that approximately half of the total transport in the model upwells within the AABW boundary current (30S to 35N). The remaining upwelling occurs in the broad region north of 35N. Following our notations, T_u is therefore equal to $1/2T_a$. We combine equations (3) and (4), to yield an expression for $\Delta\rho_a$:

$$\Delta\rho_a^2 = \frac{k_v\delta\rho}{H_a^3}A_a\frac{\beta\rho_0 L_a}{1/2gC_a} \tag{5}$$

As can be seen from (5), increase in the deep density gradient is caused both by the increase in the vertical diffusivity and decrease in the thickness of the AABW

layer. According to equation (3), this increase in the meridional density gradient is the sole reason for the intensification of the AABW transport, since the other changing quantity in the right-hand side of (3), layer thickness H_a, acts to decrease T_a. Figure 6b shows that the value of the meridional density gradient predicted by (5) are in a reasonable agreement with those computed from the OGCM. The agreement however worsens as vertical diffusivity increases. Values of H_a are computed as $2500\text{-}H$, where H is given by (2b). Experiments BL, IBL are compared against equation 5 with values of k_v taken as a depth average between 4500m and 3900m.

We then substitute expression (5) into equation (3), and finally get the relation for Ta:

$$T_a = \left\{\frac{gC_aA_a}{1/2\beta\rho_0 L_a}\right\}^{1/2}\left(\delta\rho k_v H_a\right)^{1/2} \tag{6}$$

Equation (6) suggests that the main factors affecting T_a are the vertical diffusivity, the vertical extent of the AABW flow H_a (largely controlled by the NADW cell), and the deep vertical density gradient estimated by $\delta\rho/Ha$. We take L_a to be 6,500 km, which is the distance between 30°S and 35°N; the corresponding value of A_a is then 10^{13}m^2. T_a is northward AABW transport at 30°S computed within the boundary current. To compare the predicted dependence of T_a on diffusivity with that observed in the GCM runs, we compute the constant C_a by matching equation (3) to the value of T_a in a GCM run with k_v=5×10^{-5} m^{2}s^{-1}. Values of H_a are computed as $2500\text{-}H$, where H is given by (2b). The resulting values of T_a agree reasonably well with those obtained from the OGCM runs (Figure 7). For the latter values, we integrate only the northward meridional velocities within a western boundary layer and do not include southward re-circulation. Values from the experiments BL and IBL are compared against those from the equation 6 with k_v taken as a depth average between 4500m and 3900m. The agreement with the theory is the worst for the BL run. Nevertheless, it is important to note that the numerical value of T_a in that case is still larger than that in the IBL case. According to the equation 6, T_a should be larger in a case with larger bottom diffusivity and shallower upper NADW cell. We next recall that NADW cell is shallower and deep k_v is larger in the BL than in the IBL.

4. SUMMARY AND CONCLUSIONS.

This study aims at understanding dynamical processes that control the volume and transport of the AABW in the Atlantic Ocean. We focus on the effects of the vertical mixing, and analyze results from a set of GCM experiments with different values of vertical diffusivity. Our simple model of the Atlantic basin excludes processes in the Southern Ocean that have strong influence on the density of the AABW mass and affect density balance in the deep layers. Our goal is to study the role of the dynamics of the Atlantic basin alone, which have been largely left out from the previous studies of the sensitivity of the bottom water formation.

In this study, both the AABW and NADW react to changes in vertical mixing. The response of the AABW cell to increasing diffusivity is however different from that of the NADW cell. The AABW layer in the deep Atlantic becomes thinner, because its vertical extent is controlled by the thickness of the NADW cell, which deepens. The latter effect is also observed if the upper cell intensifies due to the increased southern winds with unchanged vertical diffusivity. Nevertheless, we should not rush to the conclusion that the AABW cell has a passive role; increased surface salinity near Antarctic coast has been shown to enhance bottom water formation and push NADW cell to shallower depths [*England*, 1993; *Seidov*, 2000].

As in the NADW cell, the transport of the AABW is also proportional to the square of layer thickness and a meridional density gradient, but it intensifies for different reasons. While the increase in the T_n can be attributed to increasing thickness with unchanged density gradient, the intensification of the AABW transport T_a is explained by the strong increase in the meridional density contrast between dense water flowing from the Southern Ocean and lighter water at low latitudes. The decrease in the low-latitude density is described through the density balance in the AABW layer. The convergence of the horizontal density flux is balanced by the density loss through the vertical diffusion and upwelling. The diffusion flux increases rapidly with increasing diffusivity both due to the increase in k_v and increase in the vertical stratification. The latter effect is caused by the deepening NADW cell, which pushes isopycnal surfaces closer together.

The term "AABW" is used rather loosely in our study with no bottom topography. In the real Atlantic, the corresponding northward flowing bottom water is likely to be a combination of AABW and Circumpolar Deep Water. In addition, the bottom water flows in the middle of the South Atlantic, steered by the bottom topography, and is not always located right beneath the deep western boundary current in the upper layers. However, since this water mass is the deepest and the densest in a model at mid- and low latitudes, our scaling analysis should remain valid in more realistic configuration. The width of the AABW current is unlikely to be set by the friction alone

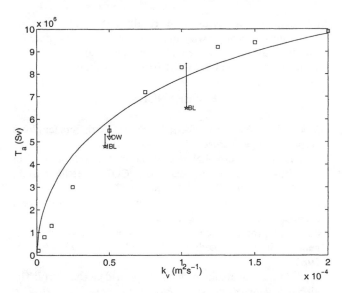

Figure 7. AABW transport T_a (Sv) vs. vertical diffusivity k_v. The squares show the values from the OGCM runs measured between 180E and 169W. The solid line shows T_a given by (6), where H_a is estimated by $2500m\text{-}H$ with H given by (2b). The stars show the values from BL and IBL runs, the triangle shows the value from the DW run. The dots connected with the corresponding experimental points show values given by (6) where H_a is estimated by $2500m\text{-}H$ with H given by (2b) and k_v averaged between 4500m and 2900m.

(as assumed in our simple model), but in large part determined by the bottom topography. Even in that case, however, the width of the bottom layer is not affected by the density structure and our derivation can be easily generalized for that case.

Based on our results, we expect the rate of AABW formation to be largely determined by the mixing processes and density distribution in the deep ocean. Realistic representation of bottom mixing is therefore necessary for proper representation of the deep circulation in an oceanic GCM. Studies of the observed distribution of mixing near the ocean bottom [e.g. *Polzin et al.*, 1997] are especially valuable in that regard. The volume of AABW present at any time is controlled by the amount of space left available for this water mass by the upper overturning cell and the bathymetry of the ocean bottom. The relationship between the AABW thickness and the depth of the upper cell may be more complicated than it is in this study with no bottom topography.

Acknowledgments. We are thankful to the two anonymous reviewers for their comments on improving the manuscript. One of the reviewers suggested additional experiments, which helped to clarify main assumptions. We thank Dr. Seidov for his advice on improving the discussion of the results in the context of other studies. This publication was supported by the Joint Institute for the Study of the Atmosphere and Ocean (JISAO) under NOAA Cooperative Agreement #NA67RJ0155, Contribution #785.

REFERENCES.

Broecker, W.S., Massive iceberg discharges as triggers for global climate change. *Nature*, 372, 421-424, 1994.

Broecker, W.S., Thermohaline circulation, the Achilles heel of our climate system: will man-made CO2 upset the current balance? *Science*, 278, 1582-1588, 1997.

Bryan, K., and L.J. Lewis, A water mass model of the world ocean, *J. Geophys. Res.*, 84, 2503-2517.

Bryan, F., Parameter sensitivity of primitive equation ocean general circulation models, *J. Phys. Oceanogr.*, 17, 970-985, 1987.

Cox, M. D., An idealized model of the World Ocean. Part I: The global-scale water masses, *J. Phys. Oceanogr.*, 19, 1730-1752, 1989.

England, M., On the formation of Antarctic Intermediate and Bottom Water in Ocean General Circulation Models, *J. Phys. Oceanogr.*, 22, 918-926, 1992.

England, M., Representing the global-scale water masses in Ocean General Circulation Models, *J. Phys. Oceanogr.*, 23, 1523-1551, 1993.

Gent, P.R., F.O. Bryan, G. Danabasoglu, S.C. Doney, W. R. Holland, W.G. Large, and J.C. McWilliams, The NCAR climate model global ocean component, *J. Climate*, 1287-1306, 1998.

Gnanadesikan, A., A simple predictive model for the structure of the oceanic pycnocline, *Science*, 283, 2077-2079, 1999.

Goodman, P., The role of North Atlantic deep water formation in an OGCM's ventilation and thermohaline circulation, *J. Phys. Oceanogr.*, 28, 1759-1785, 1998.

Goose, H., and T. Fichefet, Importance of ice-ocean interactions for the global ocean circulation: A model study, *J. Geophys. Res.*, 104, 23337-23355, 1999.

Hovine, S., and T. Fichefet, A zonally averaged, three-basin ocean circulation model for climate studies. *Clim. Dyn.*, 10, 313-331, 1994.

Maier-Reimer, E., Mikolajewicz, U., and Hasselmann, K., On the sensitivity of the global ocean circulation to changes in the surface heat flux forcing, 1991.

Manabe, S. and R.J. Stouffer, Two stable equilibria of a coupled ocean-atmosphere model. *J. Climate*, 1, 841-866, 1988.

Manabe, S. and R.J. Stouffer, Multiple-century response of a coupled ocean-atmosphere model to an increase of atmospheric carbon dioxide. *J. Climate.*, 7, 5-23, 1994.

Marotzke, J., Boundary mixing and the dynamics of three-dimensional thermohaline circulations, *J. Phys. Oceanogr.*, 27, 1713-1728, 1997.

McDermott, D. A., The regulation of Northern overturning by Southern Hemisphere winds, *J. Phys. Oceanogr.*, 26, 1234-1255, 1996.

McDougall, T.J., and J.A. Whitehead, Jr., Estimates of the relative roles of diapycnal, isopycnal and double-diffusive mixing in Antarctic Bottom Water in the North Atlantic, *J. Geophys. Res.*, 89, 10 479-10 483, 1984.

Munk, W., Abyssal recipes, *Deep-Sea Res.*, 13, 707-730, 1966.

Polzin, K.L., J.M. Toole, J.R. Ledwell, and R.W. Schmitt, Spatial variability of turbulent mixing in the abyssal ocean. *Science*, 276, 93-96, 1997.

Rahmstorf, S., On the freshwater forcing and transport of the Atlantic thermohaline circulation. *Climate Dynamics*, 12, 799-811, 1996.

Seidov, D. and B.J. Haupt, Last glacial and meltwater interbasin water exchanges and sedimentation in the world ocean. *Paleoceanography*, 14, 760-769, 1999.

Seidov, D. and M. Maslin, North Atlantic Deep Water circulation collapse during the Heinrich events, *Geology*, 27, 23-26, 1999.

Seidov, D., E. Barron, and B.J. Haupt, Meltwater and the global ocean conveyor: Northern versus southern connections, *Global and Planetary Change*, in press.

Stocker, T.F., D.G Wright, and W.S. Broecker, The influence of high-latitude surface forcing on the globalthermohaline circulation, *Paleoceanography*, 7, 529-541, 1992.

Stocker, T.F., The seesaw effect. *Science*, 282, 61-62, 1998.

Stocker, T.F., Abrupt climate changes: from the past to the future – a review. *Int. J. Earth Sciences*, 88, 365-374, 1999.

Stommel, H., The abyssal circulation, *Deep-Sea Res.*, 5, 80-82, 1958.

Stommel, H., and A.B. Arons, On the abyssal circulation of the world ocean.- II An idealized model of the circulation pattern and amplitude in oceanic basins, *Deep-Sea Res.*, 6, 217-233, 1960.

Toggweiler, J.R., and B. Samuels, On the ocean's large scale circulation near the limit of no vertical mixing. *J. Phys. Oceanogr.*, 28, 1832-1852, 1998.

Whitehead, J.A., Jr., and L.V. Worthington, The flux and mixing rates of Antarctic Bottom Water within the North Atlantic, *J. Geophys. Res.*, 87, 7903-7924, 1982.

P.J. Goodman, Lamont Doherty Earth Observatory Oceanography, 104C, 61 Rte 9W, Palisades, NY 10964 (e-mail: paulg@rosie.ldgo.columbia.edu)

I.V. Kamenkovich, JISAO, University of Washington, Box 354235, Seattle, Washington 98195-4235. (e-mail: kamen@atmos.washington.edu)

The Influence of Deep Ocean Diffusivity on the Temporal Variability of the Thermohaline Circulation

Kotaro Sakai

Institute for Global Change Research, Frontier Research System for Global Change, Yokohama, Japan

W. Richard Peltier

Department of Physics, University of Toronto, Toronto, Ontario, Canada

We investigate the influences of the diffusion coefficients for heat and salt in the deep ocean upon the stability of the Atlantic thermohaline circulation through a parameter space investigation that employs the GFDL Modular Ocean Model (MOM). The ocean general circulation model is configured to represent an idealized Atlantic basin and steadily (including an annual cycle) forced under the assumption of mixed boundary conditions. The impact on the results of the use of mixed boundary conditions is examined through additional sensitivity experiments to demonstrate that this choice of boundary conditions does not affect the main conclusions. The two primary sets of the experiments that we describe consist of sensitivity analyses for both horizontal and vertical diffusion coefficients; Bryan-Lewis type diffusion coefficient profiles are applied in such a way that one diffusion coefficient profile is varied while the other diffusion coefficient profile is fixed. We are thereby able to demonstrate that the choice of the horizontal diffusion coefficient plays a crucial role in determining the stability of the thermohaline circulation which is such that the thermohaline circulation exhibits very intense oscillations on a multi-millennium timescale when the horizontal diffusion coefficient of the deep ocean is set to a sufficiently small value. We demonstrate that the critical value required to realize an oscillatory solution of the thermohaline circulation lies well within the range of observed large-scale ocean diffusion coefficients, whereas the value of vertical diffusion coefficients affect not the bifurcation structure but the period of the millennial timescale oscillation.

1. INTRODUCTION

One of the most dramatic characteristics of climate variability during the last ice-age cycle is clearly that

associated with the so-called Dansgaard-Oeschger (D-O) oscillation. This climate oscillation is most clearly observed as an oscillation of an ice-core derived temperature proxy, namely the oxygen isotope records based on mass spectrometric measurements on the ice cores from Summit, Greenland (e.g. GRIP, 1993; Taylor et al., 1993). The large amplitude millennium timescale oxygen isotope oscillations that are most evident during oxygen isotope stage (OIS) 3, once calibrated, are believed to correspond to atmospheric temperature varia-

The Oceans and Rapid Climate Change: Past, Present, and Future
Geophysical Monograph 126

tions of amplitude 5 ~ 15 °C over the Greenland location. The timescale of each cycle of the D-O oscillation may extend to several millennia, with each cycle being characterized by a sudden warming out of the cold glacial state followed by a comparatively slow cooling out of the warm state back towards the cold state thus setting the stage for the next sudden return to the warm state.

The extent to which the impact of this millennium timescale climate variability influences the global climate system is still under active investigation. It is, however, both interesting and important that the variability of the deep North Atlantic thermohaline circulation (THC) has been closely correlated to the D-O oscillation evident in the ice core records. According to the analyses of Keigwin and Jones(1994), and of Rasmussen et al.(1996) more recently, both the relative amplitude and the period of the oscillation appear in marine records from the North Atlantic and correspond well to that of the atmospheric temperature proxy from Greenland.

Although a good correlation cannot immediately be taken to imply a direct physical connection, there is a good reason to believe that such a connection does in fact exist. The oceanic evidence analyzed by Keigwin and Jones(1994) and by Rasmussen et al.(1996) indicate that the strength of the deep circulation did indeed fluctuate on the D-O timescale during OIS 3. The thermohaline circulation (THC) in the North Atlantic is predominantly driven by North Atlantic Deep Water (NADW) production, and such NADW is formed primarily at the surface of the high latitude North Atlantic. Dense NADW naturally forms during the winter season when the surface of the North Atlantic is cooled intensely by the much colder air masses that travel across its surface from North America, and the ocean in turn releases very large amounts of heat to the atmosphere. The significance of this heat release can easily be understood by considering the relative warmth of northwest European winter climate compared with other parts of the world at the same latitude. The presence or absence of this oceanic heat source over the high latitude North Atlantic upon climate has been explicitly examined, for example, by Fawcett et al.(1997), and the shutdown and onset of the deep water formation process thereby shown to nicely explain the climate variation inferred to have occurred during the Younger-Dryas period as this is recorded in the Summit, Greenland ice core records.

The physical processes underlying the Dansgaard-Oeschger oscillation, however, have yet to be explained in an entirely satisfactory manner. Any such satisfactory explanation must clearly include identification of the properties of the climate system that support the observed "fibrillation" of THC strength that con-

trols the intensity of the oscillation as well as the reason why the observed fibrillation occurs with significant amplitude only under glacial conditions (Peltier and Sakai,2001)).

There have in fact been a few recent theoretical attempts to explain the observed millennium timescale variability. Under full glacial conditions, global average temperature was much lower than present, and extensive continental ice masses covered both northern North America and Northwestern Europe. The earliest suggestion by Broecker et al.(1990) was that coupling between the continental ice sheets and NADW formation plays the central role; the idea being that intense NADW formation would cease whenever sufficient freshwater is delivered to the surface of the North Atlantic to sufficiently reduce the density of the surface waters over the high latitude North Atlantic. These authors suggested that the meltwater delivered by continental ice sheet disintegration is the cause; when NADW formation occurs, accompanied by intense heat release to the atmosphere this enhances melting of the continental ice sheets, and the consequent increase of the freshwater discharge onto the North Atlantic then suppresses NADW formation. Reduction of NADW formation then causes cooling of the atmosphere. This cooling reduces ice sheet melting. Once the meltwater discharge is decreased, the surface salinity gradually increases until the onset of the intense NADW formation occurs once more.

This mechanism appears entirely plausible on qualitative grounds, and some paleorecords do indeed indicate the occurrence of intense episodes of freshwater outflow onto the surface of the North Atlantic from the Laurentide ice sheet complex via the creation of vast "armadas" of icebergs, events which have been recorded in deep sea sedimentary cores in the form of horizons containing large quantities of ice-rafted debris. These are currently referred to as "Heinrich events" (e.g. Heinrich,1988). This series of events is considered to occur quasi-periodically on a deca-millennial timescale. According to the analysis by Bond et al.(1993), these episodic Heinrich events serve as pacemakers of a longer timescale climate cycle, now referred to as "Bond cycles", each cycle of which consists of several cycles of the Dansgaard-Oeschger oscillation of ever decreasing amplitude but with no related Heinrich events and thus no apparent connection to ice-sheet instabilities. Another difficulty with the original Broecker et al.(1990) hypothesis is the requirement of a very close and possibly nonlinear connection between melting of the Laurentide ice sheet complex and NADW formation. Since the ice sheet complex is situated "upstream" of the source of heat release from the North Atlantic surface that occurs via NADW formation, the impact of onset

and shutdown of the THC may not affect the Laurentide ice sheet in the way required by the hypothesis. No
evidence in support of the notion that episodes of intensive melting (mass loss) of the Laurentide ice sheets are
synchronized with the Dansgaard-Oeschger oscillation
has yet been found. Recent analyses, for example by
Kreveld et al.(2000), show that minor ice rafting events
can be observed on timescales similar to the Dansgaard-
Oeschger cycles even through Holocene. These efforts
have certainly increased our knowledge on the climate
system, but we still require further information on the
role the icesheets in the millennium timescale variability
that occurred under glacial conditions.

There is another hypothesis as to the origins of the
Dansgaard-Oeschger oscillation, however, that has recently been forthcoming. This hypothesis attributes
the mechanism of the oscillation entirely to the internal
dynamics of the North Atlantic thermohaline circulation itself under conditions in which a time independent
freshwater flux anomaly is applied to the surface. The
origin of this sort of internal oscillation in the thermohaline circulation may be traced to the "flush-collapse"
type oscillation originally revealed in the THC analyses of Marotzke(1989) although no possible connection of this oscillation to any observed physical process was recognized. In our original analysis of this
problem, Sakai and Peltier(1995), we performed a series of experiments in which a two-dimensional model
of an idealized Atlantic basin was integrated under so-
called "mixed boundary conditions". These analyses
demonstrated that the Atlantic thermohaline circulation would be expected to exhibit oscillatory behavior
when the net surface freshwater flux into the basin is
sufficiently strong. Under constant anomalous freshwater forcing the period of the oscillation was shown to
vary in the range from centuries (weak fluctuation) to
millennia (very intense oscillation). Subsequent work (
Sakai and Peltier,1996; hereafter referred to as SP96)
employed a multi-two-dimensional-basin model of the
global thermohaline circulation, analysis of which revealed the existence of a hydrologically controlled bifurcation in the thermohaline circulation under a localized
but persistent freshwater flux anomaly. When a critical intensity of anomalous freshwater flux was applied
with time independent amplitude onto the high latitude
North Atlantic, the North Atlantic thermohaline circulation was shown to deliver very distinct millennium
timescale oscillations. When the anomalous freshwater
forcing was weaker than the critical intensity required to
induce the bifurcation, however, the thermohaline circulation was found to be only very weakly time dependent.
This structure has been further confirmed in Sakai and
Peltier(1997; hereafter referred to as SP97), in which
the global ocean model was coupled to an atmospheric

energy balance model. In this configuration with explicit atmospheric feedback, the model was shown to
require a somewhat stronger anomalous freshwater flux
onto the high latitude North Atlantic in order to induce the Hopf bifurcation but beyond the bifurcation
point the millennial timescale oscillation was shown to
persist. The assumption made in these analyses, based
upon the fact that the high latitude North Atlantic was
surrounded by great continental ice sheets during OIS
3, is that the hydrological cycle was significantly modified in this circumstances such that the buoyancy flux
at the surface of the high latitude North Atlantic was
much enhanced. Clear observational support for this assumption is provided by Duplessy et al.(1991) whose sea
surface salinity reconstructions for the last glacial maximum demonstrate that high latitude Atlantic sea surface salinity (SSS) was significantly reduced from modern. Nevertheless there are a number of issues that need
to be addressed before we can be entirely confident that
we have identified the correct explanation for the D-O
oscillation.

The first of these concerns the impact of coupling between the THC and the wind driven circulation. The
wind-driven circulation exerts significant influence on
the sea surface salinity distribution and can thereby
modify the impact of enhanced surface freshwater flux
on the thermohaline circulation. Furthermore sea surface temperature changes induced by variations in the
NADW formation process are likely to cause changes in
the atmospheric circulation. Properties of the gyre circulation such as the Gulf Stream and the North Atlantic
Drift, for example, are greatly influenced by the pattern
of surface wind stress. These two currents play a primary role in conveying the high salinity water from low
latitudes to higher latitudes. If these northward currents are weakened for some reason, this too results in
a decrease of high latitude SSS.

A second issue that needs to be addressed concerns
the influence of the very cold climate conditions that
were characteristic of the glacial period. The experiments described in SP97 were conducted under the assumption of "modern" surface temperature conditions
excepting those related to the hydrological cycle. The
resulting sea surface temperature distribution that obtains when intense NADW formation is maintained in
these experiments is therefore much closer to "modern"
climate and therefore inappropriate to glacial conditions. Imposing a very cold climate, however, leads
to a further enhancement of the impact of salinity
changes in determining the density of sea water (e.g.
Fofonoff,1985). Winton(1997) has recently argued that
under extremely cold climate conditions the "North Atlantic" thermohaline circulation may also be destabi-
lized and similar millennium timescale oscillation are

produced through the temperature effect as we have shown are produced by hydrological forcing alone.

A third outstanding issue, and the one that will serve as primary focus of investigation in this paper, concerns the influence of the diffusion coefficients for heat and salt upon the behavior of the thermohaline circulation in a fully three dimensional model. A typical low-resolution ocean general circulation model (resolution > 1°) cannot resolve so-called "meso-scale" eddies that are the oceanic equivalents of atmospheric cyclones and anti-cyclones which are well resolved in atmospheric general circulation models even at modest resolution because of the order of magnitude difference between the Rossby radius of deformation for the atmosphere and that for the oceans. Since these eddies play a quantitatively important role in the horizontal mixing process in the oceans, some parameterization of their influence must be included in such low-resolution ocean general circulation models.

Examinations of the sensitivity of the modeled ocean general circulation to the choice of mixing coefficients in ocean general circulation models have, of course, been repeatedly performed in the literature for various ocean general circulation models, but these have been rather sharply focused on the impact on the steady state solution. The existence of multiple equilibria of the thermohaline circulation has itself, of course, been repeatedly discussed since appearance of the paper by Stommel(1961) who employed a simple 2-box model to demonstrate this possibility. Somewhat more quantitative analyses of the same property of the large scale ocean circulation have been provided recently by Rahmstorf(1995) using a ocean general circulation model. His results show no evidence of the existence of a millennium timescale oscillation. As described earlier in the Introduction to this paper, the existence of millennium timescale oscillatory solutions for the thermohaline circulation under certain boundary conditions has been previously demonstrated mostly on the basis of two-dimensional ocean models. Equivalent results derived on the basis of full three dimensional ocean general circulation models have as yet not been produced in demonstration of the continuing existence of an intense millennium timescale oscillation that could be connected to the Dansgaard-Oeschger oscillation. This will be the focus of the present paper.

Sakai and Peltier(1999; hereafter referred to as SP99) have more recently described a dynamical systems model which also appears to capture essentially the same mechanism of thermohaline fibrillation as previously obtained in the two dimensional models of the THC, and these analyses have suggested a possible explanation of the apparent discrepancy between the predictions of the two- and three-dimensional model re-

sults. One of the results reported in this paper suggests that the influence of diffusion processes on the existence of the oscillation may be extreme. Specifically it is demonstrated therein that by imposing very large (horizontal) mixing coefficients for the deep ocean in the dynamical system, the results of the 3-box model approach those of the Stommel box model of the thermohaline circulation which exhibits only steady multiple equilibria. On the other hand, the multiple equilibria become unstable and the dynamical system allows oscillatory solutions when the mixing coefficients in the abyssal ocean component of the system are small enough.

These previous analyses, although qualitative as they are based upon a highly simplified model of the thermohaline circulation, clearly suggest why strong millennium timescale oscillations have not been clearly revealed in past integrations of three-dimensional ocean general circulation models. These model may simply be employing mixing coefficients which are larger than the critical value below which oscillatory solutions become possible. The important question then becomes how weak horizontal diffusion processes must be before oscillatory behavior begins. If the critical value is much smaller than observed large-scale mixing coefficients in the deep ocean, then the millennium timescale oscillations that have been observed in simpler models may simply be an artifact of over-simplification. The primary purpose of the present paper is to attempt to answer this question through a sequence of numerical experiments using a conventional three dimensional ocean general circulation model. The detailed design of the experiments will be discussed in the next section in which the numerical model and its specific geometric configuration are considered. The results we have obtained in these experiments are described in the third Section and our conclusions together with a brief summary of the results are offered in the concluding fourth Section.

2. OCEAN GENERAL CIRCULATION MODEL AND DESIGN OF THE NUMERICAL EXPERIMENTS

All of the experiments that were conducted in the course of the present study have been performed using a stand-alone Ocean General Circulation Model (OGCM), specifically the Modular Ocean Model (MOM) Version 2.2 β that has been developed at the Geophysical Fluid Dynamics Laboratory (GFDL) in Princeton, N.J.

For the purpose of these investigations, the numerical model has been configured in single idealized Atlantic basin mode. We specifically employ a so-called sector model that covers a rectangular domain in the longitude-latitude plane with a uniform depth of 5000 m. The longitudinal span has been held fixed to 60°, and the latitudinal extent is fixed to the range from 56°S to 80°N. In order to examine a wide range of horizontal diffusion coefficients, the horizontal resolution is chosen to be 2° in order to enable us to employ relatively small diffusion coefficients. This might be considered to constitute rather fine resolution for experiments of this kind that require very long integrations. The number of vertical levels is 16 and these are located at depths of 25, 75, 165, 265, 415, 585, 815, 1085, 1415, 1805, 2265, 2745, 3255, 3745, 4255, 4745 meters for t-cells, and 50, 120, 215, 340, 500, 700, 950, 1250, 1610, 2035, 2505, 3000, 3500, 4000, 4500, 5000 meters for w-cells.

For the purpose of these experiments, the surface boundary conditions are also idealized: prescribed seasonally varying sea surface temperature (SST) to which the time varying SST of the model is continuously restored, and the freshwater flux specified so as to include a localized high-latitude component of the forcing due to surface runoff during glacial conditions. The wind stress in both horizontal directions (longitude-latitude) is held fixed to zero for all of the analyses we shall perform in order to isolate the influence of three dimensionality from that of the wind driven circulation. A Neumann boundary condition, on the other hand, is employed for the salinity boundary condition. The latitudinal variation of the "salt flux", which is proportional to (minus) the combination of precipitation plus runoff minus evaporation, has been set to a form based upon a diagnosed "fresh water" flux at the surface by first imposing a fixed sea surface salinity using the standard parameter configuration of the OGCM. A number of different anomalous freshwater distributions have then been superimposed upon this "present salt flux" distribution, as in the previous analyses described in SP96 and SP97. The distribution shown as the solid line in Figure 1 is intended to represent the perturbed form of the flux that is assumed to obtain under ice-age conditions. As the dashed line we show the "modern"(without anomaly) distribution of freshwater flux for comparison.

The temperature conditions at the ocean surface need to be discussed in further detail in order to justify the use of a stand-alone OGCM in the study of thermohaline circulation stability. We have employed what are usually referred to as "mixed boundary conditions", a configuration that has on occasion been strongly argued against because of the possibility that the thermohaline circulation may be rendered over-sensitive under such conditions (e.g. Tziperman et al.,1994; Rahmstorf and Willebrand,1995). A conventional choice of relaxation constant for SST would be near one month, a choice

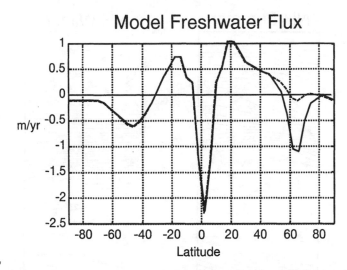

Figure 1.   Model sea surface freshwater flux. A standard anomalous freshwater has already been superimposed at the high latitude North Atlantic (solid line). "Modern" distribution of the freshwater flux is also plotted as a dashed line for comparison.

that may be motivated on the basis of the strength of the temperature feedback needed in an energy balance model that is tuned to simulate "modern climate" (e.g. North et al.,1983). Apart from the issue of realism, the literature argues that by employing such a "short" timescale for relaxation towards the reference state SST the current mode of the thermohaline circulation (and thus the deep water formation process) may become overly (unrealistically) sensitive. For this reason, use of an explicitly described active atmospheric response is clearly desirable.

We cannot at present consider employing a complete Atmospheric General Circulation Model (AGCM) as a practical candidate for this kind of investigation, because AGCMs are still prohibitively expensive. Further use of a diffusive energy balance model (EBM) as in SP97 might constitute an alternative option for the three dimensional integrations of the ocean component to be described herein. We have elected to reject this possibility in this instance, because the use of EBMs does not guarantee a realistic SST response, especially for the high-latitude Atlantic; if it were not for the large land mass (the North American continent) upstream from the ocean, the SST predicted by such an EBM may not be unreasonable. However, the air that travels over the continent becomes very much colder in winter than it would if it had traveled over the open ocean, and this cooling is not accurately captured in the EBM. The cold air mass can extract heat from the ocean much more efficiently than would a warmer air mass. This effect then reduces the effective characteristic time scale

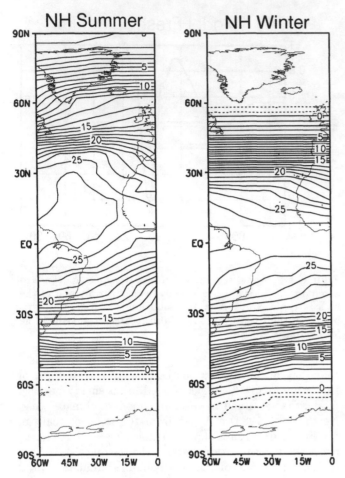

Figure 2. Model sea surface temperature distribution for (northern hemispheric) summer and winter. Seasonal variation has been obtained by superimposing the two conditions. Contour interval is 1°C. Sea shores is plotted just for convenience (MOM is configured to be a flat-bottom, sector model).

of SST relaxation to the reference SST to a value that is possibly much shorter than that which would be inferred on the basis of the EBM. The use of an EBM might be expected to overestimate the sensitivity of the THC more than one might infer through the use of the restoring boundary condition.

We have, therefore, decided to employ a very simple design for these initial three dimensional experiments on the stability of the thermohaline circulation, namely one based on the assumption of *mixed boundary conditions* and to leave investigation of the impact of more realistic surface boundary conditions to a future study.

The choice of a "best" relaxation timescale for the relaxation of SST towards some reference function of latitude is another question, and we will present the results of a set of sensitivity experiments that demonstrate the robustness of our conclusions against the vari-

ation of the time constant over a considerable range. As the "standard" restoring time constant, we will employ the value of 200 days which is approximately one order of magnitude larger than usually assumed for such restoring time constants, since this is expected to eliminate any "unphysical instability" that might be induced by employing the "short" restoring time constant often used in past experiments. In the additional experiments, results obtained when the time constant is set respectively to 50 days and 400 days will also be discussed. The reference SST field is constructed based upon the summer and winter climatology over the Atlantic sector (see Fig.2), and the reference SST pattern at every time step is obtained by superposition of the weighted mean from the two seasons.

The options employed to configure the GFDL MOM for the purpose of these experiments are those which might be considered standard for classical low-resolution OGCMs. In order to represent the vertical mixing related to the occurrence of sub-grid scale convection, a standard convective adjustment scheme has been employed. For the purpose of these experiments we will employ the simple horizontal/vertical representation of diffusion processes instead of an alternative isopycnal/diapycnal representation.

As the title of this paper indicates, our primary goal in these analyses is to understand how the choice of diffusion coefficients impacts the nature of THC time dependence that can be expected to occur under "glacial" conditions. We have adopted a simplest possible approach to this problem based upon the observational fact concerning ocean diffusion coefficients, namely that these vary most strongly as a function of depth as has been assumed, for example, in the seminal early analyses of Bryan and Lewis(1979). Although recent direct measurements of the large-scale mixing coefficients of the ocean have revealed that the range of upper ocean (e.g. Moum and Osborn,1986; Ledwell et al.,1993) vertical diffusion coefficients is $1 \sim 3 \times 10^{-5} \mathrm{m}^2/\mathrm{s}$, these direct observations cover only a very small fraction of the global oceans. Indirect measurements can help to improve this poor coverage of observational (direct) measurements of the diffusion coefficients. Olbers et al.(1985), for example, offer relatively wide coverage of such diffusion coefficient estimates for the North Atlantic. Their study presents analyses of the space-dependent mixing coefficients for both the horizontal (isopycnal) and vertical (diapycnal) directions and for two different ranges of depth, one is of the upper layer below the surface mixed layer ($100 \sim 800$m), and the other is of the intermediate layer ($800 \sim 2000$ m). By comparing the results from these analyses with those based upon direct measurements, the Olbers et al. analyses were shown to be in rather good agreement with the latter, at least for those locations from which direct observation were available for the vertical diffusion coefficients. Some may consider the analyses in Olbers et al.(1985) to be surprising. The diffusion coefficients for both directions were found to be highly space-dependent and cover a wide range extending over more than one order of magnitude. Furthermore, one finds that typical mixing coefficients that are employed in low-resolution OGCMs are rather large compared with the values obtained observationally. As opposed to the assumption in Bryan and Lewis(1979), the vertical mixing coefficient for the upper layer is found to be significantly larger than that for middle layer. For the horizontal (isopycnal) mixing coefficient, the assumption in Bryan and Lewis(1979) qualitatively agrees with the observational analyses, namely that the mixing coefficient assumes significantly larger values in the upper layer. Quantitatively, however, the values employed in global or basin-scale OGCMs may be an order of magnitude in excess of those inferred on the basis of observations. We have therefore elected to employ diffusion coefficients in the present work closer in magnitude to the observations. The actual profiles of the diffusion coefficients that we will employ in the experiments are described in the following section.

Although small scale diffusivities for heat, salt, and momentum may differ by orders of magnitude, on somewhat larger scales the effective values of these diffusion coefficients may differ significantly. In the current investigation we will simplify the experimental design (at least in part to conform to the restrictions imposed by the MOM structure) by assuming that the diffusion coefficients for heat and salt are identical in both the horizontal and vertical directions. The model allows incorporation of different values for the diffusion coefficient for momentum (spatially uniform values only). For simplicity of experimental design we have elected to employ single conservative choices for the horizontal and vertical values of this coefficient, namely $2.5 \times 10^5 \mathrm{m}^2/\mathrm{s}$ for the horizontal value (e.g. see Weaver et al.,1993) and $10^{-4}\mathrm{m}^2/\mathrm{s}$ for the vertical value. Our choice for the vertical viscosity might be considered somewhat smaller than values typically employed in OGCMs (usually in the range between $O(10^{-4}) \sim O(10^{-2})$, e.g. Bryan and Lewis,1979; Weaver et al.,1993; Danabasoglu and McWilliams,1995), but this value is still much larger than the observed vertical diffusivity for heat or salt.

3. RESULTS AND DISCUSSION

We have analyzed the results of two main sets of experiments and two subsidiary sets of experiments. In the first two subsections, we present the results obtained through analyses intended to examine the stability of

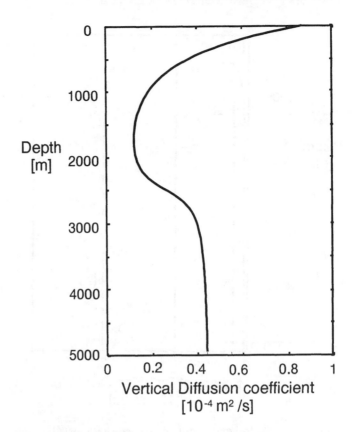

Figure 3. Standard profile of the vertical diffusion coefficient distribution for the experiments with varying horizontal diffusion coefficients.

the thermohaline circulation against changes of either horizontal or vertical diffusion coefficients separately. The final two subsections present results that are intended to demonstrate the robustness of the conclusions from the first two subsections, namely concerning the influence of lowering SST, and the influence of the use of varying relaxation times in the restoring boundary condition for SST.

3.1. The Influence of the Horizontal Diffusion Coefficients

As discussed in the previous section, choosing one specific value for the horizontal diffusion coefficient in an OGCM would constitute a poor reflections of reality. A further issue concerns the fact that typical horizontal diffusion coefficients employed in low-resolution OGCMs are generally taken to be equal to the largest value inferred anywhere on the basis of observations. We have therefore performed a first set of experiments in which the vertical profile of horizontal diffusion coefficient is varied with keeping the vertical diffusion coefficient fixed to a specific profile. Figure 3 shows the standard vertical diffusion coefficient profile for this set

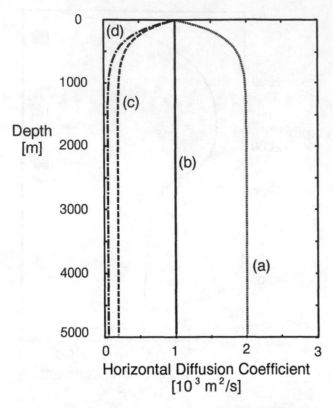

Figure 4. Horizontal diffusion coefficient distributions for the experiments with the fixed vertical diffusion coefficients distribution. Four different profiles correspond to the representative experiments to be shown herein. Labels correspond to those of the results that are later presented.

of experiments. The vertical diffusion coefficient is enhanced near the surface, which differs from the traditional "Bryan-Lewis" type profile but is based upon the observations. The influence of the choice of such a "standard" profile is directly investigated in the following subsection. Figure 4 on the other hand shows the range of horizontal diffusion coefficient profiles to be employed in our first series of experiments. We have performed more than 10 integrations employing a much wider range of profiles in a "trial and error" manner, and choose the set of profiles shown to be representative of the range of behaviors of the THC. The profiles (c) and (d) are, again, similar to the traditional "Bryan-Lewis" profile, with the exception that the depth at which the diffusion coefficient changes is here taken to be shallower than in their original study. One might refer to the value of $10^3 m^2/s$ as a "canonical" value of the horizontal diffusion coefficients (profile (b)), since many OGCMs employ a value for this parameter that is near (or even larger than) this number.

Since the nature of the THC oscillation that we intend to demonstrate that is also supported by the MOM

has already been discussed in the literature (SP96,97), no useful purpose will be served in the present paper by repeating such discussion. It will suffice to simply point out the long timescale oscillation of the thermohaline circulation, when this exists, is supported by the latitudinal contrast of upper layer North Atlantic salinity between the very saline interior that forms in low middle latitude due to excess of evaporation over precipitation and the less saline high latitude water. It will suffice for present purposes to simply present the time series of THC strength that the model delivers when the vertical profile of the horizontal diffusion coefficient is varied. In order to represent basin scale thermohaline circulation strength, we will employ time series of basin averaged overturning streamfunction. If the solution for the idealized Atlantic basin thermohaline circulation that obtains under steady seasonal forcing approaches a statistically steady state, the basin averaged overturning streamfunction should remain relatively fixed to a constant value.

Figure 5 shows such time series of THC strength for the four cases. Inspection of these results will demonstrate that there exists a bifurcation in the nature of the solutions for the idealized Atlantic thermohaline circulation that is induced simply by changing the horizontal diffusion coefficient for the deep ocean, even when there exists a time independent salinity flux variation at the surface. The model either enters an effectively steady state characterized by intense NADW formation if the horizontal diffusion coefficient is taken to be as large as those employed in typical low-resolution OGCMs, or it exhibits flush-collapse type oscillations on multi-millennia timescale when the deep ocean diffusivities are significantly reduced. The former cases appear to be similar to that of a modern state of the THC circulation, and thus the value of the basin average overturning streamfunction in these cases can be considered to correspond to that of a modern North Atlantic. This bifurcation is exactly the same as predicted in SP99. The first conclusion from these experiments, therefore, is that, as suggested on the basis of our previous work based upon the use of much simpler models, enforcing *uniformity* of the deep water mass through the use of very large horizontal mixing coefficients in the abyssal ocean eliminates the oscillatory solution that may otherwise be induced by changing steady surface boundary conditions (adding a sufficiently large fresh water flux anomaly at high northern latitudes).

Another important aspect of these results is that the period of the oscillation is rather insensitive to the magnitude of the horizontal diffusion coefficients when smaller values than those needed to induce the bifurcation are introduced. This will be clear by comparison of the results obtained for profiles (c) and (d) in Figure 5.

The deep ocean diffusivity differs by a factor of 4 in the abyss yet the period of the resulting oscillation remains approximately the same.

In the box model analyses described in SP99, one of the essential conclusions was that the behavior of the three-box model deviates significantly from the behavior of Stommel's two-box model because of the separation assumed to exist between the deep water mass in the high latitude North Atlantic and the deep water mass in the low latitude Atlantic. By increasing the efficiency of mixing between these two deep water reservoirs it was shown that the three-box model reduces to the classical Stommel model which allows only multiple equilibria of the kind further analyzed in the recent work by Rahmstorf(1995). By allowing the deep water mass to remain heterogeneous by reducing the horizontal diffusion coefficients, the analyses discussed above show that the 3-D OGCM similarly supports an oscillation in the thermohaline circulation, and that this occurs, as in the simple 2-D and box models, on the millennium timescale.

An issue which requires further discussion then concerns the question as to whether the reduction of the diffusion coefficients in the abyssal ocean that is required to allow the model to support the millennium timescale oscillation is physically plausible. In fact the required reduction of the diffusion coefficients under these boundary conditions is only to a value of $0.3 \times 10^3 \mathrm{m}^2/\mathrm{s}$, approximately one-third of the "canonical value" usually employed in low-resolution OGCMs. According to the analyses of Olbers et al.(1985), the observed value of the diffusion coefficients is in fact much less than this value throughout most of the North Atlantic Ocean, even in the intermediate water layer that extends from 800 to 2000 m depth Although the representation of diffusion processes in these model experiments is rather primitive, the important point implied by the results is simply that the requirement for the existence of the solution in which the THC strength is oscillatory is simply that the horizontal diffusion coefficients be only slightly smaller than the (overly large) value usually assumed in low-resolution OGMCs. This also explains why some low-resolution OGCMs fail to reproduce the millennium time scale oscillation; if the horizontal diffusion coefficient is taken to be larger than the critical value, then the oscillatory solutions cannot be realized. In the ensuing subsection we will focus upon the factor that most strongly controls the period of the millennium timescale variability.

A brief summary of the bifurcation structure that has already been discussed in our previous papers is as follows: the smaller value of the horizontal diffusion coefficient of salinity and temperature (and thus density) reduces northword transport of saline water in the North

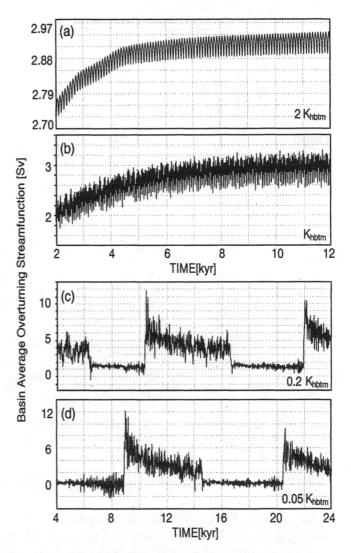

Figure 5. Time series of basin averaged overturning streamfunction. Four cases correspond to the varying horizontal diffusion coefficients respectively: From top to bottom, the largest K_h case (a), the "canonical K_h case" (b), a small K_h (c), and a smaller K_h case (d). Note that the time axis differs for lower two plates: the upper two plates show the time series of 12,000 model years, whereas the lower two plates show that of 24,000 model years. Labels denotes the diffusion coefficients shown previously.

Atlantic upper-layer and maintains the density contrast in the deeper layer beneath. These direct influences enhance the distinction between watermasses (namely, the upper-layer of the middle latitude North Atlantic, the high latitude North Atlantic, and the deep-layer of the low-middle latitudes of the Atlantic) which is necessary to drive the THC oscillation. As a result, the density of the deep water relative to that of the surface of the high-latitude North Atlantic is increased when the THC system is in the stagnant phase of an oscillation cycle.

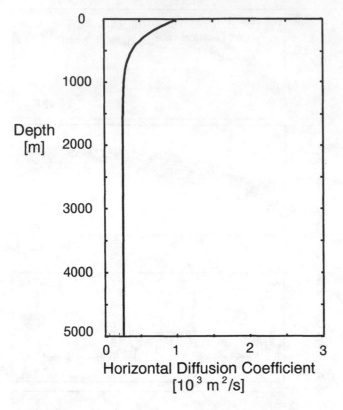

Figure 6. Standard profile of the horizontal diffusion coefficient distribution for the experiments with varying vertical diffusion coefficients.

3.2. The Influence of the Vertical Diffusion Coefficients

The existing literature on the oscillatory model of behavior of the THC, as represented for example by SP96, has already provided some insight into the factors that control the period of the oscillation in the simple two-dimensional hydrodynamic models. This timescale is primarily controlled by the time required, in the absence of motion, for diffusion to restore the density of the surface waters in the high latitude Atlantic to a convectively unstable state following the collapse phase of the oscillation. The time taken to restore the (potentially high) density of the surface water to the unstable state by increasing salinity in the low middle latitude region where evaporation exceeds precipitation will clearly involve the efficiency of vertical mixing to a significant degree, an influence upon which we will focus on this subsection.

While first keeping the vertical profile of horizontal diffusion coefficient fixed to the standard profile shown on Figure 6, we will present the results obtained from a series of experiments in which the vertical diffusion coefficients is varied. The sequence of profiles for the vertical diffusion coefficients are shown on Figure 7 where

they are labeled (a), (b) and (c). The profile of horizontal diffusion coefficients shown on Fig. 6 has been selected to be such that oscillatory solutions of the THC are allowed according to the analyses described in the last subsection and to be well within the range required by the analyses of Olbers et al.(1985)

In describing the results obtained with this sequence of models we will once again present time series of basin averaged overturning streamfunction. Figure 8 shows the results for the three cases. The central plate in this series shows the results employing the same K_v profile as that employed in the previous subsection (profile (a) on Figure 7). The upper and lower plates show the results obtained either by reducing or by increasing K_v at the surface, respectively profiles (b) and (c) on Fig.7. Inspection of these results indeed verifies our expectations based upon the results of the previously discussed 2-D models, namely that the period is very sensitive to the choice of the vertical diffusion coefficient. The smaller the vertical diffusion coefficient, the longer the period of the millennium timescale oscillation of the overturning circulation in the idealized North Atlantic basin.

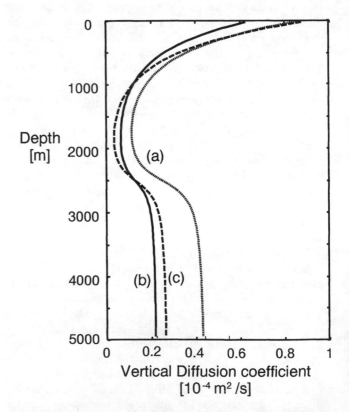

Figure 7. Vertical diffusion coefficient distributions for the experiments with the fixed horizontal diffusion coefficients distribution. Labels correspond to those of the results that are later presented.

Considering the nature of the oscillation already described in the previous series of papers (SP96,99), a simple and physical explanation of this result follows immediately. The duration of the "off state" of intense NADW formation is determined by the rate at which high salinity water accumulates in the upper layer of the low latitude North Atlantic. This process is of course very sensitive to the value of the vertical diffusion coefficient. The smaller the diffusion coefficient, the longer the time required for surface high salinity condition to diffuse into the deep halocline layer. We do not consider, on the other hand, that the duration for the "on state" of intense NADW formation is influenced in any significant way by the magnitude of K_v as will be further discussed in the following subsection.

3.3. The Impact of Varying the Restoring Time Constant for SST

In this subsection, we present the results from two experiments that are intended primarily to demonstrate the insignificance of any possible "over-sensitivity" introduced by the use of mixed boundary conditions. We need to re-emphasize here that the physically correct value of "the" restoring time constant that is employed in a mixed boundary conditions formulation is not known. Furthermore it is almost certainly the case on physical grounds that no such unique value should exist. It will suffice to provide two physical examples of circumstances that would lead one to expect that this time constant should be strongly space dependent. A first example is the previously mentioned case of the high latitude North Atlantic during winter. Very cold air outflows from North America serve to maintain the surface temperature of the high latitude North Atlantic Ocean just above the freezing point over open water. A second example concerns the equatorial Pacific Ocean. This region of the ocean experiences no similar atmospheric forcing as that of the high latitude North Atlantic. One therefore has good reason to suspect that the restoring time constant appropriate for the equatorial Pacific should much longer than that appropriate for the high latitude North Atlantic.

In order to demonstrate the insignificance of the choice of restoring time constant for SST on the existence and properties of the THC oscillation, we will present herein the results from three experiments in which the restoring time constant on SST is varied. In these experiments, the diffusion coefficients have been held fixed to the same vertical and horizontal profiles shown on Figure 9, whereas the restoring time constant is varied through the sequence 50, 200 (standard), 400 days.

Insofar as the surface boundary condition on freshwater flux is concerned, we have successfully applied the

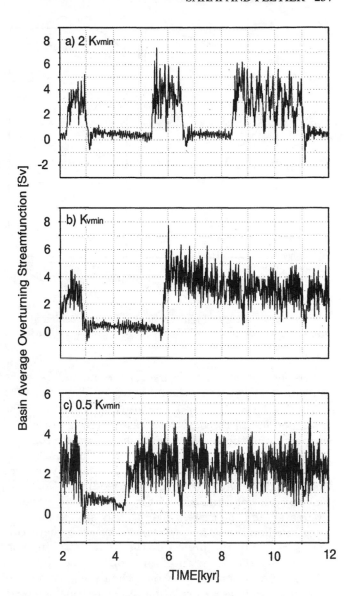

Figure 8. Time series of basin averaged overturning streamfunction. Three cases correspond to the varying vertical diffusion coefficients respectively: From top to bottom, the largest $K_{v,minimum}$ case (a), the medium $K_{v,minimum}$ case (b) which is the same as the previous subsection, and a small $K_{v,minimum}$ (c). Labels denote the diffusion coefficients shown previously.

same boundary condition shown on Fig.1 to the case when the time constant is set to 50 days as before (as well as to the 200 day base case), but we were obliged to slightly modify this surface boundary condition in order to obtain the oscillatory solution when the restoring time constant was set to 400 days (Figure 10). It is important to emphasize here that the purpose of presenting these results is to demonstrate that the millennium timescale THC oscillations may continue to exist even

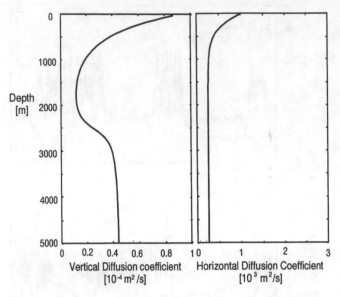

Figure 9. Horizontal and vertical diffusion coefficient profile for the experiment with varying restoring time constants for SST.

when the restoring time constant for the SST is set to an *unphysically* large value. Figure 11 shows the time series of the basin average overturning streamfunction obtained for all three of these cases. It will be clear by inspection of these results that when the time constant is reduced to 50 days from the base value of 200 days the period of the oscillation is also significantly reduced and the pulse shape substantially modified. Similarly, as the restoring time constant is increased the period is increased and the system significantly stabilized.

A further issue that warrants discussion here concerns not only the period of the oscillation but more specifically the duration of the "on" phase of intense NADW formation. In previously analyzed examples of the "flush-collapse" oscillation of the thermohaline circulation, the "on" phase was found to occupy only a very small portion of each cycle. One extreme example was produced by the three-box model discussed in SP99 which has essentially identical boundary condition to the fixed (non restoring) SST case. In this box model, the "on" phases were realized only as very sharp spikes in the time series of THC strength. For the purpose of the ensuing discussion we will exclude the time series obtained for the case that employed the longest restoring time constant, since this analysis differed not only in the restoring time constant for SST but also in the surface salinity flux. It has already been simply demonstrated that different surface salinity flux distributions may result in different periods for the oscillation (e.g. SP97).

By employing a very long restoring time constant we

influence the nature of THC time dependence in two important ways, the first concerning the stagnant phase of the oscillation and the second the intense NADW formation phase. Concerning the stagnant phase, a very long restoring time constant for SST enhances the stability of the thermohaline circulation, which leads to an increase in the threshold for the instability that is ultimately induced by low latitude salinity accumulation. Concerning the intense phase of NADW formation it is helpful to refer to SP97 in which a 2-D ocean model was coupled asynchronously to an energy balance atmospheric model. In that analysis we included atmosphere-ocean feedback, and the resulting time series of the activity of the thermohaline circulation were shown to differ significantly from those obtained by employing the stand-alone version of the ocean model in a way similar to the way we are using the 3-D OGCM in the present sequence of experiments. In this more realistic circumstance the duration of the "on" phase of intense NADW formation was shown to be stretched compared to the stagnant phase. This change in the duration of the "on" phase is an easily understood consequence of the atmospheric response. In both the long relaxation time and coupled atmosphere-ocean analyses, oceanic heat release that occurs as a result of very intense NADW formation disturbs the SST field locally over the high latitude North Atlantic where temperatures are increased. Both by employing the EBM to explicitly compute this feedback and using a longer restoring time constant for SST enhances the role of this process, by increasing the efficiency of the negative feedback that it applies on the intensity of NADW formation. When NADW formation increases, the "air" above the region warms and this reduces the density of

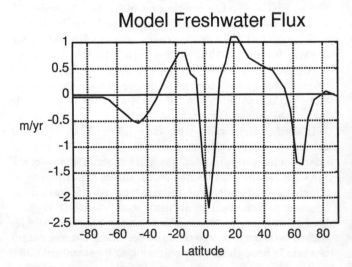

Figure 10. Model sea surface freshwater flux for the case when the restoring time constant is set to 400 days.

the sea water with which it is in contact so that further intensification of NADW formation is suppressed. This SST control on the rate NADW formation reduces the rate of consumption of saline water stored in the low-latitude upper layer. The "on" phase therefore can persist for a longer period of time then it would do in the absence of such a negative (atmospheric) feedback process as in the SP99 box model results.

The question of what might constitute a "correct" choice for the restoring time constant for SST in stand-alone ocean models remains an open question for reasons that have already been discussed in the preceding sections. To adequately address it, we will, in analyses to be described in subsequent work, employ a strategy based upon the judicious use of coupled atmosphere-ocean models in which the atmospheric component of the coupled structure is better approximated than it can be using the energy balance formalism.

3.4. The Impact of Lowering the Surface Temperature to Ice-Age Conditions

Before closing this section, we will present the results obtained from one further experiment. This was undertaken to demonstrate the impact upon the THC oscillation of lowering the sea surface temperature to the extent that occurred under glacial conditions. We do not wish to argue here for the reality of the temperature field we will employ to mimic ice-age climate. Since we seek only to develop on the basis of the experiment a qualitative argument as to how lowering the surface temperature will tend to further increase the extent to which the salinity flux (freshwater flux) will dominate the process of deep water formation.

This issue has already been addressed to some degree in Winton(1997), and so our analyses are intended to re-confirm his conclusions. The primary difference between the present experiment and those reported in previous subsections is that surface temperature will be fixed so as to better represent "cold climate" condition by uniformly reducing SST by 4°C except in those regions where surface temperature would be thereby reduced below the freezing point of sea water. The distribution of the freshwater flux is set to be identical to that in the previously performed experiments except for the high latitude anomaly. For the present purposes this high-latitude anomaly is reduced by approximately 30% *to a level such that the millennium timescale oscillation would no longer occur under modern climate conditions* (Figure 12).

Figure 13 shows the time series of basin averaged overturning streamfunction from the integration. One can clearly see that the same oscillation in thermohaline circulation intensity is produced under "glacial" bound-

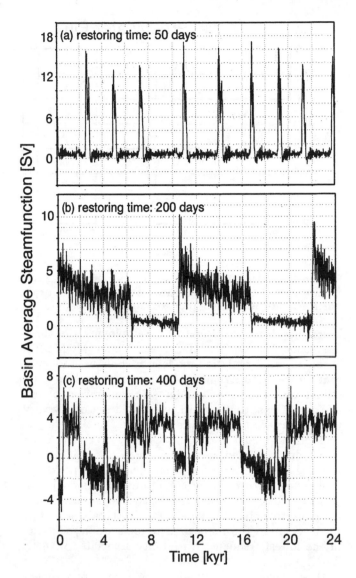

Figure 11. Time series of basin averaged overturning streamfunction. Three cases correspond to the different restoring time constant for SST. From top plate to bottom plate, the shortest case: (a) 50 days, the standard case: (b) 200 days, and the longest days: (c) 400 days.

ary conditions as that obtained previously. Of course the reduction in surface air temperature for the high latitudes during the full glacial period was likely larger than the global estimate of 4°C, so that the influence of lowering SST may play a more important role in the ice age than that which would be inferred on the basis of this experiment.

4. SUMMARY AND CONCLUSIONS

The influence of deep ocean diffusion coefficients upon the stability of the Atlantic thermohaline circu-

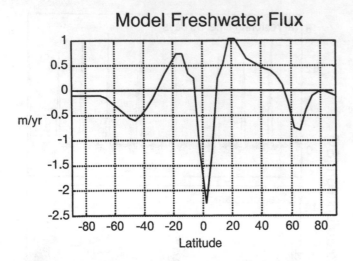

Figure 12. Model sea surface freshwater flux for a cold climate case. Anomalous freshwater in the high latitude North Atlantic is reduced approximately by 30%.

lation has been investigated by employing the three-dimensional GFDL Modular Ocean Model (MOM). The ocean general circulation model was configured to represent an idealized Atlantic basin and integrations of the model were performed under so-called mixed boundary conditions; that is with prescribed seasonally varying sea surface temperature to which the time varying SST of the model is continuously restored, and freshwater flux specified so as to include a localized high-latitude anomaly intended to mimic the influence of surface runoff during glacial conditions. The model is integrated, for each of the sensitivity analyses performed, for more than 10,000 model years in the absence of explicit wind stress forcing so as to focus upon the nature of the lowest frequency variability associated with the THC.

Depth dependent diffusion coefficients of temperature and salinity for the horizontal and vertical directions were employed in the numerical model, and a series of sensitivity experiments to investigate the impact of these coefficients on the nature of THC time dependence have been performed. The coefficients have been varied within the range of observations: $O(10^2) \sim O(10^3) \mathrm{m}^2/\mathrm{s}$ for the horizontal direction, and $O(10^{-5}) \sim O(10^{-4}) \mathrm{m}^2/\mathrm{s}$ for the vertical direction.

We derive two main conclusions on the basis of these experiments, beyond the now well established fact that the three dimensional ocean general circulation does deliver intense millennium timescale oscillation under steady boundary conditions, a fact previously demonstrated in SP97 and SP99. A first conclusion is that the horizontal diffusion coefficients for the abyssal ocean strongly control the stability of the system and thus determine the existence of the oscillatory solution, a result

that is consistent with the conclusions in SP99. When the horizontal diffusion coefficients for the deeper layer (i.e. below 2000 m) are taken to be less than a critical value, the millennium timescale oscillation is produced by the model. The critical value of the horizontal diffusion coefficient turns out to be, in the experimental configuration employed herein, approximately $3 \times 10^2 \mathrm{m}^2/\mathrm{s}$. Although this value is considerably smaller than those typically employed in low-resolution ocean general circulation models, observational estimates for the large-scale horizontal mixing coefficients, for example those by Olbers et al.(1985), clearly show that the actual value lies near this throughout most of the layer of North Atlantic intermediate water. It is of interest, and will lead to a further conclusion, that even imposing a horizontal diffusion coefficient that is less than the critical value affects the period and pattern of the oscillation insignificantly. Our second conclusion is that a factor which most strongly affects the period of the millennium timescale oscillation is the vertical diffusion coefficient at the depth of the thermocline and halocline.

Additional sets of experiments were also performed in order to demonstrate the robustness of these primary conclusions. One set was designed to investigate the influence of the lower temperatures which correspond to ice-age conditions. By lowering sea surface temperatures every where by 4°C (but maintaining the lowest temperature to -2°C), we were able to confirm the fact that lowering the temperature does enhance instability, as specifically discussed in Winton(1997), though

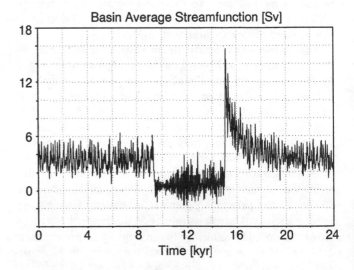

Figure 13. Time series of basin average overturning streamfunction. The result is obtained under a "cold climate" condition in which SST is set to have 4°C lower temperature than the modern SST that has been employed in this paper (the lowest temperature is still set to the freezing point of sea water).

the model continues to deliver an oscillation of the same kind as appears with somewhat weaker anomalous freshwater flux. Another set of experiments consisted of integrations with different assumed values for the restoring time constant for SST. The purpose of this set of experiments was to justify a posteriori the use of the mixed boundary conditions. We thereby confirmed that THC oscillations of the same kind are observed for the different values of the restoring time constant that range from tens of days to hundreds of days, although the period of the oscillation certainly differs: when the restoring time constant is short the THC cycle consists of a very intense and brief period of intense NADW formation followed by a relatively long period during which the THC is stagnant. With a long restoring time constant, however, the THC cycle consists of a relatively longer period of intense NADW formation followed by a shorter stage in which no NADW forms and the THC is stagnant.

Taken together the results obtained here suffice to demonstrate the robustness of the conclusion that the thermohaline circulation of the Atlantic Ocean supports millennium timescale oscillations, even when a fully configured ocean general circulation model is employed in the analysis. Our analyses have therefore further established the plausibility of our primary thesis that the millennium timescale Dansgaard-Oeschger oscillation observed in the deep ice cores from Summit, Greenland are due to an intrinsically nonlinear oscillation of the Atlantic THC that is expected to exist when the freshwater flux is in the vicinity of, but less than, the flux that would be required to shut down the circulation entirely.

According to recent analyses by Bond et al.(1997), distinct but much less significant ice-rafting events compared with the Heinrich events have been observed. Since these events are evident even after the last deglaciation, they are clearly connected to ice-rafting from the remaining icesheet, namely Greenland. In this paper we are not yet in the position to fully discuss the dynamic linkages between icebergs from Greenland and the North Atlantic THC under modern climate conditions, but it is useful to point out a few implications of the current work and previous papers: because we observe no strong oscillation in the temperature proxy for the Dansgaard-Oeschger oscillation corresponding to ice-rafting events during the Holocene, these minor ice-rafting events are unlikely to be a major factor in the mechanism underlying the Dansgaard-Oeschger oscillation. According to our previous analyses (SP99), however, the North Atlantic THC is somewhat sensitive to perturbations on the same timescales or longer than those characteristic of the primary THC oscillation. Therefore it is quite possible that such ice-rafting

events play a role as a pacemaker of the Dansgaard-Oeschger oscillation. This issue will be further investigated in future work.

In the next step of this on-going series of investigations, however, we intend to further verify the validity of the theory by exploring more explicitly the influence on the internal variability of the thermohaline circulation due to atmospheric feedback. The analyses reported in SP97 suggest that this influence is important, based upon an analysis of the coupling that employed an atmospheric EBM to represent the feedback onto the surface of the ocean. In the next step in the further development of this idea we will be obliged to use a more fully articulated model of the atmosphere that includes explicit dynamic effects. An analysis of this kind will also enable us to fully investigate the interaction between the wind driven circulation and the thermohaline circulation.

REFERENCES

Bond,G., W.Broecker, S.Johnsen, J.McManus, L.Labeyrie, J.Jouzel, and G.Bonani, Correlations between climate records from North Atlantic sediments and Greenland ice, *Nature, 365*, 143-147, 1993.

Bond,G., W.Showers, M.Cheseby, R.Lotti, P.Almasi, P.deMenocal, P.Priore, H.Cullen, I.Hajdas, and G.Bonani, A pervasive Millennial-scale cycle in North Atlantic Holocene and glacial climates, *Science, 278*, 1257-1266, 1997.

Broecker,W.S., G.Bond, and M.Klas, A salt oscillator in the glacial Atlantic? 1. the concept, *Paleoceanogr., 5*, 469-477, 1990.

Bryan,K., and L.J.Lewis, A water mass model of the world ocean, *J.Geophys.Res., 84*, 2503-2517, 1979.

Danabasoglu,G., and J.C.McWilliams, Sensitivity of the global ocean circulation to parameterizations of mesoscale tracer transports, *J.Climate, 8*, 2967-2987, 1995.

Duplessy,J.-C., L.Labeyrie, A.Juillet-Leclerc, F.Maitre, J.Dupart, and M.Sarnthein, Surface salinity reconstruction of the North Atlantic Ocean during the last glacial maximum, *Oceanologica acta, 14*, 311-323, 1991.

Fawcett,P.J., A.M.Ágústsdóttir, R.B.Alley, and C.A.Shuman, The Younger-Dryas termination and North Atlantic Deep Water formation: Insights from climate model simulations and Greenland ice cores, *Paleoceanogr., 12*, 23-38, 1997.

Fofonoff,N.P., Physical property of seawater: A new salinity scale and equation of state for seawater, *J.Geophys.Res., 90*, 3332-3343, 1985.

GRIP, Climate instability during the last interglacial period recorded in the GRIP ice core, *Nature, 364*, 203-207, 1993.

Heinrich,H., Origin and consequences of cyclic ice rafting in the Northeast Atlantic Ocean during the past 130,000 years, *Quat.Res., 29*, 142-152, 1988.

Keigwin,L.D., and G.A.Jones, Western North Atlantic evidence for millennial-scale changes in ocean circulation and climate, *J.Geophys.Res., 99*, 12397-12410, 1994.

Kreveld,S.van-, M.Sarnthein, H.Erlenkeuser, P.Grootes, S.Jung, M.J.Nadeau, U.Pflaumann, and A.Voelker, Potential links between surging ice sheets, curculation changes, and the Dansgaard-Oeschger cycles in the Irminger Sea, 60-18 kyr, *Paleoceanogr., 15*, 425-442, 2000.

Ledwell,J.R., A.J.Watson, and C.S.Law, Evidence for slow mixing across the pycnocline from an open-ocean tracer-release experiment, *Nature, 364*, 701-703, 1993.

Marotzke,J., Instabilities and multiple steady states of the thermohaline circulation, in *Oceanic Circulation Models: Combining Data and Dynamics*, edited by D.L.T.Anderson, and J.Willebrand, pp501-511, NATO ASI Ser. 284, Kluwer Academic Publishers, 1989.

Moum,J.N., and T.R.Osborn, Mixing in the main thermocline, *J.Phys.Oceanogr., 16*, 1250-1259, 1986.

North,G.R., J.G.Mengel, and D.A.Short, Simple energy balance model resolving the seasons and continents: application to the astronomical theory of the ice age, *J.Geophys.Res., 88*, 6576-6585, 1983.

Olbers,D.J., M.Wenzel, and J.Willebrand, The influence of North Atlantic circulation patterns from climatological hydrographic data, *Rev. Geophys., 23*, 313-356, 1985.

Rahmstorf,S., Bifurcations of the Atlantic thermohaline circulation in response to change in the hydrological cycle, *Nature, 378*, 145-149, 1995.

Rahmstorf,S., and J.Willebrand, The role of temperature feedback in stabilizing the thermohaline circulation, *J.Phys.Oceanogr., 25*, 787-805, 1995.

Rasmussen,T.L., E.Thomsen, T.C.E.van Weering, and L.Labeyrie, Rapid changes in surface and deep water conditions at the Faeroe Margin during the last 58,000 years, *Paleoceanogr., 11*, 757-771, 1996.

Peltier,W.R., and K.Sakai, Dansgaard-Oeschger oscillations: a hydrodynamic theory, in the North Atlantic, *A Changing Environment*, P.Schräfer, W.Rizrau, M.Schüter and J.Threde (eds.), Springer-Verlag, Berlin, 45-66, 2001.

Sakai,K., and W.R.Peltier, A simple model of the Atlantic thermohaline circulation: internal and forced variability with paleoclimatological implications, *J.Geophys.Res.-Oceans, 100*, 13455-13479, 1995.

Sakai,K., and W.R.Peltier, A multi-basin reduced model of the global thermohaline circulation: paleoceanographic analyses of the origins of ice-age climate variability, *J.Geophys.Res.-Oceans, 101*, 22535-22562, 1996.

Sakai,K., and W.R.Peltier, Dansgaard-Oeschger oscillations in a coupled atmosphere-ocean climate model, *J.Clim, 10*, 949-970, 1997.

Sakai,K., and W.R.Peltier, A dynamical systems model of the Dansgaard-Oeschger oscillation and the origin of the Bond cycle, *J.Clim., 12*, 2238-2255, 1999.

Stommel,H.M., Thermohaline convection with two stable regimes of flow, *Tellus, XIII(2)*, 224-230, 1961.

Taylor,K.C., G.W.Lamorey, G.A.Doyle, R.B.Alley, P.M.Grootes, P.A.Mayewski, J.W.C.White, and L.K.Barlow, The 'flickering switch' of late Pleistocene climate change, *Nature, 361*, 432-436, 1993.

Tziperman,E., J.R.Toggweiler, Y.Feliks, and K.Bryan, Instability of the thermohaline circulation with respect to mixed boundary conditions: is it really a problem for realistic model?, *J.Phys.Oceangr, 24*, 217-232, 1994.

Weaver,A.J., J.Marotke, P.F.Cummins, and E.S.Sarachik, Stability and variability of the thermohaline circulation, *J.Phys.Oceanogr., 23*, 39-60, 1993.

Winton,M., The effect of cold climate upon North Atlantic deep water formation in a simple ocean-atmosphere model, *J.Clim., 10*, 37-51, 1997.

W. Richard Peltier, Department of Physics, University of Toronto, 60 St George St., Toronto, ONT, M5S 1A7, Canada. (e-mail: peltier@atmosp.physics.utoronto.ca)

K. Sakai, Institute for Global Change Research, Frontier Research System for Global Change, 3175-25 Showamachi, Kanazawa-ku Yokohama City, Kanagawa 236-0001, Japan. (e-mail: kotaro@jamstec.go.jp)

The Climatic Influence of Drake Passage

H. Bjornsson[1]

Program in Atmospheric and Ocean Sciences, Princeton University, Princeton, New Jersey

J. R. Toggweiler

Geophysical Fluid Dynamics Laboratory, Princeton, New Jersey

The influence of Drake Passage on the earth's temperature distribution is explored in an idealized coupled model. In a version of the model without Drake Passage the temperature distribution is symmetric about the equator, due in large part to the fact that the meridional overturning in the ocean is symmetric about the equator with deep water formation in both hemispheres. With Drake Passage open, the overturning takes on the form of an interhemispheric conveyor with deep water formation primarily in the opposite (northern) hemisphere. Surface temperatures rise in the northern hemisphere and fall in the southern hemisphere as the ocean transports a large amount of heat northward across the equator. The magnitude of the thermal asymmetry between the hemispheres depends on the depth of the circumpolar channel and the strength of the winds over the channel. The opening of Drake Passage also leads to the formation of a low-salinity intermediate water mass reminiscent of Antarctic Intermediate Water. The intermediate water mass is associated with a warming and thickening of the thermocline that extends from the circumpolar channel into the highest latitudes of the northern hemisphere. Paleoclimatic implications of this work are discussed.

1. INTRODUCTION

A prominent feature of the oceanic circulation is the Atlantic *conveyor*, an interhemispheric overturning circulation that links a northward flow of warm upper ocean water into the North Atlantic with a southward flow of cold deep water out of the Atlantic. The conveyor is a significant factor in the earth's climate because it transports a very large amount of heat into the high latitudes of the northern hemisphere. A well-known feature of the upper northward branch of the conveyor is its high salt content, a property that allows the water in the northward flow to become especially dense when it is cooled. Heat losses in the northern North Atlantic thereby lead to mass sinking and the formation of a major ocean water mass, North Atlantic Deep Water (NADW), that spreads far and wide through the ocean's interior. The standard view is that buoyancy forces drive the Atlantic conveyor, i.e. the cooling and salinification that lead to the formation of NADW are thought to initiate the conveyor's flow.

The high salinities observed in the North Atlantic are due in part to high evaporation rates and the atmospheric transport of water vapor out of the Atlantic drainage basin.

[1] Now at: Icelandic Meteorology Office, Reykavik, Iceland.

The Oceans and Rapid Climate Change: Past, Present, and Future
Geophysical Monograph 126
Copyright 2001 by the American Geophysical Union

243

Water-Planet Model

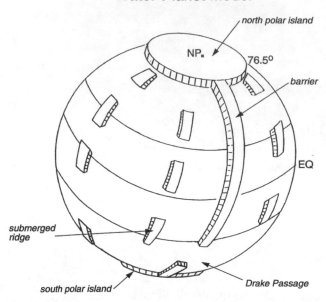

Figure 1. Schematic of ocean basin setup.

Broecker [1991] has suggested that this loss of freshwater is the ultimate driver for the conveyor. Numerous experiments with Ocean General Circulation models (OGCMs) show that the conveyor is indeed sensitive to the freshwater forcing in the North Atlantic. If the amount of added freshwater is increased the conveyor can be reduced in intensity, or even turned off completely [*Hughes and Weaver*, 1994; *Rahmstorf*, 1996; *Fanning and Weaver*, 1997; *Manabe and Stouffer*, 1999a].

There are a number of reasons to think that the conveyor may not be entirely buoyancy driven. A series of OGCM experiments starting with *Gill and Bryan* [1971] have shown that the existence of a continuous band of open ocean beyond the perimeter of Antarctica, and the Antarctic Circumpolar Current (ACC) flowing through through the open channel, have a major impact on global-scale circulations of which the conveyor is a part. The open channel is often characterized geographically by the span of open water between South America and Antarctica known as Drake Passage. *Cox* [1989] and *England* [1993] have shown that the opening of Drake Passage in an ocean GCM leads to major changes in the location and properties of the main oceanic water masses, including NADW. *Toggweiler and Samuels* [1995] have shown that the strength of the wind stress in the latitude band of Drake Passage may directly affect the overturning rate of the Atlantic conveyor and the rate of formation of NADW. This body of work suggests that mechanical forces in the circumpolar channel play an important role in the conveyor circulation by "setting the stage" for the work of buoyancy forces in the North Atlantic.

In the OGCM studies cited above, the impact of an open Drake Passage was limited by the fact that the same restoring boundary conditions were used before and after Drake Passage was opened, or before and after the southern wind stress was changed. This means that surface temperatures and salinities were basically fixed and were not allowed to respond to circulation changes. In this study, the impact of a Drake Passage-like gap in the south is examined in an idealized coupled model. Air temperatures and ocean temperatures in this model are free to change. A field of freshwater fluxes is applied to the model that allows salt to move within the ocean in response to changes in the circulation. We find under these circumstances that the simple act of opening of Drake Passage is able to initiate an interhemispheric overturning circulation much like the observed conveyor. No special forcing features are included to enhance the process of deep-water formation in the model. These results suggest that an open circumpolar channel in the south does more than just "set the stage"; it seems in our opinion to be an indispensable part of the conveyor circulation.

Two paleoclimatic implications are taken up below. The first concerns paleoclimatic changes associated with the opening of Drake Passage some 35 million years ago and the gradual deepening of the circumpolar channel over time. Aspects of this work are discussed in more detail in *Toggweiler and Bjornsson* [2000]. The second is concerned with the sensitivity of the global climate to processes that operate within and around Drake Passage, e.g. the thermal response to changing wind stresses in the Drake Passage channel.

2. THE WATER PLANET MODEL

The model used in this paper describes a water-covered earth in which land is limited to two polar islands and a thin barrier that extends from one polar island to the other (see Fig. 1 and top panel in Fig. 2). The polar islands cover each pole out to 76.5° latitude, roughly the latitude of the Weddell and Ross Sea embayments on Antarctica's perimeter. Figure 1 shows a schematic of the model's land-sea configuration with a segment of the barrier (Drake Passage) removed near the south polar island. The barrier between the polar islands is very thin, 7.5° of longitude wide, or 2% of a latitude circle.

The coupled model consists of a three-dimensional ocean general circulation model coupled to a one-dimensional energy balance model of the overlying atmosphere. As in other coupled models, the model used here finds a latitudinal temperature distribution in which the flux of outgoing long-wave energy is in balance with a specified input of short-wave solar energy. The model does not have interactive winds or an interactive hydrological cycle. Instead, a latitudinally varying wind stress field is imposed on the ocean along with a latitudinally varying salt flux. No attempt is made to include an ice-albedo feedback due to snow cover or ice on the polar islands, and there is no sea ice on the ocean.

The atmospheric model, loosely based on the model of *North* [1975] solves a one-dimensional equation describing the latitudinal variation of the atmospheric heat budget. It is discretized into 40 grid cells which are aligned with the meridional grid in the ocean. Each grid cell receives a specified input of shortwave solar energy which is symmetric about the equator. Each grid cell then radiates longwave energy to space as a function of the local temperature. Atmospheric cells diffuse heat meridionally and exchange heat with all ocean grid cells in the same latitude band. The exchange of sensible and latent heat between the ocean and atmosphere is parameterized in terms of a linear damping of ocean-atmosphere temperature differences. The energy balance model is described in more detail in *Toggweiler and Bjornsson* [2000]

The ocean model is based on the GFDL (Geophyscial Fluid Dynamics Laboratory) MOM 2 (Modular Ocean Model) code [*Pacanowski*, 1996]. The model solves a three-dimensional equation for momentum and a pair of conservation equations for temperature and salinity. The model used here is built on a coarse grid (4.5° latitude by 3.75 ° longitude with 12 vertical levels). The maximum depth is 5000 m. Drake Passage in these experiments is quite wide, extending from 38.25S to 65.25S. The rationale for such a wide gap is based on the fact that the Antarctic Circumpolar Current (ACC) in coarse models tends to be strongly influenced by lateral friction near the tip of South America. A wide gap helps eliminate lateral friction as a leading order term in the momentum budget for the water planet's ACC.

The ocean floor is mostly flat except for a series of submerged ridge segments that extend up from the bottom. The ridge segments are staggered across the ocean floor as shown in the top panel of Figure 2. The ridges serve as a sink for wind generated momentum. Pressure gradients across the submerged ridges take up more than 90% of the momentum added to the ocean by the wind in the latitude band of Drake Passage. The remainder (less than 10%) is taken up by lateral friction.

The model uses a modified version of the *Bryan and Lewis* [1979] vertical mixing scheme in which vertical mixing varies with depth from $0.15 cm^2 s^{-1}$ in the upper kilometer to $1.3 cm^2 s^{-1}$ in the lowest kilometer. The model includes the Gent-McWilliams (GM) parameterization for tracer mixing as implemented in MOM by *Griffies* [1998]. The coefficients governing GM thickness mixing and diffusion along isopycnals have been set to $0.6 \times 10^7 cm^2 s^{-1}$. The model uses flux corrected transport (FCT) to advect tracers, rather than centered differences, and has no background horizontal mixing. There is no freezing point limitation on the

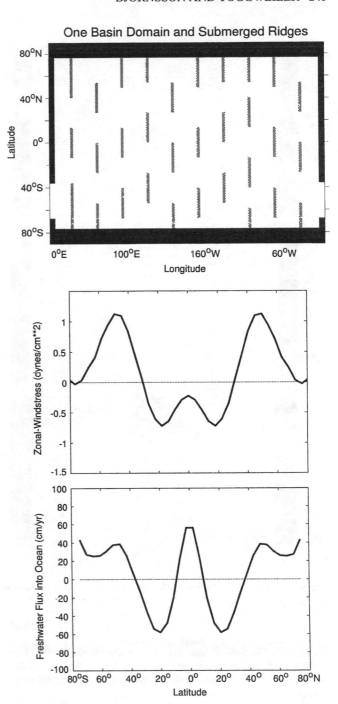

Figure 2. Experimental setup. Top panel: The model domain, the submerged ridges are lightly shaded. Middle panel: The applied wind forcing. Bottom panel: The applied virtual freshwaterflux to the ocean.

temperatures generated by the model, thus ocean temperatures are allowed to fall below -2°C.

The wind forcing on the model is based on the annual-mean wind-stress climatology of *Hellerman and Rosenstein*

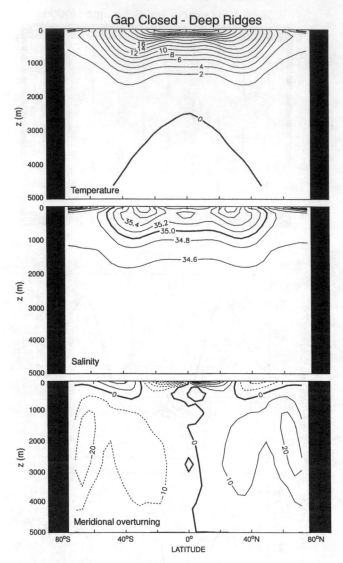

Figure 3. Results obtained with the Drake Passage gap closed and deep ridges. Shown are meridional sections through the ocean showing zonally averaged temperatures and salinities (top and mid panels) and the meridional overturning (bottom panel). The meridional overturning is a streamfunction with units of Sv $(= 10^6 m^3 s^{-1})$ derived from the zonally integrated meridional and vertical flow. The effects of the Gent-McWilliams parameterization on tracers is not included. Dashed Contours indicate overturning in a counter-clockwise direction.

[1983]. Zonal wind stresses in Hellerman and Rosenstein were first zonally averaged and then northern and southern stresses at the same latitude were then averaged to create a wind-stress field that is symmetric about the equator. The salinity forcing used by the model is derived from the surface water balance (precipitation + runoff - evaporation) determined during the spin-up of the GFDL R30 coupled model [*Knutson and Manabe*, 1998]. The net water flux was first

converted to an equivalent salt flux and zonally averaged. It was then averaged between the hemispheres to create a forcing field that is hemispherically symmetric like the wind field. Wind stresses and water fluxes used to force the ocean are plotted as a function of latitude in the mid and bottom panels of Figure 2.

Two sets of experiments were performed to examine the climatic impact of the opening of Drake Passage. The two sets differ only in the depth of the ridge segments that extend up from the bottom. One set has the ridge tops raised up to 742 m below the surface (the base of model level 5) in an attempt to simulate conditions during the time period shortly after Drake Passage was opened. In the second set the ridge tops were deepened to 2768m (the base of level 9), a depth more characteristic of the topographic obstacles that the modern ACC encounters. The two sets of experiments will be referred to as the "shallow ridges" and "deep ridges" experiments, respectively. Each set of experiments includes a pair of model runs, one with the Drake Passage closed and the other with Drake Passage open. These are referred to as the "closed gap" and "open gap" experiments, respectively. Selected pairs of closed gap and open gap experiments have also been run without any salinity forcing. Differences between these temperature-only model runs and the standard runs with salinity forcing show how salinity differences influence the circulation.

Each experiment was run out for a minimum of 5000 years, with select experiments run on to 6000 years to ascertain that the models had converged properly. The results of these experiments are described below. For a more complete discussion the reader is referred to *Toggweiler and Bjornsson* [2000].

3. THE EFFECTS OF OPENING UP DRAKE PASSAGE

The local effects of opening up Drake Passage are well known from previous work. Geostrophically balanced meridional flows are inhibited within a zonally re-entrant circumpolar channel. This means that the relatively warm surface waters north of Drake Passage cannot easily pass into or across the channel. Hence the possibility of a significant transport of oceanic heat into and across the ACC is much reduced. The Ekman transport within the channel also brings up cold water from the interior. The cold upwelled water takes up solar and sensible heat from the atmosphere and exports some of this heat northward. These local effects make the southern Hemisphere distinctly different from the northern Hemisphere.

Figure 3 shows the zonally averaged temperatures and salinities and the zonally integrated meridional overturning obtained for a control run with a closed gap and deep ridges. The meridional overturning in the interior consists of a circulation cell in each hemisphere with deep water formation (sinking) against each polar island. The net meridional flow in the upper 1000 m of each hemisphere is poleward as expected. The surface water close to the polar islands has a temperature just below the freezing point. Sea surface temperatures in the tropics reach $28°C$ on either side of the equator. As expected from the forcing, the circulation, temperature and salinity fields are symmetric about the equator. When this experiment is repeated with the topographic ridges raised to the shallow position a very similar result is obtained.

Figure 4 shows the same set of results for a model run with an open Drake Passage gap and shallow ridges. One notices immediately how the open gap breaks down the symmetry in the circulation and tracer fields. Deep water formation (sinking) next to the south polar island is strongly suppressed while the sinking next to the north polar island increases by about 50%. The circulation associated with northern sinking extends across the equator down to the open gap in the south. The streamlines indicate that there is a continuous flow of mid-depth water from the north polar island to the gap of about 5 Sv. Deep water from the northern hemisphere becomes incorporated into the shallow wind driven overturning cell within the latitude band of the gap and is brought to the surface as part of the Ekman divergence south of 50S. The surface water flowing northward within the gap is pumped downward into the main thermocline and flows all the way up to the sinking region next to the north polar island. The net flow in the upper kilometer between the equator and 40S is now northward instead of southward.

Associated with this interhemispheric circulation is a near surface transport of heat and salt from the southern hemisphere into the northern hemisphere. The northern hemisphere becomes saltier as seen by much larger volumes of water enclosed by the 35.2 psu contour in the northern hemisphere. At the same time, a strong halocline develops next to the south polar island with a minimum surface salinity of 32.8 psu. The surface water next to the southern polar island cools to less than $-2°C$ while the waters next to the northern polar island warm to more than $2°C$. Opening the gap decreases the average SST south of 40S by $2.4°C$ while it increases the average SST north of 40N by $1.8°C$. Thus, the high latitudes of the southern hemisphere cool, as expected, while the high latitudes of the northern hemisphere warm up.

The asymmetric circulation set up by the open gap is enhanced when the ridges are in their deep position (see Figure 5). The wind driven overturning in the latitude band of the gap deepens significantly as it extends downward to the top of the deeper submerged ridges. The overturning in the northern hemisphere and the southward outflow of northern deep water to the open gap both become significantly

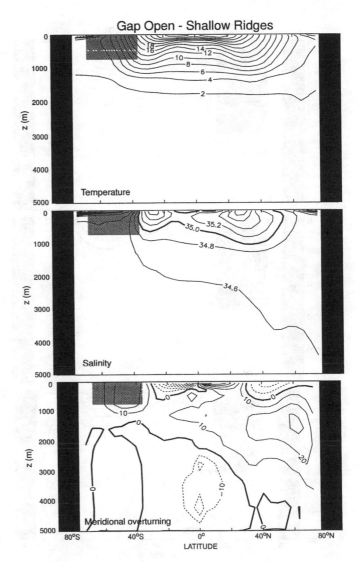

Figure 4. Same meridional sections as in Fig. 3 but for the shallow ridges model with the Drake Passage gap open. The position of the gap is shaded.

stronger. Opening the gap with deep ridges increases the average SST north of 40N by $2.8°C$, while it decreases the average SST south of 40S by $3.4°C$. A warming of the interior ocean is also observed. Individual isotherms in Fig. 5 are substantially deeper than in Fig. 3. This thickening of the thermocline leads to a subsurface maximum in temperature change. Figure 6 shows that the region of maximum temperature increase lies at or above the depth of maximum overturning in both sets of experiments. Thus, the maximum temperature increase tends to occur in a water mass that is flowing northward.

Fig. 7 shows the changes in atmospheric temperature that occur in response to an open Drake Passage. Results for both

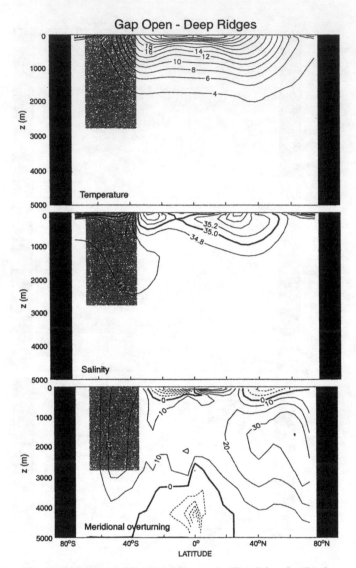

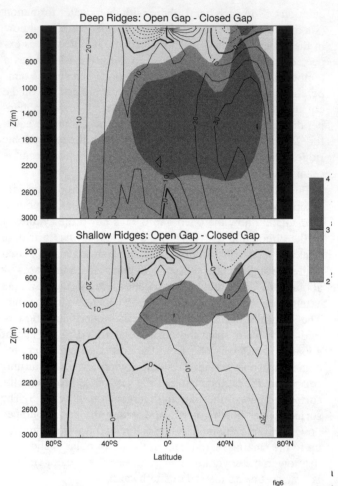

ing circulation (contours). Upper panel shows results from the deep ridges model. Lower panel shows results with shallow ridges. The shading in the figure highlights only the regions of maximum temperature increase (2 to $4°C$). The upper ocean cools in latitudes near the gap.

Figure 5. Same meridional sections as in Fig. 3 but for the deep ridges model with the Drake Passage gap open.

the deep ridges (dot-dashed) and shallow ridges (solid) experiments are shown. The opening of Drake Passage cools the high latitudes of the southern hemisphere and warms the high latitudes of the northern hemisphere. Maximum warming in the north is concentrated in the latitude band adjacent to the north polar island where deep water is being formed. Deeper ridges produce larger air temperature changes, the peak-to-peak difference is just over $4°C$ with shallow ridges, but just over $6°C$ with deep ridges.

It is noteworthy that air temperature changes in the tropics are minimal. A conveyor circulation that is entirely buoyancy driven would be expected to transport tropical heat poleward. In such a case a switch-on of a buoyancy-driven conveyor would cause the North Atlantic region to warm

at the expense of cooler tropical regions. The switching on of an interhemispheric conveyor linked to Drake Passage warms the high-latitudes of the northern hemisphere at the expense of a cooler south. The deep ridges results in Fig. 7 suggest that surface temperatures at 60N should be some $6.0°C$ warmer than at 60S. This is a large effect which should be obvious in ocean data. Indeed, Levitus SST observations show that SSTs between 50 and 60N within the Atlantic basin are about $6.0°C$ warmer than at 50 to 60S (see Fig. 11 in Toggweiler and Bjornsson, 2000).

As in the temperature field the asymmetries in the salinity field become more pronounced with deep ridges (Fig. 5). The salinity field now features a prominent tongue of relatively fresh water that extends northward at intermediate depths north of the gap. This is the water planet model's version of the salinity minimum associated with Antarctic Inter-

mediate Water (AAIW) seen in ocean observations. Intermediate and deep water in the northern hemisphere, in contrast,
becomes a bit saltier in the deep ridges model. These two
features - a northward penetration of fresh intermediate water, and a strong transequatorial overturning circulation - are
prime hallmarks of the conveyor circulation seen in the real
Atlantic. These features appear in the deep ridges model as
a consequence of an open Drake Passage. Nothing special
is done with the surface forcing to help the model generate
these features.

The virtual freshwater flux applied to the model is hemispherically symmetric. This means that the implied atmospheric water vapor transport across the equator is zero. It
follows that at steady state the net oceanic salinity transport
across the equator is also zero. Yet, both Figs. 4 and 5 show a
strong asymmetry in salinities, with the northern hemisphere
being saltier. This salinity difference must be built up during
the initial transient phase of the model run. Figure 8 shows
the average hemispheric salinity difference obtained in the
deep ridges, open gap experiment. From the figure, it is clear
that most of the salinity difference is built up during the first
1000 years of the model run. The point here is simply that
the hemispheric differences in salinity are a consequence of
conveyor circulation, they are not the cause of the conveyor
circulation.

This argument is strengthened by the fact that interhemispheric salinity differences are not even necessary for the
maintainance of a conveyor circulation. The experiments in

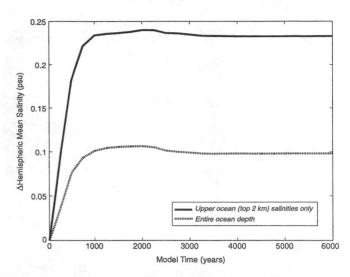

Figure 8. Buildup of hemispheric differences in salinity for the
deep-ridges/open-gap experiment. Shown is the northern hemispheric average salinity minus the southern hemispheric average
salinity. The solid line shows the result obtained when the average
was restricted to the upper 2 km of the ocean, for the dashed line
the average extended throughout the entire depth of the ocean.

Figures 4 and 5 (open gap with shallow and deep ridges)
were repeated as temperature-only runs without an imposed
salinity flux (i.e. all model salinities remained at the initial value of 34.7 psu). A conveyor circulation with sinking in the northern hemisphere was still obtained in the
temperature-only mode. This is not to say that the interhemispheric overturning circulation is unaffected by interhemispheric salinity differences. Indeed, deep water in the northern hemisphere is made relatively dense because of interhemispheric salinity differences. Thus, a well-known feedback operates that builds up interhemispheric salinity differences in conjunction with a stronger overturning. This issue
is explored in more detail in the following section.

In summary, an open Drake Passage leads directly to the
development of an overturning circulation which favors deep
water formation in the northern hemisphere. The northern hemisphere becomes saltier and the southern hemisphere
fresher. These effects are clear in the shallow ridges experiment, but become more pronounced with the deep ridges.
In our deep ridges experiment a tongue of low salinity water stretches northwards at intermediate depths in the southern ocean, reminiscent of the AAIW tongue in the Atlantic.
An open gap causes the northern hemisphere to warm considerably while the upper ocean and atmosphere in the gap
latitudes cools. The thermocline deepens leading to a subsurface warming maximum, with the greatest warming occurring in northward flowing intermediate waters.

The overturning with Drake Passage is also a new kind
of overturning in which the relative work done by buoyancy

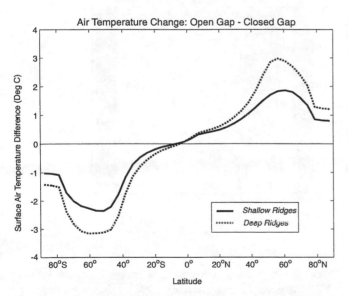

Figure 7. The change in air temperature as a result of opening the
gap, i.e., the air temperature obtained with the gap open minus the
air temperature obtained with the gap closed. The solid line shows
the results obtained with the shallow ridges, and the dot-dashed line
shows the results obtained with the deep ridges.

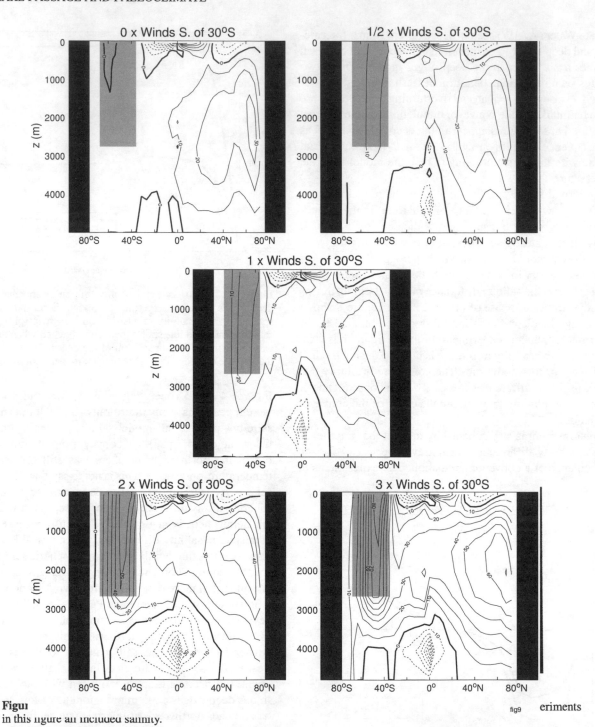

Figur
in this figure an included salinity.

fig9

eriments

forces and wind forces is reversed. The energy that drives the interior overturning in the model without Drake Passage comes entirely from work done by heating, cooling, and vertical mixing to maintain horizontal density differences. With the opening of Drake Passage, the winds driving the circumpolar current add a new source of energy that exceeds the input from heating, cooling, and vertical mixing [*Wunch*, 1998; *Toggweiler and Samuels*, 1998]. It has been suggested that an interhemispheric overturning linked to Drake Passage can be maintained by the circumpolar winds alone with relatively little input from vertical mixing [*Toggweiler and Samuels*, 1998; *Webb and Suginohara*, 2001].

4. THE EFFECTS OF WINDS IN DRAKE PASSAGE LATITUDES

The deep ridges experiment with an open gap was used as starting point for four different "wind experiments" in which the zonal windstress south of 30S (the latitude where the windstress goes to zero, see figure 2b) was multiplied by constant factors of $0, \frac{1}{2}, 2$, and 3. Each of the wind experiments was integrated for 3000 years beyond the end of the baseline deep-ridges run. For comparison purposes each of these experiments was repeated with the model run in a temperature-only mode.

Figure 9 shows the overturning obtained for the five different wind strengths. The overturning for the baseline 1x winds is shown in the middle. The wind-driven overturning in the latitude band of the gap is clearly enhanced as the windstress south of 30S is enhanced. The maximum value of the overturning stream function in the northern hemisphere and the outflow of deep water from the north polar island into the latitude band of the gap also increases. The examples in Fig. 9 were obtained from simulations run with an imposed salinity flux. Temperature-only versions of the model also show a similar response.

The change in the strength of the interhemispheric overturning with the strength of the applied wind stress is summarized in Fig. 10. The strength of the interhemispheric overturning, given here by the outflow of northern deep water across 20S, increases linearly with the wind stress south of 30S for both temperature-only and salinity-forced versions of the model. The slope of the two lines in Fig. 10

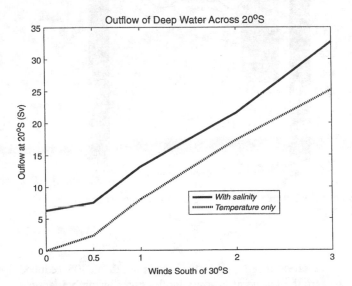

Figure 10. The sensitivity of the outflow across 20S to changes in the windstress south of 30S. The solid line shows the results obtained in experiments that include salinity, the dashed line shows the results obtained using the temperature-only model.

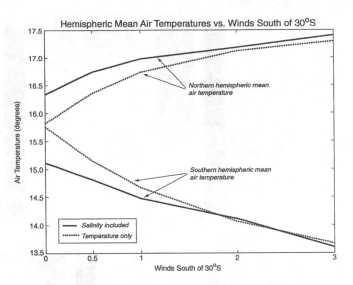

Figure 11. Hemispheric means of surface air temperature as a function of windstress south of 30S. The solid line shows the results obtained in experiments that include salinity, the dashed line shows the results obtained using the temperature-only model

is the same. The main effect of salinity differences in the model is to provide a nearly constant upward offset in the level of overturning for a given wind stress. When the winds south of 30S are turned off in the temperature-only model the interhemispheric overturning at 20S goes to zero and the overturning of northern deep water is confined to the northern hemisphere. When the winds south of 30S are turned off in the model with salinity differences the northern hemisphere overturning cell continues to extend into the southern hemisphere (see top-left panel in Fig. 9) and an outflow of 7 Sv at 20S remains. The latter effect is a consequence of interhemispheric salinity differences that develop in response to an gap open in the absence of winds (Fig. 8). The salinity differences that develop when the gap is opened thus feed back on the circulation to strengthen it. In this sense the upward offset in Fig. 10 is the salinity driven component of the interhemispheric conveyor.

In the previous section it was noted that the interhemispheric conveyor circulation transfers heat from the southern hemisphere to the northern hemisphere (see Fig. 7). Stronger winds south of 30S should enhance this effect. Figure 11 shows mean northern and southern hemisphere air temperatures as a function of wind stress south of 30S. In the standard case with deep ridges and salinity differences (wind stress factor = 1 in Fig. 11) there is a $2.5°C$ difference between mean air temperatures in the northern and southern hemispheres. In the temperature-only experiment the mean interhemispheric air temperature difference is about $2°C$. As the wind stress is increased beyond the standard value, the northern hemisphere becomes warmer and the southern

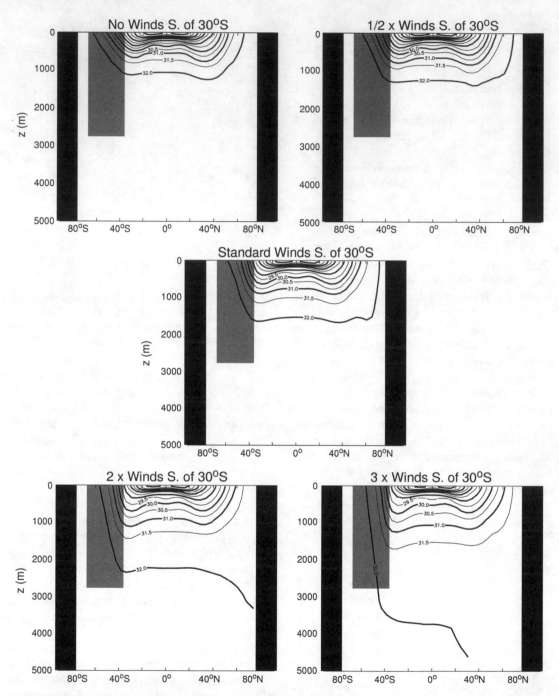

Figure 12. The zonally averaged density field (σ_1) obtained when the windstress south of 30S is varied. The experiments in this figure were obtained with the temperature-only model.

hemisphere becomes cooler. The change in temperature for a given increase in wind stress is fairly similar for the experiments including salinity and in the temperature-only experiments. With weaker winds, temperature differences between the hemispheres are reduced and the role of salinity becomes more pronounced. With the winds south of 30S reduced to zero the temperature-only model gives virtually identical mean air temperatures in the two hemispheres; the model with salinity differences retains a $1.3°C$ offset in mean air temperature.

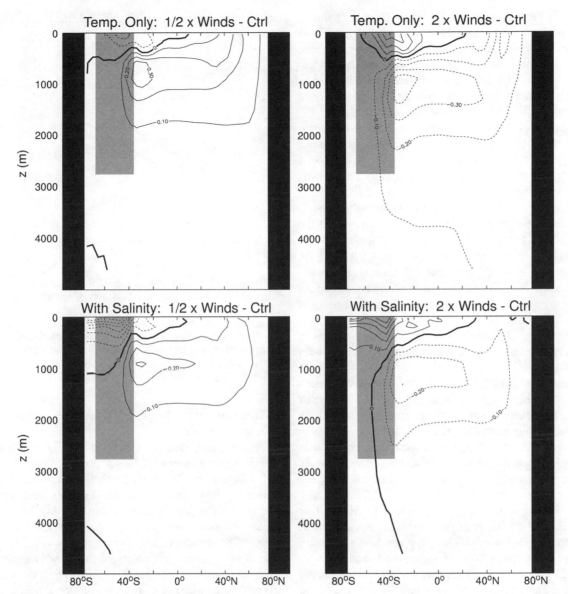

Figure 13. Density change from the control run with winds south of 30S reduced by half (left column) or doubled (right column). The upper panels show the results obtained in the temperature-only experiments, the bottom panels show the results obtained when salinity is included.

4.1. The Density Structure of the Ocean and the Wind Strength in the Gap

The previous subsection showed that the strength of the interhemispheric overturning circulation is sensitive to the wind strength in the latitude band of the open gap. In this section we explore this connection further and examine the mechanism behind it.

Figure 12 shows the zonally averaged density (σ_1) distribution obtained in the temperature-only model with each of the five different wind stress fields used in the previous section. It shows that the downward penetration of light upper ocean water is enhanced with progressively stronger winds. With no winds south of 30S (top left panel in Fig. 12) the 31.5 isopycnal is found at 740 m at the equator. With $\frac{1}{2} \times$ winds (top right panel) the same isopycnal is at 860 m. In the control run (middle panel) the corresponding depth is 1000 m. In the double winds case (lower left panel) this depth is 1260 m and with $3 \times$ winds the depth of the 31.5 isopycnal is 1525 m.

The changes in density can be split into two regimes. In the upper ocean above 500 m isopycnal depths are relatively insensitive to changes in the winds south of 30S. Most of the changes in density seen in Fig. 12 is found in the lower pycnocline-to-intermediate depth range. Figure 13 shows the difference in σ_1 between the $\frac{1}{2}\times$ and $2\times$ winds cases and the standard case. The top two panels show results from the temperature-only model. The lower two panels show results from the model with salinity differences. Changes in density in response to the winds have a subsurface maxima around 1 km depth that is strongest just north of the gap.

Isopycnal layers in this depth range outcrop in the northern half of the gap, in a region of Ekman flux convergence and active subduction. Figure 14 shows the thickness of the isopycnal layer that outcrops in the northern half of the gap. The isopycnal thickness is measured at 20S; experiments with and without salinity are included in the figure. In both cases the density difference across the gap at the ocean surface is fairly uniform, close to 2 kg/m^3 in the temperature-only experiments, and close to 1.5 kg/m^3 for all windstrengths in the experiments with salinity (except the no-winds case where the surface density difference across the gap is very small). In both cases the layer thickening at 20S it almost linearly related to the windstress south of 30S, as would be expected if the thickening were due to enhanced Ekman flux convergence, which also increases linearly with the wind strength.

The density changes in Fig. 13 illustrate a somewhat paradoxical result: when the winds south of 30S and the rate of deep-water formation in the north are stronger, the sinking water in the north is actually less dense. When the winds in the south and northern sinking are weakest, the sinking water in the north is densest. One almost reflexively expects denser water in the north to lead to stronger overturning, not weaker. Looked at another way, the results in Fig. 13 are actually consistent with expectations. The changes in thermocline density in Fig. 13 are larger in the southern hemisphere than in the northern hemisphere. Thus, the north-south density contrast becomes larger with stronger winds and stronger overturning. This result is consistent with *Rahmstorf* [1996] who found a linear relationship between the strength of the Atlantic conveyor and the north-south density difference in the thermocline between 50-55N and 35-40S. In *Rahmstorf* [1996] the Atlantic conveyor was weakened through the introduction of freshwater anomalies in the North Atlantic.

Figure 15 shows the average steric height at 50N and at 40S as a function of wind stress for models run with salinity differences and with temperature-only. It is very clear that the steric height is linearly related to the wind stress south of 30S both at 40S and at 50N. The curves for 40S and 50N diverge slightly with stronger winds leading to a larger north-south difference. The bottom panels of the figure show the correspondance between the average north-south steric height difference between 40S and 50N and the outflow of northern water across 20S. A stronger outflow is associated

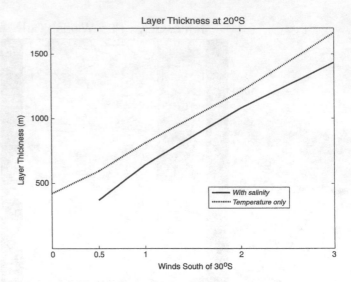

Figure 14. The thickness at 20S of the layer that outcrops in the northern half of the gap, as a function of windstress south of 30S. The thickness is defined as depth at 20S to the σ_1 density surface that outcrops in the middle of the gap minus the depth (also at 20S) to the σ_1 surface that outcrops at the northern end of the gap. The data for the no-winds case with salinity is omitted, since in this case the density difference at the surface is very small and there is not a clear density layer outcroping at the surface.

with a larger steric height difference, but the relationship is not exactly linear.

The steric height is calculated by integrating twice the steric volume anomaly (the reciprocal of the density anomaly) from a specific depth level to the surface [*Hughes and Weaver*, 1994]. The depth level of interest here is the level of no motion that separates southward flowing deep water from the northward flowing water above. To obtain the results in Fig. 15 the depth to the 31.5 σ_1 surface was used as a level of no motion, since this density surface tends to coincide with the depth level of maximum overturning. If the steric height is calculated with the level-of-no-motion at a fixed depth, say 1500m, the relationship between steric height difference and northern outflow vanishes. Similar results were obtained by *Hughes and Weaver* [1994] who found an approximate linear relationship between the Atlantic conveyor strength and steric height differences between the northern and southern hemispheres. The results here show that the existence of this relationship is linked to the deepening pycnocline and the deepening level-of-no motion; it is not just due to changing north-south density differences above some fixed depth.

The water type most affected by changes in southern hemisphere winds is found in at intermediate depths in the lower thermocline. This is the water mass that exhibits the greatest warming when the gap is opened (see Fig. 6), and is further characterized by low salinities at intermediate depth just

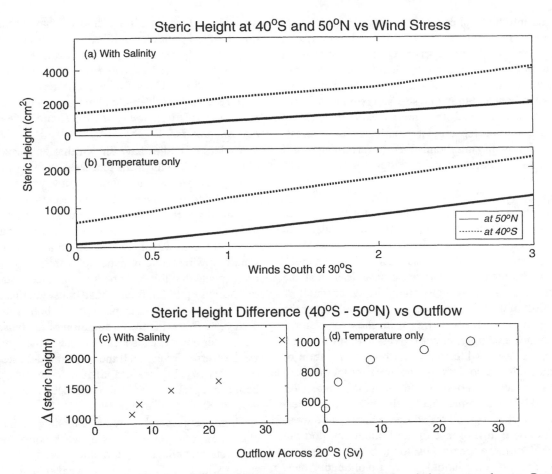

Figure 15. Upper two panels: The dependence of steric height at 40S and 50N on the windstress over the gap. Panel a) for experiments where salinity is included, panel b) for the temperature only experiments. The lowest two panels show the north-south difference in steric height, experiments including salinity in (c) and temperature only experiments in d).

north of the gap (the "AAIW water" tongue of Fig. 5). The results here are consistent with those of *Vallis* [2000] who documented the formation of a thermostad below the main wind-driven thermocline in an idealized GCM with a Drake Passage-like gap. They are also consistent with the box model of the thermocline in *Gnanadesikan* [1999] which relates the thickness of the ocean's thermocline to the wind-driven conversion of high-density deep water in the south to low-density thermocline water. In all three studies the volume of lower thermocline/intermediate water is enlarged by the Ekman flux convergence in the latitude band north of the maximum westerlies. It makes sense that this water mass would expand northward and that its growth as a water mass should scale with the windstress south of 30S.

5. DISCUSSION

5.1. Drake Passage and Paleoclimate

According to geologic records, the thermal isolation of Antarctica began about 35 million years ago near the Eocene-

Oligocene boundary with the accumulation of glacial ice on Antarctica and an abrupt cooling of the ocean's deep water *Kennett* [1977]. This would appear to mark the initiation of a circumpolar flow around Antarctica. Kennett identifies this event with the separation of Australia from Antarctica rather than the opening of Drake Passage. *Lawver and Gahagan* [1998] claim that a middle depth opening through the present day Palmer Peninsula (a proto Drake Passage) was already in place at the time when Antarctica and Australia separated. The opening of a deep passage between South America and Palmer Peninsula occurred sometime later [*Barker and Burrell*, 1977] . Technically speaking, the experiments described here are more relevant to the separation of Antarctica and Australia. Comparison of results from experiments with the two different ridge heights (see Fig. 7) suggest that the opening of a middle depth channel in a latitude band of westerly winds should be more than sufficient to initiate significant cooling on Antarctica.

The experiments carried out in this paper explore the sensitivity of the interhemispheric overturning circulation to a wide range of wind stresses. It is unlikely, however, that the circumpolar ocean ever experienced the range of wind stresses imposed here. On the other hand, rather clear evidence exists that meridional shifts in the location of the Antarctic polar front and the wind stress maximum have occurred [*McCulloch et al.*, 2000; *Howard and Prell*, 1992]. From a paleoclimate perspective it is perhaps more interesting to consider how the interhemispheric overturning would respond to a meridional shift in the westerlies. A northward shift in southern hemisphere winds has been implicated in the glacial CO_2 cycle [*Sigmund and Boyle*, 2000].

The wind stress maximum associated with the southern hemisphere westerlies today is up to 10° of latitude north of the open channel passing through Drake Passage. Much of the zonal flow of the ACC is also north of the channel. It is only in the vicinity of Drake Passage itself that the wind stress maximum is actually found in the channel. A northward shift of the wind stress maximum would move the stress maximum even further away from the open channel and would almost certainly reduce the eastward stress within the channel. It is possible that a northward migration of the westerlies might also draw the polar easterlies out from Antarctica into the southern part of the channel. *Toggweiler and Samuels* [1995] imagined that the link between the interhemispheric overturning and the southern hemisphere westerlies comes about in response to the Ekman transport out of the channel, or more specifically, to the zonally integrated northward transport at the latitude of the tip of South America (56S). In this case a northward shift of the westerlies would reduce the overturning in much the same way that an overall reduction in wind stress would reduce the overturning.

The existence of a link between winds in the southern hemisphere, an open Drake Passage, and the interhemispheric overturning seems to be a robust effect in models, but it is not entirely clear at the present time how the wind-overturning link works in models with a less idealized geometry. This uncertainty is very much in play when one considers how the overturning should respond to a northward shift in the winds. Furthermore, given the idealized nature of the experiments, one must be careful when discussing the relevance of the "closed gap" experiments to the ocean circulation that existed prior to the opening of Drake Passage. Obviously, real world asymmetries in the structure of ocean basins, may have resulted in an ocean circulation different from the very symmetric two cell circulation of Fig. 3.

5.2. Multiple Equilibria and Climate Variability

As discussed in the Introduction, there is a substantial body of work on the buoyancy driven aspects of conveyor dynamics. One constant theme of that discussion is the possibility of multiple equilibria of the conveyor circulation [*Bryan*, 1986; *Manabe and Stouffer*, 1988; *Marotzke and Willebrand*, 1991; *Power and Kleeman*, 1993; *Hughes and Weaver*, 1994]. The stability properties of these solutions have received considerable attention, both in zonally averaged models [*Bjornsson et al.*, 1997] and in OGCMs [*Manabe and Stouffer*, 1999b].

Modelling studies have shown that enhanced freshening of the North Atlantic leads to a weak conveyor (e.g. *Weaver et al.* [1998]) or even a collapsed conveyor (e.g. *Seidov and Maslin* [1999]). However, neither of these particular studies examined whether a weakened conveyor might return back to its original state once the anomalous forcing was terminated. If the conveyor is to remain in a weak or a collapsed state following the termination of a freshwater pulse, it is a prerequisite that this state be a stable steady state. *Manabe and Stouffer* [1999b] found that a conveyor that is weakened by a transient freshwater pulse can rebound in a few hundred years following the termination of the pulse. The same freshwater pulse can also lead to a permanent collapse and a weak reverse circulation if the ocean model uses a low vertical diffusivity. Low vertical diffusivity levels are thought to be more representative of the real ocean [*Watson and Ledwell*, 2000; *Ledwell et al.*, 1998].

There is considerable evidence to suggest that during glacial times conveyor episodes of weak conveyor circulation did occur and climatic ramifications were significant. Consistent with Fig. 11 an episodic weakening of the conveyor gives rise to a "seesaw signal" in Atlantic ocean temperatures [*Broecker*, 1998; *Stocker*, 1998]: A weak conveyor circulation cools the North Atlantic and warms the South Atlantic. However, there is no evidence to suggest that the conveyor was ever in a reverse state for any duration of time.

The above can be summarized as follows: First, ocean models do have multiple equilibria, and in some of these a reverse conveyor is stable. Second, there is no observational evidence for a stable reverse conveyor, rather the evidence is consistent with episodic variations in the strength of a forward conveyor. While this may seem contradictory, there are several ways of reconciling the above two statements. First, the real ocean may not have a stable reverse conveyor, and will thus always return to the forward conveyor once external disruptions (such as a freshwater pulse) terminate. A second possibility is that the real ocean has a stable reverse conveyor, but the freshwater pulses into the Atlantic Ocean during the last glacial were not of sufficient magnitude to "kick" the conveyor into a reverse state.

The results of the recent study by *Ganapolski and Rahmstorf* [2001] suggest that the first of the above alternatives is the one that applies to the glacial Atlantic. They used a zonally averaged ocean model coupled to a simplified atmospheric model. When they prescribed holocene boundary conditions they found two stable solutions, a forward conveyor and a reverse conveyor. However, when they prescribed glacial conditions, they could not obtain the reverse conveyor state. They found instead a stable weak forward conveyor and a weakly unstable state with a strong forward conveyor. They suggest that the glacial climatic record can be explained by the system alternating between the the stable weak conveyor (cold state) and the weakly unstable strong conveyor (warm state).

This paper argues that Southern Ocean winds also influence the Atlantic conveyor. The winds can be thought of as an "added driver" that makes the forward conveyor more resilient in response to freshwater perturbations in the North Atlantic. The added driver idea is consistent with observations that show that the glacial conveyor recovered from massive disruptions such as Heinrich Events. Since this added driver is independent of buoyancy, it would not be affected by interglacial-to-glacial changes of the large scale buoyancy fluxes across the ocean surface, and thus it is possible that during glacial times it made the forward conveyor more stable and the reverse conveyor less stable.

Although this paper focuses on the influence of the southern ocean winds, it is quite possible that other southern ocean processes affect the Atlantic conveyor. Recently *Seidov et al.* [2001] examined the impact of freshwater pulses into the southern ocean on the Atlantic conveyor and found that a freshening of the southern ocean can lead to a stronger forward conveyor. These results and those from this paper suggest that a "seesaw" could be forced from the high latitudes of either hemisphere.

5.3. Alternate Model Configurations

The illustrations above show that there is a clear link between the interhemispheric overturning and the strength of the Ekman convergence north of the westerly wind stress maximum. The Ekman convergence is shown to have a direct impact on north-south pressure differences and density differences within the thermocline, factors which have been shown previously to correlate very strongly with rates of deep-water formation in the North Atlantic [*Hughes and Weaver*, 1994; *Rahmstorf*, 1996]. The circumpolar channel in the water planet model is unrealistically wide. As such, it covers most of the westerly wind band. In the real world there is a land barrier, South America, that extends into the latitude band of the westerlies. Does the Ekman convergence north of the wind stress maximum operate in the same way in a latitude band that is blocked by land? The circumpolar belt is also divided among three ocean basins whereas the interhemispheric conveyor operates only in the Atlantic. How

do wind effects in the Indian and Pacific basins influence the overturning in the Atlantic Ocean?

The first question in the previous paragraph can be addressed in the context of the present model. The response of the interhemispheric overturning to changing winds was examined in a model in which the land barrier was extended to cover the northern half of the gap. The open channel is thereby restricted to the latitude band between 51.75S to 65.25S. The model was run in a temperature only mode with an open gap using the standard winds and the 3x winds of section 4. These experiments revealed that the overturning circulation retains more-or-less the same sensitivity to winds south of 30S as the model with a wide gap. A similar deepening of isopycnals was obtained with the narrower gap.

Further experiments with separate "Atlantic" and "Pacific" basins are needed to answer the second question. An identical pair of basins with an open gap in the south would presumably each acquire half the overturning seen in the present single basin model. One can then ask what sort of changes in configuration and/or interbasin atmospheric water vapor transport is required to allow one of the two basins to capture the interhemispheric overturning circulation. Is the maximum northward extent of the two basins important? What effect does a different southward extent of "South American" and "African" land barriers have on the relative overturning of the two basins?

5.4. Fully Coupled Models and Models with Intermediate Levels of Coupling

As pointed out in the introduction, the first studies to examine the impact of an open Drake Passage and the sensitivity of the Atlantic overturning to the wind stress in the latitude band of Drake Passage [*Cox*, 1989; *Toggweiler and Samuels*, 1995] used simple restoring boundary conditions for temperature and salinity. *Rahmstorf and England* [1997] examined the sensitivity of the Atlantic overturning to southern hemisphere wind stresses in a model with an intermediate level of coupling. They replaced the restoring of salinity with a fixed salt flux, as is done here, and adopted a hybrid thermal boundary condition which allows surface temperatures to respond to changes in circulation subject to a slow restoring of surface temperatures to a set of fixed atmospheric temperatures. The present study is also carried out in a model with an intermediate level of coupling, because wind stresses and salt fluxes are imposed, but it is a thermally coupled model in the sense that the model's atmospheric temperatures are prognostic variables of the system. No one has yet "opened Drake Passage" in a fully coupled model in which wind stresses, evaporation and precipitation are also prognostic variables.

Rahmstorf and England found that the sensitivity of the

North Atlantic overturning circulation to winds in the southern hemisphere was reduced in their intermediate coupled model in comparison with the results of *Toggweiler and Samuels* [1995]. A simple temperature feedback was identified that is responsible for the reduced sensitivity. With this feedback, an initial enhancement of the overturning (in response to increased southern hemisphere winds) leads to warmer temperatures in the North Atlantic which in turn leads to reduced deep-water formation and a slower overturning. The net result is a reduced overturning sensitivity. The same mechanism is presumably at work in the water planet model where northern hemisphere temperatures increase in response to stronger overturning. From a climate perspective, however, it is the increase in northern temperatures which is of interest. Thus, the results of Rahmstorf and England (1997) actually reinforce the idea that wind stresses in the latitude band of Drake Passage are important for northern hemisphere climate.

6. CONCLUSIONS

The model results presented in this paper show that the opening of Drake Passage initiates a conveyor-like overturning that transports heat and salinity across the equator leading to a warmer northern hemisphere and colder southern hemisphere. Once an open channel is established interhemispheric temperature differences are sensitive to the strength of the winds blowing over the channel.

These effects can be understood in terms of the water mass formation that occurs when Drake Passage is open. The opening of a Drake Passage-like gap leads to the vertical expansion of the density class that outcrops in the band of Ekman flux convergence north of the wind stress maximum. Water of this density class flows northward within the lower thermocline into the northern hemisphere. The volume transported by this water mass is shown to be quite sensitive to the Ekman flux convergence in the outcroping region and hence to changes in southern hemisphere winds generally. Enhancement of the volume transport leads to a deepening of the thermocline and an increase in the volume of water which is exposed to the atmosphere and sinks in the northern hemisphere.

As discussed earlier, there is a considerable volume of literature on the buoyancy driven aspects of the conveyor circulation. In this paper we have shown that there is a whole suite of additional processes acting within the southern hemisphere that influence the conveyor and thereby influence the climate of the northern hemisphere. We conclude that the Atlantic conveyor circulation is a hybrid circulation, part thermohaline, part wind driven, which is very dependent on the existence of an open circumpolar channel in the high latitudes of the southern hemisphere.

REFERENCES

Barker, P. F., and J. Burrell, The opening of Drake Passage, *Mar. Geol.*, 25, 15–34, 1977.

Bjornsson, H., L. A. Mysak, and G. A. Schmidt, Mixed boundary conditions versus coupling with an energy moisture balance model for a zonally averaged ocean climate model, *J. Climate*, 10, 2412–2430, 1997.

Broecker, W. S., The Great Ocean Conveyor, *Oceanography*, 4, 79–89, 1991.

Broecker, W. S., Paleocean circulation during the last deglaciation: A bipolar seesaw?, *Paleoceanography*, 13, 119–121, 1998.

Bryan, F., High-latitude salinity effects and interhemispheric thermohaline circulations, *Science*, 323, 301–304, 1986.

Bryan, K., and L. Lewis, A water mass model of the World Ocean, *J. Geophys. Res.*, 84, 2503–2517, 1979.

Cox, M. D., An idealized model of the world ocean. Part I: the global scale watermasses, *J. Phys. Oceanogr.*, 19, 1730–1752, 1989.

England, M., Representing the global-scale water masses in ocean general circulation models , *J. Phys. Oceanogr.*, 23, 1523–1553, 1993.

Fanning, A. F., and A. J. Weaver, Temporal-geographical meltwater influences on the North Atlantic Conveyor: Implications for the Younger Dryas, *Paleoceanography*, 12, 307–320, 1997.

Ganapolski, A., and S. Rahmstorf, Rapid changes of glacial climate simulated in a coupled climate model, *Nature*, 409, 153–158, 2001.

Gill, A. E., and K. Bryan, Effects of geometry on the circulation of a three-dimensional southern hemisphere ocean model, *Deep-Sea Res.*, 18, 685–721, 1971.

Gnanadesikan, A., A simple predictive model for the structure of the oceanic pycnocline, *Science*, 283, 2077–2079, 1999.

Griffies, S. M., The Gent-McWilliams skew flux, *J. Phys. Oceanogr.*, 28, 831–841, 1998.

Hellerman, S., and M. Rosenstein, Normal monthly wind stress over the world ocean with error estimates, *J. Phys. Oceanogr.*, 13, 1093–1104, 1983.

Howard, W. R., and W. L. Prell, Late Quaternary surface circulation of the Southern Indian Ocean and its relationship to orbital variations, *Paleoceanography*, 7, 79–117, 1992.

Hughes, T. M. C., and A. J. Weaver, Multiple equilibria of an asymmetric two-basin ocean model., *J. Phys. Oceanogr.*, 24, 619–637, 1994.

Kennett, J. P., Cenozoic evolution of Antarctic glaciation, the circum-Antarctic ocean, and their impact on global paleoceanography, *J. Geophys. Res.*, 82, 3843–3860, 1977.

Knutson, T. R., and S. Manabe, Model assessment of decadal variablility and trends in the tropical Pacific Ocean, *J. Climate*, 11, 2273–2296, 1998.

Lawver, L. A., and L. M. Gahagan, The initiation of the Antarctic Circumpolar Current and its impact on Cenozoic Climate, in *Tectonic Boundary Conditions for Climate Model Simulations*, edited by T. Crowley and K. Burke, pp. 213–226, Oxford University Press, 1998.

Ledwell, J. R., A. J. Watson, and C. S. Law, Mixing of tracers the pycnocline, *J. Geophys. Res.*, 103, 21,499–21,529, 1998.

Manabe, S., and R. J. Stouffer, Two stable equilibria of a coupled ocean-atmosphere model, *J. Climate*, 1, 841–866, 1988.

Manabe, S., and R. J. Stouffer, The role of thermohaline circulation in climate, *Tellus*, 51A, 91–109, 1999a.

Manabe, S., and R. J. Stouffer, Are two modes of the thermohaline circulation stable?, *Tellus, 51A*, 400–411, 1999b.

Marotzke, J., and J. Willebrand, Multiple equilibria of the global thermohaline circulation, *J. Phys. Oceanogr., 21*, 1372–1385, 1991.

McCulloch, R. D., M. J. Bentley, R. S. Purves, N. R. J. Hulton, D. E. Sugden, and C. M. Clapperton, Climatic inferences from glacial and paleoecological evidence at the last glacial termination, southern South America, *J. Quat. Sci., 15*, 409–417, 2000.

North, G. R., Analytical solution to a simple climate model with diffusive heat transport, *J. Atmos. Sci., 32*, 1301–1307, 1975.

Pacanowski, R., MOM 2: Documentation, users guide and reference manual, *Tech. Rep. 3.2*, GFDL Ocean Technical Report, Princeton, New Jersey, 1996.

Power, S. B., and R. Kleeman, Multiple equilibria in a global ocean general circulation model, *J. Phys. Oceanogr., 23*, 1670–1681, 1993.

Rahmstorf, S., On the freshwater forcing and transport of the Atlantic thermohaline circulation, *Climate Dynamics, 12*, 799–811, 1996.

Rahmstorf, S., and England, Influence of southern hemispheric winds on North Atlantic Deep Water flow, *J. Phys. Oceanogr., 27*, 2040–2054, 1997.

Seidov, D., and M. Maslin, North Atlantic deep water circulation collapse during heinrich events, *Geology, 27*, 23–26, 1999.

Seidov, D., E. Barron, and B. J. Haupt, Meltwater and the global ocean conveyor: Northern versus southern connections, *Global and Planetary Change, (in press)*, 2001.

Sigmund, D. M., and E. A. Boyle, The Mystery of Glacial/Interglacial Variations in Atmospheric Carbon Dioxide, *Nature*, 2000.

Stocker, T., The seasaw effect, *Science, 282*, 61–62, 1998.

Toggweiler, J. R., and H. Bjornsson, Drake Passage and paleoclimate, *J. Quat. Sci., 15*, 319–328, 2000.

Toggweiler, J. R., and B. Samuels, Effect of Drake Passage on the global thermohaline circulation, *Deep-Sea Res., 42*, 477–500, 1995.

Toggweiler, J. R., and B. Samuels, On the ocean's large-scale circulation near the limit of no vertical mixing, *J. Phys. Oceanogr., 28*, 1832–1852, 1998.

Vallis, G. K., Large-scale circulation and production of stratification: effects of wind, geometry and diffusion, *J. Phys. Oceanogr., 30*, 933–954, 2000.

Watson, A. J., and J. R. Ledwell, Oceanographic experiments using sulphur hexafluoride, *J. Geophys. Res., 105*, 14,325–14,337, 2000.

Weaver, A. J., M. Eby, A. F. Fanning, and E. C. Wiebe, Simulated influence of carbon dioxide, orbital forcing and ice sheet s on the climate of the Last Glacial Maximum, *Nature, 394*, 847–853, 1998.

Webb, D. J., and N. Suginohara, Vertical mixing in the ocean, *Nature, 409*, 37, 2001.

Wunch, C., The work done by the wind on the oceanic general circulation, *J. Phys. Oceanogr., 28*, 2332–2340, 1998.

H. Bjornsson, Icelandic Meteorology Office, Bustadavegur 9, IS-150, Reykavik, Iceland (halldor@vedur.is).

J.R. Toggweiler GFDL/NOAA, P.O. Box 308, Princeton, NJ 08542, USA (jrt@gfdl.noaa.gov).

Stability and Variability of the Thermohaline Circulation in the Past and Future: a Study with a Coupled Model of Intermediate Complexity

Andrey Ganopolski and Stefan Rahmstorf

Potsdam Institute for Climate Impact Research, Potsdam, Germany.

We present the results of a study of the stability properties of the thermohaline ocean circulation (THC) and the possible mechanisms of rapid climate changes using the climate system model of intermediate complexity, CLIMBER-2. We consider two climate states, the warm modern climate and the cold glacial climate, and compare the stability properties of the THC in both cases. While for modern climate there are two stable circulation modes: the "warm" and the "off" mode, under glacial conditions we find only one ("cold") stable mode of Atlantic ocean circulation. This mode is characterized by deep water formation south of Iceland and has a shallower overturning with weaker outflow to the southern Atlantic. Another mode of the THC similar to modern "warm" mode is marginally convectively unstable under glacial conditions and a relatively small negative perturbation in the freshwater flux to the Nordic Seas can trigger rapid transitions between "cold" and "warm" modes. After being excited, the "warm" mode is metastable on the time scale of a few hundred years. We speculate that the stable "cold" mode corresponds to stadial periods with cold climate over the Northern Atlantic, while temporal transitions between "cold" and "warm" modes could explain the observed Dansgaard-Oeschger oscillations. The warm modern climate does not possess this type of instability, but rapid global warming can trigger a complete collapse of the THC. We discuss the mechanisms and uncertainties of future changes in the THC.

1. INTRODUCTION

The understanding that the Earth system, as a complex and strongly nonlinear object, can posses a typically nonlinear features, such as multiple equilibria and rapid transitions between them, becomes one of the cornerstones of the Earth sciences. The first example of multiple equilibria in the climate system due to positive albedo feedback, was demonstrated by *Budyko* [1969] using an energy balance climate model. *Stommel* [1961] with a simple box model find two equilibrium solutions for the ocean thermohaline circulation. It took few decades before convincing example

of existence of different climate states for the same external boundary conditions was obtained using complex climate models (AOGCMs). *Manabe and Stouffer* [1988] with the GFDL coupled climate model find two considerably different climate states, corresponding to two modes of the ocean thermohaline circulation. The paleodata from ice cores and other proxies analyzed in the beginning of 90th [e.g., *Dansgaard et al.*, 1993; *Bond et al.*, 1993] showed numerous and vigorous millennia-scale oscillations of the temperature in the Northern Atlantic and surrounding areas, indicating the possible relation with the instability of the thermohaline circulation [*Broecker*, 1994]. This and other results provoked growing interest in the problem of stability and variability of the THC. Furthermore, the results of simulations of future climate change indicated the possibility of a complete collapse of the THC due to global warming and changes in the hydrological cycle [*Manabe and Stouffer*, 1994]. This

The Oceans and Rapid Climate Change: Past, Present, and Future
Geophysical Monograph 126

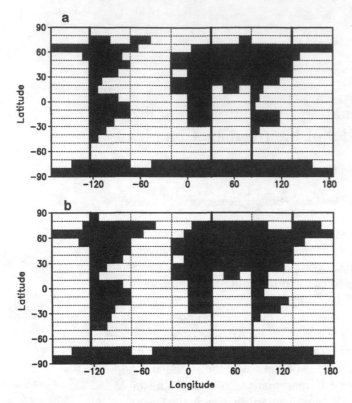

Figure 1. Representation of the Earth's geography in the model. (a) The area of land for present conditions (b) The area of land (including shelf ice) during glacial conditions. Dashed lines show the atmospheric grid, solid lines separate ocean basins.

finding made the problem of stability of the THC not only of scientific but also of great practical importance.

While the importance of a better understanding of the nonlinear Earth system dynamics in general, and of the THC in particular, became widely recognized, it also became apparent that existing scientific tools were not very suitable to tackle these questions. The typical time scales of the ocean range from a hundred to thousand years, which is at the upper limit of the applicability of GCMs. Some other important processes in the climate system, involving the biosphere, geosphere and cryosphere, have even longer time scales. Simple models, in contrast, can easily operate on these time scales, but the lack of geographical realism (usually they are zonally averaged or box-type models), small number of processes considered, and necessity to make a large number of prescriptions compared to the number of the model's degrees of freedom, limit the usefulness of such models. As a result of these limitations a new class of models, called Earth system models of intermediate complexity (EMICs), appeared aiming to fill the growing gap between simple and complex models. By design EMICs are simpler and more spatially aggregated or coarse-resolution compared to GCMs. This is why they are at least a few orders of magnitude less computationally expensive than GCMs, but being based on an improved understanding of the major processes in the Earth system, these models are able to realistically simulate a considerable number of processes and characteristics. Due to their low computational cost models of intermediate complexity provide a unique opportunity to perform transient paleoclimate simulations on multimillenial and even million-year time scales. In respect of future centennial time-scale climate projections, EMICs "play" on the same ground as comprehensive climate models, but they are still very useful because they allow us to perform large series of experiments and help to understand the role of different processes, uncertainties and possible reasons for differences between the results of complex models.

2. MODEL AND EXPERIMENTAL DESIGN

2.1 Model

For the experiments presented here we used Earth system model of intermediate complexity CLIMBER-2 (Version 3). This version is similar to the previous one described in detail in *Petoukhov et al.* [2000], apart from enhanced horizontal and vertical resolution in the ocean. CLIMBER-2 is a coarse resolution geographically explicit model designed for long-term climate simulations. Geographically the model resolves only individual continents (subcontinents) and ocean basins (Figure 1). Atmosphere, land surface, and vegetation models have the same spatial resolution: 10° in latitude and 51° in longitude. The ocean and sea ice models in longitudinal direction resolve only three ocean basins and in latitudinal direction have a resolutions of 2.5°. Vertically the ocean model has 20 unequal levels with an upper mixed layer of 50m thickness. Each oceanic grid cell communicates with either one, two or three atmosphere grid cells, depending on the width of the ocean basin. Energy, water and momentum exchanges between atmosphere and ocean are calculated on a complementary grid, with a resolution of 2.5° in latitude and 51° in longitude. In latitudinal direction the atmospheric characteristic are interpolated to the ocean grid, while in longitudinal direction (if one ocean grid corresponds to more than one atmosphere grid cell) an additional parameterization is employed to account for the longitudinal gradient of SST due to the ocean gyre circulation.

The atmosphere model POTSDAM (POTsdam Statistical-Dynamical Atmosphere Model) is based on a statistical-dynamical approach and describes the large-scale climatological (i.e. ensemble averaged) dynamics and thermodynamics of the atmosphere. Synoptic process are not described explicitly but instead parameterized in terms of large-scale diffusion with a calculated diffusion coefficient derived from the theory of baroclinic instability. Dynamics are based on the quasigeostrophic approximation and a special parameterization for the zonally averaged meridional circulation. The model's prognostic equations are derived using an assumption about the universal vertical structure of temperature and humidity. The vertical profile of the temperature is assumed to be linear in the troposphere and specific humidity is exponential with a constant vertical scale. The model employs multi-level radiation schemes for the calculation of short-wave and long-wave radiation fluxes, which take into account two types of clouds (stratus and cumuli), water vapor, carbon dioxide, and aerosols. The model does not resolve the diurnal cycle and has a time step of one day.

Land surface processes are described by the Atmosphere-Surface Interface (ASI) model, which calculates surface fluxes, temperature, soil moisture and snow cover. The model is based on BATS scheme [*Dickinson et al.*, 1986]. Each grid cell is considered as a mixture of up to six different surface types: open ocean, sea ice, forest, grassland, desert and glaciers. The last surface type differs in that it always has the albedo of snow. If the model predicts annual net accumulation of snow at any grid cell, this net accumulation is added to the runoff from this grid cell. The fractions of different vegetation types are calculated using a simple global dynamical vegetation model [*Brovkin et al.*, 1997].

The ocean model is based on the zonally averaged multi-basin model of *Stocker and Wright* [1991]. The meridional component of velocity is calculated in the model via the meridional density gradients and the zonal component of wind stress, computed in the atmosphere module. The ocean is represented by three sectors of variable width, stretching from the North Pole to Antarctica. Within the latitudinal belt where the oceans are separated by continents there is no horizontal mass exchange, while in the Arctic and Southern Oceans the oceanic sectors are connected with each other and the averaged zonal velocity is prescribed. Sea ice is described by a one-layer thermodynamic model with a simple treatment of horizontal advection and diffusion. It predicts thickness and concentration of sea ice. The model does not resolve the gyre circulation, so that meridional heat and salt transports are solely due to the meridional overturning circulation and diffusion, except for the Northern Atlantic, where freshwater transport due to the subpolar gyre circulation is parameterized [see *Ganopolski and Rahmstorf*, 2001]. The rationale for employing this parameterization is the following. At present more than half of total surface freshwater to the Arctic and Nordic seas, estimated as 0.18 Sv, escapes from the Arctic via the Canadian archipelago and the East Greenland current in the form of low salinity surface currents (0.08 Sv) and sea ice transport (0.02 Sv) without mixing with the incoming salty Atlantic water [*Aagaard and Carmack*, 1989]. This helps to sustain convection and to form dense NADW in this area. Since these processes cannot be explicitly described in the zonally averaged ocean model, we implemented special parameterizations for freshwater bypass and ice export. These parameterizations are tuned to produce observed present-day

fluxes of freshwater and sea ice and allow us to obtain stability properties of the THC which agree with the results of GCMs.

2.2 *Experimental Design and Background Climates*

Below we will discuss the stability properties of the THC under two radically different climate conditions: modern and glacial. Under modern climate we understand hereafter equilibrium preindustrial climate state - i.e. climate corresponding to the present-day orbital parameters, preindustrial (280 ppm) CO_2 concentration and potential vegetation cover. To obtain this equilibrium state the model was integrated for 5000 years without using flux correction. This modern climate state simulated by the model is very similar to that described in detail in *Petoukhov et al.* [2000], where the performance of the previous model version was presented. CLIMBER-2 is able to simulate the spatial and temporal (seasonal) variability of different climatological characteristics in good agreement with empirical data. In particular, the model simulate strong overturning in the Atlantic (Figure 2a) with formation of deep water between 60°N and 70°N.

Glacial climate corresponds to LGM (21 KyBP) boundary conditions. In this case CO_2 concentration was set to 200 ppm, the annual cycle of insolation was changed to that of 21 KyBP, continental ice sheets, their elevation and changes in area of continents were prescribed using the reconstruction of Peltier (1994). The elevation of ice-free land was uniformly increased by 120m and the global salinity increased by 1 p.s.u. to account for sea level drop. For glacial conditions, the efficiency of freshwater bypass was reduced by 50% to take into account the closing of the Canadian archipelago by ice sheets and the partial closure and shallowing of Denmark Strait. However, this is not crucial for the results. Vegetation cover, as a part of climate system, interactively adjusts to the new climate conditions and produces a positive feedback for global cooling caused by a considerable reduction of forest area [*Ganopolski and Claussen*, 2000]. The results of the LGM climate simulations in major details agree well with the results of the previous CLIMBER-2 version presented in *Ganopolski et al.* [1998]. In particular, the new version of the model simulates substantial changes in the THC in the Atlantic - namely a southward shift in the formation area and shallowing of NADW compared to the modern climate (Figure 2d). It is important to stress that the new version has a four-time higher latitudinal resolution in the ocean model, and thus this shift corresponds to 6 grid points and cannot be considered as a numerical artifact of coarse resolution. This southward shift was recently been confirmed in the LGM simulation of the coupled GCM HadCM3 [*Hewitt et al.*, 2001]. These changes in the Atlantic circulations also agree with the results of OGCM simulations performed using surface boundary conditions derived from proxy data [*Seidov*

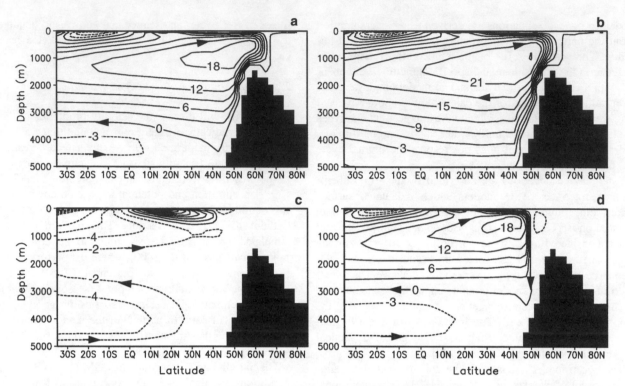

Figure 2. Modes of Atlantic thermohaline circulation in the coupled model. (a) Holocene warm mode. (b) Glacial warm (or interstadial) mode. (c) Holocene "off" mode. (d) Glacial cold (stadial) mode. Figure from Ganopolski and Rahmstorf [2001].

and Maslin, 1999; *Seidov et al.*, this issue]. Comparison of CLIMBER-2 with the results of LGM simulations performed with different AGCMs [e.g. *Kageyama et al.*, 2001] shows that our results fall within the range of complex climate model simulations and are in agreement with many terrestrial and marine paleodata.

In the last part of the paper we consider future climate change scenarios. In these experiments the model was driven by prescribed CO_2 scenarios. The performance of the model in global warming simulations is described in detail in *Ganopolski et al.* [2001].

3. ANALYSIS OF STABILITY OF THE OCEAN THERMOHALINE CIRCULATION FOR DIFFERENT CLIMATE

In general the THC is defined as a component of the ocean circulation which "is driven by changes in sea water density arising from changes in temperature versus salinity" [*Houghton et al.*, 1996] and is complimentary to the wind-driven component of the ocean circulation. In reality, it is difficult to distinguish between these components, since density fields are strongly affected by wind forcing. In a more narrow sense the THC can be defined as a meridional ocean circulation driven by large-scale meridional density gradients. This com-

ponent of the ocean circulation is typically depicted by meridional overturning diagrams and is closely linked to the area of deep water formation [*Rahmstorf*, 1995]. The topology and intensity of this component of the ocean circulation is primarily defined by large scale surface heat and freshwater fluxes. Only this part of the THC can be simulated with the zonally averaged ocean model.

The stability properties of the THC can be presented in the form of stability diagrams (Figure 3). The figure shows the quasi-equilibrium dependence of the intensity of the THC (maximum of meridional overturning) on freshwater flux for modern and glacial conditions. To obtain stability diagrams for different climates we apply a slowly changing freshwater flux anomaly in different regions of the Atlantic ocean similar to *Rahmstorf* [1996]. The shape of the stability diagram depends on whether freshwater is applied south of the NADW formation area or directly in this area. Otherwise, the results are not very sensitive to the precise position and the details of the applied perturbations. The points E (rectangles) on the stability diagram, corresponding to zero anomalies of freshwater flux, represent the equilibrium THC for unperturbed climate states.

Figure 3 shows that the stability properties of the THC for modern and glacial climates differ considerably. For modern climate, like in *Manabe and Stouffer* [1988], *Stocker and*

Wright [1991] and *Rahmstorf* [1996], there are two stable and fundamentally different modes of operation of the THC. The first mode the is so-called "conveyor-on" or "warm" mode with deep water formation in the Northern Atlantic (Figure 2a corresponding to point E1 in Figure 3a). Another mode is the "conveyor-off" mode (Figure 2c corresponding to point E2 in Figure 3a). When the freshwater flux to the Atlantic exceeds some critical threshold (point C in Figure 3a) the THC becomes advectively unstable [see *Rahmstorf*, 2000] and collapses on a time scale of about a thousand years. When freshwater inflow is applied directly in the NADW area the hysteresis loop is narrower because a shutdown (point A) of convection is triggered before the advective stability threshold is reached. The collapse of the THC due to convective instability occurs on a time scale of about a hundred years.

The glacial climate has a rather different shape of the stability diagram (Figure 3b). In this case for unperturbed conditions ($\Delta F_s=0$) there is only one stable climate state corresponding to the "cold" mode of the THC with deep water formation in the area around 50°N. Overturning is shallower than for modern conditions, although its maximum intensity is almost the same. Second, in the case of glacial climate hysteresis behavior is much less pronounced than for the modern climate. Finally, the stability diagram differs not only quantitatively (in respect of the position of transition points between different modes) but also qualitatively between the cases where freshwater flux anomaly is applied north or south of 50°N. When the freshwater flux anomaly is applied southward of 60°N (present-day area of deep water formation) the evolution of the THC can be described as a gradual transition between the "cold" and "conveyor-off" modes. Unlike the modern climate Figure 3b does not reveal a clear threshold for the stability of the "cold" mode. With the increase of freshwater flux the overturning becomes weaker and shallower and the area of deep convection migrates to the south. Since this shift in the area of convection leads to a decrease of the density of formed deep water, AABW gradually fills the deep and intermediate Atlantic and the overturning becomes very similar to the modern "conveyor-off" mode (Figure 2c). When the freshwater flux to the Northern Atlantic is decreased, overturning gradually recovers, following almost the same path. If the freshwater flux anomaly is applied north of 50°N, the response of THC is similar to the previous case, but already for a relatively small negative anomaly of the freshwater flux (about -0.02 Sv) deep convection is triggered in the Nordic seas and a transition of the THC occurs from the "cold" glacial mode to a "warm" mode similar to the modern "on" mode of the THC (Figure 2d). This transition is accompanied by a rapid decrease of sea ice cover in the northern Atlantic and an increase of the air temperature over the Atlantic sector in the latitudinal belt between 50°N-70°N by more than 6°C, and the whole Northern Hemisphere

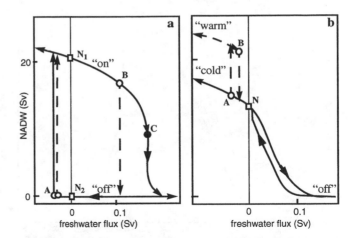

Figure 3. Schematic stability diagram for the Atlantic thermohaline circulation for the present climate (a) and the glacial climate (b) based on Ganopolski and Rahmstorf [2001]. Solid lines correspond to freshwater perturbation added in the latitude belt 20°-50°N, and dashed lines - in the latitude belt 50°- 70°N. Points N represent the equilibrium climate states for unperturbed conditions. C is a bifurcation point for advective instability. Points A and B represent transitions between "on" and "off" modes for present climate and between "warm" and "cold" for glacial climate due to start-up and collapse of convection in Nordic seas.

warms by 1K. At the same time, due to the increase of the interhemispheric oceanic heat transport, the Southern Hemisphere is colder in "warm" mode by almost the same amount. When the freshwater flux increases again (i.e., the applied negative anomaly is reduced) transition between "warm" and "cold" modes occurs (point B in Figure 3b) almost for the same negative freshwater flux anomaly as transition A.

The proximity of the points A and B in freshwater space for glacial conditions can be explained by the fact that the width of the hysteresis loop A-B is proportional to the oceanic heat transport, as it is the negative buoyancy flux related to releasing heat to the atmosphere which has to overcome the positive buoyancy flux from freshwater input in this area to sustain convection. For glacial conditions even in the "warm" mode the oceanic heat transport to the Nordic Seas is much smaller than for the modern climate (Figure 4) and this is why the convective hysteresis loop is much narrower in the former case. In the real climate system there are several feedbacks which can additionally affect the stability of the "warm" mode and the width of A-B hysteresis. A transition from the "cold" to the "warm" mode has a pronounced effect on climate over the Laurentide and especially the Fennoscandian ice sheet. An increase of air temperature should increase ablation near the southern margins of these ice sheets and increase the freshwater flux to the northern Atlantic (destabilizing feedback), while an increase of precipitation should lead to an increase of the accumulation at high elevations, and thus at least temporarily increase the freshwater losses from

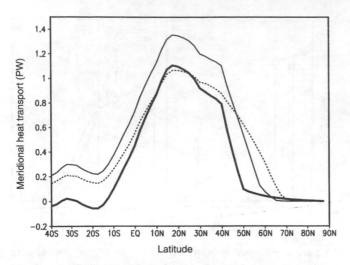

Figure 4. Meridional heat transport in the Atlantic. Glacial cold mode - thick solid line; glacial warm mode - thin solid line; present-day climate - dashed line.

the northern Atlantic (stabilizing feedback). The combined impact of these feedbacks can be time-dependent since ice sheets have long time scales. In the present version of CLIMBER-2 ice sheets are assumed to be in equilibrium and thus these feedbacks are not taken into consideration.

The fact that both points A and B are very close to the unperturbed glacial climate is crucial for the proposed mechanism of rapid glacial climate changes discussed below. This proximity is due to the fact that the terms of the high-latitude freshwater budget, especially runoff, become much smaller in a cold glacial climate, which therefore has a near-zero net freshwater input on the scale of Figure 2. Since modeling of climates considerably different from the present one involves many uncertainties (in particular in defining the boundary conditions and model parameters) and difficulties in the validation of model results, it is important to prove that the results are robust with respect to these uncertainties. One of the possible source of uncertainty is the efficiency of the freshwater bypass (discussed above) during glaciation. To prove that our results are robust with respect to the details of the parameterizations of the freshwater bypass, we performed three additional experiments for glacial conditions. The only differences between these experiments are related to the freshwater bypass parameterizations. In the first experiment (E1) we used the same numerical parameters as for modern conditions; in the second one (E2), both freshwater bypass and ice export parameters were set to one half of their present-day values, and in the third one (E3) both parameterizations were completely switched off. These three experiments cover the whole possible range of uncertainties with respect to freshwater bypass and ice export parameterizations; actually the first and third experiment represent two rather ex-

treme cases. The results of the stability analysis for these experiments are shown in Figure 5. Here, unlike in Figure 3, we used the annual surface temperature over the northern Atlantic to trace the transition between "cold" and "warm" modes, because this characteristic is the most sensitive measure of these transitions. As seen on Figure 5, the stability diagrams for all three experiments are qualitatively similar, and are characterized by sharp transitions from cold to warm modes (B-transitions). Bifurcation points are shifted between the experiments by values comparable to but smaller than the differences in freshwater transport by the freshwater bypass. The latter is a consequence of the fact that part of the freshwater exported to the south then returns back in the upper ocean layer to the area of convection in the "warm" mode. Both bifurcation points A and B are very close to the "0" perturbation point in experiments E1 and E2, while in the experiment without any export of freshwater (E3) the "cold" mode is more stable and a larger amplitude of negative freshwater forcing is needed to trigger convection in the Nordic Seas. These experiments show that the features shown in Figure 2 are indeed robust. In addition, the sensitivity study showed that the colder the climate, the more the points A and B move to the left, i.e., the more stable the "cold" mode becomes and the larger the freshwater perturbation required to trigger a transition to the "warm" mode. The three qualitatively different circulation modes found in our model coincide with those deduced from paleoclimatic data [*Alley et al,* 1999] and obtained in OGCM simulations [e.g. *Seidov and Maslin,* 1999].

Why is the response of the THC to freshwater forcing for

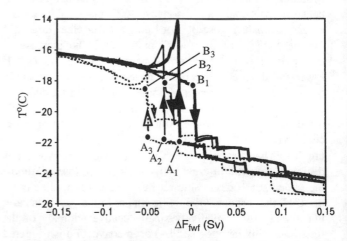

Figure 5. Stability diagram for glacial conditions in terms of North Atlantic sector air temperature (60°-70°N). Thick solid line corresponds to experiment with present parameters of the freshwater bypass, thin solid line - parameters of the freshwater bypass are half of present values, dashed line - without freshwater bypass. Dots indicate the positions of convective transitions from "warm" to "cold" (A) and from "cold" to "warm" (B) modes.

glacial condition so robust, and why does the ocean circulation operate in two distinctly different modes? The reason for this discrete structure is the glacial surface freshwater flux, which differs substantially from the present. Figure 6 shows the annual surface freshwater flux into the Atlantic ocean for glacial conditions in comparison with modern conditions. The cold glacial climate is characterized by a strong reduction of precipitation in middle and high latitudes of the Northern Hemisphere [see e.g. *Ganopolski et al.*, 1998]. The reduction of precipitation (directly, and via a reduction of the Eurasian river runoff) as well as an increase of sea ice export from this area lead to a strong decrease of the surface freshwater flux to the Arctic and Nordic seas. At the same time in the latitudinal belt 40°-60°N the freshwater flux is higher than at present due to a reduction of evaporation, a southward shift of the storm-track and the melting of sea ice in this area (Figure 6). The reduced ocean meridional heat transport compared to modern conditions cannot sustain deep water formation in this area (area B in Figure 6). Deep convection can operate only either to the south of this area (area A corresponding to the "cold" mode) or to the north (area C corresponding to the "warm" mode). This explains the existence of two distinct modes of operation of the glacial THC. Figure 6 also shows that the contribution to the freshwater flux due to ice export is relatively small compared to background climatological flux. Ice export only reduces the freshwater flux in the Nordic seas and thus makes it easy to trigger convection in the area C, but it does not affect the major features of freshwater forcing. Thus, although freshwater export from the Arctic is important both for the glacial and the modern climate conditions to obtain quantitatively correct stability properties of the THC, qualitatively the results are independent of a specific selection of model parameters.

4. MILLENNIA SCALE VARIABILITY OF THE CLIMATE DURING GLACIATION

4.1 Dansgaard-Oeschger Events

Previous attempts to understand the mechanisms of rapid climate changes have focused on modelling cold events triggered by freshwater inflow into the North Atlantic [*Schiller et al.*, 1997; *Manabe and Stauffer*, 1997; *Fanning and Weaver*, 1997]. All these experiments were performed for modern climate conditions and cannot be directly applied for the explanation of the instability of glacial climate. The results of our simulations suggest that the stability properties of the THC under glacial climate conditions were rather different from the modern. In particular, we have shown that the "cold" mode is the only stable mode of the glacial Atlantic ocean circulation. This implies that cold climate conditions over the northern Atlantic during stadial periods correspond

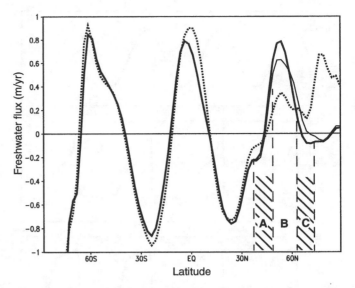

Figure 6. Annual surface freshwater flux into the Atlantic sector for glacial (solid lines) and modern (dashed line) conditions. These fluxes include precipitation, evaporation, runoff and melting/freezing of ice (i.e. explicitly include effect of ice export parameterization). Thick lines correspond to standard model set-up; thin line corresponds to experiment in which freshwater bypass parameterizations were switched off. The strong maximum in the Arctic for modern conditions is due to Eurasian rivers runoff.

to the normal mode of the glacial THC and do not require additional external forcings for their explanation. The actual problem is to explain the mechanism of "warm" glacial events - Dansgaard-Oeschger events (D/O). Here we speculate that D/O oscillations represent rapid transitions between the "warm" and "cold" modes of the THC, triggered by small variations in freshwater flux to the northern Atlantic.

Spectral time series analysis [*Grootes and Stuiver*, 1997] of Greenland ice core data reveals a dominant peak at a periodicity of 1470 years associated with the D/O events. A cycle with similar periodicity was also traced in ocean sediments in the northern Atlantic [*Bond et al.*, 1999]. During glaciation this cycle is well correlated with Greenland temperature, but it is also detected during the Holocene, albeit without a clear counterpart in Greenland temperatures. The mechanism which could cause such a periodicity is still unclear. It was suggested that millenial scale solar variability could be a driving force, but the correlation between solar variability and millenia-scale climate variability is not proven so far. Another possible mechanism, self-sustained oscillation in the climate-ice sheet system, was proposed by *Paillard and Labeyrie* [1994]. *Alley et al.* [this issue] speculate that D/O oscillations can be explained by stochastic resonance in the climate system caused by random fluctuations in freshwater flux. The exploration of different mechanisms will be the subject of a forthcoming paper. Here we would merely like to

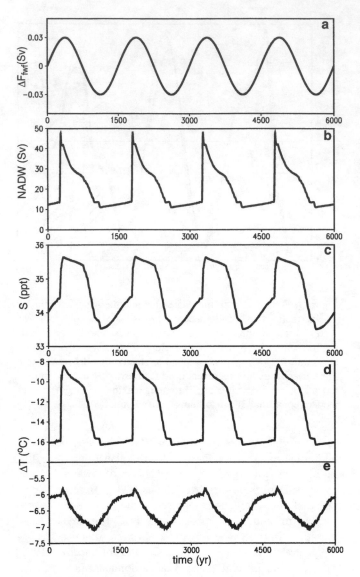

Figure 7. Simulated D/O events. (a) Forcing, (b) Atlantic overturning, (c) Atlantic salinity at 60°N, (d) air temperature in the northern North Atlantic sector (60°-70°N), and (e) temperature in Antarctica (temperature values are given as the difference to the present-day climate)

illustrate that a weak periodic freshwater forcing (irrespective of its origin) applied to the high latitudes of the North Atlantic can trigger rapid reorganizations of the Atlantic THC and thereby strong variations in climate consistent with those recorded in paleoclimate proxy data in both hemispheres.

To test this hypothesis, we impose a periodic variation in the freshwater forcing of the Atlantic in the latitude belt 50°-70°N (Figure 7a). The amplitude of this forcing is small (0.03 Sv) compared to the other components of the freshwater bal-

ance of the Arctic and Nordic seas. The response to the imposed forcing is shown in Figures 7b-e. During a period of negative anomaly in freshwater, convection is triggered in the area north of 60°N and a transition from the "cold" to the "warm" mode of the THC occurs (transition A in Figure 2b). This transition is accompanied by a vigorous flush in the strength of overturning and an increase of northward oceanic heat and salt transport. Both these factors temporarily stabilize the "warm" mode, which otherwise could not be sustained when the anomaly of freshwater flux is increasing. As the ocean circulation and density gradually (on a time scale of a few hundred years) adjust to each other, overturning is weakening until at some moment the combined effect of a decrease of the oceanic heat transport and an increase of freshwater forcing leads to a complete shut-down of convection in the Nordic Seas and a rapid transition from the "warm" to the "cold" mode of the THC. The time evolution of Greenland temperatures during these warm events has three phases (Figure 7d): an abrupt initial warming on a time scale of a decade, then a gradual cooling trend during five hundred years, terminated by a rapid (about hundred years) temperature drop back to stadial conditions. All these features of simulated warm events resemble the observed D/O events. The region of maximum surface air temperature response is centered on the northern North Atlantic, where temperature changes reach 8°C. This is less than the observed temperature changes in Greenland (up to 15°C, [*Dahl-Jensen et al.*, 1998]), but one has to take into account that the temperature changes shown in Figure 7d represent the average air temperature anomalies over the whole Atlantic sector between 60°N and 70°N, but not the maximum changes tween 60°N and 70°N, but not the maximum changes which can occur over a smaller area. In Antarctica, temperatures are increasing during the stadial phase and decreasing during the warm event with an amplitude of about 1°C (Figure 7e).

It is important that the characteristics of the simulated D/O events are rather insensitive to the amplitude of the imposed forcing cycle once this amplitude exceeds some critical threshold. To demonstrate this threshold behavior we performed a series of experiments with different parameters for the freshwater bypass described in the previous section. We forced the system with periodic freshwater input with a time-varying amplitude (Figure 8a). As shown in Figure 8b the threshold amplitude depends on model parameters. For both experiments E1 (present day freshwater bypass) and E2 (half of the present day freshwater bypass parameters) the threshold is about 0.02 Sv, while in E3 (no freshwater bypass) full-scale D/O oscillations start only when amplitude of freshwater forcing exceeds 0.05 Sv. When the amplitude is above the threshold, in all experiment the amplitude and the shape of D/O oscillations are very similar and rather insensitive to the amplitude of applied forcing.

4.2 Heinrich Events and Bond-cycle

To simulate climate the impact of Heinrich events (large iceberg discharges to the northern Atlantic, [*Heinrich*, 1988]) a much larger freshwater perturbation with an amplitude of 0.15 Sv and a duration of 2000 years was added to the 1500-year periodic forcing (Figure 9a). This additional perturbation was applied in the same area of the northern North Atlantic as the periodic forcing described above. In response to this freshwater release the conveyor belt gradually shuts down and at the time of maximum freshwater inflow becomes very similar to the modern conveyor-off mode. Since, as seen from Figure 3, the "conveyor-off" mode is not an equilibrium state for unperturbed glacial climate, the THC gradually recovers after the freshwater release starts to decrease. Compared to a stadial mode (cold phase of D/O events), Heinrich events cause only 1-2°C additional cooling over the northern part of the North Atlantic (Figure 9b), while maximum of cooling is located between 40°N-50°N, where the temperature drops by 4°C. This pattern of climate response to Heinrich events is in agreement with paleo records, which in Greenland show similar stadial temperatures irrespective of the occurrence of Heinrich events, while in mid-latitudes the

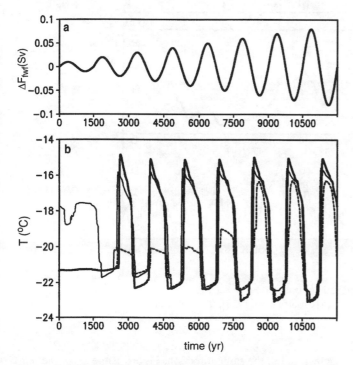

Figure 8. Triggering of rapid changes by freshwater forcing of variable amplitude. (a) Forcing, (b) air temperature in the northern North Atlantic sector (60°-70°N). Thick solid line corresponds to experiment with present parameters of the freshwater bypass (E1), thin solid line - parameters of the freshwater bypass are half of present values (E2), dashed line - without freshwater bypass (E3).

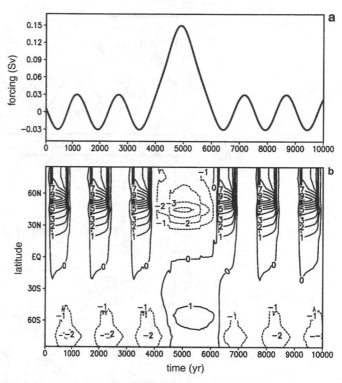

Figure 9. Simulated D/O events and Heinrich event. (a) Forcing, (b) annual surface air temperature anomalies (differences from equilibrium LGM state) for the Atlantic sector.

Atlantic temperature response to Heinrich events is much stronger [*Paillard and Cortijo*, 1999; *Bard et al.*, 2000]. Compared with D/O oscillations (Figure 7), Heinrich events show a much more pronounced bipolar see-saw [*Stocker*, 1998] with a strong Antarctic response (2°C warming compared to interstadials). This is due to the larger changes in interhemispheric heat transport and longer duration of these events. In agreement with *Blunier et al.* [1998] the temperature in Antarctica gradually increases during Heinrich events and decreases during Greenland interstadials.

Although these experiments explain many features of the observed glacial millenial-scale climate variability, there is still one important aspect known from palerecords, which is not captured by the model. Greenland ice cores indicate that a series of warm events always starts after a relatively long cold phases, usually associated with a Heinrich event, and that the series of individual D/O oscillations shows a decline in amplitude and duration (so-called Bond-cycle). This is especially clearly seen between Heinrich events H4 and H2, when the 1500-year cycle is most pronounced in the Greenland ice cores. To simulate these features observed in the paleodata, we superimposed on the freshwater scenario described above a linear trend in freshwater flux of 0.01 Sv/Kyr (Figure 10a). The parameters of the applied freshwa-

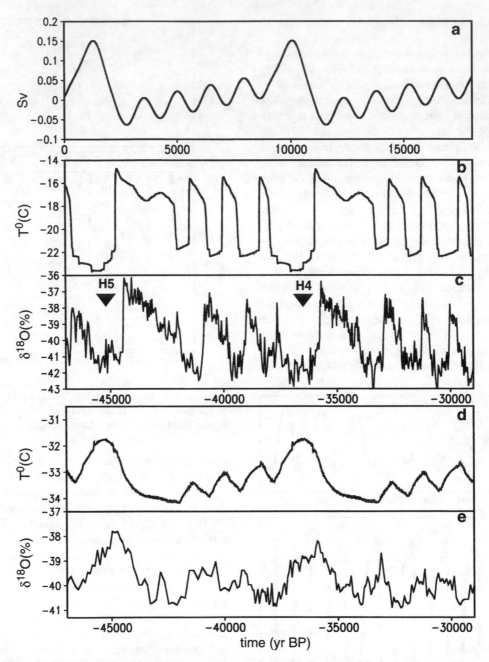

Figure 10. Simulated Bond cycle in comparison with paleoclimatic records. (a) Freshwater forcing, (b) air temperature in the northern North Atlantic sector (60°-70°N), (c) GRIP d[18]O data [Grootes et al, 1993], (d) temperature over Antarctica, (e) Byrd d[18]O data [Blunier et al., 1998].

ter flux were selected in such a way that the total freshwater input between two Heinrich events is equal to zero. This trend could represent either a gradual increase in iceberg discharge from the major ice sheets following a Heinrich event, or a gradual decrease in intensity of the freshwater bypass due to advances of ice sheets and a closing of Denmark Strait. It is also possible that both mechanisms worked to-

gether and in the same direction. Figure 10b-e show the temporal response of the Northern Atlantic and Antarctic temperatures to this forcing in comparison with GRIP and Byrd data. A characteristic nonlinear response is seen in the figure. Greenland temperature responds less to the strong Heinrich events than to the much weaker 1500-years forcing. Moreover, it does not respond to each 1500-year cycle, staying longer in the warm mode directly after Heinrich events. Between Heinrich events, the temperature over the Northern Atlantic shows a pronounced cooling trend. The Antarctic response, in contrast, is much stronger to Heinrich events than to individual D/O events. All these features are consistent with paleodata, showing that changes in freshwater forcing applied only in the Northern Atlantic could fairly realistically explain global-scale changes in temperature. In particular, our results could explain why the Southern Hemisphere appears to lead the Northern Hemisphere by 1-2 kyr, while the driver of these changes is located in the Northern Hemisphere.

5. CENTENIAL TIME-SCALE RESPONSE OF THERMOHALINE CIRCULATION TO THE GLOBAL WARMING

Based on the past instability of the Atlantic THC and on physical considerations, concerns have been raised that anthropogenic climate change might trigger another instability of the ocean circulation [e.g., *Broecker*, 1987]. Results of model simulations [*Manabe and Stouffer*, 1994] have confirmed this possibility. Most simulations performed during the last years with GCMs show a reduction of the strength of the THC during the next century, although they disagree over the amplitude of the changes [e.g., *Rahmstorf*, 1999]. In particular, a recent experiment with the HadCM-3 coupled model [*Wood et al.*, 1999] shows only moderate reduction in the strength of the THC even for a quadrupling of CO_2 concentration. *Dixon et al.* [1999] and *Mikolajewicz and Voss* [2000] analyzed the role of temperature and salinity changes in slowing down the THC and came to qualitatively different conclusions. Simplified models were extensively used to assess uncertainties in the future evolution of the THC. *Stocker and Schmittner* [1997] studied the critical thresholds for CO_2 concentration and its rate of change beyond which the THC collapses. In *Rahmstorf and Ganopolski* [1999] we have shown that temperature rise alone only causes a weakening of the THC but cannot lead to a complete shutdown of the ocean conveyor belt. To compare different climate models we proposed a new diagnostic quantity – "North Atlantic hydrological sensitivity" - which is defined as the change in the freshwater flux to the North Atlantic (north of 50°N) per degree of annual warming in the Northern Hemisphere. The long-term evolution of the THC crucially depends on changes in the freshwater flux to the northern Atlantic. If the hydrological sensitivity exceeds some critical value, the THC rapidly collapsed after the year 2100 and a substantial long-lasting cooling over the North Atlantic was simulated. In *Ganopolski et al.* [2001] we extended this study by an analysis of different CO_2 scenarios and model parameters. Here we report further global warming experiments with the newest CLIMBER-2 version to analyze the impact of increased freshwater flux to the northern Atlantic and the role of convective and advective instabilities of the THC.

Figure 11 shows results of the model experiments for two standard 1% increase CO_2 scenarios. Comparison of Figures 11b and 11c show that changes in freshwater flux to the northern Atlantic (north of 50°N) are well correlated with a warming in the Northern Hemisphere. The hydrological sensitivity of CLIMBER-2 is 0.013 Sv/K, while the GFDL model has hydrological sensitivity of 0.030 Sv/K. The latest version of the Hadley Centre model, HadCM-3 [*Wood et al.*, 1999], has a hydrological sensitivity 0.020 Sv/K. Moreover, different climate models have a different temperature sensitivity to CO_2. As a result, the total increase of freshwater flux to the northern Atlantic in the GFDL model for the same CO_2 scenario is 3.5 time large than in CLIMBER-2, and in ECHAM-3/LSG model is almost twice as larger [*Mikolajewicz and Voss*, 2000]. To cover a possible range of uncertainties of the changes in hydrological cycle, we performed experiments with artificially enhanced hydrological sensitivity by applying an additional freshwater flux proportional to the annual warming in the Northern Hemisphere as described in *Rahmstorf and Ganopolski* [1999]. Although some part of the additional freshwater flux to the northern Atlantic could come from the melting of the Greenland ice sheet and glaciers, the larger part of this increase is due to changes in atmospheric moisture transport. This is why an increase of freshwater flux to the northern Atlantic has to be compensated in the other parts of the oceans. Here we consider two cases: increase of the freshwater input to the northern Atlantic is compensated in the Pacific (1) or in the tropical Atlantic (2). Results of GCM experiments suggest that the second case is more realistic.

The temporal evolution of the Atlantic THC for different CO_2 scenarios and hydrological sensitivities is shown in Figure 11d-f. For convenience we will use acronyms for all these experiments in the form C^kH^n, where "C" refers to the CO_2 scenario and "H" refers to the hydrological sensitivity. In these abbreviations k=2 corresponds to a doubling of CO_2 and k=4 to a quadrupling. Index "n" shows the enhancement factor of hydrological sensitivity (compared to CLIMBER-2 hydrological sensitivity). The letter "A" or "P" after this number refers to the area where anomalous flux is compensated ("A" corresponds to the Atlantic, and "P" corresponds to the Pacific). In these abbreviations our standard experiments are denoted as C^2H^1 and C^4H^1. In the experiments with unaltered hydrological sensitivity (Figure 11d) overturning gradually weakens with the increase of CO_2 and reaches its minimum soon after CO_2 is stabilized. For doubling of CO_2 the model shows a 20% decrease of the THC, and for a quadrupling a 35% decrease. In both cases, after stabilizing of the CO_2 concentration, the THC gradually recovers. If we increase the effective hydrological sensitivity by a factor of four the THC responds strongly to the increase of CO_2 but

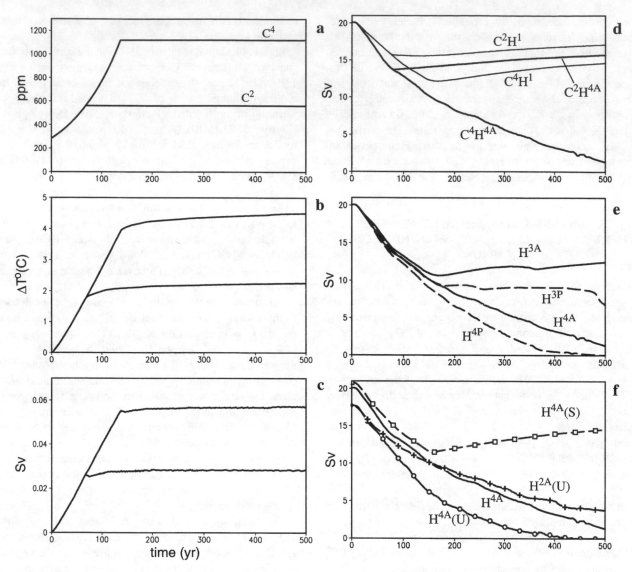

Figure 11. Time series of forcing scenarios and model variables. (a) CO_2 scenarios; (b) Northern Hemisphere temperature changes; (c) model simulated changes in freshwater flux to the Northern Atlantic; (d-e) NADW. Thin solid lines on (d) correspond to the standard runs, thick solid lines on (d-f) correspond to experiments with "Atlantic compensation" and dashed lines correspond "Pacific compensation". Only CO_2 quadrupling experiments (C^4 experiments) are shown in (e) and (f), this is why the first index in experiment's acronyms is omitted in these figures.

still survives in a doubling of CO_2 (experiment C^2H^{4A}). In a quadrupling CO_2 experiment (C^4H^{4A}) the THC completely stops after several hundred years. If we compare the differences between compensation of additional freshwater flux in the Atlantic and in Pacific (Figure 11e) then we see that interoceanic transport of moisture is more effective than a redistribution of freshwater within one basin, but there is no qualitative difference between C^4H^{4A} and C^4H^{4P}. The results of experiments C^4H^{3A} and C^4H^{3P} show that the threshold value of the hydrological sensitivity which separates the cases

of recovery and shutdown of the THC for a quadrupling CO_2 experiment is close to 0.04 Sv/K.

The rapidity of the shutdown, and the fact that it occurs even if freshwater is only redistributed within the Atlantic, suggests that the convective instability mechanism plays a major role. To test this hypothesis we performed two additional sets of experiments. In the first one, denoted by "U", the threshold of convective instability was lowered by a reduction of the intensity of the freshwater bypass by a factor of two. In the second one, denoted by "S", the threshold of convective instability was increased by an increase of the intensity of the freshwater bypass by 50%. Results shown in Figure 11e indicate that during initial phase of the thermohaline slowdown changes in convection do not play an important role and the rate of decline of the THC is almost the same in all experiments. However, the long-term evolution of the THC after stabilization of the CO_2 concentration is different. In experiments with lower stability of convection the THC declines more rapidly and a lower input of freshwater is needed for a complete shutdown. Figure 11e shows that in this case the shutdown of the THC occurs already for a doubled hydrological sensitivity ($C^4H^{2A}(U)$). In the case of more stable convection ($C^4H^{4A}(S)$) the THC starts to rise after stabilization of CO_2 concentration. Thus our results suggest that simulations of rapid changes in the THC need accurate representation of both changes in the freshwater balance of the Northern Atlantic and convective stability properties.

6. SUMMARY AND CONCLUSIONS

Using a climate system model of intermediate complexity we performed a stability analysis of the THC for modern and glacial climate conditions. We show that the stability properties of glacial climate differ fundamentally from those of the modern climate. For modern climate the "warm" and the "off" modes represent the two stable Atlantic circulation states. For glacial condition the situation is rather different. Our stability analysis suggests that the "cold" (stadial) mode represents the only stable mode of the glacial Atlantic ocean circulation, with NADW formation south of Iceland. Another mode of glacial circulation similar to the modern "warm" mode is marginally unstable for glacial conditions. The fact that transitions between "cold" and "warm" modes of the THC can be triggered by small changes in the freshwater flux to the Northern Atlantic is explained by a strong reduction both of the freshwater flux to the Arctic and meridional oceanic heat transport to high latitudes.

Our simulations show that warm events similar to the observed D/O events in time evolution, amplitude and spatial pattern can be triggered in the model as a temporary flip to the "warm" mode with NADW formation in the Nordic Seas. We demonstrated that a weak climate cycle can trigger large-amplitude episodic warm events due to the non-linear threshold response of the Atlantic ocean circulation. The positions of deep water formation areas for "cold" and "warm" modes are separated by a strong maximum in surface freshwater flux and this explains why both transitions are rapid and the amplitude of simulated D/O events is rather insensitive to the details of the model parameterizations and amplitude of applied forcing. Another result of our sensitivity study is that the glacial "cold" mode becomes more stable the colder the climate conditions. This could explain why the THC stays predominantly in the cold mode during the coldest parts of the glacial while D/O events occur almost each 1,500 year cycle during more moderate glacial conditions (50-30 kyr b.p.).

Heinrich events, simulated in the model as large freshwater pulses, lead to a collapse of NADW formation but no major further cooling in Greenland. The temperature response of Antarctic temperature to Heinrich events is much stronger than to individual D/O events. Both these results are in agreement with the paleoclimatic records. When an additional linear trend is applied to the freshwater flux, the model simulates a sequence of events resembling the one recorded in Greenland and Antarctic ice cores. Thus our results suggest that relatively small variations in freshwater input to the Northern Atlantic could have caused many of the observed features of glacial climate.

Our results suggest that the THC for the modern climate is more stable than for the cold glacial climate with respect to a small perturbation of the freshwater flux, but a sufficiently large perturbation could still destabilize the modern THC and lead to a complete and irreversible shut down of the conveyor belt. The threshold value of the freshwater flux depends on the area where anomalous freshwater flux is applied. This is due to different mechanisms of the instability of the THC.

We performed a series of simulations for two CO_2 scenarios and a range of different hydrological sensitivities. A significant reduction of the strength of the THC in the Atlantic occurs in all experiments. While the initial reduction of the THC is largely due to surface warming, the long-term response of the THC depends on the increase of freshwater flux to the northern North Atlantic. Our results suggest that there is a threshold value of hydrological sensitivity beyond which the Atlantic THC breaks down. The mechanism of rapid (time scale of about one hundred years) collapse of the THC is a convective instability. This is why this threshold value depends on the stability properties of deep convection in the northern North Atlantic. Smaller hydrological sensitivity is needed in the case of reduced export of freshwater from Nordic Seas.

REFERENCES

Aagaard, K., and E.C. Carmack, The role of sea ice and other fresh water in the Arctic circulation, *J. Geophys. Res., 94,* 14,485-14,498, 1989.

Alley, R.B., and P.U. Clark, P.U. (1999) The deglaciation of the Northern Hemisphere: a global perspective, *Annual Rev. Earth Planetary Sci., 27,* 149-182, 1999.

Alley, R.B., P.U. Clark, L.D. Keigwin, and R.S. Webb, Making sense of millennial scale climate change, in *Mechanisms of global climate change at millennial time scales,* edited by P.U. Clark, R.S Webb, and L.D. Keigwin, pp. 385-394, American Geophysical Union, Washington, D.C., 1999.

Bard, E., F. Rostek, J.-L. Turon, and S. Gendreau, Hydrological impact of Heinrich events in the subtropical Northheast Atlantic, *Science, 289*, 1321-1324, 2000.

Blunier, T., Chappellaz, J. Scwander, A. Dallenbach, B. Stauffer, T.F. Stocker, D. Raynaud, J. Jouzel, H.B. Clausen, C.U. Hammer, and S.J. Johnsen, Asynchrony of Antarctic and Greenland climate climate change during the last glacial period, *Nature, 394*, 739-743, 1998.

Bond, G., W. Broecker, S., Johnsen, J. McManus, L. Labeyrie, J. Jouzel, and G. Bonani, Correlations between climate records from North Atlantic sediments and Greenland ice, *Nature, 365, 143-147*, 1993.

Bond, G., W. Showers, M. Elliot, M. Evans, R. Lotti, I. Hajdas, G. Bonani, and S. Johnson, The North Atlantic's 1-2 kyr climate rhythm: Relation to Heinrich events, Dansgaard/Oeschger cycles and the Little Ice Age. in *Mechanisms of global climate change at millennial time scales*, edited by P.U. Clark, R.S Webb, and L.D. Keigwin, pp. 35-58, American Geophysical Union, Washington, D.C., 1999.

Broecker, W.S., Unpleasant surprise in the greenhouse? *Nature, 328*, 123, 1987.

Broecker, W.S., Massive iceberg discarges as triggers for global climate change, *Nature, 372*, 421-424, 1994.

Brovkin, V., A. Ganopolski, and Y. Svirezhev, A continuous climate-vegetation classification for use in climate-biosphere studies, *Ecological Modelling*, 101,251-261, 1997.

Budyko, M.I., The effect of solar radiation variations on the climate of the earth, *Tellus, 21*, 611-619, 1969.

Daansgard, W., S.J. Johnsen, H.B. Clausen, D. Dahljensen, N.S. Gundestrup, C.U. Hammer, C.S Hvidberg, J.P. Steffensen, A.E. Sveinbjornsdottir, J. Jouzel, and G. Bond, Evidencies for general instability of past climate from a 250-Kyr icecore record, *Nature, 364*, 218-220, 1993.

Dahl-Jensen, D., K. Mosegaard, N. Gundestrup, G.D. Clow, S.J. Johsen, A.W. Hansen, and N. Balling, Past temperatures directly from the Greenland ice sheet, *Science, 282*, 268-279, 1998.

Dixon, K.W., T.L. Delworth, M.J. Spelman, and R.J. Stouffer, The influence of transient surface fluxes on North Atlantic overturning in a coupled GCM climate change experiment, *Geophys. Res. Let., 26*, 2749-2752, 1999.

Fannig, A.F., and A.J. Weaver, Temporal-geographical meltwater influence on the North Atlantic conveyor: Implication for the Younger Dryas, *Paleocean., 12*, 307-320, 1997.

Ganopolski, A, and M. Claussen, Simulation of Mid-Holocene and LGM climates with a climate system model of intermediate complexity, in *Proceedings of the third PMIP Workshop*, WCRP-111, 201-204, 2000.

Ganopolski, A., V. Petoukhov, S. Rahmstorf, V. Brovkin, M. Claussen, A. Eliseev, and C. Kubatzki, CLIMBER-2: A climate system model of intermediate complexity. Part II: Model sensitivity. *Clim. Dyn.* (in press), 2001.

Ganopolski, A., and S. Rahmstorf, Rapid changes of glacial climate simulated in a coupled climate model, *Nature, 409*, 153-158, 2001.

Ganopolski, A., S. Rahmstorf, V. Petoukhov, and M. Claussen, Simulation of modern and glacial climates with a coupled model of intermediate complexity, *Nature, 391*, 351-356, 1998.

Grootes, P.M., M. Stuiver, J.W.C. White, S. Johnsen, and J. Jouzel, Comparison of oxygen isotope records from the GISP2 and GRIP Greenland ice cores, *Nature, 366*, 552-554, 1993.

Grootes, P.M., and M. Stuiver, Oxygen 18/16 variability in Greenland snow and ice with 103- to 105-year time resolution, *J. Geophys. Res., 102*, 26455-26470, 1997.

Heinrich, H., Origin and consequences of cyclic ice rafting in the northeast Atlantic ocean during the past 130,000 years, *Quaternary Res., 29*, 143-152, 1988.

Hewitt, C.D., A.J. Broccoli, J.F.B. Mitchell, R.J. Stouffer, A coupled model study of the last glacial maximum: Was part of the North Atlantic relatively warm? *Geophys. Res. Let.*, 2001 (in press).

Houghton, J.T., L.G. Meira Filho, B.A. Callander, N. Harris, A. Kattenberg, K. Maskell, *Climate Change 1995 - The science of climate change*, Cambridge University Press, 572 pp, 1996.

Kageyama, M., O. Peyron, S. Pinot, P. Tarasov, J. Guiot, S. Joussame, and G. Ramstein,The Last Glacial Maximum climate over Europe and western Siberia: a PMIP comparison between models and data, *Clim. Dyn.*, 17, 23-43, 2001

Manabe, S., and R.J. Stouffer, Two stable equilibria of a coupled ocean-atmosphere model, *J. Climate*, 1, 841-866, 1988.

Manabe, S., and R.J. Stouffer, Multiple-century response of a coupled ocean-atmosphere model to an increase of atmospheric carbon dioxide, *J., Climate*, 7, 5-23, 1994.

Mikolajewicz, U., and R. Voss, The role of the individual air-sea flux components in CO2-induced changes of the ocean's circulation and climate. *Clim. Dyn.,16*, 627-642, 2000.

Paillard, D., and E. Cortijo, A simulation of the Atlantic meridional circulation during Heinrich event 4 using reconstructed sea surface temperatures and salinities, *Paleoceanography, 14*, 716-724, 1999.

Paillard, D., and L. Labeyrie, Abrupt climate warming after Heinrich events: the role of the thermohaline circulation, *Nature*, 372, 162-164, 1994.

Petoukhov, V., A. Ganopolski, V. Brovkin, M. Claussen, A. Eliseev, C. Kubatzki, S. Rahmstorf, CLIMBER-2: a climate system model of intermediate complexity. Part I: Model description and performance for present climate, *Clim. Dyn., 16*, 1-17, 2000.

Rahmstorf, S., Bifurcations of the Atlantic thermohaline circulation in response to changes in the hydrological cycle, *Nature, 378*, 145-149, 1995.

Rahmstorf, S., On the freshwater forcing and transport of the Atlantic thermohaline circulation, *Clim. Dyn., 12*, 799-811, 1996.

Rahmstorf, S., Shifting seas in greenhouse? *Nature, 399*, 523-524, 1999.

Rahmstorf, S., The thermohaline ocean circulation: a system with dangerous thresholds? *Climatic Change*, 46, 247-256, 2000.

Rahmstorf, S., and A. Ganopolski, Long-term global warming simulations with efficient climate model, *Climatic Change, 43*, 353-367, 1999.

Schiller, A., U. Mikolajewicz, and R. Voss, The stability of the North Atlantic thermohaline circulation in a coupled ocean-atmosphere general circulation model, *Clim. Dyn., 13*, 325-347, 1997.

Seidov, D., and M. Maslin, North Atlantic deep water circulation collapse during Heinrich events, *Geology, 27*, 23-26, 1999.

Seidov D., B. J. Haupt, and M. Maslin, Ocean bi-polar seesaw and climate: southern versus northern meltwater impact, (this issue), 2001.

Stocker, T.F., The seesaw effect, *Science, 282*, 61-62, 1998.

Stocker, T.F., and A. Schmittner, Influence of CO_2 emission rates on the stability of the thermohaline circulation, *Nature, 388*, 862-865.

Stocker, T.F., and D.G. Wright, Rapid transitions of the ocean's deep circulation induced by changes in surface water fluxes, *Nature, 351*, 729-732, 1991.

Stocker, T.F., and D.G. Wright, A zonally averaged ocean model for the thermohaline circulation. Part II: Interocean circulation in the Pacific Atlantic basin system, *J. Phys. Oceanogr*, 21, 1725-1739, 1991.

Stommel, H.M., Thermohaline convection with two stable regimes of flow. *Tellus*, 13, 529-541, 1961.

Wood, R., A. Keen, J.F.B. Mitchel, and J.M. Gregory, Changing spatial structure of the thermohaline circulation in response to atmospheric CO_2 forcing in a climate model, *Nature, 399*, 572-575, 1999.

Andrey Ganopolski and Stefan Rahmstorf, Potsdan Institute for Climate Impact Research, D-14412 Potsdam, Germany

The Future of the Thermohaline Circulation – A Perspective

Thomas F. Stocker, Reto Knutti, and Gian-Kasper Plattner

Climate and Environmental Physics, Physics Institute, University of Bern

Evidence from paleoclimatic archives suggests that the ocean atmosphere system has undergone dramatic and abrupt changes with widespread consequences in the past. Climatic changes are most pronounced in the North Atlantic region where annual mean temperature can change by 10°C and more within a few decades. Climate models are capable of simulating some features of abrupt climate change. These same models also indicate that changes of this type may be triggered by global warming. Here we summarize what is known about such future changes and discuss the state of our knowledge about these potential threats to the stability of the Earth System.

1. INTRODUCTION

In the discussion of future climate change, a new issue has caught the attention of scientists and policymakers alike: the possibility of non-linear changes in the Earth System. Non-linearity has many characteristics: non-linear changes are not easily extrapolated from ongoing observed changes, they may have large amplitudes and they may occur as surprises. Some of these changes may even be irreversible in the sense that they occur in response to perturbations and persist long after the perturbations have stopped to influence the climate system. Inherent to such changes is their reduced predictability. Among such non-linear changes are the collapse of large Antarctic ice masses and rapid sea level rise, the desertification of entire land regions, the thawing of permafrost and associated release of large amounts of radiatively active gases, and the collapse of the large-scale Atlantic thermohaline (i.e., the temperature and salinity driven) circulation (THC). The latter has clearly caught most of the attention and has spurred

much research in the last decade. With the availability of high-resolution paleoclimatic records from the polar ice sheets and from marginal and deep basins in the ocean, a detailed picture of sequences of abrupt climate changes during the last glaciation and the glacial-interglacial transition emerges. The last of these dramatic coolings and warmings occurred about 8200 years before present. Climate modeling has also produced important insights into the properties and role of the Atlantic THC, and model simulations indicate that such changes could lie ahead.

The purpose of this article is to summarize our current understanding of future changes of the Atlantic thermohaline circulation, and to give an assessment of the uncertainties which are associated with this phenomenon. This necessitates a discussion of stabilizing and destabilizing feedback mechanisms associated with the thermohaline circulation.

The paper presents a discussion of the representation of the THC in models in section 2, and a brief summary of past evidence for changes of the THC in section 3. Dynamical concepts responsible for the limited stability of the THC and associated feedbacks are presented in section 4. Section 5 discusses the implications of a breakdown of the Atlantic THC on air temperature and

The Oceans and Rapid Climate Change: Past, Present, and Future
Geophysical Monograph 126

sea level. A catalog of feedback mechanisms is presented in section 6. The question of a possible "runaway greenhouse effect" due to a collapse of the THC is addressed in section 7; conclusions follow in section 8.

2. THERMOHALINE CIRCULATION AND THEIR REPRESENTATION IN MODELS

The thermohaline circulation of the world ocean is driven by differences in buoyancy caused by heat and freshwater fluxes at the surface of the ocean [*Warren*, 1981; *Gordon*, 1986]. These fluxes lead to the formation of dense water masses preferentially in the Greenland-Iceland-Norwegian (GIN) Seas, in the Labrador Sea and around Antarctica [*Killworth*, 1983; *Marshall and Schott*, 1999]. The dense waters from the Nordic seas flow in deep western boundary currents southward into the Southern Ocean from where they are distributed into the deep Indian and Pacific Oceans [*Schmitz*, 1995]. The return flow takes various paths through the Indonesian Passage and around Africa into the Atlantic [*de Ruijter et al.*, 1998], and through the Drake Passage. The popular view of this global circulation is a 'conveyor belt' [*Broecker*, 1991], but this is somewhat misleading because the pathways are not continuous, and observations and inverse calculations indicate a much more complicated structure of boundary currents and recirculations [*Macdonald and Wunsch*, 1996].

The thermohaline circulation strongly influences the climate on regional-to-hemispheric scales. In the Atlantic Ocean, the meridional heat transport is mostly carried by the THC and is due to the surface and deep western boundary currents: warm waters flow northward in the Gulf Stream/Transatlantic Drift system and the cold deep waters flow southward. In combination this yields a northward meridional heat flux in the Atlantic at all latitudes with a maximum of about 10^{15}W [*Macdonald*, 1998; *Ganachaud and Wunsch*, 2000]. Associated with the presence of the warm waters at the western mid-latitudes in the North Atlantic is an intense storm system whose transient eddies also transport substantial heat northward. Changes in both the ocean THC and the atmospheric storm tracks would seriously affect the climate in northwestern Europe.

The large-scale dynamics of the thermohaline circulation can be characterized on the basis of the conservation of angular momentum [*Stommel and Arons*, 1960a; *Stommel and Arons*, 1960b]. Stommel and Arons assumed that localized deep water formation is compensated by uniform deep upwelling. Water on the rotating

Earth must flow poleward in the deep interior in order to conserve angular momentum. This mass flux is compensated by an opposite flow in deep western boundary currents which connect the three ocean basins [*Stommel*, 1958]. In recent years, it became clear that water masses move preferentially along isopycnals (surfaces of constant density) and that uniform deep upwelling is inconsistent with tracer distributions [*Toggweiler and Samuels*, 1998]. Upwelling appears much more localized and is likely associated with topographic features in the deep ocean [*Ledwell et al.*, 2000]. This will make the current structure in the deep ocean more complicated and fragmented than the present understanding suggests. Because deep currents are difficult to measure, emerging high-resolution ocean models will augment our knowledge significantly. However, the present generation of highest resolution models is not yet simulating deep ocean processes in a prognostic mode due to computational constraints [*Smith et al.*, 2000].

Over the last few decades a hierarchy of models has been developed with which the variability of the thermohaline circulation was investigated (see review by *Weaver et al.*, 1999). Most of these models use coarse resolution and require various degrees of parameterizations of convection, deep water formation and mixing. This still poses a limitation on the accuracy with which issues like the stability of the thermohaline circulation, or natural variability of the THC can be addressed. Furthermore, atmospheric processes associated with air-sea heat and freshwater fluxes are often crudely accounted for, especially in ocean-only models. In spite of these limitations, a number of important physical mechanisms have been identified and described.

The general structure of the thermohaline circulation is simulated in 3-dimensional ocean general circulation models of relatively coarse resolution. Deep western boundary currents can be identified in these models, and the water mass distribution that they produce is consistent with observations [*Semtner and Chervin*, 1992; *Maier-Reimer*, 1993; *Drijfhout et al.*, 1996]. However, many of these models still require rather unrealistic forcing in the high latitudes either by prescribing artificially high values of salinity (e.g., around Antarctica to promote deep water formation) or by restoring to observed values of temperatures and salinity at entire depth sections. Global models of the THC have still many deficiencies in simulating key processes thought to be important for the THC. There is hope that increasing resolution will alleviate these problems to a large extent; the most recent simulations at 1/10 degree resolution exhibit encouraging details of the surface flows along western boundaries, and the statistics of eddies [*Smith et al.*, 2000].

Notwithstanding, the following processes will require continued attention in the simulation of the THC. Interbasin exchanges such as the Agulhas current system [*de Ruijter et al.*, 1998] or the Indonesian Passages, are poorly captured by coarse-resolution models. Of particular importance are water mass transformation processes in the marginal basins of the high latitudes. In the Atlantic, the GIN and Labrador Seas are known to influence strongly the water mass structure of the intermediate and deep waters from which global waters such as North Atlantic Deep Water (NADW) derive [*Dickson et al.*, 1996; *Dickson et al.*, 2000]. The presence and formation of sea ice is known to dominate the fluxes of heat and freshwater in the high latitudes, but this component is often not included in current global ocean circulation models. A few potentially important feedback mechanisms associated with sea ice will be discussed below.

3. THE PALEO-THERMOHALINE CIRCULATION AND ITS CHANGES

Without the paleoclimatic records, our knowledge about ocean changes would be limited to theoretical insight from models. Indeed, evidence from such records hinted at the importance of ocean circulation changes for climate change. Since the reconstruction of the rapid movement of the North Atlantic polar front during the last termination [*Ruddiman and McIntyre*, 1981] and the bold proposal by the late Hans Oeschger that carbonate isotope changes in Lake Gerzensee (Switzerland) and isotopic changes in Greenland ice have the same common origin [*Oeschger et al.*, 1984], changes in ocean circulation have moved to center stage for the explanation of rapid climate change. Oeschger proposed that the ocean may be a "flip-flop" system, and *Broecker et al.* [1985] and *Broecker and Denton* [1989] have collected and synthesized evidence from the paleoclimatic records which point to the ocean as one of the key elements of abrupt change.

During the last decade an unprecedented growth of evidence for abrupt climate change during the last glaciation has occurred. This was enabled by a significant increase in temporal resolution of marine and polar ice core records, as well as new parameters and smaller analytical uncertainties in various proxy parameters. The most important manifestations of abrupt climate changes are the remarkable sequences of abrupt warmings and slower coolings registered in various ice cores

from the Greenland ice sheet. They serve as model events although it should be noted that they are signals at a very remote and special location on the planet.

Dansgaard et al. [1993] counted 24 of these abrupt events that are now referred to as Dansgaard/Oeschger (D/O) events. Their evolution bears remarkable similarity among each other: the warming is always abrupt and measurements of stable isotopes of gases enclosed in the ice demonstrate that in Greenland the annual mean temperature warmed by about 16°C within a few decades [*Lang et al.*, 1999; *Severinghaus and Brook*, 1999]. The last of these warming events was the transition from the cold Younger Dryas (YD) to the warm Preboreal about 11,650 years ago and was accompanied by an abrupt increase in accumulation [*Alley et al.*, 1993] and a sudden decrease of the dust load within less than 10 years [*Taylor et al.*, 1993]. The coolings, in general, are more gradual and evolve over a timescale of 1-3 thousand years, punctuated by some shorter cold events (e.g., during the Bølling/Allerød).

The wide spread nature of these abrupt climate changes has been confirmed in many paleoclimatic records from different archives and different locations provided that the temporal resolution of these archives is sufficient. This evidence is reviewed in several recent papers (*Broecker*, 1997; *Stocker*, 2000; *Alley and Clark*, 1999). High-resolution marine sediment records point at the central role of the ocean, in particular the North Atlantic. The first clear case of abrupt change of sea surface temperature (SST) that correlated surprisingly well with the Greenland ice core was presented by *Lehman and Keigwin* [1992] who showed that SST off the Norwegian shelf changed rapidly by more than 5°C within a few decades. The changes in SSTs during deglaciation correlate strongly with those inferred from the Greenland ice cores. A further important finding was that each D/O event was associated with a layer of ice rafted debris originating from icebergs from the circum-North Atlantic ice sheets [*Bond et al.*, 1999]. The thickest of these layers in the marine sediments are referred to as Heinrich events and they tend to occur close to some of the longest of the D/O events. This was an indication that the THC may have been disrupted by massive input of freshwater from melting icebergs. A still open problem, however, is the exact timing of the layers of ice rafted debris and the climate changes. Clearly, the meltwater input around the last major cold event, the YD, seems to occur about 1000 years too early to be directly responsible for the break-down of the Atlantic THC [*Bard et al.*, 1996; *Clark et al.*, 1996]. As this meltwater input is inferred from the rise in sea level, this does not exclude that there are smaller but crucial inputs of meltwater around the onset of YD. There are indeed indications of ice rafted debris at that time [*Bond et al.*, 1999]. The case is clearer for the 8,200-yr cold event [*Alley et al.*, 1997] which apparently occurred in response to a huge outflow of proglacial lakes [*Barber et al.*, 1999]; the cooling in Greenland was estimated at about 7°C [*Leuenberger et al.*, 1999] and most likely due to a slow-down of the Atlantic THC. It is clear that both a better synchronization of these meltwater records with the Greenland ice cores, as well as detailed investigation of the provenance of the ice rafted debris are necessary to solve this problem. The first is needed to compare the temporal evolution simulated by coupled models, the second will be the basis of more realistic freshwater perturbations used in coupled models.

Irrespective of possible mechanisms, an SST record from Bermuda traces all D/O events and indicate that these were climate phenomena of large, at least Atlantic, if not hemispheric extent. Sea surface temperatures varied in concert with Greenland by about 4°C for each D/O event [*Sachs and Lehman*, 1999]. These climate signals are also recorded in remarkable detail in the tropical Atlantic (Cariaco Basin off Venezuela, *Peterson et al.*, 2000), in the East Atlantic (off the coast of Portugal, *Shackleton and Hall*, 2000) and in the Santa Barbara Basin (eastern Pacific, *Behl and Kennet*, 1996). Moreover, *Oppo and Lehman* [1995] found that deep water mass properties co-vary with surface properties in the same core confirming that the Atlantic thermohaline circulation is indeed involved in these large climatic changes. Rapid deep water changes were also found by *Adkins et al.* [1998] who analyzed a coral that grew in deep water and registered an abrupt change from nutrient-rich to nutrient-low waters during the last deglaciation when the North Atlantic went through a series of dramatic climatic changes.

Indications of ocean changes also come increasingly from regions outside the North Atlantic, and probably these contribute now most to an understanding of the underlying mechanisms. Recent studies have found that in the South Atlantic large Heinrich events, which are cold events in the North Atlantic, appear as warmings [*Vidal et al.*, 1999; *Clark et al.*, 1999]. This is consistent with what one would expect if the THC collapsed in the North Atlantic: as the Atlantic THC is associated with a cross-hemispheric heat transport, a THC collapse would result in excess heat in the South Atlantic. This is the bi-polar seesaw [*Broecker*, 1998; *Stocker*, 1998] suggested earlier by model simulations [*Crowley*, 1992; *Stocker et al.*, 1992; *Seidov et al.*, 2001].

Apart from paleoclimatic evidence directly from regions influenced by the Atlantic THC, there is now a growing body of circumstantial evidence that the THC plays a dominant role for abrupt climate change. The synchronization of climatic events identified in ice cores from Greenland and Antarctica during the last deglaciation [*Blunier et al.*, 1997], as well as during the glacial [*Blunier et al.*, 1998] is consistent with the concept of the bi-polar seesaw. It is remarkable that only the larger and longer of the D/O events have a counterpart in the Antarctic record, suggesting that only those are associated with a full collapse of the THC which would result in a strong inter-hemispheric coupling [*Stocker and Marchal*, 2000]. A further indirect indicator for THC changes is the radiocarbon concentration (^{14}C) in atmospheric CO_2, because a cessation of the THC would lead to an increase in ^{14}C. This was reconstructed from radiocarbon-dated varved sediments [*Hughen et al.*, 2000] at the beginning of the YD cold event. Recent simulations suggest that the millennial changes of ^{14}C during deglaciation were a combination of changes in the production rate of ^{14}C and changes in the THC [*Muscheler et al.*, 2000; *Marchal et al.*, 2001].

In spite of this wealth of high-quality paleoclimatic information, there are major limitations that seriously hamper progress. The most important is a relatively poor data base outside the North Atlantic region. Especially the tropics are not well represented although it is clear that probably important processes for abrupt change may originate there. Practically nothing is known about the long-term variability in the tropics; it is not known whether the El Niño-Southern Oscillation (ENSO) cycle persisted in the glacial, or whether the tropics may have resided in a phase-locked state similar to La Niña. It would also be extremely important to know whether the statistics of the ENSO cycle is subject to long-term changes over millennia, as recent modelling suggests [*Clement et al.*, 1999].

The Southern Ocean is severely undersampled and yet many records (e.g., the CO_2 during the glacial and during deglaciation) point to its importance in pacing climate change. Reliable proxies at high temporal resolution that inform us about water mass composition and origin would be extremely valuable and important progress in this respect was made [*Yu et al.*, 1996; *Marchal et al.*, 2000].

Polar ice cores have contributed significantly to progress in the last decade. However, it turned out that the interpretation of classical proxies such as the stable isotopes of ice had to be revised [*Cuffey et al.*, 1995; *Johnsen et al.*, 1995; *Lang et al.*, 1999]. Likewise, the regionality of the isotopic signal in Antarctica is practically unknown, but there are indications that it may be important in correctly interpreting the climate signals recorded in these cores [*Steig et al.*, 1998; *Mulvaney et al.*, 2000].

Even if the apparent gaps in coverage of paleoclimatic records will be closed in the future, the synchronization of records from various archives remains a top priority. New techniques have been established to synchronize polar ice cores based on methane [*Blunier et al.*, 1998] and stable isotopes in oxygen of air [*Sowers and Bender*, 1995]. However, changes in the enclosure processes and the gas age-ice age difference still limit the synchronization to a few centuries.

4. THRESHOLDS IN THE OCEAN-ATMOSPHERE SYSTEM

The freshwater balance in the North Atlantic is one of the major components that governs the strength of the THC. The THC is driven by atmosphere-ocean fluxes of heat and freshwater: in high latitudes surface waters are cooled and lose buoyancy. On the other hand, there is excess precipitation which increases buoyancy. The competition of these two effects, which activate very different feedback mechanisms in response to a change, gives rise to the possibility of large changes in the thermohaline circulation. *Stommel* [1961] showed that different removal times of sea surface temperature and salinity anomalies result in two very different circulation modes. This can be described as feedback mechanisms which influence the THC (Fig. 1). The two feedback mechanisms are due to the advection of surface waters from the low to the high latitudes via the wind-driven and thermohaline circulation [*Bryan*, 1986]. A stronger THC results in an increased meridional heat flux which tends to warm the surface waters in the high northern latitudes. This decreases density and therefore acts as a negative feedback mechanism for the THC. Negative feedback mechanisms may give rise to oscillations. *Delworth et al.* [1993] have shown that this feedback mechanism causes an interdecadal oscillation of the THC. Similarly, with increasing THC, more saline waters are transported northward which tends to increase density and speed up the THC – a positive feedback which may cause instability. The interplay between the temperature and the salinity feedbacks are the origin of multiple equilibria of the THC [*Stommel*, 1961].

Following the pioneering work of *Bryan* [1986], many ocean only models [*Stocker and Wright*, 1991; *Mikolajewicz and Maier-Reimer*, 1994; *Rahmstorf*, 1994; *Sei-*

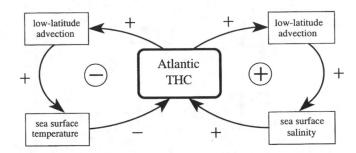

Figure 1. Principal advective feedback mechanisms influencing the Atlantic THC. The signs attached to the arrows indicate the correlation between changes in the quantity of the outgoing box with that of the ingoing box, e.g., warmer sea surface temperatures (SST) lead to weaker THC. Resulting correlations of a loop are circled and they indicate whether a process is self-reinforced (positive sign) or damped (negative sign). A stabilizing loop (left) is associated with changes in SST due to changes in the advection of heat. This loop may give rise to oscillations. The second loop (right) is due to the influence of advection of low-latitude salty waters into the areas of deep water formation. The resulting correlation is positive and the loop may therefore cause instabilities.

dov and Maslin, 1999] and coupled climate models [*Manabe and Stouffer*, 1988] have confirmed earlier hypotheses that the ocean-atmosphere system has more than one stable mode of operation [*Broecker et al.*, 1985]. The obvious statement is that, like many nonlinear physical systems, the ocean-atmosphere system may exhibit hysteresis behavior [*Stocker and Wright*, 1991]: for certain values of a control variable more than one stable state is permissible. Numerous modelling studies have demonstrated that the Atlantic surface freshwater balance is a key control variable for the THC which can assume different modes of operation.

For a simplified ocean-atmosphere model [*Knutti and Stocker*, 2000], this hysteresis behaviour is illustrated in Fig. 2a. Starting in a present-day steady state of the ocean-atmosphere system with active overturning, an anomalous freshwater flux, ΔF, into the North Atlantic is slowly increased (0.1 Sv/1000 yr; 1 Sv$=10^6$ m^3/s) until the North Atlantic becomes too fresh and deep water formation stops (transition from state 1 to state 2). The freshwater input is then decreased until the deep water formation restarts again. In this model experiment, two equilibrium states of the THC are possible for present day conditions (zero freshwater anomaly), one with active deep water formation in the North Atlantic (state 1) and one without (state 3). This classical picture can be extended by applying the freshwater

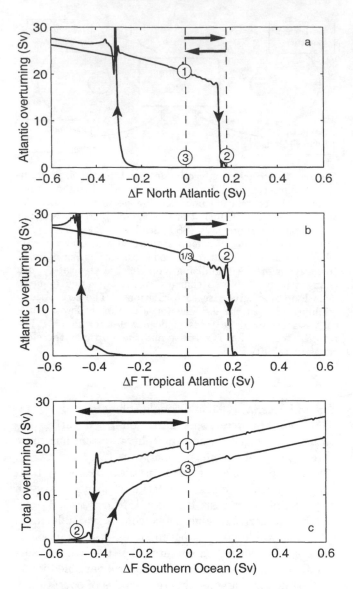

Figure 2. Hysteresis behaviour of the thermohaline circulation. The three panels show the dependence of the Atlantic overturning (total global overturning in panel c) on a slowly changing freshwater anomaly ΔF applied (a) in the North Atlantic, (b) in the tropical Atlantic or (c) in the Southern Ocean. Depending on the initial location of the state on the hysteresis curve (indicated by the circled numbers) and the amplitude of the perturbation (indicated by the horizontal arrows), three qualitatively different response types can occur: (a) non-linear/irreversible, (b) linear/reversible or (c) non-linear/reversible.

flux in the tropical Atlantic (Fig. 2b) or in the Southern Ocean (Fig. 2c) instead of the North Atlantic. When the freshwater input is applied in the tropical Atlantic, a very similar picture is observed, except that the hys-

teresis loop is stretched, because only part of the freshwater anomaly is transported to the regions of deep water formation in the North Atlantic. The perturbation must therefore be larger, or last longer, to induce a transition. When extracting freshwater from the Southern Ocean, southern sinking strengthens and the deep oceans are increasingly dominated by Antarctic Bottom Water, until a certain threshold is reached where deep water formation in the north stops. When the freshwater loss of the Southern Ocean is reduced again, the Antarctic Bottom Water retreats from the deep oceans and sinking in the northern hemisphere starts again. In this specific model experiment, intermediate water in the North Pacific evolves, but other model versions may also show Atlantic deep water formation. As the perturbation is applied in a zonally uniform way in the Southern Ocean, there is no direct control on the location where deep water formation in the north starts first. Fig. 2c indicates that changes in the THC of the northern hemisphere can also be remotely triggered from ocean changes around Antarctica.

The existence of hysteresis implies three fundamentally different responses to perturbations, which depend on the initial location in phase space and the amplitude of the perturbation. The response may be linear/reversible, non-linear/reversible, or non-linear/irreversible. Figure 2 can be used to illustrate the three different response types. When a temporary freshwater flux of about 0.18 Sv is applied to the tropical Atlantic, the overturning is slightly reduced but recovers when the perturbation stops (Fig. 2b). The system shows a linear/reversible response. Applying the same perturbation in the North Atlantic, the threshold for a linear response is crossed and the system moves to a different circulation mode (Fig. 2a). The response is non-linear and irreversible. Considering the total amount of overturning in the northern hemisphere, a non-linear but almost reversible response is observed when a strong temporary perturbation is applied in the Southern Ocean (Fig. 2c).

Hysteresis behavior is well known in climate models and considered to be a robust feature, but its structure is highly model dependent. A critical question is: where are we now on the hysteresis, what is its structure and how close is the threshold?

As the freshwater supply to the North Atlantic is influencing the THC, any forcing modifying the freshwater fluxes directly or indirectly may move the THC beyond a threshold. Model simulations with different climate models suggest that the maximum concentration of CO_2 constitutes a threshold provided the THC has a second equilibrium state [*Manabe and Stouffer*, 1993]. In these simulations the threshold lies between 2× and 4×CO_2 concentration, but the existence of the threshold and its value strongly depend on the climate sensitivity of the coupled model, the details of the hydrological cycle and other parameterizations. Because of the interplay between the buoyancy uptake rate, which is limited by vertical mixing processes, and the warming of the atmosphere, also the rate of CO_2-increase is determining whether or not a threshold is crossed: the ocean-atmosphere system appears less stable under faster perturbations [*Stocker and Schmittner*, 1997].

The structure of the oceanic reorganization beyond the threshold, modeled in models of different complexity, is very similar: the Atlantic THC ceases and deep ocean ventilation stops. This leads to a reduction in the meridional heat transport in the Atlantic, and hence a regional cooling is superimposed on the global warming. It depends on the model's climate sensitivity whether the combined effect leads to a net warming or net cooling in the regions most affected by the meridional heat transport of the Atlantic THC.

5. FUTURE CHANGES COULD LIE AHEAD

5.1. Short-term Evolution of the Atlantic THC

Given its relative stability over many millennia, one may wonder why the Atlantic THC is of any interest today. It turns out that anticipated global warming, caused by anthropogenic emissions of greenhouse gases, is another process that influences significantly the surface freshwater balance of the Atlantic ocean [*Manabe and Stouffer*, 1994]. When air temperature rises, surface waters in the high latitudes also tend to warm up. This decreases surface density and reduces the THC; high-latitude amplification of the warming due to the snow-albedo feedback adds to this effect. In addition, the hydrological cycle may be enhanced in a warmer atmosphere because of increased evaporation, larger atmospheric moisture capacity and increased meridional transport of latent heat [*Dixon et al.*, 1999]. Both these effects tend to reduce the THC because they decrease surface water density. While the relative strength of these two mechanisms is debated [*Dixon et al.*, 1999; *Mikolajewicz and Voss*, 2000], a general reduction of the Atlantic THC in response to global warming appears to be a robust result found by the entire hierarchy of current climate models: The majority of coupled climate models indicates a reduction of the THC from 10% to 50% under increasing CO_2 concentration in the atmosphere for the next 100 years [*IPCC*, 2001]. This is illustrated in Fig. 3 which shows that the spread of

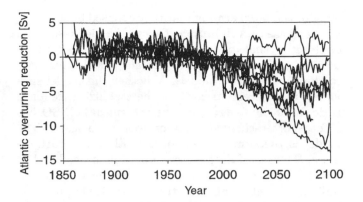

Figure 3. Anomaly of the maximum meridional overturning streamfunction in the Atlantic for a series of coupled model simulations up to year 2100 in Sv ($1\,\text{Sv} = 10^6\text{m}^3/\text{s}$). All models assume the same scenario of increasing CO_2 but each model has an individual response with respect to the changes in heat and freshwater fluxes. Nevertheless, most models indicate a reduction of the overturning of up to 50%. One model appears to be stable; this provides an indication of potentially important new feedback mechanisms. [Modified from *IPCC*, 2001]

the simulated reduction increases significantly after year 2000. This may partly result from the very large differences in maximum Atlantic overturning from which these models start. Values range from weak overturning of about 16 Sv to very strong overturning of over 25 Sv. It is known that the stability of the THC depends on the strength of the THC itself [*Tziperman*, 2000], and this may directly influence the amplitude of the transient response. In addition, each coupled model has a different climate sensitivity, i.e., warming and freshwater balance anomalies, that are realized over the 100 years of integration. A more quantitative assessment of these results requires more simulations and systematic comparisons in which these differences in control state and climate sensitivity are taken into account.

While all these models point towards a reduced THC under global warming, there are two exceptions. A stabilizing process for the THC has recently been suggested based on a coupled climate model [*Latif et al.*, 2000]. According to this model, and in agreement with reanalysis data [*Schmittner et al.*, 2000], El Niños are associated with an increased atmospheric freshwater export from the tropical Atlantic which tends to make surface waters flowing northward in the Atlantic saltier and stabilize the Atlantic THC. Because the coupled model shows a tendency to more El Niño phases under global warming, the tropical Atlantic tends to become saltier and thus compensates, in this model, the gain of buoyancy in the northern high latitudes caused by warming and increased precipitation. Due to the high resolution of the isopycnal model in the tropics, the ENSO simulation in the model of *Latif et al.* [2000] is superior to that in the other models. Whether this model simulated the correct strength of the stabilizing feedback is doubtful, however, since the present-day ENSO has a too short recurrence time of about two years. The model is therefore biased towards El Niño. Furthermore, the area of deep water formation is simulated at a rather poor resolution due to the use of isopycnal coordinates, suggesting that the high-latitude processes influencing the THC are probably not well captured. Nevertheless, this model points to a potentially important stabilizing feedback for the THC. This warrants focused investigation with other comprehensive climate models.

Gent [2001] presents a new simulation with a comprehensive coupled AOGCM in which the THC remains stable under global warming. In this model, the warming in the western Atlantic and stronger winds cause an increase in evaporation which is not compensated by local precipitation, i.e., there is net export of water vapor from the North Atlantic region. This helps to offset the buoyancy gain by warming and stabilizes the THC. However, it should be noted that the meridional overturning in the control simulation is unrealistically large which tends to stabilize the THC [*Tziperman*, 2000]. Also, the coupled model does not yet dispose of a river routing scheme which would certainly influence the control state of the THC.

At this point it may be concluded that due to such stabilizing feedback mechanisms, a collapse of the THC may no longer be of concern. As discussed above, however, most 3-dimensional simulations indicate a reduction of the THC by the year 2100, but not a complete collapse [*IPCC*, 2001]. A collapse appears therefore unlikely to occur by 2100 but it can not be ruled out for later. Recent simulations indicate that modes of natural variability and their future changes may also influence the strength and stability of the THC [*Delworth and Dixon*, 2000]. Important in this context is to emphasize that a reduction of the THC moves the system closer to the threshold and the likelihood of a collapse may well increase [*Tziperman*, 2000]. Concern for this potentially irreversible phenomenon may thus increase in the future. In any case, it should be a key priority to better understand the relative strength of different positive and negative feedback mechanisms with respect to the THC.

5.2. Long-term Evolution of the Atlantic THC

In the preceding section we discussed the fate of the THC under greenhouse warming for the next 100 years based on simulations with comprehensive 3-dimensional coupled general circulation models. A longer perspective and a careful investigation of parameter space is still difficult to achieve with such models. Here, simplified models of intermediate complexity fill an important void. Such models, which have been verified in various contexts, e.g., for tracer studies of paleoclimatic simulations, allow us to make progress regarding the question of robust results.

As ocean models indicate, an initial reduction of deep water formation in the northern North Atlantic due to a gain in buoyancy from a stronger meridional transport of moisture results in a freshening of the surface waters. A reduction of sea surface density is also caused by the increased surface air temperatures further reducing the thermohaline circulation. The question now is whether this reduction leads to a permanent shut-down, i.e., will certain thresholds be crossed in the process of a slow changing of the forcing? There is presently only one simulation using a comprehensive AOGM extending over many centuries [*Manabe and Stouffer*, 1993; *Manabe and Stouffer*, 1994]. They showed that a complete shut-down can indeed occur in response to a sufficiently large perturbation. In those simulations the critical threshold lies between 2× and 4×CO_2.

In order to investigate systematically on which quantities this threshold depends, only models of reduced complexity can be used, because they allow for a large number of long-term simulations. *Stocker and Schmittner* [1997] found that besides the stabilization level of greenhouse gases in the atmosphere, the rate of increase of greenhouse gas concentration also determines the threshold. This is illustrated in Fig. 4. The climate sensitivity is set at 3.7°C for a doubling of CO_2 in agreement with *Manabe and Stouffer* [1993]. The standard rate of CO_2 increase is 1%/yr compounded; experiments with a fast rate of 2%/yr (denoted F) and a slow rate of 0.5%/yr (denoted S) are also performed. The maximum CO_2 values are 560 ppmv (exp. 560), 650 ppmv for experiments 650 and 650F, and 750 ppmv for experiments 750 and 750S. Once the maximum value is reached, CO_2 is held constant (Fig. 4a).

Simulated global mean surface air temperature changes do not depend on the emission history for a given maximum CO_2 concentration (Fig. 4b). However, there exists a bifurcation point for the maximum meridional

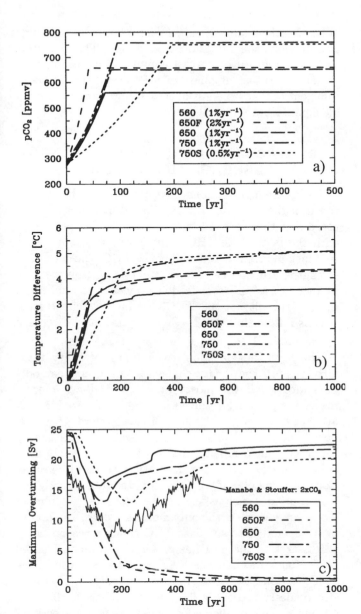

Figure 4. (a) Prescribed evolution of atmospheric CO_2 for five global warming experiments. (b) Simulated global mean surface air temperature changes. The climate sensitivity for a doubling of CO_2 was set at 3.7°C in agreement with the simulation of *Manabe and Stouffer* [1993]. (c) Evolution of the maximum meridional overturning of the North Atlantic in Sverdrup (1 Sv= $10^6 m^3/s$). Note the good agreement between the overturning simulated in the simplified model with that of the 3D AOGCM of *Manabe and Stouffer* [1993]. [Modified from *Stocker and Schmittner*, 1997]

overturning of the North Atlantic (Fig. 4c). In all cases, a reduction is obtained with an amplitude depending on the values of maximum atmospheric CO_2 and of the rate of CO_2 increase. The circulation collapses perma-

nently for a maximum concentration of 750 ppmv with an increase at a rate of 1%/yr (exp. 750). It recovers, however, and settles to a reduced value if the increase is slower (0.5%/yr, exp. 750S) or if the final CO_2 level is reduced to 650 ppmv (exp. 650). Similarly, for a fast increase (exp. 650F) at a rate of 2%/yr the circulation collapses. All experiments have been integrated for 10,000 years and no further changes have been observed. In other words, once the THC collapses it settles to a new equilibrium and changes are hence irreversible. Even if CO_2 concentrations return to preindustrial levels many centuries after emissions are exhausted, the Atlantic THC may remain shut off [*Rahmstorf and Ganopolski*, 1999].

The few model simulations suggest that the critical level for THC collapse is somewhere between double and fourfold preindustrial CO_2 concentration. Extensive parameter studies show that the position of thresholds depends critically on various model parameterizations [*Schmittner and Stocker*, 1999]. A crucial process determining stability is the representation of vertical mixing in ocean models. *Manabe and Stouffer* [1999] argued that the vertical diffusivity in ocean models determines the number of equilibria of the THC: models with higher vertical diffusivity appear to be more stable and exhibit fewer equilibrium states. It is thus crucial to improve ocean mixing schemes and investigate their effect on the stability of the THC. A recent parameter study with a simplified ocean model indicates that the THC continues to have limited stability when modern mixing schemes are used, but that the values of thresholds depend quantitatively on the mixing [*Knutti et al.*, 2000].

The Atlantic THC is an important transport mechanism from the surface to the deep ocean. The amount of heat mixed into the interior of the ocean therefore also depends on the strength of the THC. A stronger THC would represent a more efficient downward transport of heat. More than half of the projected sea level rise is due to the thermal expansion of the water column [*IPCC*, 1996]. Because the vertical distribution of excess heat affects sea level, changes of the THC have the potential to influence the rate of sea level rise and its final value.

Simplified coupled models permit the construction of well-defined experiments. *Knutti and Stocker* [2000] placed their model on a bifurcation point of the Atlantic THC: just beyond the bifurcation point the THC collapses, otherwise it recovers. The point is that the equilibrium atmospheric global warming in these two simulations is nearly identical but the internal distribution of heat may differ substantially. This is shown in Fig. 5.

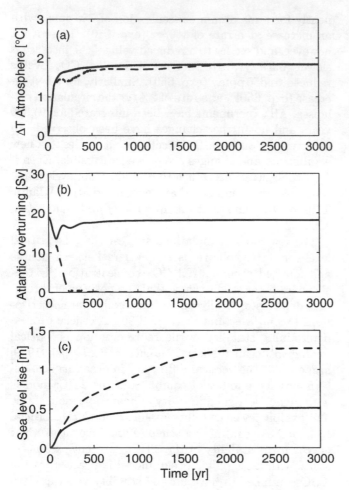

Figure 5. (a) Global mean atmospheric temperature increase, (b) Atlantic deep overturning, and (c) global mean sea level rise versus time in two almost identical global warming experiments using the model version with a Gent&McWilliams mixing scheme. If the Atlantic deep overturning collapses (dashed lines), sea level rise due to thermal expansion is much larger for the same atmospheric temperature increase than if the overturning recovers (solid lines). [From *Knutti and Stocker*, 2000]

Global mean warming is about 1.8°C with an equilibrium sea level rise of about 0.5 m if the THC remains active (solid lines, Fig. 5). However, if the circulation collapses, the equilibrium sea level rise is about 0.7 m larger, although the equilibrium atmospheric warming is identical (dashed lines, Fig. 5).

This result is counterintuitive at first sight because one might argue that a collapsed THC prevents heat from mixing into the interior and the warming now takes place only in the uppermost layers of the ocean. However, when the THC slows down, the surface waters

in the North Atlantic tend to cool relative to a simulation in which the THC does not change. The increased air-sea temperature contrast enhances the heat uptake during the transient phase of a few centuries. This is sufficient to take up additional heat and more than double equilibrium sea level rise.

6. STABILIZING AND DESTABILIZING FEEDBACK MECHANISMS

The preceding sections have illustrated the importance of the THC for future climate change and indicated that there are still major uncertainties associated with the fate of the THC. This is due to a number of feedback mechanisms involving the THC whose relative strengths are poorly known. In this section we give an overview of different feedback mechanisms influencing the stability of the THC. Various feedback mechanisms influencing the THC were previously discussed by *Marotzke* [1996] and *Rahmstorf et al.* [1996].

The important question regarding the stability of the THC to perturbations or changes in the forcing can only be addressed if the most important feedback mechanisms are properly resolved in models and if their relative strength is simulated realistically. Even if we believe that a particular model succeeds in representing the important feedbacks, verification is extremely difficult. High-resolution paleoclimatic records in conjunction with model components that simulate directly observed proxies such as, e.g., stable isotopes, greenhouse gases, are currently the only means to assess the ability of these models to simulate THC changes reasonably.

Research in the past has probably too much focused on feedback mechanisms that lead to a collapse of the THC. Beyond the two principal feedback mechanisms already described in Fig. 1 above, a number of other stabilizing and destabilizing feedbacks have been described in model simulations. In the following figures the signs attached to the arrows indicate the correlation between changes in the quantity of the outgoing box with that of the ingoing box. Resulting correlations of a loop are circled and they indicate whether a process is self-reinforced (positive sign) or damped (negative sign).

An example of two mechanisms associated with the response of the atmospheric circulation to changes in the THC are given in Fig. 6. A stabilizing feedback is due to increased Ekman divergence in the area of deep water formation if the THC reduces. This brings up more saline waters from depth, thus helping maintain the THC [*Fanning and Weaver*, 1997]. Conversely, *Marotzke and Stone* [1995] used a simple box model representing ocean-atmosphere interaction and suggested a destabilizing feedback mechanism due to the meridional transport of moisture. A weaker THC cools the high latitudes and thus increases meridional temperature gradients. This leads to a stronger meridional circulation in the atmosphere and stronger meridional moisture flux. The stronger import of moisture to the high latitudes decreases sea surface salinity and hence enhances the reduction of the THC.

The influence of Arctic sea ice on the THC has hardly been studied because ocean models are often used without sea ice components. However, the influence of sea ice on the THC is potentially important because in the high latitudes, sea ice formation and sea ice export constitutes an important contribution to the freshwater balance [*Dickson et al.*, 2000]. Here we propose two feedback mechanisms associated with sea ice which have not yet been quantitatively studied with models (Fig. 7). A local effect arises through the process of brine rejection when sea ice forms. Increased sea ice formation thus tends to increase sea surface salinity (SSS) locally which promotes deep water formation. A weakening THC results in enhanced Arctic sea ice for-

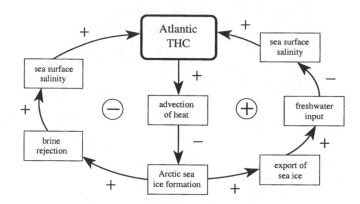

Figure 7. Schematics of a local (left loop) and a non-local (right loop) feedback mechanims associated with changes in Arctic sea ice in response to THC changes.

mation which tends to increase SSS and therefore the THC. On the other hand, sea ice is exported from the Arctic reducing SSS upon melting. This far-field effect in the form of a freshwater flux represents a destabilizing feedback mechanism for the THC. Beyond the feedback mechanisms associated with the salt balance, sea ice also influences the atmosphere-ocean heat transfer. Such feedback mechanisms have been discussed and simulated by *Yang and Neelin* [1993].

Finally, we speculate that there may also exist an interhemispheric feedback mechanism which involves deep water mass characteristics (Fig. 8). A stronger Atlantic THC extracts more heat from the Southern Ocean thereby cooling it [*Crowley*, 1992; *Stocker*, 1998]. Cooling promotes sea ice formation and, via the process of brine rejection, enhances the formation of Antarctic Bottom Water (AABW). AABW is the densest large-scale water mass in the world ocean with an influence as far north as the North Atlantic. Ocean models suggest that if the density of southern component waters increases, NADW and with it the Atlantic THC tends to reduce [*Stocker et al.*, 1992; *England*, 1992]. This constitutes a negative feedback mechanism.

It is clear that the present "gallery" of feedback mechanisms is not complete and that further processes will be investigated as models increase their resolution and completeness. For example, the far-field effects of changes in the tropics and how these may influence the THC [*Latif et al.*, 2000; *Schmittner et al.*, 2000] are still poorly studied and only very few experiments with comprehensive models exist todate. It is, however, evident that we are not yet in the position to assess the overall stability of the THC to perturbations with sufficient confidence because the strength of the individual feed-

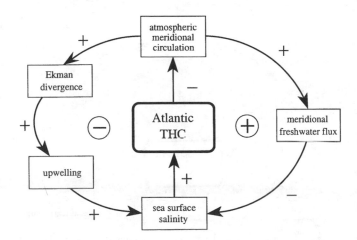

Figure 6. Schematics of two feedback mechanisms associated with changes in the atmospheric circulation in response to THC changes. The signs attached to the arrows indicate the correlation between changes in the quantity of the outgoing box with that of the ingoing box, e.g., increased sea surface salinity (SSS) leads to stronger THC. Resulting correlations of a loop are circled and they indicate whether a process is self-reinforced (positive sign) or damped (negative sign). A stabilising loop (left) is associated with changes in SSS due to wind stress changes. This loop may give rise to oscillations. The second loop (right) is due to the influence of the meridonal flux of freshwater whose resulting correlation is positive; the loop may therefore cause instabilities.

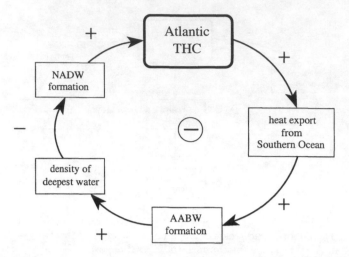

Figure 8. Schematics of a possible interhemispheric feedback mechanism.

back mechanisms is still poorly known. In particular, the strength of those feedback mechanisms that involve high-latitude processes such as sea ice, deep water formation and flow over sills, are fraught with the largest uncertainties. Progress in this area can only come from models which resolve the scales of these processes realistically.

7. A POSSIBLE RUNAWAY GREENHOUSE EFFECT?

Given the insights about the workings of the climate system, we pose an important question: Do these massive ocean reorganisations have the potential to trigger a runaway greenhouse effect? The reasoning goes as follows (Fig. 9). A warming atmosphere clearly leads to increasing sea surface temperatures which, in turn, reduce the solubility of CO_2 in the surface waters. Warmer waters hold less dissolved carbon and warming thus causes an outgassing of this greenhouse gas. This constitutes a positive feedback loop (top in Fig. 9) enhancing the initial increase of atmospheric CO_2. A further positive feedback loop (left in Fig. 9) is associated with the effect of downward transport of carbon by the THC. If the THC collapses, much less carbon will be buried in the deep sea, again reinforcing accumulation of CO_2 in the atmosphere. There is a third feedback loop added in Fig. 9. This is associated with the reaction of the marine biosphere as described first by *Siegenthaler and Wenk* [1984]. Its strength and even sign are very uncertain, but model simulations suggest that it may be a negative feedback which partly compensates the in-

crease in atmospheric CO_2 caused by the left loop [*Joos et al.*, 1999].

Model simulations using 3-dimensional ocean general circulation models with prescribed boundary conditions predicted a minor [*Maier-Reimer et al.*, 1996] or a rather strong [*Sarmiento and Le Quéré*, 1996] feedback between the circulation changes and the uptake of anthropogenic CO_2 under global warming scenarios. However, the complete interplay of the relevant climate system and carbon cycle components was only recently taken into account using a physical-biogeochemical model of reduced complexity [*Joos et al.*, 1999; *Plattner et al.*, 2001].

With this model, different experiments with the ocean carbon pumps operating or suppressed can be performed. Such experiments are essential for a better understanding of the various processes influencing ocean uptake of CO_2 and hence atmospheric concentration (Fig. 10). By placing the model on a bifurcation point of the Atlantic THC, the impact of a breakdown of NADW formation on atmospheric CO_2 can be examined. Fig. 10a shows a set of simulations where the NADW formation weakens but recovers almost completely; the case of a complete collapse of the Atlantic

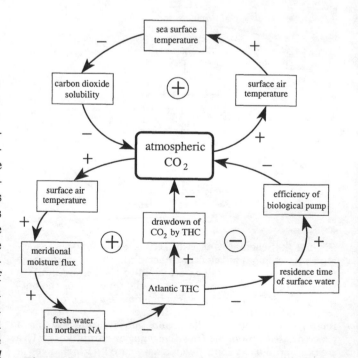

Figure 9. Possible feedback mechanisms that influence the atmospheric CO_2 concentration. In two cases, a positive feedback occurs with the potential of reinforcing the warming. NA denotes North Atlantic.

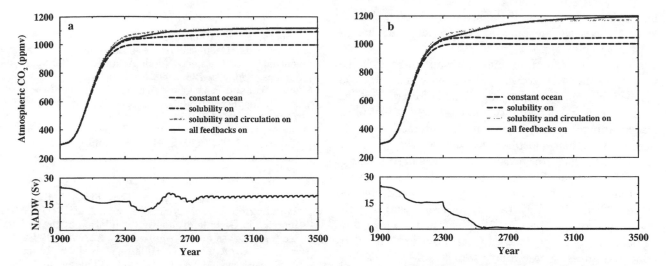

Figure 10. Evolution of atmospheric CO₂ and the formation rate of North Atlantic Deep Water (NADW, lower panels) in a model in which the emissions of carbon are prescribed based on a previous CO₂-stabilization simulation with a constant ocean. The stabilization target is 1000 ppmv. The model was placed on a bifurcation point: (a) NADW formation weakens and recovers thereafter; (b) NADW formation breaks down completely. The global mean temperature increase for a doubling of CO₂ is about 2.6°C, identical in both sets of simulations.

THC is given in Fig. 10b. The CO₂ stabilization scenario is realized in a simulation with prescribed carbon emissions and a constant climate (long-dashed lines). The full simulation including all feedbacks (solubility, circulation and biota) shows a larger atmospheric CO₂ concentration for the same carbon emissions (the global warming at CO₂-doubling is 2.6°C). The additional increase in atmospheric CO₂ of about 11% in Fig. 10a is almost entirely associated with the effect of decreased solubility due to the warming. The breakdown of the THC in the Atlantic further reduces ocean CO₂ uptake significantly, resulting in an almost 20% increase in atmospheric CO₂ (Fig. 10b). In general, in this model the circulation and marine biota feedback nearly compensate each other, at least until year 2500, and the solubility effect remains the only significant global warming feedback [*Joos et al.*, 1999]. However, if the THC in the Atlantic collapses, the circulation feedback becomes dominant with minor contributions by the solubility effect and the marine biota feedback. The reduction of strength of the ocean as a major carbon sink due to global warming appears to be a robust result, but the model also shows that dramatic feedback effects (such as a runaway greenhouse effect) are very unlikely. The maximum increase of atmospheric CO₂ in the case of a collapsed Atlantic THC is estimated at about 20%.

However, a recent study indicates that the terrestrial biosphere may constitute a more important feedback on

the century time scale. In an integration using a coupled ocean-atmosphere-biosphere general circulation model, *Cox et al.* [2000] showed that by 2100 the terrestrial biosphere acts as a strong source of carbon due to increased soil respiration in a warmer and wetter world and the saturation of the fertilization effect at high CO₂. The effect on atmospheric CO₂ is estimated at more than 35%, significantly more than the oceanic effects in the year 2100. But as with the ocean simulations, one should emphasize that these results are still highly uncertain [*Joos et al.*, 2000]. Nevertheless, they indicate that significant, but not catastrophic positive feedback mechanisms are associated with the carbon cycle.

8. CONCLUSIONS

Since *Broecker* [1987] first drew attention to the possibility of large and abrupt changes associated with the THC, the joint effort of paleoclimatic reconstruction and climate modeling has provided tremendous detail to our understanding of the dynamics of non-linear THC changes in the Atlantic and its climatic impact. High-resolution records have told us that changes can be completed within a few years with amplitudes that are hard to imagine. In Greenland, mean annual temperature changed by up to 16°C within just a decade or so. Various records have indicated that these changes are wide spread and influence regions at least in the northern hemisphere. For some of the strongest events, changes are also seen as far south as Antarctica, but changes appear to be in opposite phase.

As the collection of findings of abrupt climate change in various records grows, paleoclimate modeling becomes increasingly important. This is for two reasons. First, only few paleoclimatic indices have a direct physical meaning; in most cases complex interactions in the ocean-atmosphere-biosphere system set the variations of variables that are measured. Models are needed to quantify the contribution of individual climate-relevant components to the observed signal. Second, models serve to quantitatively test hypotheses regarding the causes of abrupt change. Whereas in the early days of paleo-science, hypotheses were often formulated by educated guessing, nowadays we are able to put these hypotheses to a quantitative test. Clearly, the models do not replace further data but often they help us focus the data and embed them better into a large-scale context.

These same models send us a clear message: future changes of the THC in response to global warming are likely. Although the magnitude of the change is highly uncertain, the models agree that the THC in the Atlantic will reduce due to the gain of buoyancy associated with the warming and a stronger hydrological cycle. A few models, however, show a weak response and suggest that additional feedback mechanisms are at work. This apparent disagreement is the seed for new research. We need a better understanding of the individual feedback mechanisms, positive and negative, that are collectively influencing the THC. A few of these feedbacks are discussed in this article, others will be discovered with the next generation of climate models which will include better coupling to the atmospheric components, higher resolution and improved representation of deep water formation processes.

It is clear that future changes of the THC remain an important issue, even though there are few indications that an abrupt shut-down, a surprise, is a likely occurrence in the near future. It is also known that a slow-down of the THC moves the system closer to thresholds – and this should be of sufficient concern to warrant intensified research into this topic.

Acknowledgments. The efforts and patience of the editor, Dan Seidov, are appreciated. The thoughtful review of Peter Clark is gratefully acknowledged. This work is supported by the Swiss National Science Foundation and the Swiss Federal Office of Science and Education through the EC-project GOSAC.

REFERENCES

Adkins, J. F., H. Cheng, E. A. Boyle, E. R. Druffel, and L. Edwards, Deep-sea coral evidence for rapid change in ventilation of the deep North Atlantic 15,400 years ago, *Science, 280*, 725–728, 1998.

Alley, R. B., and P. U. Clark, The deglaciation of the northern hemisphere: a global perspective, *Ann. Rev. Earth Plan. Sci., 27*, 149–182, 1999.

Alley, R. B., P. A. Mayewski, T. Sowers, M. Stuiver, K. C. Taylor, and P. U. Clark, Holocene climatic instability: A prominent, widespread event 8200 yr ago, *Geology, 25*, 483–486, 1997.

Alley, R. B., D. A. Meese, C. A. Shuman, A. J. Gow, K. C. Taylor, P. M. Grootes, J. W. C. White, M. Ram, E. D. Waddington, P. A. Mayewski, and G. A. Zielinski, Abrupt increase in Greenland snow accumulation at the end of the Younger Dryas event, *Nature, 362*, 527–529, 1993.

Barber, D. C., A. Dyke, C. Hillaire-Marcel, A. E. Jennings, J. T. Andrews, M. W. Kerwin, G. Bilodeau, R. McNeely, J. Southon, M. D. Morehead, and J.-M. Gagnon, Forcing of the cold event of 8,200 years ago by catastrophic drainage of Laurentide lakes, *Nature, 400*, 344–348, 1999.

Bard, E., B. Hamelin, M. Arnold, L. Montaggioni, G. Cabioch, G. Faure, and F. Rougerie, Deglacial sealevel record from Tahiti corals and the timing of global meltwater discharge, *Nature, 382*, 241–244, 1996.

Behl, R. J., and J. P. Kennet, Brief interstadial events in the Santa Barbara basin, NE Pacific, during the past 60 kyr, *Nature, 379*, 243–246, 1996.

Blunier, T., J. Chappellaz, J. Schwander, A. Dällenbach, B. Stauffer, T. F. Stocker, D. Raynaud, J. Jouzel, H. B. Clausen, C. U. Hammer, and S. J. Johnsen, Asynchrony of Antarctic and Greenland climate change during the last glacial period, *Nature, 394*, 739–743, 1998.

Blunier, T., J. Schwander, B. Stauffer, T. Stocker, A. Dällenbach, A. Indermühle, J. Tschumi, J. Chappellaz, D. Raynaud, and J.-M. Barnola, Timing of temperature variations during the last deglaciation in Antarctica and the atmospheric CO_2 increase with respect to the Younger Dryas event, *Geophys. Res. Let., 24*, 2683–2686, 1997.

Bond, G. C., W. Showers, M. Elliot, M. Evans, R. Lotti, I. Hajdas, G. Bonani, and S. Johnson, The North Atlantic's 1–2 kyr climate rhythm: Relation to Heinrich events, Dansgaard/Oeschger cycles and the Little Ice Age, in *Mechanisms of Global Climate Change at Millennial Time Scales*, edited by P. U. Clark, R. S. Webb, and L. D. Keigwin, Volume 112 of *Geophysical Monograph*, pp. 35–58, Am. Geophys. Union, Washington, D. C., 1999.

Broecker, W. S., Unpleasant surprises in the greenhouse?, *Nature, 328*, 123–126, 1987.

Broecker, W. S., The great ocean conveyor, *Oceanography, 4*, 79–89, 1991.

Broecker, W. S., Thermohaline circulation, the Achilles heel of our climate system: will man-made CO_2 upset the current balance?, *Science, 278*, 1582–1588, 1997.

Broecker, W. S., Paleocean circulation during the last deglaciation: a bipolar seesaw?, *Paleoceanogr., 13*, 119–121, 1998.

Broecker, W. S., and G. H. Denton, The role of ocean-atmosphere reorganizations in glacial cycles, *Geochim. Cosmochim. Acta, 53*, 2465–2501, 1989.

Broecker, W. S., D. M. Peteet, and D. Rind, Does the ocean-atmosphere system have more than one stable mode of operation?, *Nature, 315*, 21–25, 1985.

Bryan, F., High-latitude salinity effects and interhemispheric thermohaline circulations, *Nature, 323*, 301–304, 1986.

Clark, P. U., R. A. Alley, and D. Pollard, Northern hemisphere ice-sheet influences on global climate change, *Science, 286*, 1104–1111, 1999.

Clark, P. U., R. B. Alley, L. D. Keigwin, J. M. Licciardi, S. J. Johnsen, and H. Wang, Origin of the first global meltwater pulse, *Paleoceanogr., 11*, 563–577, 1996.

Clement, A., R. Seager, and M. Cane, Orbital controls on the El Niño/Southern Oscillation and the tropical climate, *Paleoceanogr., 14*, 441–455, 1999.

Cox, P. M., R. A. Betts, C. D. Jones, S. A. Spall, and I. J. Totterdell, Acceleration of global warming due to carbon-cycle feedbacks in a coupled climate model, *Nature, 408*, 184–187, 2000.

Crowley, T. J., North Atlantic deep water cools the southern hemisphere, *Paleoceanogr., 7*, 489–497, 1992.

Cuffey, M. K., G. D. Clow, R. B. Alley, M. Stuiver, E. D. Waddington, and R. W. Saltus, Large Arctic temperature change at the Wisconsin-Holocene glacial transition, *Science, 270*, 455–458, 1995.

Dansgaard, W., S. J. Johnsen, H. B. Clausen, D. Dahl-Jensen, N. S. Gundestrup, C. U. Hammer, C. S. Hvidberg, J. P. Steffensen, A. E. Sveinbjornsdottir, J. Jouzel, and G. Bond, Evidence for general instability of past climate from a 250-kyr ice-core record, *Nature, 364*, 218–220, 1993.

de Ruijter, W. P. M., A. Biastoch, S. S. Drijfhout, J. R. E. Lutjeharms, R. P. Matano, T. Pichevin, P. J. van Leeuwen, and W. Weijer, Indian-Atlantic interocean exchange: Dynamics, estimation and impact, *J. Geophys. Res., 104*, 20885–20910, 1998.

Delworth, T., S. Manabe, and R. J. Stouffer, Interdecadal variations of the thermohaline circulation in a coupled ocean-atmosphere model, *J. Clim., 6*, 1993–2011, 1993.

Delworth, T. L., and K. W. Dixon, Implications of the recent trend in the Arctic/North Atlantic Oscillation for the North Atlantic thermohaline circulation, *J. Clim., 13*, 3721–3727, 2000.

Dickson, R., J. Lazier, J. Meincke, P. Rhines, and J. Swift, Long-term coordinated changes in the convective activity of the North Atlantic, *Prog. Oceanogr., 38*, 241–295, 1996.

Dickson, R. R., T. J. Osborn, J. W. Hurrell, J. Meincke, J. Blindheim, B. Adlandsvik, T. Vinje, G. Alekseev, and W. Maslowski, The Arctic Ocean response to the North Atlantic Oscillation, *J. Clim., 13*, 2671–2696, 2000.

Dixon, K. W., T. L. Delworth, M. J. Spelman, and R. J. Stouffer, The influence of transient surface fluxes on North Atlantic overturning in a coupled GCM climate change experiment, *Geophys. Res. Let., 26*, 2749–2752, 1999.

Drijfhout, S. S., E. Maier-Reimer, and U. Mikolajewicz, Tracing the conveyor belt in the Hamburg large-scale geostrophic ocean general circulation model, *J. Geophys. Res., 101*, 22563–22575, 1996.

England, M. H., On the formation of Antarctic intermediate and bottom water in ocean general circulation model, *J. Phys. Oceanogr., 22*, 918–926, 1992.

Fanning, A. F., and A. J. Weaver, Temporal-geographical meltwater influences on the North Atlantic conveyor: implications for the Younger Dryas, *Paleoceanogr., 12*, 307–320, 1997.

Ganachaud, A., and C. Wunsch, Improved estimates of global ocean circulation, heat transport and mixing from hydrographic data, *Nature, 408*, 453–457, 2000.

Gent, P. R., Will the North Atlantic Ocean thermohaline circulation weaken during the 21st century?, *Geophys. Res. Let.,* (in press), 2001.

Gordon, A. L., Interocean exchange of thermocline water, *J. Geophys. Res., 91*, 5037–5046, 1986.

Hughen, K. A., J. R. Southon, S. J. Lehman, and J. T. Overpeck, Synchronous radiocarbon and climate shifts during the last deglaciation, *Science, 290*, 1951–1954, 2000.

IPCC, *Climate Change 1995, The Science of Climate Change,* Intergovernmental Panel on Climate Change, Cambridge University Press, 572 pp., 1996.

IPCC, *Third Assessment Report of Climate Change,* Intergovernmental Panel on Climate Change, Cambridge University Press, in preparation, 2001.

Johnsen, S. J., D. Dahl-Jensen, W. Dansgaard, and N. Gundestrup, Greenland palaeotemperatures derived from GRIP bore hole temperature and ice core isotope profiles, *Tellus, 47B*, 624–629, 1995.

Joos, F., G.-K. Plattner, T. F. Stocker, O. Marchal, and A. Schmittner, Global warming and marine carbon cycle feedbacks on future atmospheric CO_2, *Science, 284*, 464–467, 1999.

Joos, F., I. C. Prentice, S. Sitch, R. Meyer, G. Hooss, G.-K. Plattner, and K. Hasselmann, Global warming feedbacks on terrestrial carbon uptake under the IPCC emission scenarios, *Global Biogeochem. Cyc.,* (submitted), 2000.

Killworth, P. D., Deep convection in the world ocean, *Rev. Geophys. Space Phys., 21*, 1–26, 1983.

Knutti, R., and T. F. Stocker, Influence of the thermohaline circulation on projected sea level rise, *J. Clim., 13*, 1997–2001, 2000.

Knutti, R., T. F. Stocker, and D. G. Wright, The effects of sub-grid-scale parameterizations in a zonally averaged ocean model, *J. Phys. Oceanogr., 30*, 2738–2752, 2000.

Lang, C., M. Leuenberger, J. Schwander, and S. Johnsen, 16°C rapid temperature variation in Central Greenland 70,000 years ago, *Science, 286*, 934–937, 1999.

Latif, M., E. Roeckner, U. Mikolajewicz, and R. Voss, Tropical stabilization of the thermohaline circulation in a greenhouse warming simulation, *J. Clim., 13*, 1809–1813, 2000.

Ledwell, J. R., E. T. Montgomery, K. L. Polzin, L. C. St. Laurent, R. W. Schmitt, and J. M. Toole, Evidence for enhanced mixing over rough topography in the abyssal ocean, *Nature, 403*, 179–181, 2000.

Lehman, S. J., and L. D. Keigwin, Sudden changes in North Atlantic circulation during the last deglaciation, *Nature, 356*, 757–762, 1992.

Leuenberger, M., C. Lang, and J. Schwander, $\delta^{15}N$ measurements as a calibration tool for the paleothermometer and gas-ice age differences. A case study for the 8200 B.P. event on GRIP ice, *J. Geophys. Res., 104*, 22163–22170, 1999.

Manabe, S., and R. J. Stouffer, Two stable equilibria of a coupled ocean-atmosphere model, *J. Clim., 1*, 841–866, 1988.

Macdonald, A. M., The global ocean circulation: a hydrographic estimate and regional analysis, *Prog. Oceanogr.*, *41*, 281–382, 1998.

Macdonald, A. M., and C. Wunsch, An estimate of global ocean circulation and heat fluxes, *Nature*, *382*, 436–439, 1996.

Maier-Reimer, E., Geochemical cycles in an ocean general circulation model. Preindustrial tracer distributions, *Global Biogeochem. Cyc.*, *7*, 645–677, 1993.

Maier-Reimer, E., U. Mikolajewicz, and A. Winguth, Future ocean uptake of CO_2: interaction between ocean circulation and biology, *Clim. Dyn.*, *12*, 711–721, 1996.

Manabe, S., and R. J. Stouffer, Century-scale effects of increased atmospheric CO_2 on the ocean-atmosphere system, *Nature*, *364*, 215–218, 1993.

Manabe, S., and R. J. Stouffer, Multiple-century response of a coupled ocean–atmosphere model to an increase of atmospheric carbon dioxide, *J. Clim.*, *7*, 5–23, 1994.

Manabe, S., and R. J. Stouffer, Are two modes of the thermohaline circulation stable?, *Tellus*, *51A*, 400–411, 1999.

Marchal, O., R. François, T. F. Stocker, and F. Joos, Ocean thermohaline circulation and sedimentary $^{231}Pa/^{230}Th$ ratio, *Paleoceanogr.*, *15*, 625–641, 2000.

Marchal, O., T. F. Stocker, and R. Muscheler, Changes in radiocarbon during Younger Dryas, *Earth Planet. Sci. Lett.*, (in press), 2001.

Marotzke, J., Analysis of thermohaline feedbacks, in *Decadal Climate Variability, Dynamics and Predictability*, edited by D. L. T. Anderson and J. Willebrand, Volume I 44 of *NATO ASI*, pp. 333–378, 1996.

Marotzke, J., and P. H. Stone, Atmospheric transport, the thermohaline circulation, and flux adjustments in a simple coupled model, *J. Phys. Oceanogr.*, *25*, 1350–1364, 1995.

Marshall, J., and F. Schott, Open-ocean convection: observations, theory and models, *Rev. Geophys.*, *37*, 1–64, 1999.

Mikolajewicz, U., and E. Maier-Reimer, Mixed boundary conditions in ocean general circulation models and their influence on the stability of the model's conveyor belt, *J. Geophys. Res.*, *99*, 22633–22644, 1994.

Mikolajewicz, U., and R. Voss, The role of the individual air-sea flux components in CO_2-induced changes of the ocean's circulation and climate, *Clim. Dyn.*, *16*, 627–642, 2000.

Mulvaney, R., R. Röthlisberger, E. W. Wolff, S. Sommer, J. Schwander, M. A. Hutterli, and J. Jouzel, The transition from the last glacial period in inland and near-coastal Antarctica, *Geophys. Res. Let.*, *27*, 2673–2676, 2000.

Muscheler, R., J. Beer, G. Wagner, and R. C. Finkel, Changes in deep-water formation during the Younger Dryas event inferred from ^{10}Be and ^{14}C records, *Nature*, *408*, 567–570, 2000.

Oeschger, H., J. Beer, U. Siegenthaler, B. Stauffer, W. Dansgaard, and C. C. Langway, Late glacial climate history from ice cores, in *Climate Processes and Climate Sensitivity*, edited by J. E. Hansen and T. Takahashi, Volume 29 of *Geophysical Monograph*, pp. 299–306, Am. Geophys. Union, 1984.

Oppo, D. W., and S. J. Lehman, Suborbital timescale variability of North Atlantic Deep Water during the past 200,000 years, *Paleoceanogr.*, *10*, 901–910, 1995.

Peterson, L. C., G. H. Haug, K. A. Hughen, and U. Röhl, Rapid changes in the hydrologic cycle of the tropical Atlantic during the last glacial, *Science*, *290*, 1947–1951, 2000.

Plattner, G.-K., F. Joos, T. F. Stocker, and O. Marchal, Feedback mechanisms of ocean carbon uptake under global warming scenarios in a zonally averaged climate model, *Tellus*, (in press), 2001.

Rahmstorf, S., Rapid climate transitions in a coupled ocean-atmosphere model, *Nature*, *372*, 82–85, 1994.

Rahmstorf, S., and A. Ganopolski, Long-term global warming scenarios computed with an efficient coupled climate model, *Clim. Change*, *43*, 353–367, 1999.

Rahmstorf, S., J. Marotzke, and J. Willebrand, Stability of the thermohaline circulation, in *The Warmwatersphere of the North Atlantic Ocean*, edited by W. Krauss, Berlin, pp. 129–157. Bornträger, 1996.

Ruddiman, W. F., and A. McIntyre, The mode and mechanism of the last deglaciation: oceanic evidence, *Quat. Res.*, *16*, 125–134, 1981.

Sachs, J. P., and S. J. Lehman, Subtropical North Atlantic temperatures 60,000 to 30,000 years ago, *Science*, *286*, 756–759, 1999.

Sarmiento, J. L., and C. Le Quéré, Oceanic carbon dioxide in a model of century-scale global warming, *Science*, *274*, 1346–1350, 1996.

Schmittner, A., C. Appenzeller, and T. F. Stocker, Enhanced Atlantic freshwater export during El Niño, *Geophys. Res. Let.*, *27*, 1163–1166, 2000.

Schmittner, A., and T. F. Stocker, The stability of the thermohaline circulation in global warming experiments, *J. Clim.*, *12*, 1117–1133, 1999.

Schmitz, W. J., On the interbasin-scale thermohaline circulation, *Rev. Geophys.*, *33*, 151–173, 1995.

Seidov, D., E. Barron, and B. J. Haupt, Meltwater and the global ocean conveyor: northern versus southern connections, *Global Plan. Change*, (in press), 2001.

Seidov, D., and M. Maslin, North Atlantic deep water circulation collapse during Heinrich events, *Geology*, *27*, 23–26, 1999.

Semtner, A. J., and R. M. Chervin, Ocean general circulation from a global eddy-resolving model, *J. Geophys. Res.*, *97*, 5493–5550, 1992.

Severinghaus, J. P., and E. J. Brook, Abrupt climate change at the end of the last glacial period inferred from trapped air in polar ice, *Science*, *286*, 930–934, 1999.

Shackleton, N. J., and M. A. Hall, Phase relationships between millennial-scale events 64,000–24,000 years ago, *Paleoceanogr.*, *15*, 565–569, 2000.

Siegenthaler, U., and T. Wenk, Rapid atmospheric CO_2 variations and ocean circulation, *Nature*, *308*, 624–626, 1984.

Smith, R. D., M. E. Maltrud, F. O. Bryan, and M. W. Hecht, Numerical simulation of the North Atlantic at $1/10°$, *J. Phys. Oceanogr.*, *30*, 1532–1561, 2000.

Sowers, T., and M. Bender, Climate records covering the last deglaciation, *Science*, *269*, 210–213, 1995.

Steig, E. J., E. J. Brook, J. W. C. White, C. M. Sucher, M. L. Bender, S. J. Lehman, D. L. Morse, E. D. Waddington, and G. D. Clow, Synchronous climate changes in Antarctica and the North Atlantic, *Science*, *282*, 92–95, 1998.

Stocker, T. F., The seesaw effect, *Science, 282*, 61–62, 1998.

Stocker, T. F., Past and future reorganisations in the climate system, *Quat. Sci. Rev., 19*, 301–319, 2000.

Stocker, T. F., and O. Marchal, Abrupt climate change in the computer: is it real?, *Proc. US Natl. Acad. Sci., 97*, 1362–1365, 2000.

Stocker, T. F., and A. Schmittner, Influence of CO_2 emission rates on the stability of the thermohaline circulation, *Nature, 388*, 862–865, 1997.

Stocker, T. F., and D. G. Wright, Rapid transitions of the ocean's deep circulation induced by changes in surface water fluxes, *Nature, 351*, 729–732, 1991.

Stocker, T. F., D. G. Wright, and W. S. Broecker, The influence of high-latitude surface forcing on the global thermohaline circulation, *Paleoceanogr., 7*, 529–541, 1992.

Stommel, H., The abyssal circulation, *Deep Sea Res., 5*, 80–82, 1958.

Stommel, H., Thermohaline convection with two stable regimes of flow, *Tellus, 13*, 224–241, 1961.

Stommel, H., and A. B. Arons, On the abyssal circulation of the world ocean - I. Stationary planetary flow patterns on a sphere, *Deep Sea Res., 6*, 140–154, 1960a.

Stommel, H., and A. B. Arons, On the abyssal circulation of the world ocean - II. An idealized model of the circulation pattern and amplitude in oceanic basins, *Deep Sea Res., 6*, 217–233, 1960b.

Taylor, K. C., G. W. Lamorey, G. A. Doyle, R. B. Alley, P. M. Grootes, P. A. Mayewski, J. W. C. White, and L. K. Barlow, The 'flickering switch' of late Pleistocene climate change, *Nature, 361*, 432–436, 1993.

Toggweiler, J. R., and B. Samuels, On the ocean's large-scale circulation near the limit of no vertical mixing, *J. Phys. Oceanogr., 28*, 1832–1852, 1998.

Tziperman, E., Proximity of the present-day thermohaline circulation to an instability threshold, *J. Phys. Oceanogr., 30*, 90–104, 2000.

Vidal, L., R. Schneider, O. Marchal, T. Bickert, T. F. Stocker, and G. Wefer, Link between the North and South Atlantic during the Heinrich events of the last glacial period, *Clim. Dyn., 15*, 909–919, 1999.

Warren, B. A., Deep circulation of the world ocean, in *Evolution of Physical Oceanography – Scientific Surveys in Honor of Henry Stommel*, edited by B. A. Warren and C. Wunsch, pp. 6–41. MIT Press, 1981.

Weaver, A. J., C. M. Bitz, A. F. Fanning, and M. M. Holland, Thermohaline circulation: high-latitude phenomena and the difference between the Pacific and Atlantic, *Ann. Rev. Earth Planet. Sci., 27*, 231–285, 1999.

Yang, J., and J. D. Neelin, Sea-ice interaction with the thermohaline circulation, *Geophys. Res. Let., 20*, 217–220, 1993.

Yu, E.-F., R. François, and M. P. Bacon, Similar rates of modern and last-glacial ocean thermohaline circulation inferred from radiochemical data, *Nature, 379*, 689–694, 1996.

T. F. Stocker, R. Knutti and G.-K. Plattner, Climate and Environmental Physics, Physics Institute, University of Bern, 5 Sidlerstrasse, CH-3012 Bern, Switzerland. (stocker@climate.unibe.ch; knutti@climate.unibe.ch; plattner@climate.unibe.ch)